Small Satellites Missions and Applications

Small Satellites Missions and Applications

Editors

Simone Battistini

Filippo Graziani

Mauro Pontani

Basel • Beijing • Wuhan • Barcelona • Belgrade • Novi Sad • Cluj • Manchester

Editors

Simone Battistini	Filippo Graziani	Mauro Pontani
MBDA Italia S.p.A.	GAUSS Srl	Sapienza University of Rome
Rome	Roma	Rome
Italy	Italy	Italy

Editorial Office
MDPI
St. Alban-Anlage 66
4052 Basel, Switzerland

This is a reprint of articles from the Special Issue published online in the open access journal *Applied Sciences* (ISSN 2076-3417) (available at: https://www.mdpi.com/journal/applsci/special_issues/Small_Satellites).

For citation purposes, cite each article independently as indicated on the article page online and as indicated below:

Lastname, A.A.; Lastname, B.B. Article Title. *Journal Name* **Year**, *Volume Number*, Page Range.

ISBN 978-3-0365-8550-5 (Hbk)
ISBN 978-3-0365-8551-2 (PDF)
doi.org/10.3390/books978-3-0365-8551-2

Contents

About the Editors

Simone Battistini

Simone Battistini is currently the Head of the Simulation & Modeling and Performance Department at MBDA Italy. He has a Masters Degree in Control System Engineering (2009) and a PhD in Aerospace Engineering (2013), both from Sapienza University of Rome (Italy). He has held academic positions as an Associate Professor at the University of Brasilia (Brazil, 2013–2018) and as a Senior Lecturer in Aerospace Engineering at Sheffield Hallam University (UK, 2019–2022). He also served as a Visiting Researcher at Technion in Haifa (Israel, 2012) and as a Visiting Professor at the University of Vigo (Spain, 2017).

His research interests span across aerospace and control system engineering, with an emphasis on missiles and spacecraft, which have found application in many fields, such as flight mechanics; simulation; guidance, navigation, and control of aerospace vehicles; Kalman filtering; small satellites missions; and tracking. He is coeditor and coauthor of the book *Cubesat Handbook*, published by Academic Press.

Filippo Graziani

Filippo Graziani was a Professor of Astrodynamics at Aerospace Engineering School of Sapienza University of Roma for 35 years until 2012, when he retired, and was the Dean of the School from 2004 to 2010. He is a Member of the International Academy of Astronautics (IAA) and a Member of the IAA Trustees Board. His didactical and research activity has been mainly directed towards the "hands-on" space educational programs.

He participated in the main Italian space programs starting with the San Marco satellites in 1970 and he was the team leader of the Italian University Satellites Program (UNISAT) with the aim of designing, manufacturing, and launching microsatellites with his students. Ten microsatellites have been launched since 2000.

In 2012, he founded the company GAUSS (Group of Astrodynamics for the Use of Space Systems) as a spin-off of the Aerospace Engineering School, active in the space technology field, and for which he is the President and CEO. He is the author of more than 200 technical papers on astrodynamics and space systems. He has been the Co-Editor of *Acta Astronautica* since 2009. He received the "Utkin Golden Medal" for international relationship between Russia and Italy for University Satellite Launches and the "M.K.Yangel—100 years Golden Medal" for the contribution to the development of space science in the world. Since 1975, he has participated in the IAF Conferences.

Mauro Pontani

Mauro Pontani is currently an Associate Professor of Aerospace Engineering at the Department of Astronautical, Electrical, and Energy Engineering, Sapienza University of Rome. He is a corresponding member of the International Academy of Astronautics. He was Visiting Scholar and Visiting Researcher at the University of Illinois at Urbana-Champaign, IL, and at Rice University, Houston, TX, several times from 2004 through 2017. He earned the Italian National Qualification for the position of Full Professor of Aerospace Engineering in 2017. He is a member of the Academic Board of the Ph.D. course "Energy and Environment" at Sapienza.

His research interests are in the field of astrodynamics and aerospace trajectory optimization, and specifically include aerospace mission analysis and design, analytical and numerical methods for trajectory optimization, guidance and control of aerospace vehicles, dynamic game theory applied to aerospace trajectories, satellite constellations, launch, ascent and descent vehicles, and satellite release systems. He is the author of more than 100 scientific publications (including 42 journal papers), and currently serves as Associate Editor for the *Journal of Optimization Theory and Applications, Acta Astronautica, Aerotecnica Missili e Spazio, Mathematical Problems in Engineering*, and the *International Journal of Aerospace Engineering*. He has taken part in several research projects as researcher or principal investigator.

Preface

In the last 20 years, small satellites have created a paradigm shift in the world of space missions. While, at first, they were deemed suitable only for academic and amateur projects, today, they are employed in large-scale industrial projects that have previously made use of larger satellites. Smaller dimensions, lower costs, use of standards, faster and in-series production are some of the elements that have contributed to the success of these platforms. Small satellites provide a robust and agile alternative to traditional satellites, which has enabled innovative concepts of space missions. Remote sensing, telecommunications, scientific exploration, and data collection systems are some of the applications where small satellites have been massively introduced due to the short revisit times and global coverage that can be obtained if large numbers of small satellites are used.

The aim of this book is to present new insights to the reader and to bring together the state of the art in the field of small satellites. The two main topics of this book are the applications of small satellites in commercial and scientific missions, and the spacecraft subsystem technologies that are being proposed to enable the aforementioned concepts and missions. The reader will find analyses of real missions and in-orbit data in the small satellites missions section, as well as examples of the design of advanced technological solutions in the spacecraft subsystem section.

The challenges posed by modern space missions require novel solutions in space systems. The contributions of this book indicate a growing awareness of the importance of small satellites to meet these challenges, both from researchers and from engineers. We hope that this book will contribute to increase the knowledge of the scientific and industrial space communities.

Simone Battistini, Filippo Graziani, and Mauro Pontani
Editors

Article

Transient Attitude Motion of TNS-0#2 Nanosatellite during Atmosphere Re-Entry

Danil Ivanov [1,*], **Dmitry Roldugin** [1], **Stepan Tkachev** [1], **Yaroslav Mashtakov** [1], **Sergey Shestakov** [1], **Mikhail Ovchinnikov** [1], **Igor Fedorov** [2], **Nikolay Yudanov** [2] and **Artem Sergeev** [2]

[1] Keldysh Institute of Applied Mathematics RAS, 125047 Moscow, Russia; rolduginds@gmail.com (D.R.); stevens_l@mail.ru (S.T.); yarmashtakov@gmail.com (Y.M.); shestakov.sa@gmail.com (S.S.); ovchinni@keldysh.ru (M.O.)

[2] JSC Russian Space Systems, 111250 Moscow, Russia; tm016@rniikp.ru (I.F.); kolyan2606@mail.ru (N.Y.); vorchun@yandex.ru (A.S.)

* Correspondence: danilivanovs@gmail.com

Abstract: Attitude motion reconstruction of the Technological NanoSatellite TNS-0 #2 during the last month of its mission is presented in the paper. The satellite was designed to test the performance of the data transmission via the Globalstar communication system. This system successfully provided telemetry (even during its atmosphere re-entry) up to an altitude of 156 km. Satellite attitude data for this phase is analyzed in the paper. The nominal satellite attitude represents its passive stabilization along a geomagnetic field induction vector. The satellite was equipped with a permanent magnet and hysteresis dampers. The permanent magnet axis tracked the local geomagnetic field direction with an accuracy of about 15 degrees for almost two years of the mission. Rapid altitude decay during the last month of operation resulted in the transition from the magnetic stabilization to the aerodynamic stabilization of the satellite. The details of the initial tumbling motion after the launch, magnetic stabilization, transition phase prior to the aerodynamic stabilization, and subsequent satellite motion in the aerodynamic stabilization mode are presented.

Keywords: nanosatellite; attitude determination; passive magnetic stabilization; aerodynamic stabilization

Citation: Ivanov, D.; Roldugin, D.; Tkachev, S.; Mashtakov, Y.; Shestakov, S.; Ovchinnikov, M.; Fedorov, I.; Yudanov, N.; Sergeev, A. Transient Attitude Motion of TNS-0#2 Nanosatellite during Atmosphere Re-Entry. *Appl. Sci.* **2021**, *11*, 6784. https://doi.org/10.3390/app11156784

Received: 9 June 2021
Accepted: 20 July 2021
Published: 23 July 2021

Publisher's Note: MDPI stays neutral with regard to jurisdictional claims in published maps and institutional affiliations.

1. Introduction

Passive attitude control systems are quite popular for nano, pico, and femto-satellites due to strict limitations in terms of mass, size, cost, and energy. These systems utilize natural magnetic, gravitational, or aerodynamic torques to provide the passive stabilization required for missions. Each type of passive stabilization provides different satellite attitudes. Gravitational torque aligns the minimum moment of inertia axis along the nadir direction. Magnetic torque provides tracking of the local geomagnetic field direction by the onboard magnetic dipole. Each type of passive attitude control system is effective on different altitudes. The ratio between the magnitudes of these major environmental torques changes as the satellite orbit evolves. This transforms into changes of the attitude motion regimes during a satellite's lifetime. These regimes of the passively controlled satellites have well-studied both theoretically and experimentally onboard a large number of satellites with gravitational [1–4], magnetic [5–10] and aerodynamic [11–15] control systems. The transient motion between the magnetic and aerodynamic attitude stabilization regimes during the orbit altitude degradation has not been investigated experimentally yet to the best of our knowledge. This is due to the extremely short time between the entering of the dense upper atmospheric layers and the satellite operation termination due to heating. Telemetry gathering is a difficult task for this period. Therefore, the sensor measurements necessary for the attitude motion reconstruction are rarely available for the last weeks of the satellite operation. Technological NanoSatellite TNS-0 #2 utilized GlobalStar antennas [16] as its main data transfer device. This provided frequent communication sessions even during the last days of the mission in the upper atmosphere.

The nominal satellite attitude was a passive magnetic stabilization. The longitudinal axis of the satellite was approximately directed along the local geomagnetic induction vector with a deviation no greater than 12 degrees. The satellite had a relatively elongated geometry with significant displacement between the centers of mass and pressure. The influence of the aerodynamic torque rapidly increased as the satellite entered the dense layers of the atmosphere. On 10 August 2019, the magnetic stabilization changed to an almost chaotic motion with a high angular velocity. Aerodynamic stabilization was achieved in about two weeks. This motion was maintained until its breakdown in the atmosphere on 6 September 2019. Chaotic tumbling was observed during the breakdown.

The structure of the paper is as follows. First, the TNS-0 #2 description is presented. Then, the attitude motion equations (taking into account the magnetic and aerodynamic torques) are provided and a short explanation of the motion reconstruction technique is given. This technique is then applied to process the telemetry received during the magnetic stabilization, transient motion, and aerodynamic stabilization.

2. TNS-0 #2 Description

The TNS-0#2 nanosatellite was developed by JSC Russian Space Systems. It was successfully launched on 17 August 2017 from the International Space Station during the spacewalk of Russian cosmonauts. TNS-0 #2 nanosatellite is a hexagonal prism with a 26.4 cm height and an 18.7 cm diameter. The sides are covered with solar panels. Globalstar antennas, a set of Sun sensors, and a radio link antenna are located on the upper side of the satellite. A handle is attached at the bottom for the cosmonaut to hold the satellite during the spacewalk. The satellite is presented in Figure 1.

Figure 1. TNS-0 #2 nanosatellite (Credit: JSC Russian Space Systems).

The mass of the satellite is 4.8 kg and its center of mass is located almost on the axis of geometric symmetry of the satellite body, as shown in Figure 2. The inertia tensor in the reference frame with the origin in the center of mass of the satellite is

$$
J = \begin{bmatrix} 0.06153 & -0.00013 & -0.00033 \\ -0.00013 & 0.06669 & -0.00012 \\ -0.00033 & -0.00012 & 0.01287 \end{bmatrix} \text{kg·m}^2.
$$

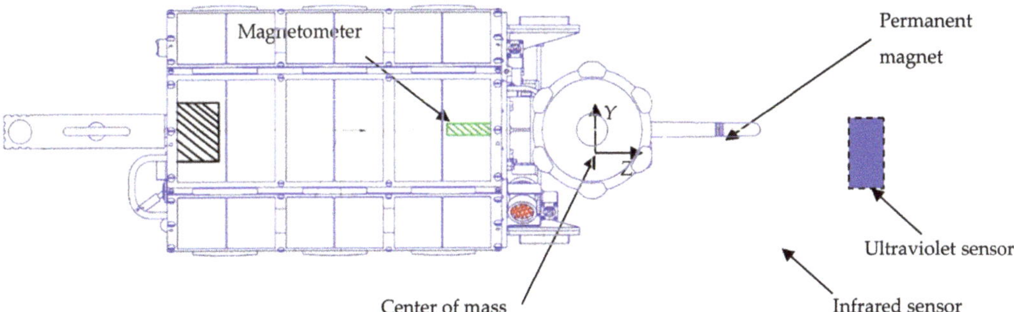

Figure 2. Center of mass location, permanent magnet, magnetometer and Sun sensors positions in the satellite body.

The satellite is equipped with the passive magnetic attitude control system developed by the Keldysh Institute of Applied Mechanics of RAS. The system consists of a set of hysteresis rods for the angular velocity damping and a permanent magnet located along the axis of symmetry to stabilize this axis along the local geomagnetic field induction vector. The dipole moment magnitude of the permanent magnet is 2.2 A·m^2. Grids of hysteresis rods were installed on the upper and lower sides of the nanosatellite body. The location of the permanent magnet, magnetometer, and Sun sensors is shown in Figure 2.

Three-axis magnetometer and a set of Sun sensors were installed onboard. Attitude motion was reconstructed using their measurements. The measurements root-mean-square error of the magnetometer is $\sigma = 100$ nT. The non-orthogonality of the measuring axes is not worse than 1 degree. Eight optical sensors are installed on the nanosatellite THS-0 #2: six photodiode sensors, one ultraviolet Sun sensor, and one infrared horizon sensor.

More details on the passive attitude control system can be found in [17].

3. Attitude Motion Reconstruction Technique

In this section, a technique for the sensors' measurements processing of TNS-0 #2 is presented. A mathematical model of the angular motion of the nanosatellite with relevant assumptions is provided.

3.1. TNS-0#2 Attitude Motion Equations

Consider the motion of a satellite with hysteresis rods and a permanent magnet, taking into account both gravitational and aerodynamic torques. Assume that the satellite is a rigid body moving along a circular orbit around the Earth. The geomagnetic field model is IGRF [18]. The parallelogram model is used to describe the effect of the hysteresis in rods [19].

The following reference frames are used in the paper. $OX_OY_OZ_O$ is the orbital reference frame with the origin placed in the satellite center of mass. The axis OZ_O is directed along the satellite radius-vector from the Earth center. OY_O is perpendicular to the orbital plane, and axis OX_O complements these axes. $OXYZ$ is the body-fixed reference frame, and its axes are directed as shown in Figure 2.

The attitude motion is described using the Euler equations and kinematic relations based on the quaternion. The satellite state vector consists of the absolute angular velocity vector Ω and the attitude quaternion $\Lambda = (\mathbf{q}, q_0)$. Here $\mathbf{q}$ is the vector part of the quaternion and q_0 is the scalar part. Additionally, the direction cosines matrix $\mathbf{A}$ is used for the torques formulation and for the attitude representation.

The dynamical motion equations are as follows:

$$\mathbf{J}\dot{\Omega} + \Omega \times \mathbf{J}\Omega = \mathbf{M}_{mag} + \mathbf{M}_{grav} + \mathbf{M}_{hyst} + \mathbf{M}_{aero}$$

where $\mathbf{J}$ is the inertia tensor, $\mathbf{M}_{mag}$, $\mathbf{M}_{grav}$, $\mathbf{M}_{hyst}$, and $\mathbf{M}_{aero}$ are the magnetic torque caused by the permanent magnet, gravitational torque, magnetic torque due to the hysteresis rods, and aerodynamic torque, respectively. The gravitational torque is

$$\mathbf{M}_{grav} = 3\omega_0^2 (\mathbf{A}\mathbf{e}_3) \times \mathbf{J}(\mathbf{A}\mathbf{e}_3)$$

where $\mathbf{e}_3 = \begin{bmatrix} 0 & 0 & 1 \end{bmatrix}^T$ is a local vertical vector written in the orbital reference frame, and ω_0 is the orbital angular velocity vector $\omega_0 = \begin{bmatrix} 0 & \omega_0 & 0 \end{bmatrix}^T$. The torque due to the permanent magnet is as follows:

$$\mathbf{M}_{mag} = \mathbf{m} \times \mathbf{B}$$

where $\mathbf{m}$ is the dipole moment of the permanent magnet, $\mathbf{B}$ is the geomagnetic induction vector. The torque caused by the hysteresis rods is

$$\mathbf{M}_{hyst} = \mathbf{m}_{hyst} \times \mathbf{B}$$

where $\mathbf{m}_{hyst}$ is the resulting dipole magnetic moment of all the rods. The dipole moment of one rod is

$$\mathbf{m}_{hyst}^k = \mu_k V_k H_0 W \mathbf{e}_k / \mu_0$$

where μ_k is the relative magnetic permeability of the k-th rod, V_k is its volume, H_0 is the mean magnitude of the geomagnetic $\mathbf{H}$-field in the current point of the orbit, $W(H_\tau)$ is a dimensionless function describing the dependence of the induction of the rod related to H_0 according to the parallelogram hysteresis model, μ_0 is the magnetic constant, and $H_\tau = \mathbf{H}\mathbf{e}_k$, $\mathbf{e}_k$ is the unit vector directed along the rod in the body reference frame.

The aerodynamic torque is

$$\mathbf{M}_{aero} = -\sum \mathbf{d} \times \mathbf{f}_a.$$

Vector $\mathbf{d}$ determines the position of the satellite centre of mass relative to the centre of pressure, and $\mathbf{f}_a$ is the aerodynamic drag force acting on the side in the body-fixed reference frame. The sum over all sides facing the incoming flow is performed.

Dynamic equations are supplemented with kinematic relations. The attitude quaternion is used in the numerical simulation. Its kinematic equation is as follows:

$$\dot{\Lambda} = \tfrac{1}{2}\mathbf{C}\Lambda$$

$$\mathbf{C} = \begin{bmatrix} 0 & \omega_3 & -\omega_2 & \omega_1 \\ -\omega_3 & 0 & \omega_1 & \omega_2 \\ \omega_2 & -\omega_1 & 0 & \omega_3 \\ -\omega_1 & -\omega_2 & -\omega_3 & 0 \end{bmatrix}.$$

Here $\omega = [\omega_1, \omega_2, \omega_3]^T$ is the angular velocity vector relative to the orbital reference frame,

$$\omega = \Omega - \mathbf{A}\omega_0$$

These equations are used for the satellite motion reconstruction with the onboard sensors measurements.

3.2. Measurements Processing Technique

The algorithms of the attitude motion determination using on-board sensors measurements are well studied in the literature. There are two main approaches: real-time on-board motion estimation using recursive or local algorithms [20–22] and post-flight measurements processing for motion reconstruction [23–25] and for satellite parameters characterization [26,27]. The first approach is commonly used for active attitude control systems, and the second is more suitable for motion estimation of satellites with passive attitude control systems. The magnetometer measurements processing technique was used for the TNS-0 #2 nanosatellite attitude motion reconstruction. Sun sensors were not used since they are sensitive to the albedo and therefore have very low accuracy. Also, Sun sensors measurements are not available in the shadowed part of the orbit. Nevertheless, Sun sensors are used for the verification of the motion estimation established using magnetometer measurements.

The problem of the attitude motion reconstruction is formulated as follows. It is necessary to obtain such initial conditions for the attitude motion equations that the difference between the predicted measurements calculated using the measurements model and the actual measurements from the on-board sensor achieves a minimum by the mean square criterion.

Consider an initial conditions vector consisting of the quaternion vector part $\mathbf{q}(t = 0)$ and angular velocity vector $\boldsymbol{\omega}(t = 0)$:

$$\boldsymbol{\xi} = [\mathbf{q}(t = 0), \boldsymbol{\omega}(t = 0)]^T$$

For the given vector $\boldsymbol{\xi}$ the attitude quaternion $\Lambda(t = t_k)$ for any time t_k is obtained by the integration of the motion equations. The quaternion provides the prediction of the magnetometer measurements

$$\tilde{\mathbf{b}}_{\mathrm{model}}^k = \mathbf{A}(\Lambda_k)\mathbf{b}_o^k$$

where $\tilde{\mathbf{b}}_{\mathrm{model}}^k$ is a predicted unit vector of the geomagnetic induction field in the orbital reference frame, and $\mathbf{b}_o^k$ is calculated according to IGRF model for a given position of the satellite (obtained using either GPS/GLONASS receiver or TLE and SGP4 model). The problem of the vector $\boldsymbol{\xi}$ determination reduces to the problem of the following function minimization

$$F(\boldsymbol{\xi}) = \sum_{k=1}^{N} \left(\left\| \tilde{\mathbf{b}}_{\mathrm{model}}^k - \mathbf{b}_{meas}^k \right\| \right)^2$$

where $\mathbf{b}_{meas}^k$ is a unit magnetometer measurements vector excluding the constant bias. The minimization of the function $F(\boldsymbol{\xi})$ is carried out using the nonlinear optimization methods.

4. Measurements Processing Results and Passive Attitude Motion Regimes Analysis

4.1. Attitude Stabilization After The Launch

The first communication session that included the sensors measurements telemetry occurred on 19 August at 17 h 11 min UTC. Figure 3 shows the measurements of the magnetometer.

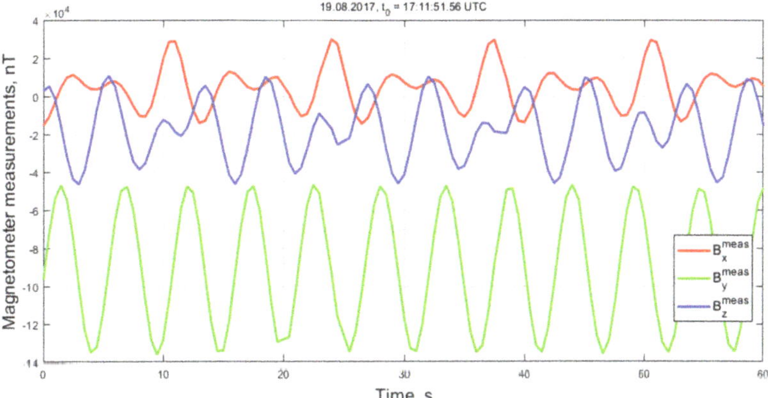

Figure 3. The first measurements from the magnetometer.

Magnetometer measurements include the permanent magnet and hysteresis rods influence. The main contribution is due to the magnet. The corresponding constant bias in the magnetometer measurements was estimated using the least squares method. The technique involves the functional of the difference between the field magnitudes according to the measurements and according to the IGRF model. Figure 4 shows the obtained geomagnetic field corrected for the constant bias of the magnetometer measurements $\mathbf{B}_{bias} = \begin{bmatrix} 5.8 & -90.8 & -20.9 \end{bmatrix} \cdot 10^3 \, \text{nT}$.

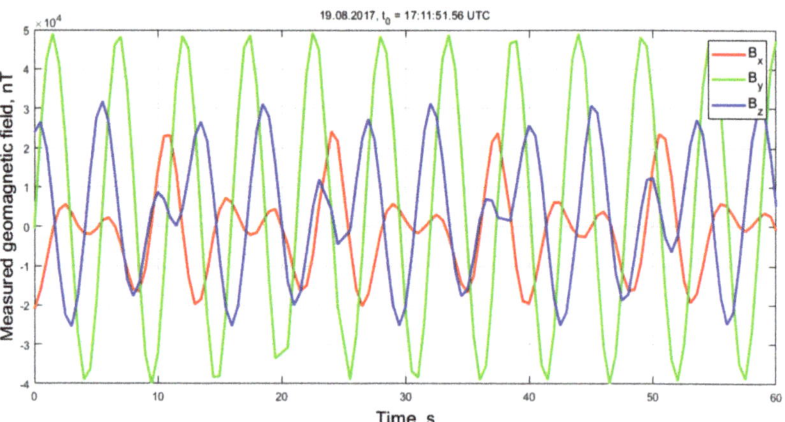

Figure 4. Measured geomagnetic field excluding bias.

Figure 5 depicts the measured and predicted unit geomagnetic induction vector according to the motion reconstruction scheme discussed in previous section. The magnetometer measurements and their predicted magnitudes are close. This indicates relatively accurate satellite attitude reconstruction.

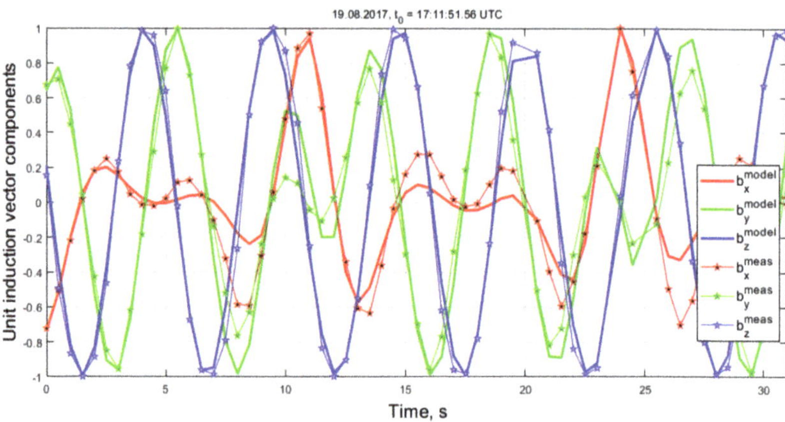

Figure 5. Measured and predicted unit vector of the geomagnetic field.

Figure 6 shows the angular velocity reconstructed according to the found initial conditions and the equations of motion. Note that the aerodynamic torque is not taken into account for the motion after the launch. The initial orbit altitude is about 420 km and the aerodynamic torque small and can be neglected.

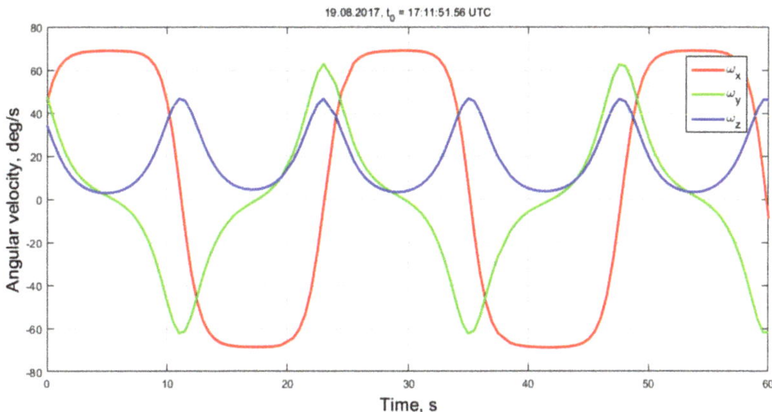

Figure 6. Angular velocity components.

Measurements from the Sun sensors were predicted and compared with actual measurements. It is assumed that the photodiode sensors have a sensitive area in the form of a cone with an opening angle of 120 degrees. The sensor measurements within this region are considered to have a cosine-like dependence on the incidence angle. The measurements are zero outside this region. The measurements were normalized by the maximum value according to the procedure described in [17]. Figure 7 presents a comparison of sensor measurements and their predicted values using the satellite motion model with initial conditions obtained from the magnetometer data. There is some correlation between the corresponding curves. Periodic significant discrepancy is due to the unaccounted influence of the Earth's albedo. Its contribution can reach up to 30% of the Sun's influence. In particular, non-zero measurements from Sun sensors № 4 and 5 can be explained by the effect of the light reflected from the Earth. These sensors are located at opposite sides of the satellite. This indicates the clearly low accuracy of the Sun sensors. Therefore, they are not used directly in the attitude reconstruction process. Their measurements are utilized only to validate the results of the reconstruction with the magnetometer measurements.

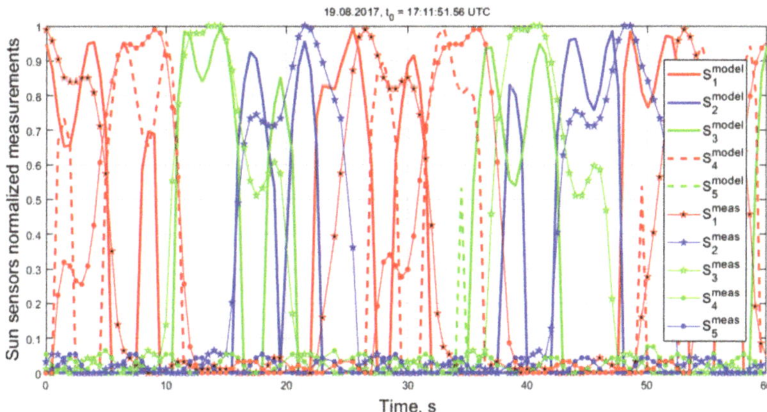

Figure 7. Sun sensors normalized measurements and its predicted values.

Figure 8 shows the time history of the angular velocity magnitude after the launch of the satellite. The curve is close to linear, which corresponds to the model of the hysteresis dampers. The damping finished and the satellite achieved the required magnetic attitude after about 36 days. This is due to a very high initial angular velocity of about 79 deg./s.

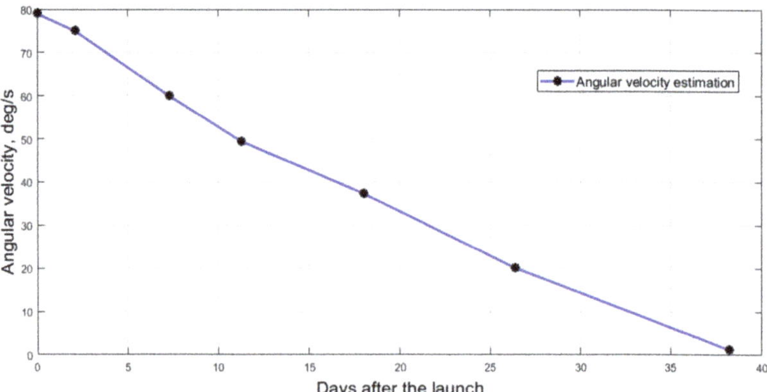

Figure 8. Angular velocity damping history.

4.2. Magnetic Stabilization of TNS-0 # 2 Nanosatellite

Telemetry for approximately two orbits was stored and downloaded on 2 October for the proper assessment of the nominal magnetic stabilization motion. Figure 9 shows the deviation of OZ axis from the local magnetic field vector. This angle does not exceed 12 degrees, and the average deviation along the orbit is about 5 degrees. This is typical for a passive magnetically stabilized satellite. The characteristic period of the forced oscillations (about 9 min) caused by the uneven rotation of the local geomagnetic field induction vector is clearly seen. The frequency is close to the natural frequency of the satellite's oscillations as a rigid body with a permanent magnet in a constant external magnetic field. The period of comparatively slow amplitude variation of these oscillations (a time equal to half the satellite's revolution in orbit around the Earth) is associated with a change in the magnitude of the induction vector of the local geomagnetic field. Figure 10 depicts the angular velocity during the motion. The satellite slowly rotates around the longitudinal axis with angular velocity of 0.4 deg./s. This rotation cannot be damped by the hysteresis rods installed perpendicular to the rotation axis.

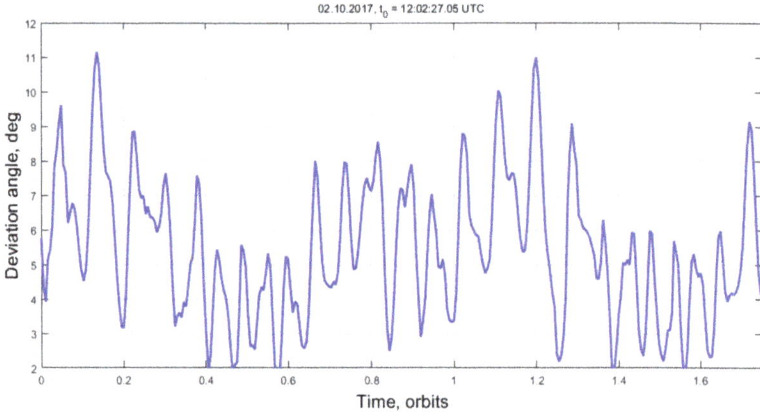

Figure 9. Deviation from the direction of the local magnetic field.

The described passive magnetic stabilization maintained since October 2017 till the last several months of the mission. The deviation of the permanent magnet axis from the local geomagnetic field for a number of months in 2018 is shown in Figure 11.

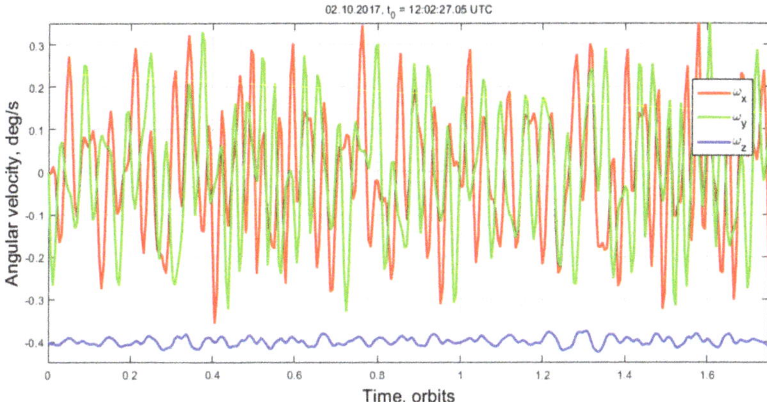

Figure 10. Angular velocity in the steady-state motion.

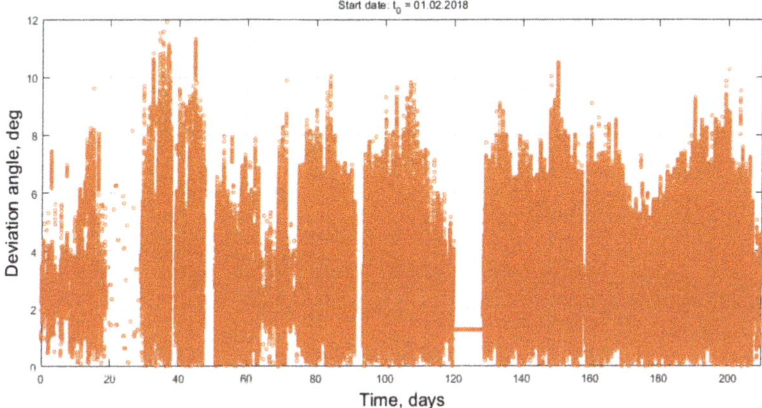

Figure 11. Deviation of OZ axis from the direction of the local magnetic field for 210 days starting from 1 February 2018.

The deviation eventually increased due to the increasing influence of the aerodynamic torque as the satellite orbit decayed. The satellite altitude for the whole mission duration is shown in Figures 12 and 13 shows the magnet axis deviation from the local geomagnetic field starting from 1 July 2019. The deviation increased up to 40 degrees in July. The magnetic stabilization was completely lost in August; the satellite was tumbling.

The satellite achieved aerodynamic stabilization in about two weeks. Then the satellite lost this attitude on 6 September, the last day of the mission. The transient motion and aerodynamic stabilization are discussed in the following section.

4.3. Transient Attitude Motion

The telemetry obtained on 20 August 2019 is considered below. The satellite altitude was 240 km, the average atmosphere density value is 4×10^{-11} kg/m^3. The aerodynamic torque value prevailed over the magnetic torque. The aerodynamic torque is strongly affected by the center of mass shift relative to the center of pressure. This displacement was included in the vector of the estimated parameters in the least squares method. The displacement estimation provided approximately 7 cm, along the longitudinal axis.

The satellite lost the magnetic stabilization on August. However, the aerodynamic stabilization was not achieved as indicated by Figure 14. The satellite rotation rate is relatively high as presented in Figure 15. This represents the transient motion from the magnetic to the aerodynamic stabilization. The aerodynamic torque is large enough to disturb the magnetic stabilization. However, the magnetic torque is still strong enough to prevent the aerodynamic stabilization.

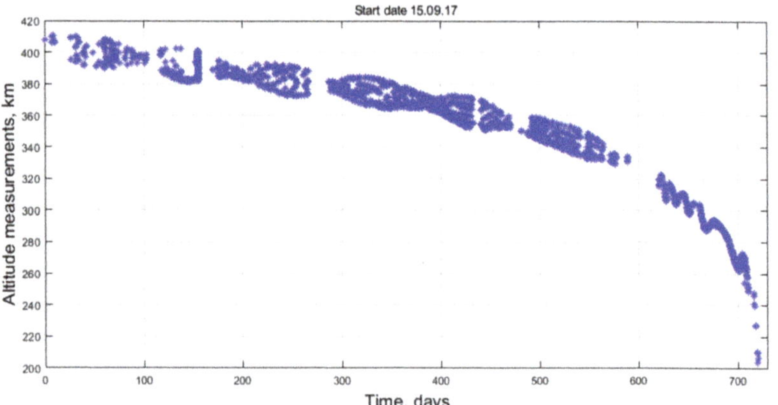

Figure 12. Altitude for the whole mission lifetime according to the onboard GPS/GLONASS receiver.

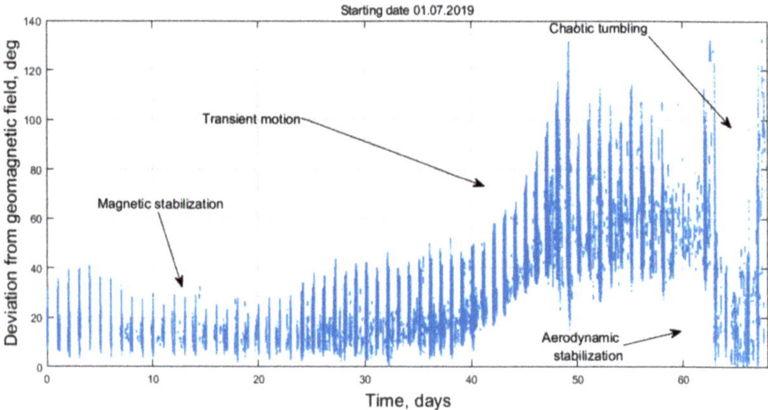

Figure 13. Deviation of OZ axis from the direction of the local magnetic field.

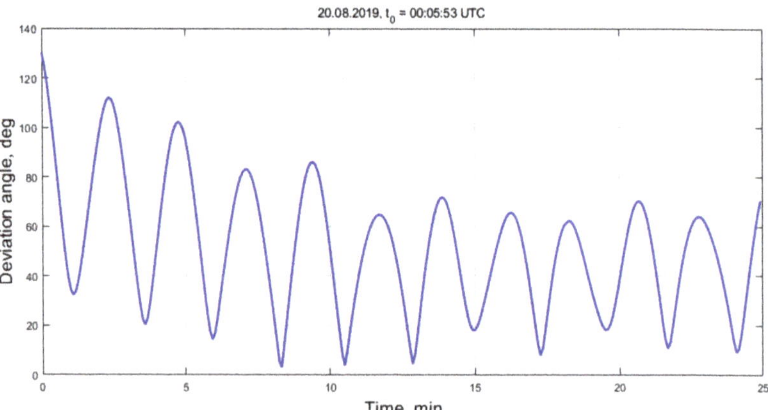

Figure 14. Deviation of OZ axis of the satellite from the flow direction.

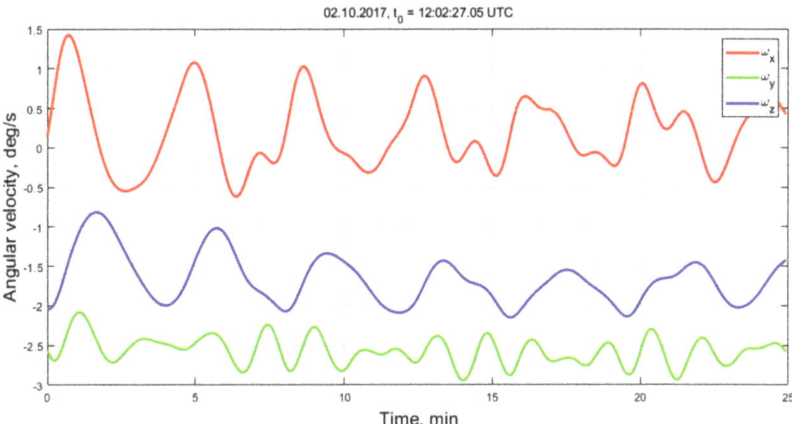

Figure 15. Estimated angular velocity.

4.4. Aerodynamic Attitude Stabilization

The satellite exhibited the tumbling transient motion for about two weeks. The telemetry analysis for 5 September 2019 is provided below. The satellite altitude is approximately 180 km and the atmosphere density is 3×10^{-10} kg/m^3. The aerodynamic torque is larger than the magnetic one by an order of magnitude. The deviation of the axis of the approximate dynamical symmetry from the direction of the incoming airflow does not exceed 30 degrees according to Figure 16. The angular velocity decreased to 0.5 deg/s (Figure 17) and a slow rotation around OZ axis is observed.

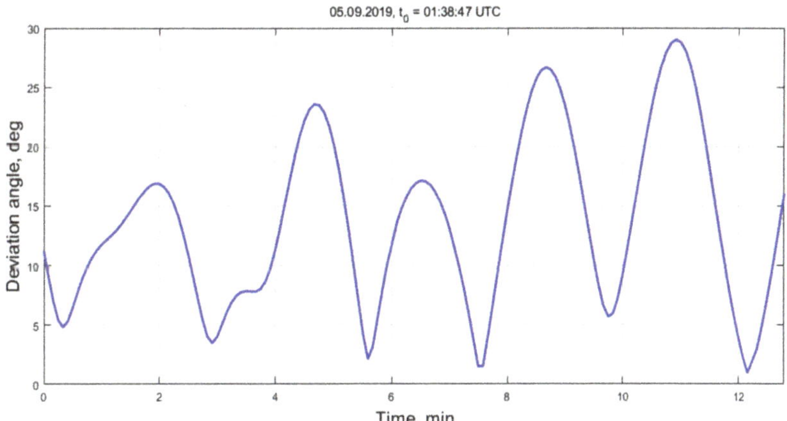

Figure 16. Deviation of OZ axis from the direction of the incoming air flow.

4.5. Tumbling During The Very Last Day

6 September 2019 was the last day of the mission. The satellite altitude was 153 km and the atmosphere density was 1.5×10^{-9} kg/m^3. The satellite lost the aerodynamic stabilization, as seen in Figures 18 and 19.

The satellite tumbling rate increased up to 5 degrees per second before the contact was eventually lost. The reason of this motion type cannot be established reliably but one of the reasons can be the loss of the hull integrity.

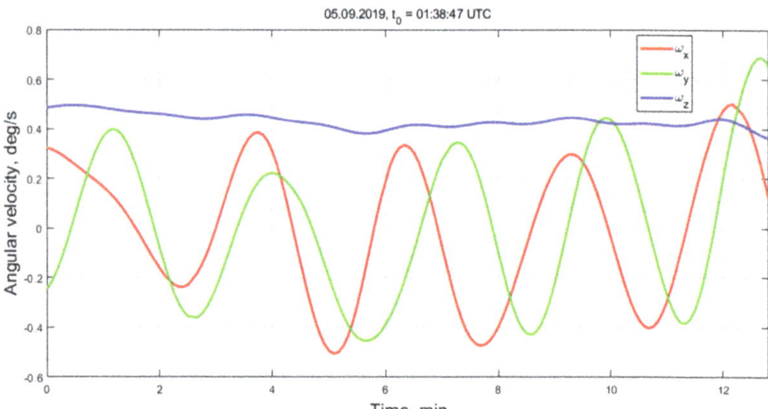

Figure 17. Angular velocity in the body reference frame.

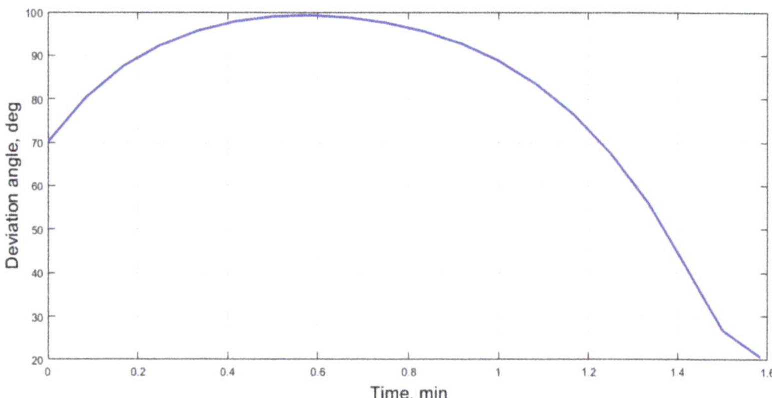

Figure 18. Deviation of OZ axis from the direction of the incoming air flow.

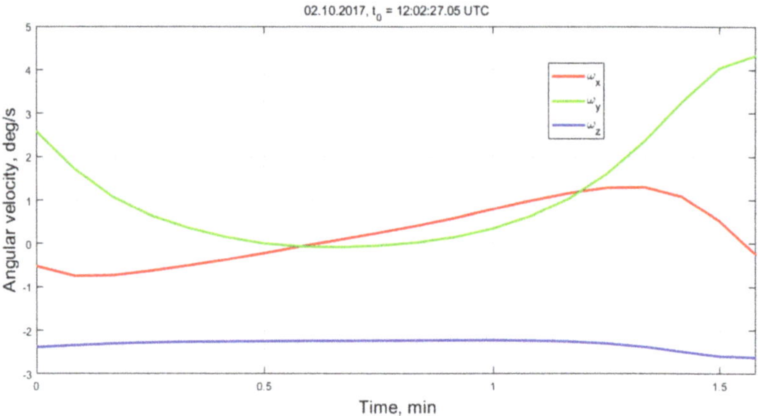

Figure 19. Angular velocity in the body reference frame.

5. Conclusions

The attitude motion of TNS-0 #2 satellite was analyzed. The aerodynamic torque influence gradually increased compared to the magnetic torque due to the orbit decay in last months of the mission. This caused the transition from the passive magnetic stabilization of about 12 degrees of accuracy to the aerodynamic stabilization with the deviation from the incoming airflow of about 30 degrees. The communication via the GlobalStar satellites provided enough onboard sensor measurements for the attitude motion reconstruction up the altitude of 153 km, probably just before the breakdown of the satellite in the dense atmospheric layers.

Author Contributions: Data processing methodology, D.I.; manuscript preparation, D.R.; obtained data analysis, S.T.; data processing, Y.M. and S.S.; work coordination, M.O.; communication with satellite, telemetry preprocessing, I.F., N.Y., and A.S. All authors have read and agreed to the published version of the manuscript.

Funding: This research received no external funding.

Institutional Review Board Statement: Not applicable.

Informed Consent Statement: Not applicable.

Data Availability Statement: Not applicable.

Conflicts of Interest: The authors declare no conflict of interest.

References

1. Ovchinnikov, M.Y.; Shargorodskiy, V.D.; Pen'kov, V.I.; Mirer, S.A.; Guerman, A.D.; Nemuchinskiy, R.B. Nanosatellite REFLECTOR: Choice of parameters of the attitude control system. *Cosm. Res.* **2007**, *45*, 60–77. [CrossRef]
2. Sarychev, V.A.; Mirer, S.A.; Sazonov, V.V. Plane oscillations of a gravitational system satellite-stabilizer with maximal speed of response. *Acta Astronaut.* **1976**, *3*, 651–669. [CrossRef]
3. Sarychev, V.A.; Gutnik, S.A. Gravity-oriented satellite dynamics subject to gravitational and active damping torques. *Cosm. Res.* **2018**, *56*, 68–74. [CrossRef]
4. German, A.D.; Gutnik, S.A.; Sarychev, V.A. Satellite dynamics due to gravity and constant torques. *J. Comput. Syst. Sci. Int.* **2017**, *56*, 125–136. [CrossRef]
5. Battagliere, M.L.; Santoni, F.; Ovchinnikov, M.; Graziani, F. Hysteresis Rods in The Passive Magnetic Stabilization System for University Micro and Nanosatellites. In Proceedings of the 59th IAC, Glasgow, UK, 29 September–3 October 2008; p. 10.
6. Long, M.; Lorenz, A.; Rodgers, G.; Tapio, E.; Tran, G.; Jackson, K.; Twiggs, R.; Bleier, T. A Cubesat Derived Design for a Unique Academic Research Mission in Earthquake Signature Detection. In Proceedings of the 16th Annual/USU Conference on Small Satellites, Logan, TX, USA, 12–15 August 2002; p. 17.
7. Tsuda, Y.; Sako, N.; Eishima, T.; Ito, T.; Arikawa, Y.; Miyamura, N.; Tanaka, A.; Nakasuka, S. University of Tokyo's CubeSat Project-Its Educational and Technological Significance. In Proceedings of the 15th Annual AIAA/USU Conference on Small Satellites, Logan, TX, USA, 13–16 August 2001; p. 8.
8. Battagliere, M.L.; Santoni, F.; Piergentili, F.; Ovchinnikov, M.; Graziani, F. Passive magnetic attitude stabilization system of the EduSAT microsatellite. *Aerosp. Eng.* **2010**, *224*, 1097–1107. [CrossRef]
9. Hennepe, F.T.; Zandbergen, B.T.C.; Hamann, R.J. Simulation of the Attitude Behaviour and Available Power Profile of the Delfi C3 Spacecraft with Application of the OpSim Platform. In Proceedings of the Paper at the 1st CEAS European Air and Space Conference, Berlin, Germany, 10–13 September 2007; p. 9.
10. Burton, R.; Rock, S.; Springmann, J.; Cutler, J. Online attitude determination of a passively magnetically stabilized spacecraft. *Acta Astronaut.* **2017**, *133*, 269–281. [CrossRef]
11. Miguel, N.; Colombo, C. Deorbiting spacecraft with passively stabilised attitude using a simplified quasi-rhombic-pyramid sail. *Adv. Sp. Res.* **2020**, *67*, 2561–2576. [CrossRef]
12. Xuegang, Z.; Zhencai, Z.; Hongyu, C. Aerodynamic passive stabilization design and flight data analyses for transitional regime satellite LX-1. *Acta Astronaut.* **2020**, *167*, 232–238. [CrossRef]
13. Sarychev, V.A.; Gutnik, S.A. Satellite dynamics under the influence of gravitational and aerodynamic torques. A study of stability of equilibrium positions. *Cosm. Res.* **2016**, *54*, 388–398. [CrossRef]
14. Psiaki, M.L. Nanosatellite attitude stabilization using passive aerodynamics and active magnetic torquing. *J. Guid. Control. Dyn.* **2004**, *27*, 347–355. [CrossRef]
15. Berthet, M.; Yamada, K.; Nagata, Y.; Suzuki, K. Feasibility assessment of passive stabilisation for a nanosatellite with aeroshell deployed by orbit-attitude-aerodynamics simulation platform. *Acta Astronaut.* **2020**, *173*, 266–278. [CrossRef]
16. Ovchinnikov, M.; Ivanov, D.; Pansyrnyi, O.; Sergeev, A.; Fedorov, I.; Selivanov, A.; Khromov, O.; Yudanov, N. Technological NanoSatellite TNS-0 #2 Connected Via Global Communication System. *Acta Astronaut.* **2020**, *170*, 1–5. [CrossRef]
17. Ivanov, D.S.; Ovchinnikov, M.Y.; Pantsyrnyi, O.A.; Selivanov, A.S.; Sergeev, A.S.; Fedorov, I.O.; Khromov, O.E.; Yudanov, N.A. Nanosatellite TNS-0 No.2 Attitude Motion After The Launch From ISS. *Cosm. Res.* **2019**, *57*, 272–288. [CrossRef]

18. Thébault, E.; Finlay, C.C.; Beggan, C.D.; Alken, P.; Aubert, J.; Barrois, O.; Bertrand, F.; Bondar, T.; Boness, A.; Brocco, L.; et al. International Geomagnetic Reference Field: The 12th generation. *Earth Planets Sp.* **2015**, *67*, 79. [CrossRef]
19. Sarychev, V.A.; Penkov, V.I.; Ovchinnikov, M.Y. Mathematical model of hysteresis based on magneto-mechanical analogy. *Math. Simul.* **1989**, *1*, 122–133.
20. Ovchinnikov, M.Y.; Ivanov, D.S.; Ivlev, N.A.; Karpenko, S.O.; Roldugin, D.S.; Tkachev, S.S. Development, integrated investigation, laboratory and in-flight testing of Chibis-M microsatellite ADCS. *Acta Astronaut.* **2014**, *93*. [CrossRef]
21. Ivanov, D.; Ovchinnikov, M.; Ivlev, N.; Karpenko, S. Analytical study of microsatellite attitude determination algorithms. *Acta Astronaut.* **2015**, *116*, 339–348. [CrossRef]
22. Kim, O.-J.; Shim, H.; Yu, S.; Bae, Y.; Kee, C.; Kim, H.; Lee, J.; Han, J.; Han, S.; Choi, Y. In-Orbit Results and Attitude Analysis of the SNUGLITE Cube-Satellite. *Appl. Sci.* **2020**, *10*, 2507. [CrossRef]
23. Beuselinck, T.; Van Bavinchove, C.; Abrashkin, V.I.; Kazakova, A.E.; Sazonov, V.V. Determination of attitude motion of the Foton M-3 satellite according to the data of onboard measurements of the Earth's magnetic field. *Cosm. Res.* **2010**, *48*, 246–259. [CrossRef]
24. Abrashkin, V.I.; Voronov, K.E.; Piyakov, Y.Y.; Sazonov, V.V.; Semkin, N.D.; Chebukov, S.Y. Attitude motion of the Photon M-4 satellite. *Cosm. Res.* **2014**, *54*, 315–322. [CrossRef]
25. Abrashkin, V.I.; Voronov, K.E.; Piyakov, I.V.; Puzin, Y.Y.; Sazonov, V.V.; Semkin, N.D.; Chebukov, S.Y. Rotational motion of Foton M-4. *Cosm. Res.* **2016**, *54*, 296–302. [CrossRef]
26. Kramer, A.; Bangert, P.; Schilling, K. UWE-4: First Electric Propulsion on a 1U CubeSat—In-Orbit Experiments and Characterization. *Aerospace* **2020**, *7*, 98. [CrossRef]
27. Belokonov, I.V.; Kramlikh, A.V.; Lomaka, I.A.; Nikolaev, P.N. Reconstruction of a Spacecraft's Attitude Motion Using the Data on the Current Collected from Solar Panels. *J. Comput. Syst. Sci. Int.* **2019**, *58*, 286–296. [CrossRef]

applied sciences

Article

Passive Thermal Control Design Methods, Analysis, Comparison, and Evaluation for Micro and Nanosatellites Carrying Infrared Imager

Shanmugasundaram Selvadurai [1,*], Amal Chandran [1,2], David Valentini [3] and Bret Lamprecht [2]

1 Satellite Research Centre, School of Electrical and Electronics Engineering, Nanyang Technological University, Singapore 639798, Singapore; achandran@ntu.edu.sg
2 Laboratory for Atmospheric and Space Physics, University of Colorado, Boulder, CO 80309, USA; bret.lamprecht@lasp.colorado.edu
3 Thales Alenia Space, 06150 Cannes, France; david.valentini@thalesaleniaspace.com
* Correspondence: sselvadurai@ntu.edu.sg

Abstract: Advancements in satellite technologies are increasing the power density of electronics and payloads. When the power consumption increases within a limited volume, waste heat generation also increases and this necessitates a proper and efficient thermal management system. Mostly, micro and nanosatellites use passive thermal control methods because of the low cost, no additional power requirement, ease of implementation, and better thermal performance. Passive methods lack the ability to meet certain thermal requirements on larger and smaller satellite platforms. This work numerically studies the performance of some of the passive thermal control techniques such as thermal straps, surface coatings, multi-layer insulation (MLI), and radiators for a 6U small satellite configuration carrying a mid-wave infrared (MWIR) payload whose temperature needs to be cooled down to 100K. Infrared (IR) imagers require low temperature, and the level of cooling is entirely dependent on the infrared wavelengths. These instruments are used for various applications including Earth observations, defence, and imaging at IR wavelengths. To achieve these low temperatures on such instruments, a micro-cryocooler is considered in this study. Most of the higher heat dissipating elements in the satellite are mounted to a heat exchanger plate, which is thermally coupled to an external radiator using thermal straps and heat pipes. The effects of the radiator size, orbital inclinations, space environments, satellite attitude with respect to the sun, and surface coatings are discussed elaborately for a 6U satellite configuration.

Keywords: thermal control systems; nanosatellite; micro-satellite; heat pipes; radiators; thermal straps; infrared imagers

Citation: Selvadurai, S.; Chandran, A.; Valentini, D.; Lamphrecht, B. Passive Thermal Control Design Methods, Analysis, Comparison, and Evaluation for Micro and Nanosatellites Carrying Infrared Imager. *Appl. Sci.* **2022**, *12*, 2858. https://doi.org/10.3390/app12062858

Academic Editors: Simone Battistini, Filippo Graziani and Mauro Pontani

Received: 6 February 2022
Accepted: 8 March 2022
Published: 10 March 2022

Publisher's Note: MDPI stays neutral with regard to jurisdictional claims in published maps and institutional affiliations.

1. Introduction

Satellite thermal control methods that are currently used by conventional space missions are well established and proven. The term "NewSpace" is being adopted by many agencies and industries in recent days. According to experts, NewSpace is an approach focused on lowering the barriers of entry into space, by providing cheaper access to space and by commercializing the space sectors [1]. This has also further increased the interests among many small satellite manufacturers, research organizations, universities, and start-ups to build complex missions at a lower cost using nanosatellites and CubeSat platforms, especially for technology demonstration, science, research, educational projects, and commercial applications [2]. These nanosatellites typically weigh 1 to 10 kg and follow CubeSat size standards where 1U is a $10 \times 10 \times 10$ cm^3 cube. There are other challenges in the new era of space technologies but this article focuses only on thermal management challenges on small satellites. Due to miniaturized electronics, high power components, and payloads, the satellite becomes heavily packed within the smaller satellite volume, and this leads to thermal problems as the available radiative surface area is reduced [3,4]. There is a necessity to solve thermal challenges for high power, scientific, and cryogenic small satellite missions. Taking advantage of small satellites, earth exploration missions are now being carried out using nano/microsatellites and are expected to grow exponentially [5–7]. Earth observation (EO) refers to the use of remote sensing technologies to monitor land, marine and atmosphere. Most of the earth observations are done in the infrared region of the electromagnetic spectrum. Infrared detectors that are designed to operate in the wavelength ranging from 0.76 μm to 103 μm need to be cooled down for better performance. The relationship between infrared wavelength and temperature is defined by Wien's displacement law in Equation (1) as follows,

$$\lambda T = b \tag{1}$$

where, b is Wien's constant (2898 μmK), λ—Detection wavelength, and T—Low temperature. From this relation, it is clear that the radiation wavelength shifts toward the shortwave direction as the temperature rises. In general, the longer the infrared wavelength is, the lower the operating temperature of the detector will be [8].

Infrared detectors operating at a non-zero temperature are subjected to dark current (DC) noise which is directly related to pixel area, detector material properties and detector temperature. DC noise is the dominant limitation in many detectors and a relationship to calculate the DC is given in Equation (2) [9].

$$I_{DC} = C \times T^2 \times A_d \times e^{\frac{\Delta Q}{kT_d}} \tag{2}$$

where, I_{DC}—Dark Current intensity (A); C = Constant $(1.2 \times 10^6$ A·m^{-2}·K$^{-2})$; T_d—Pixel Temperature (K); ΔQ—Energy band gap (eV); A_d—Pixel area (m^2); Q—Electron charge $(-1.6022 \times 10^{-19}$ J); k—Boltzmann constant $(1.38064852 \times 10^{-23}$ J/K); T—Detector temperature (K).

Dark current is significantly reduced when operating at lower temperatures, and hence most of the infrared detectors are required to be cooled to lower temperatures for certain applications. The low temperature requirement becomes of utmost importance for the detectors operating at the cryogenic temperature regime [8]. This study considers a mid-wave infrared (MWIR) instrument whose detector is required to be maintained between 0 to -100 °C for a few specific mission operations. Traditional passive thermal control methods are studied both numerically and analytically to analyze the effectiveness of the overall thermal control system in maintaining the payloads and components at required temperature levels.

For missions with such infrared (IR) payloads and stringent thermal requirements, active thermal control systems are used. Satellite active thermal control systems (ATCS) rely on input power for operation and have been shown more effective in maintaining the required temperature within required limits [10] but this increases the total power budget of the satellite and eventually the cost. In general, small satellites cannot generate more power due to limited solar array size. Some of the active thermal systems are thermo electric coolers, electric heaters, and pumped fluid loop systems. Besides cost and input power, these methods require additional volume which is limited in small satellites. Unlike active systems, a passive thermal control system (PTCS) does not require any additional input power from the satellite and it also can be implemented at lower cost. Some of the passive thermal control systems are, multi-layer insulation (MLI), thermal straps, radiators, thermal louvers, heat pipes, phase change materials, and thermal switches [11]. No small satellite missions with thermally sensitive IR instruments (operating at cryogenic range of temperatures) have flown and it is only due to lack of thermal control systems. NASA's JPL has developed an active cryocubesat (ACCS) [12] for such a thermally sensitive IR instrument, which is a breakthrough technology for many future cryogenic small satellite missions. To save cost, power, and volume inside the satellite, a few PTCS are studied for a nano/micro satellite carrying a an IR instrument.

2. Satellite and Payload Considered

Technological advancements and miniaturization have led to the use of infrared (IR) instruments in nanosatellite platforms [13]. Infrared instruments used for space imaging, Earth observation (EO), astronomy, surveillance, and hyper- spectral remote sensing at higher IR wavelengths come with thermally sensitive detectors, and these detectors are to be operated in a controlled temperature environment for better performance [14].

Table 1 lists all of the instruments in different IR wavelength categories and their detector temperature requirements. Cooled detectors are proven to have higher sensitivity with an extremely reduced signal-to-noise ratio [15]. In some literature, the term thermal infrared (TIR) is used and it is a combined form of both MWIR and SWIR.

Table 1. Infrared instrument—Detector temperature requirements [16,17].

IR Wavelength Category	Wavelength (μm)	Detector Temperature (*K*)
Near-Infrared (NIR)	0.75–1.4	300
Short Wavelength Infrared (SWIR)	0.75–1.4	300
Mid Wavelength Infrared (MWIR)	3–8	50–80
Long Wavelength Infrared (LWIR)	8–15	50–80
Far Infrared (FIR)	15–1000	0.05–20

Figure 1 shows how the heat flows from each element of the system to a deep space radiator. Passive thermal control methods analyzed in this study are mainly for MWIR detector's thermal requirement and other mission requirements. For this payload as seen in Figure 1, it is assumed that the detector is lying inside the instrument and it is coupled to a micro-cryocooler through a high thermal conductivity strap. Heat generated from active micro-cryoccoler parts and associated electronics is carried away to radiator then radiated to space.

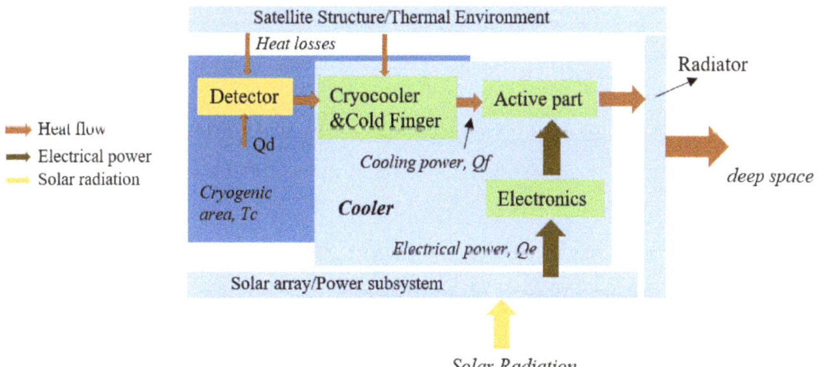

Figure 1. Payload configuration.

This study initially elaborates the theoretical background of the passive thermal control methods such as multi-layer insulation (MLI), thermal straps, radiators, heat pipes, and surface coatings. This is then supported with numerical analyses and performance evaluation studies carried out for a nanosatellite configuration.

3. Passive Thermal Control Methods

3.1. Multi-Layer Insulation

Thermal insulation is crucial for cryogenic and infrared satellite missions. Multi-layer insulation (MLI) blankets are the most efficient insulation material for space applications [18]. The MLI blanket is in general comprised of a number of low-emittance sheets combined with low-conduction netting layers to control the heat transfer for low temperature applications such as cryogenic instruments. Thermal insulation performance is quantified as effective emissivity (e^*) and is dependent on the number of inner layers as well as geometric considerations (number of bends, holes, etc.). Effective emissivity is defined as in Equation (3) [19]

$$e^* = \frac{Q_{Total}}{\sigma(T_{hot}^4 - T_{cold}^4)} \tag{3}$$

where, e^*—Effective emissivity; Q—Total incoming heat flux (W/m^2) ; σ—Stefan-Boltzmann constant (5.670374419 $\times$ 10^{-8} W/m^2·K^4); T_{hot}—Temperature of the MLI hot side (K); T_{cold}—Temperature of the MLI cold side (K). Typical values for e^* are closer to in the range 0.0150 to 0.03 [19].

MLI in satellites is widely used for the following reasons [20].

1. To prevent excessive thermal flux from/to various satellite components
2. To minimize thermal gradients through out the components
3. To reduce temperature variations due to change in orbital environment conditions.

Thermal conduction across the thickness of the MLI is very sensitive to the layer compression. To minimize the MLI conduction heat transfer, any compressive pressure or bending of blankets must be avoided [19]. This conduction term is defined as the total temperature difference (ΔT) between the outer blanket layer and the inner blanket layer divided by the total number of layers. Assuming that the temperature distribution is uniform throughout the blankets, the conduction heat transfer per unit area is given by Equation (4) [21]. MLI efficiency reduces as the size decreases because heat transfer at the blanket edges increases and hence MLI generally does not perform well on small satellite platforms [10].

$$q_c = \frac{k_c \Delta T}{n} \tag{4}$$

where, q_c—MLI conduction heat load per unit area (W/m^2), k_c—Conductance of a single MLI layer (W/m^{2}K), ΔT—Temperature difference across the MLI layer (K), n—Number of MLI layers. Heat flux due to conduction can be reduced by increasing the number of inner layers and this reduction is linear with increasing layer numbers. Conductive heat transfer is typically negligible and ignored in MLI blankets. Unlike conductive heat transfer, radiation heat transfer is considered and is given in Equation (5) [21].

$$q_r = 2.835 \times 10^{-5} \epsilon \frac{T_{hot}^4 - T_{cold}^4}{n} \tag{5}$$

q_r—MLI radiative heat flux (W/m^2), ϵ—Emissivity of MLI outer layers, T_{hot}—Temperature of the MLI hot side (K), T_{cold}—Temperature of the MLI cold side (K), n—Number of MLI layers. From both conduction and radiation terms, it can be proven that the total heat flux is minimized when the number of MLI layers are increased.

3.2. Surface Paints

The satellite's passive thermal control system mainly uses specially prepared thermal coatings to maintain the subsystem temperature within safe operating ranges. In space, external satellite surface paints or coatings are greatly influenced by adverse environmental effects, namely, Atomic Oxygen (ATOX), molecular contamination, and Ultraviolet radiation (UV) [22]. Energy absorbed by the satellite surfaces depends on the external surface characteristics and area. The major concern in using paints is the degradation of physical and thermo-optical properties of the components and this degradation is dependent on mission duration and orbit altitude. Thermal radiation heat transfer on satellites can be controlled using materials that have specific thermo-optical surface properties, which are solar absorptivity (α) and infrared emissivity (ϵ). These properties are dependent on materials and processing techniques [23,24]. α governs how much solar incident heating a spacecraft absorbs, while ϵ determines how much heat a spacecraft emits to space [10]. By altering α and ϵ, the overall temperature of the spacecraft can be controlled to some extent. There are numerous space qualified paints and coatings that are commercially available and most of them have space heritage. Radiative tapes also provide better performance. For example, second-surface fluorinated ethylyne propylene (FEP) provides better performance as radiator coatings [10] and for most small satellites, adhesive tapes or surfaces finishes (polishing, anodize, alodine) have been considered. End-of-life (EOL) properties are considered in the early design phase as the absorbed solar energy will gradually increase over the years due to degradation of surface properties. In general, satellite components will run cooler in the early phase of the mission life and additional heaters may be used to avoid temperature drops for the critical components [25].

3.3. Thermal Straps

Thermal straps are excellent passive heat transfer devices that are commonly used on space missions to conduct the heat from inaccessible regions of the spacecraft and radiate into space. Thermal straps come in various shapes and lengths according to the requirement. They are comprised of thin wires of high thermal conductivity metals or foils which make them flexible and efficient thermal links. Thermal straps are made of different metals and some of them are listed in Table 2 with their thermal properties.

Table 2. Thermal strap materials.

Material Name	Thermal Conductivity, k, W/m·K
Graphene	$\sim$3500
Pyrolytic graphene	$\sim$1500
Graphite fiber	$\sim$800
Copper	$\sim$450
Aluminium	$\sim$225

Note: $\sim$:- Actual values may change but they remain close to the tabulated values.

Thermal contact conductance (TCC) is a property of heat conduction between solid bodies in thermal contact. This property is highly dependent on contact pressure and surface flatness. TCC influences the overall performance of the thermal strap and the relationship is given in Equation (7) [26].

$$Q_{strap} = \frac{\lambda A}{l} \Delta T \tag{6}$$

$$C = \frac{Q_{strap}}{\Delta T} = \frac{Q_{htr} - Q_{leak}}{\Delta T} = \frac{Q_{htr} - Q_{rad} - Q_{htrleads} - Q_{TCleads}}{\Delta T} \tag{7}$$

where, λ—Thermal conductivity (W/mK), A and l are, respectively, the cross-section (m^2) and the length (m) of the heat path inside the medium, C—Thermal contact conductance (W/K), Q_{strap}—applied heat load into the thermal strap (W), Q_{htr}—heat load from the heater pad, Q_{leak}—total heat lost from the heater pad due to radiation (Q_{rad}), from the heater wire leads ($Q_{htrleads}$), and thermocouple leads ($Q_{TCleads}$). ΔT is the difference in temperature measured at the ends (°C). Variation in thermal strap conductance is shown in Figure 2. The lower the thermal strap conductance, the higher the difference in temperature between two ends of the thermal strap will be.

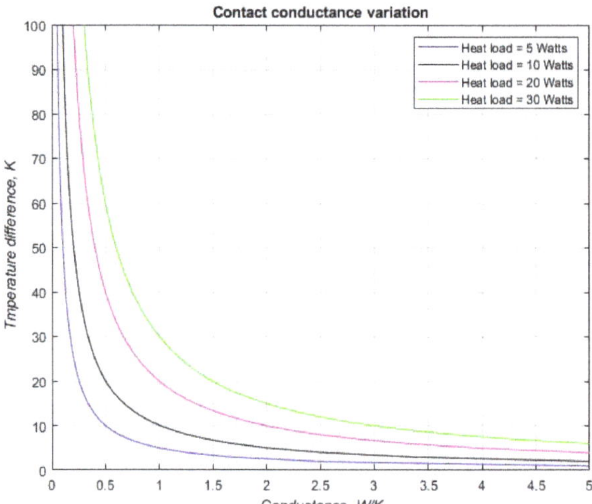

Figure 2. Thermal strap conductance.

3.4. Thermal Straps Conductance Characterization

The test unit consists of a copper thermal strap of the dimensions as described in Figure 3, a thin polyimide heater pad, and two thermocouples for temperature measurement. The entire unit was kept inside the thermal vacuum chamber (TVAC) which contains a temperature-controlled high emissivity shroud and baseplate. The main idea of this test is to find out the thermal conductance of a custom made thermal strap. Prior to the actual test, a preliminary start-up test was performed at a vacuum condition, at 10^{-5} mbar, to analyze how conduction heat transfer plays a major role in transferring the heat from one end to another end. One end of the thermal strap was heated sufficiently using a thin polyimide heaterpad (from minco, HK6903) and a thermal interface material (TIM) is used between the heaterpad and the copper block to minimize the resistance as the entire unit is kept at complete vacuum condition where there is no medium. Two T-Type thermocouples were attached to both hot and cold ends of the thermal strap to measure the actual temperature. The custom designed flexible copper braid that is cold pressed at each of its ends into copper blocks is covered with MLI to block the heat from getting radiated out.

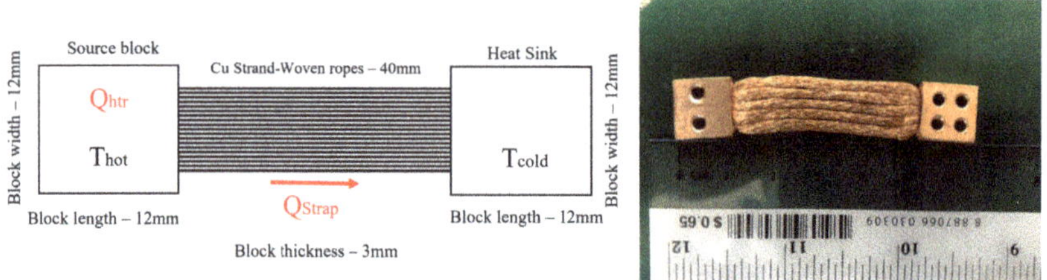

Figure 3. Thermal Strap—test sample.

Incremental heat loads were applied using an external power supply to the hot end, and the cold end is connected to an another copper block (cold sink) to generate a thermal gradient across the strap. Temperatures were measured from both the ends of the strap. It is assumed that the radiation loss is negligible due to MLI and the entire test sample is thermally isolated from the TVAC's baseplate (a bare printed circuit board and a PEEK sheet) using thermal insulators and thus conduction heat loss is also considered to be negligible.

In the test setup, hot and cold end blocks of the copper strap were exposed to the test chamber due to the fact that the radiation from the copper blocks is assumed to be negligible as the surface emissivity of bare copper is 0.02 [27]. However, other measurement uncertainties, such as TVAC system thermocouple reading uncertainties of <0.5 °C, and external power supply error of 0.005 percent, are taken care of and ensured to be within the minimum allowable limits. The stray heat leaks from the sources such as wire harnesses (no harnesses used), and PC104 connectors (a standard interboard connector with 104 metal pins for board-to-board communication and it is made up of black high temperature, glass filled nylon) are neglected in this analysis. When the heater was turned ON with 0.5 W initially, the temperature started to increase and then stabilized after approximately 50 min. Temperature readings after stabilization were recorded from both ends and the same sequence was followed for the rest of the incremental heat loads. For varying heat loads from 0.5 W to 3 W, hot and cold end temperature from the TVAC system were recorded and used to compute the thermal conductance using Equation (7) and plotted as shown in Figure 4. With increasing heat load, the temperature difference between hot and cold ends is increased, but the strap conductance of the strap remains close to a constant value. Temperature gradient (ΔT) for every incremental heat load (Q_{strap}) increases proportionally and this is the reason why the strap conductance is nearly constant. From the experiment, it is observed that the strap conductance varied between 5.17 W/K to 5.010 W/K for the given heat load, and thus an average value of 5 W/K is used for all the thermal simulations carried out in this study. This custom made thermal strap is expected to operate only under 3 W of heat load and hence, higher heat load (>5 W) test was considered to be unnecessary.

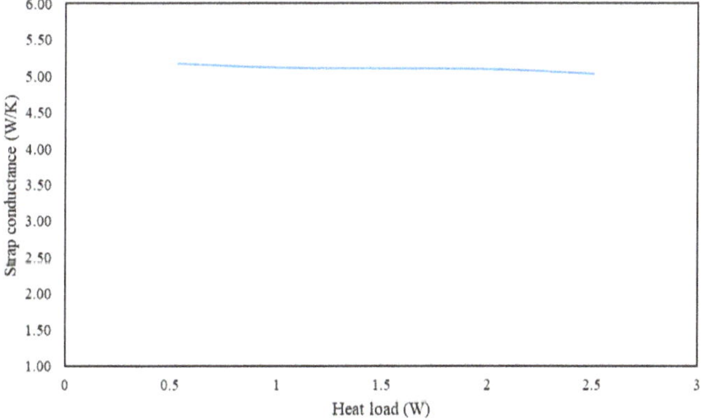

Figure 4. Thermal strap conductance.

3.5. Passive Radiator

Radiators used on the spacecraft are passive radiating elements. In space, there is no medium present to convect or conduct away the waste heat from the spacecraft components to deep space, and it has to be lost only through radiative heat transfer. Radiator panels are specially designed for this purpose, and they come in various configurations such as single active face (Body-mount radiators, BMR) and double active face (deployable radiators, DR) [28,29]. Radiator design and its mounting location on the spacecraft body greatly influences the performances of the radiating panel. Most importantly, radiators are mounted in a location where external fluxes are minimized. Body-mount radiators are designed as an integral part of the spacecraft (S/C) or can be designed separately and mounted to the external surface of the S/C structural body [30]. Secondly, optical properties of the radiators are considered important as the radiating power is dependent on surface coating. Radiating power (Q_{rad}) of the radiator is defined by the Stefan-Boltzmann equation as in (8).

$$Q_{rad} = A_{Rad}\epsilon\sigma F_v(T_{rad}^4 - T_s^4) \tag{8}$$

where, A_{Rad}—Radiator surface area, ϵ—emissivity of the radiator surface, σ—Stefan Boltzmann constant ($5.670374419 \times 10^{-8}$ W/m^2·K^4), F_v—view factor and it will normally be close to unity and it can be lower if the respective surface is partially occulted by other spacecraft components, T_{rad}—Temperature of the radiator surface and T_s—deep space temperature (assumed 3 K). Q_{rad} is radiating power of the radiator and it is strongly dependent on the surface temperature.

Thermal equilibrium condition is defined as the total incoming heat (Q_{in}) equal to heat leaving (Q_{out}) the system and it is given by Equation (9). The following governing equations are used for calculating the radiator area for any given environmental conditions [11,31].

$$Q_{in} = Q_{out} \tag{9}$$

On comparing,

$$q_s \alpha A_s \cos(\theta) = \epsilon \sigma T^4 A \tag{10}$$

Steady state radiating surface temperature is computed by,

$$T = \left[\frac{q_s \alpha \cos(\theta)}{\epsilon \sigma} \right]^{\frac{1}{4}} \tag{11}$$

At LEO, a satellite's thermal equilibrium condition is given by Equation (12),

$$Q_{Internal} + Q_{Solar} + Q_{Albedo} + Q_{Infrared} = Q_{rad} \tag{12}$$

Radiator sizing depends on the total external heat load which is a sum of the solar flux (Q_{Solar}), planet albedo (Q_{Albedo}), planet IR flux ($Q_{Infrared}$), and internal heat load ($Q_{Internal}$).

Equation (13) shows the heat balance between the radiator and the space environment in steady state.

$$(A_s q_s + A_A q_A)\alpha + A_E q_E \epsilon + Q_{internal} = A\sigma\epsilon F_v T_{rad}^4 \tag{13}$$

Equation (13) can be further simplified as follows,

$$Q_{external} + Q_{internal} = A\sigma\epsilon F_v T_{rad}^4 \tag{14}$$

Alternatively,

$$\dot{q}_{absorbed} \cdot A + Q_{internal} = A\sigma\epsilon F_v T_{rad}^4 \tag{15}$$

where, $Q_{external}$—Total external flux and it is a sum of Q_{solar}, Q_{Albedo}, and $Q_{Infrared}$. A_s, A_A and A_E are the projected areas receiving, respectively, solar, albedo and planetary radiation. F_v is the view factor from the radiator to the space environment (assumed perfect view into space, $F_v = 1$) and A_E is the projected area of the earth. q_s, q_a, and q_E are solar flux, albedo flux, and earth IR flux, respectively. $\dot{q}_{absorbed}$ is the total flux or heat absorbed from solar, albedo and infrared radiation per unit area.

Radiator area is calculated using Equation (16) [32],

$$A_{rad} = \frac{Q_{internal}}{\epsilon\sigma F_v T_{rad}^4 - \dot{q}_{absorbed}} \tag{16}$$

And the surface temperature of the radiator is calculated using Equation (17).

$$T_{rad} = \left[\frac{q_s \alpha \cos(\theta) + \frac{Q_w}{A_r}}{\sigma\epsilon} \right]^{\frac{1}{4}} \tag{17}$$

where α is absorptivity of the radiator surface, q_s is the incident solar flux on the radiator panel, Q_w is the heat to be rejected. By knowing $Q_{internal}$ and other environmental heat loads, radiator area can be calculated using Equation (16) assuming that the view factor is 1 for preliminary calculations. If the radiator design calculations are carried out for GEO satellites, Q_{albedo} and Q_{IR} are generally ignored as the effect of earth's albedo and IR flux are insignificant (approximately <0 [11]) because as the altitude increases, environmental loads from earth decreases rapidly. One exception to this is the case of cryogenic systems, which operate at very low temperatures that even small environmental heat loads from earth are significant to the thermal design [11].

The radiator panel for the 6U satellite platform is optimized for two design parameters.

- Radiator size: Radiator panel is allowed to vary in size for maximum radiating capacity. Length and width of the radiator are varied while keeping the thickness constant.
- Radiator Thickness: Radiator thickness is varied while keeping the surface area constant.

Radiating temperature of the radiator for both of these conditions are varied as the input heat load remains same. To account for varying orbital thermal loads, performance studies were carried out for two orbital worst-case conditions. A few assumptions made for this performance study were,

- The radiator loses heat only by radiation mode of heat transfer
- Calculations are done for steady state conditions with constant properties
- Radiator temperature is assumed to be constant across thickness
- Heat loss from the edges is negligible.
- Incoming solar flux is negligible as the radiator looks into deep space all the time in the orbit and mounted below the solar panel.
- Radiator plate is considered to be at a uniform temperature initially.
- Material properties do not change with respect to the temperature.
- Earth shine and planetary radiation are taken into consideration.

Firstly, a body-mount radiator is studied. Based on the advanced standard for Cube-Sats [4], the maximum surface area of the body-mount radiator on a 6U satellite is not more than 200 mm × 300 mm. However, full surface area cannot be utilized due to a few mechanical constraints, and hence the maximum area is assumed to be 290 mm × 180 mm (0.0522 m^2). Similarly, the size of the radiator panel for a 27U satellite is considered to be 300 mm × 300 mm (0.09 m^2). The Figures 5 and 6 illustrate the orientation of the body-mount and deployable radiators for 6U and 27U satellites with respect to the sun.

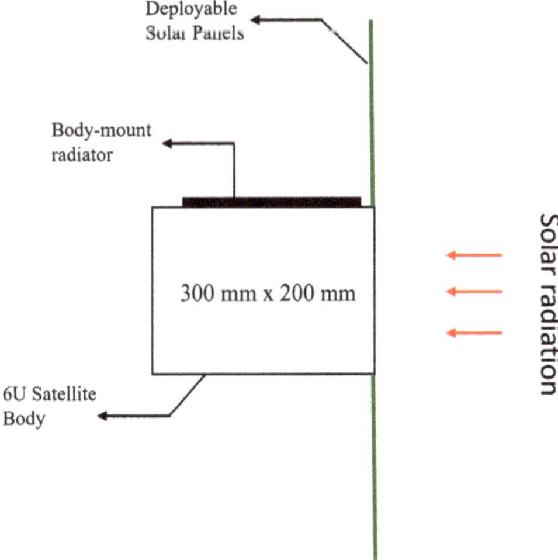

Figure 5. BMR-Satellite Orientation with respect to Sun.

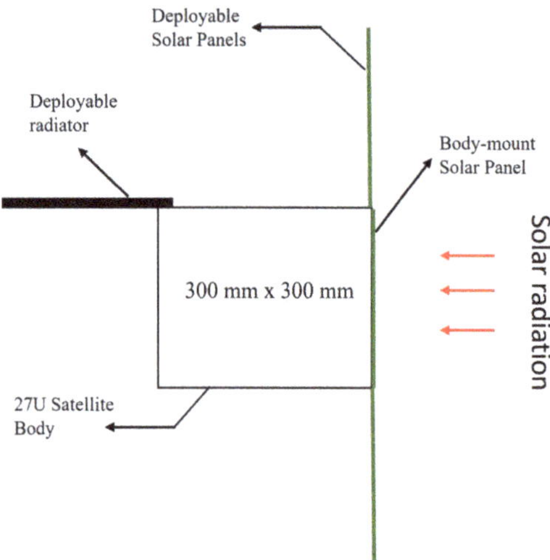

Figure 6. DR-Satellite Orientation with respect to Sun.

As can be seen in Figure 7, total heat rejected by both the body-mount (BMR) and deployable radiators (DR) are increasing exponentially (due to fourth power of T) when the temperature of the radiator is increased for a constant radiating surface area of 0.2 m^2. Total heat rejected for the deployable radiator is higher because of double-active radiating surfaces, whereas the BMR has a single-active radiating surface. For the above calculations, it is assumed that the radiating surfaces are coated with black paint which has an emissivity of 0.85. The radiator is mounted below the solar panel for the 6U configuration, so they do not absorb much heat if exposed to the sun but, typically the radiators are painted white. Figure 8 illustrates the relationship between radiator area (A_{rad}), temperature (T_{rad}), and total heat to be dissipated ($Q_{internal}$). For a constant temperature, the total heat loss into deep space increases with increase in radiator surface area, and this variation is plotted for different radiating temperatures. Placement of the radiator panels on the satellite body is another important decision as the radiator should not be exposed to solar radiation, which is a major heat source. A typical configuration for BMR radiator on a 6U satellite body is shown in Figure 5. In the same way, deployable radiator configuration is also implemented on the satellite body. These panels are thermally isolated from rest of the satellite body to minimize thermal conduction but a strong thermal coupling is established for the detector/component of interest which needs to be cooled.

Thermal storage devices can be used to reduce the surface area of the radiator panels, which indirectly reduces the peak load if there is a transient heat load in the system. However, it is still challenging for small satellite thermal control technologies. Thermal dissipation for high power small satellites is challenging using only a body-mount radiator panel, and this is best addressed by increasing the radiating area by means of deployable panels. If deployable radiator panels are used, it is very important to ensure that the thermal conductivity is higher for the hinges used. Flexible thermal straps along with proper mechanical hinges, which have variable bending angles, can be used to deploy radiators [33]. Flexible thermal straps can be considered for thermal connection between the heat dissipating element and the deployable radiator.

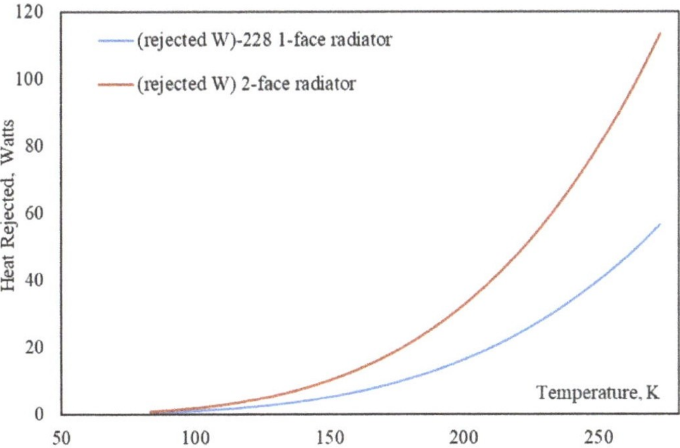

Figure 7. Heat rejection with temperature.

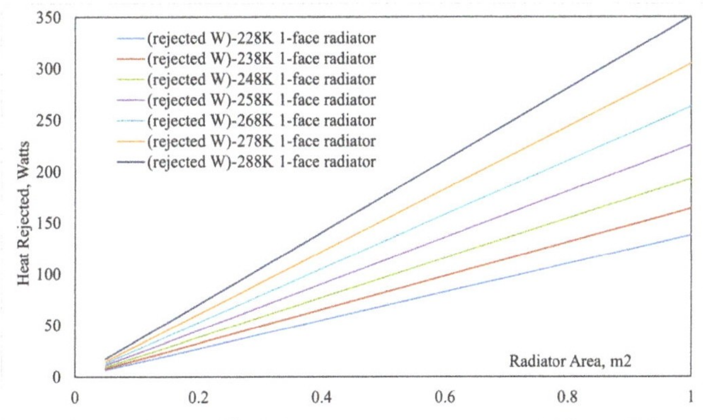

Figure 8. Heat rejection capacity with surface area

Besides radiator surface area, thickness also has a considerable effect on its temperature. According to [34], Mackay and Leventhal et al. derived relationships for heat transfer from an uniform plate heated at one edge. For a thin rectangular deployable radiator radiating in free space, it is assumed that the heat enters uniformly at the fin base and then it passes from the fin faces by radiation [33]. Heat dissipated by the fin faces is given by Equation (20) and this relationship takes thickness parameter (L) into consideration.

$$q_{actual} = 2k\delta L \left[\frac{\sigma\epsilon}{5k\delta}\right]^{\frac{1}{2}} \left[T_b^5 - T_a^5\right]^{\frac{1}{2}} \tag{18}$$

where, q_{actual}—Actual heat input at the base of the fin, δ—width of the fin, k—Thermal conductivity, L—Radiator thickness, T_b—Radiator base temperature, T_a—Radiator tip temperature.

The fin efficiency is calculated from Equation (20). It is defined as the ratio of the actual heat dissipation to the ideal dissipation [35]. In ideal heat dissipation, the entire fin is at the base temperature considered.

$$q_{Ideal} = 2\sigma\epsilon\delta L T_b^4 \tag{19}$$

$$\eta = \frac{q_{actual}}{q_{ideal}} \tag{20}$$

3.6. Heat Pipes

Heat pipes are two-phase heat transfer devices that are categorized as one of the efficient passive thermal control methods for space applications because they do not need any additional power for their operation [36]. With smaller cross sections and high thermal conductance, these devices are capable of transferring heat over a long distance within the satellite. Heat pipes are broadly divided into three sections, namely: evaporator, adiabatic, and condenser. A typical heat pipe as illustrated in Figure 9 [37] has one evaporator section that takes heat from a heat source. The heat causes change of phase of the working fluid (water is used in this study) from liquid to vapor and moves to the adiabatic or transportation section due to increased vapor pressure at the evaporator section and then the vapor reaches the condenser section where condensation rejects the latent heat of the fluid to the sink. The condensed liquid then moves back to the evaporator section due to capillary pumping action [11,31]. The size of the heat pipe entirely depends on the heat load that it is subjected to. Constant Conductance heat Pipes (CCHP) are the basic and standard isothermal heat pipes whose thermal conductivity also can be varied by changing the physical properties such as pore diameter, porosity, and permeability [38].

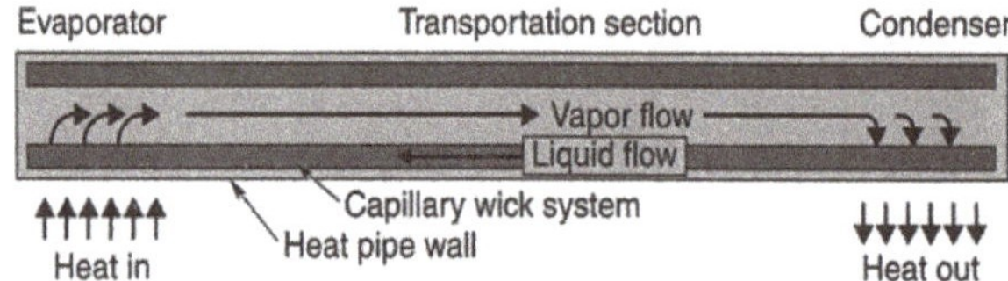

Figure 9. Heat Pipe—schematic.

Depending on the working fluid, wick structure, size, and operational temperature, heat pipes undergo various heat transfer limitations [39]. Total and effective heat pipe lengths are given in Equations (21) and (22).

$$l_{total} = l_e + l_{ad} + l_c \tag{21}$$

$$l_{eff} = \frac{1}{2}(l_e + l_c) + l_{ad} \tag{22}$$

Heat pipe parameters used in this study are given in Table 3 and these values are derived based on available volume within the satellite.

Table 3. Heat pipe design parameters used.

Parameters	Symbol	Values
Vapor core diameter	r_v	0.003
Wall material	-	Copper
Working fluid	-	Water
Wall area (m²)	A_w	0.0000026×10^{-6}
Wall shape	-	circular
Evaporation length of heat pipe (m)	l_e	0.08
Adiabatic length of heat pipe (m)	l_{ad}	0.08
Condensation length of heat pipe (m)	l_c	0.12
Effective length of heat pipe (m)	l_{eff}	0.18
Thermal conductivity of heat pipe material (W·m⁻¹·K⁻¹)	λ_m	389

In this study, the heat pipes are considered to be directly embedded onto both heat exchanger (HX) and radiator aluminum panels to maximize the heat transfer rate. Embedded heat pipes typically increase the effective thermal conductivity by several factors [40].

4. Mechanical Design and Thermal Modelling of 6U CubeSat

4.1. Mechanical Design Considerations

Figure 10 shows a generic 6U satellite configuration with most heat-dissipating internal components such as batteries, avionics, actuators, communication module, payload and a cryocooler. A deployable radiator designed for both of these configurations are analyzed numerically. Radiator panels are mounted to one of the larger satellite surfaces in these configurations.

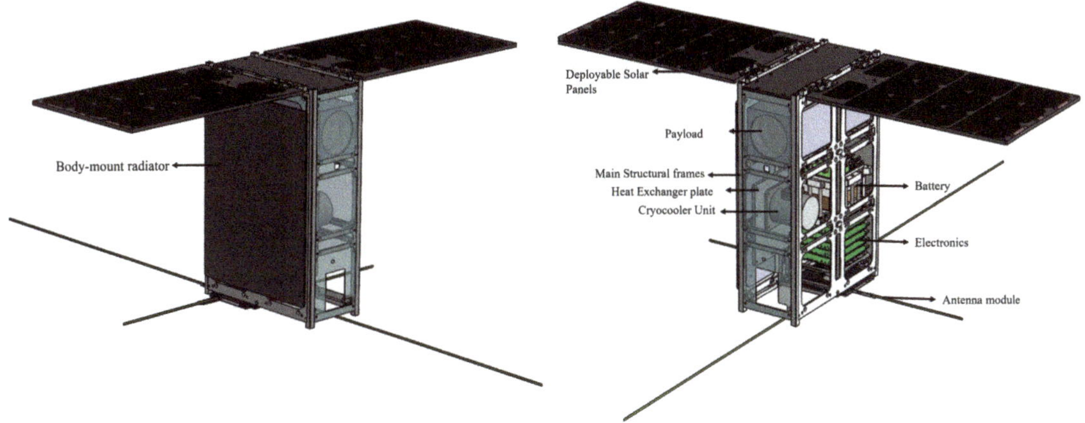

Figure 10. 6U—Mechanical design.

The most heat dissipating elements are directly attached to a baseplate manufactured from aluminum 6061-T6. For the deployable radiator, the hinge mechanism keeps the radiator at stowed position below the deployable solar panel. Radiator is deployed into deep space after the solar panels are released from Hold Down and Release mechanism (HDRM). The base of the radiator is connected to the heat exchanger using high thermal conductive links such as thermal straps and thermal interface materials in order to improve the rate of heat transfer. The baseplate in this design accommodates a micro cryocooler and payload electronics. The avionics subsystem in the satellite is separately linked to the satellite external structure to keep them under operating temperature limits.

The proposed mass distribution for two satellite design configurations are listed in Table 4.

Table 4. Mass distribution—6U satellite.

Components	6U Mass (kg)	27U Mass (kg)
Payload	3.5	6.2
Avionics	0.50	0.75
Battery	0.5	0.9
Heat Exchanger	0.40	0.75
Solar Panels	0.325	0.65
Satellite Structure	3.2	6.9
Radiator	0.35	0.9

Thermo-optical properties of some of the components that are considered in the proposed 6U satellite design are listed in Table 5.

Table 5. Surface properties—Used.

Components	Materials	Coating	Absorptivity	Emissivity
Solar Panels	FR4	FR4 default	0.6	0.7
Payload external	Al	Yellow Chrome	0.2	0.37
Radiator	Al 6061-T6	Silver Teflon	0.09	0.95
Battery	Al 6061-T6	Polyimide + Al + PEEK	0.43	0.52
Cryocoooler Cold finger	Copper	Default copper	0.4	0.05
Heat Exchanger	Al 6061-T6	Black anodize	0.73	0.86

4.2. Environmental Load Studies and Orbit Parameters

Optimization is carried out on the satellite such that the design is able to handle the environmental heat loads for any orbital inclination and earth seasons. Orbital inclination varying from $0 < i < 97.3$ degrees is studied to confirm the design is optimum for all the external heat load conditions. The variation of beta angles for every inclination is extracted from Systems Tool Kit (STK) software and verified with a commercial thermal analysis software, Thermal Desktop®.

Thermal simulations are carried out for four Earth seasons and seasonal environmental heat loads and the Earth's seasonal variations are shown in Figure 11. All the Earth-bound Low Earth Orbits (LEO) will experience the same environmental thermal loads during its mission period, and they affect the thermal behaviour of the satellite components. Hence, the thermal design should consider all these fluctuations of the thermal loads. For the given orbital parameters, worst case mission conditions will be identified for the given satellite configurations and analyzed numerically. In general, for satellite thermal design, two extreme cases are the worst conditions.

1. Worst cold condition: Satellite at LEO will experience coldest temperature in orbit once a year, and it will happen during the Northern hemisphere summer where the satellite will receive lesser flux from the sun. Earth is located at aphelion, and most of the electronics and payloads are turned off. Beginning of the life (BOL) thermo-optical parameters are used.
2. Worst hot condition: Satellite at LEO receives 10% higher than the satellite would receive during summer. This happens during winter conditions. Earth is located at perihelion, and all the electronics and payloads are operating at their peak power. End of the life (EOL) thermo-optical parameters are used.

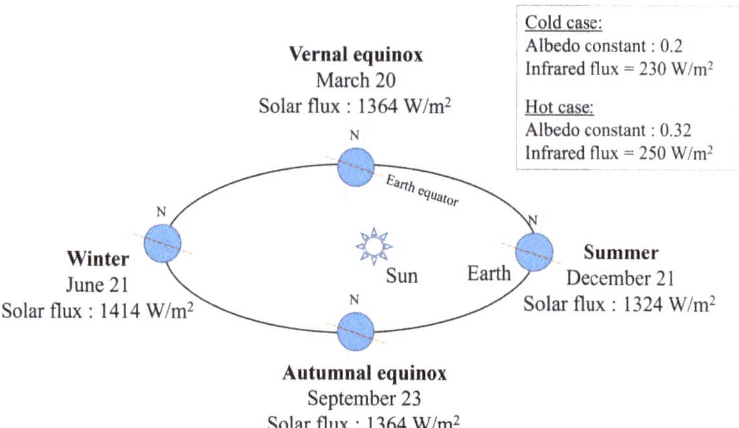

Figure 11. Variation in environmental heat loads due to Earth's season.

The orbital properties are illustrated pictorially in Figure 11. In addition to these properties, other important properties that should be considered for thermal calculations are beta angle, orbital inclination, and orbit type. Beta angle (β) is defined as the angle between the solar vector and satellite orbital plane. This angle varies as the orbit precesses around the Earth and as the Earth moves around the sun. This angle correlates to how much sun light falls onto the satellite surfaces. For a higher beta angle, the satellite spends more of the orbit in the sun. This angle varies with the inclination of the orbit. If the orbit is not sun synchronous, β will move through the range as given in Equation (23) [41], but it has some limitations for higher inclination orbits.

$$\beta = \pm(23.5 + |i|) \tag{23}$$

where, i—orbital inclination, varies from $-90° \leq \beta \leq +90°$. Beta angles where the sun is north of the orbit plane are considered positive. However, due to orbit precession, the right ascension of ascending node varies with time and hence the β equation is modified as in Equation (24) [11].

$$\beta = \sin^{-1} \cos \Gamma \sin \Omega \sin i - \sin \Gamma \cos \epsilon' \cos \Omega \sin i + \sin \Gamma \sin \Gamma \cos i \tag{24}$$

where, Γ—Ecliptic true solar longitude, Ω—Right ascension of ascending node (RAAN), ϵ' is Obliquity of the ecliptic (currently the tilt is at 23.5), and i is the orbit's inclination. RAAN and i are dependent on the satellite's orbit but Γ is a function of earth's position in orbit around the sun [42]. Two factors that affect the beta angle are the change of seasons and perturbation of the orbit due to earth oblateness. As the beta angle varies, there are two consequences of interest to be considered [41].

- Planet shadow varies: Fraction of sunlight reaching the satellite reduces. This is illustrated in Figure 12.
- The incident solar intensity varies. Variation of solar intensity ranges from 0 to 90 degrees and it is clearly seen from the variations plotted using System Tools Kit (STK) tool. The relationship between the sun season and beta angles is also shown in Figure 13.

Satellite in low earth circular orbits spends approximately 37% in eclipse and 63% in sunlight for low inclination orbits. Satellite at 408 km altitude, eclipse duration gradually reduces until when the beta angle approximately equals to 70 degrees as shown in Figure 12 [41].

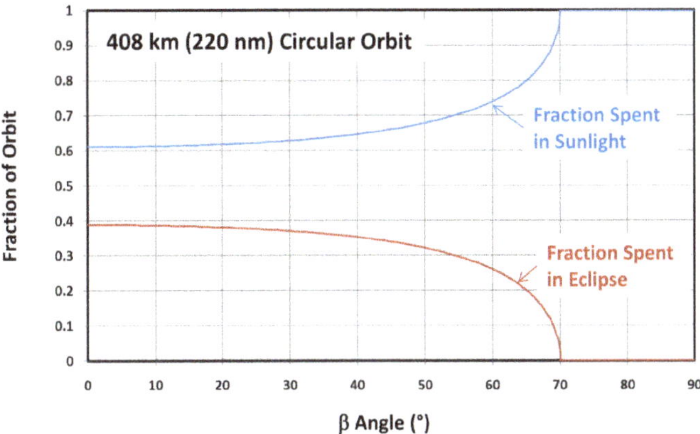

Figure 12. Fraction of time spent in sunlit and eclipse.

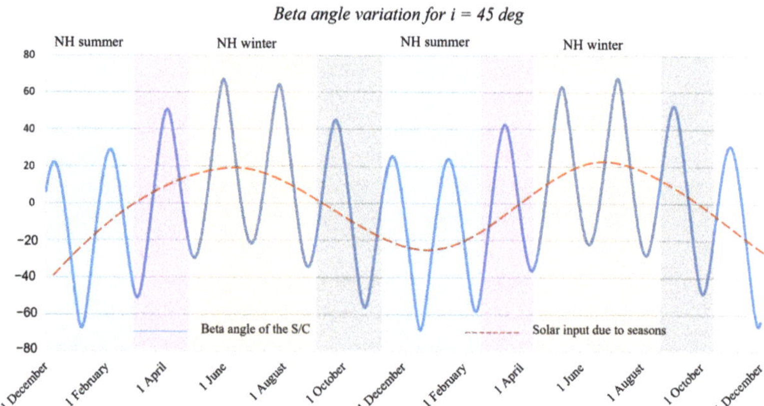

Figure 13. Beta angle variation—STK.

4.3. Geometrical and Thermal Mathematical Model

The Geometrical Mathematical Model (GMM) is a mathematical representation of physical surfaces of the components. It is used to calculate radiation coupling between surfaces as well as heating rates due to environmental heat fluxes. The radiation interchange couplings and environmental heat fluxes calculated by the GMM are used as inputs to the Thermal Mathematical Model (TMM). The TMM consists of a resistor-capacitor network of the spacecraft to calculate heat flows and temperatures.

A thermal model of the 6U satellite was developed using the Thermal Desktop (TD) tool, which performs both the GMM and TMM components of the calculation. All the satellite components such as outer metal plates of the satellite, internal components, payload, and avionics are modelled as solid blocks and solar panels are modelled as 2D surfaces to reduce computational time. Moreover, the variation in temperature across the thickness is insignificant. However, the mass of each elements in both satellite configurations are kept the same as provided in Table 4. In the process of developing thermal mathematical model, thermal couplings were introduced to simulate the physical contact between various satellite components.

Convergence studies with different finite elements were carried out on the thermal model and confirmed that the change in temperature (ΔT) is negligible(<1%). Thermal models for these satellites are built such that it reflects all the physical connections, thermal behaviour, and to have the same actual mass. The 6U satellite thermal model has a mass of 9 kg.

Thermal contact conductance (TCC) properties are generally empirically derived. However, a table of standard values for bolted interfaces from the Spacecraft Thermal Control Handbook are used in this study. All the major contacts with TCC are listed in Table 6 [11].

Table 6. Thermal contact conductance.

Contacting Bodies	TCC Values [W/K]
Aluminium parts to aluminium parts (Structures)	10
Solar Panels to Structure	2
Payload to Structure	3
Cryocooler to HX	5
Radiator to Structure	1.5
HX to Structure	1.2
HX to Radiator (Thermal straps)	5

4.4. Thermal Analysis—Heat Load Parameters

Thermal analyses were performed for the given mission parameters and heat loads for both extreme temperature conditions. The 6U satellite with all the major heat dissipating elements such as a mini cryocooler, battery pack, avionics and payload are considered in the thermal model. Passive thermal control methods include heat pipes, thermal straps, and body-mount radiator are analyzed initially. Later, radiator design for the same 6U satellite configuration is analyzed for various design parameters.

5. Thermal Analysis and Results Discussion

5.1. Thermal Analysis of a 6U Satellite with Heat Pipes

For the satellite configuration shown in Figures 14 and 15 and heat load conditions as listed in Table 7, thermal analyses were carried out for both worst hot and worst cold conditions with constant conductance heat pipes (CCHP). The thermal model of the 6U satellite with cryogenic payload is shown in Figure 16. This thermal model has a micro cryocooler modelled as a cylinder with a copper cold finger, generic cryogenic payload also modelled as a cylinder, and heat dissipating elements such as electronics, heaters, and batteries. The thermal contact conductance values used are provided in Table 6. For this study, it is considered that the satellite is orbiting around the earth at 535 km equatorial orbit with 5° inclination. As can be seen from the thermal model, one of the larger surfaces of the 6U satellite (maximum radiator area considered for BMR is 290 mm × 250 mm) is used as the radiator plate to which the heat pipes are attached. This study considers a generic mid-wave infrared instrument whose detector must be kept around 95K as reference. To provide this cryogenic cooling, a miniature stirling cryocooler (Ricor K508) is considered as it provides better performance at low input power.

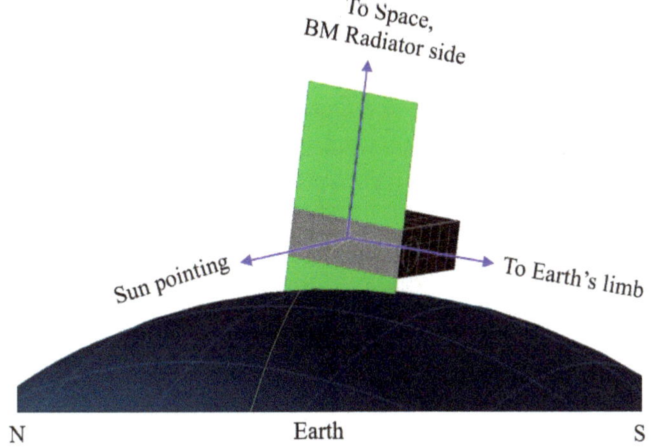

Figure 14. Satellite's attitude in orbit—BMR.

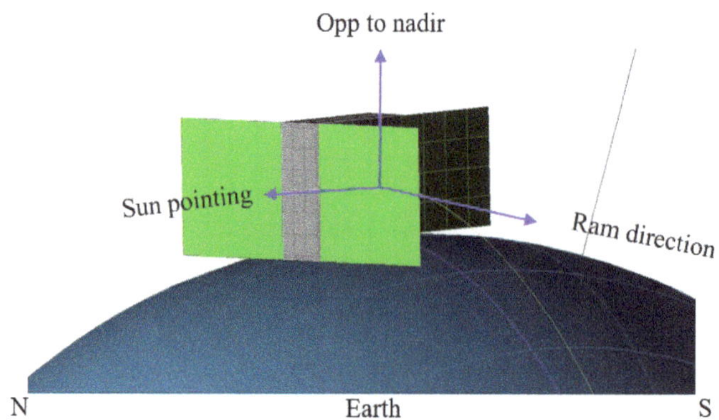

Figure 15. Satellite's attitude in orbit—DR.

Table 7. Thermal Model—Heat loads.

Components	Peak Operation (W)—Hot	Nominal (W)—Cold
Electrical Power System (EPS)	3.5	3.5
ADCS	0.65	0.65
OBDH	0.5	0.5
Payload	12.0	0.0
Camera	1.75	0.0
Battery	3.75	3.75
cryocooler	17	0

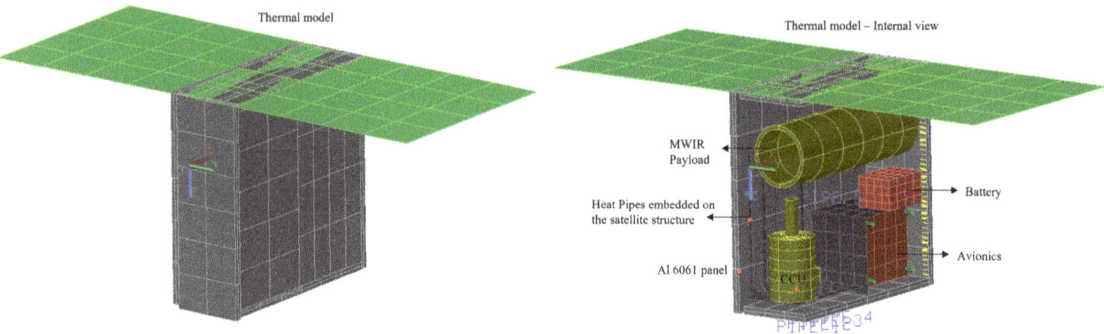

Figure 16. 6U Satellite Body-mount radiator.

Ricor K508N comes with a built-in compressor unit, and it consumes a maximum of 20W during its operation. The cryocooler and heaters are mounted to the Heat Exchanger (HX) plate that is firmly attached to the satellite structure. The cryocooler cold head is assumed to be connected to the IR instrument's detector using high conductance copper thermal straps covered with MLI to minimize the radiation loss and to aid the strap conductance. Two different radiator configurations are analyzed to dissipate the waste heat from the HX and the instrument.

1. Body-mount radiator: One of the large surfaces of the 6U satellite is modelled as a body-mount radiator with high emissivity paint. CCHP are modelled such that it connects both HX and the radiator panel. From the calculation, it was found that the heat pipe with 300 mm long can carry up to 5 W of power. Hence, four heat pipes are used and each has a condenser length of 60mm and an evaporator length of 120 mm. The mass of the satellite's structural panel with embedded heat pipes is

860 g. The temperature summary of the body-mount radiator with embedded heat pipes is shown in Figure 16. All the internal components are exchanging heat between themselves and the external plates of the satellite exchange the heat with external space environment. The amount of heat absorbed or emitted from any component is strongly dependent on its surface thermo-optical properties.

The 6U satellite with BMR is analyzed for worst hot case orbital condition. The MWIR instrument in this study is assumed to take science measurement all the time in the orbit and hence the micro cryocooler considered to be running continuously to meet the instrument's thermal requirement at the detector. Cryocooler is mounted to the HX and heat pipes are connected to both HX and BMR. The main objective of this arrangement is to keep the temperature within the system interface temperature of 25 °C to 30 °C. However, the simulation results show that the body-mounted radiator panel with four heat pipes is not able to provide sufficient thermal performance as the temperature of the Cryocooler unit (CCU) exceeds its maximum operating temperature of 85 °C as seen in Figure 17. Since both CCU and HX are placed inside the satellite, their temperature should be maintained within the desirable limit in order to ensure the rest of the satellite components are safe for continuous satellite operation. An alternative thermal solution, deployable radiator, is studied for the same satellite configuration.

2. Deployable radiator: For the same 6U satellite thermal model, the radiator is changed to deployable configuration to compare the radiating capacity with BMR panel. As with the BMR, the deployable radiator is made from aluminum 6061-T6 with a thickness of 6 mm. CCHP are embedded to both HX and radiator separately and are linked together using thermal straps along the edges of the radiator panel to make it flexible. The surface area of the deployable radiator shown in Figure 18 is assumed to be 150 mm × 270 mm with 620 g of mass including heat pipes.

   The maximum operational range of CCU is −40 °C to +85 °C and the CCU temperature profile shown in Figure 19 indicates 35 °C of difference from its maximum limit. Although the temperature of CCU is within its operational limit, the overall system interface temperature, HX temperature in this case, remains high at 43 °C which is still higher for other components. When compared to BMR panel configuration, deployable radiator panel radiates two times higher for the same heat load.

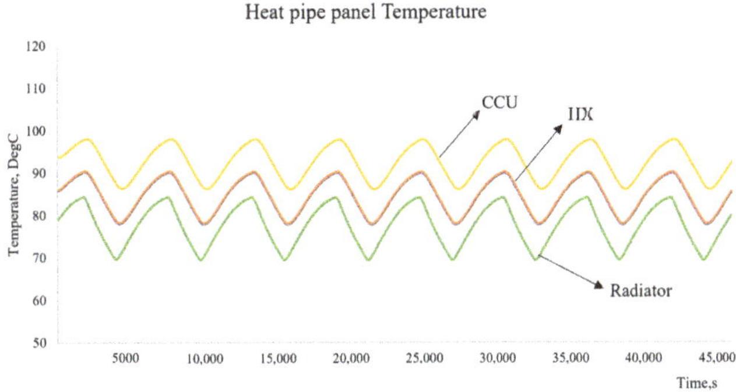

Figure 17. 6U Satellite body-mount radiator with heat pipes—Temperature.

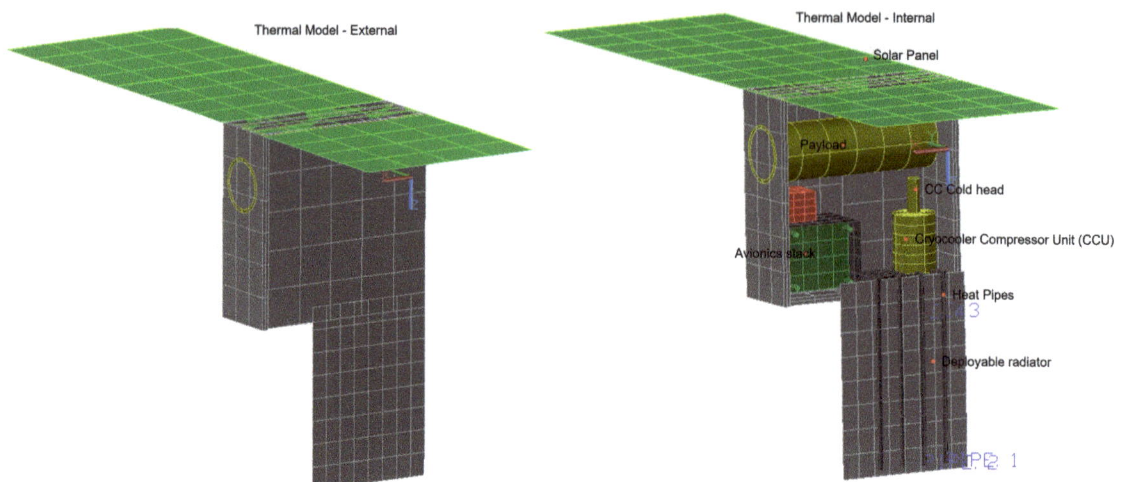

Figure 18. 6U Satellite deployable radiator with heat pipes—Thermal model.

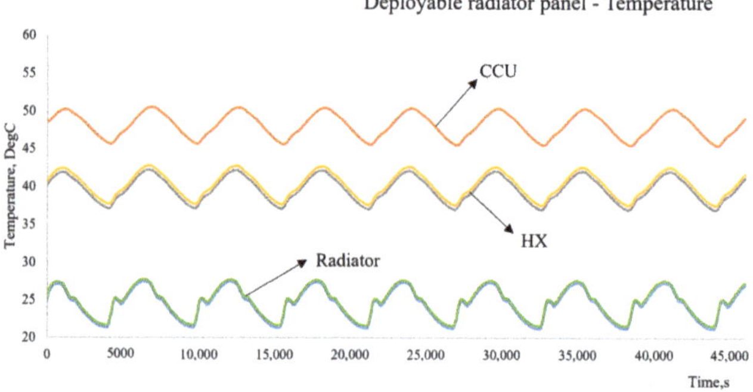

Figure 19. 6U Satellite deployable radiator with heat pipes.

5.2. Thermal Analysis of 6U satellite with Thermal Straps

In this case, a flexible copper thermal strap of the dimensions as indicated in Figure 3 is considered to be thermally coupled to the HX and the deployable radiator panel. With higher applied joint pressure, thermal coupling between the radiator and HX plate results from a large surface area attachment and both HX and the radiator panel are aluminum which has a relatively higher thermal conductivity. Since the thermal strap is coupled to the HX plate, it is assumed that the thermal strap is isothermal with HX temperature. The radiator panel is thermally isolated from the structure so that no parasitic heat loads flow into the radiator panel except through the deployable mechanism and only the thermal strap is the effective thermal path between the HX and the radiator panel for heat conduction. Considering the length of the thermal strap in this configuration, a large temperature gradient is expected between the end points of the strap. It is also assumed that deep space is 0 K and the radiator has a full view factor to space, a thermal simulation with the thermal strap and the radiator was carried out. Figures 20 and 21 show the temperature profiles of the CCU, HX and the radiator panel. HX needs to be maintained between 10 °C to 20 °C and the simulation shows that the HX exceeds its operational limit with thermal straps. As the temperature of the radiator panel varies across its surface area, a few locations including radiator edges and middle regions were picked to plot the distributions and it is depicted in Figure 21.

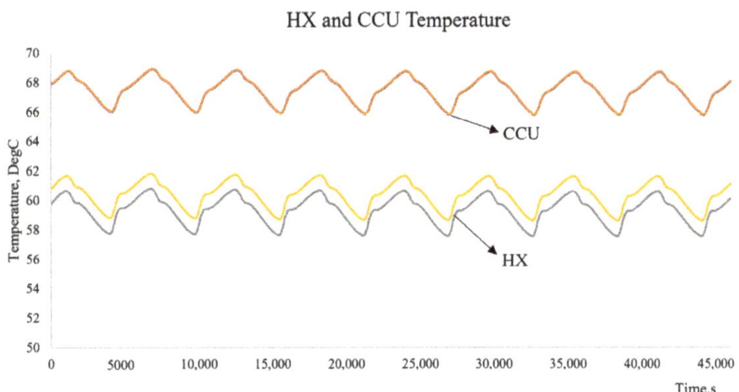

Figure 20. 6U Satellite deployable radiator with Thermal Straps—CC and HX Temperature.

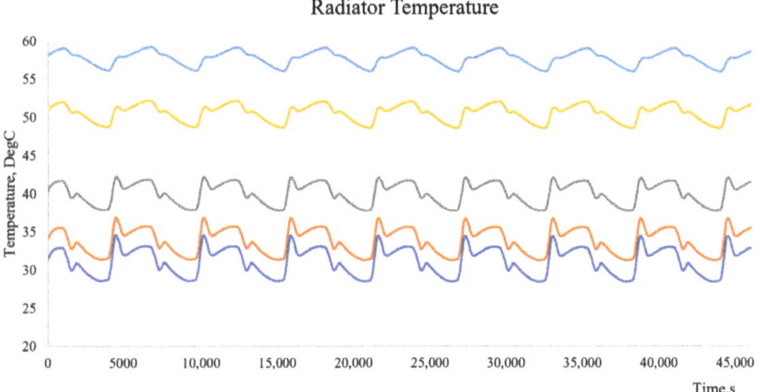

Figure 21. 6U Satellite deployable radiator with Thermal Straps—Radiator Temperature.

Passive TCS studies carried out for a typical high powered 6U satellite platform using thermal straps and heat pipes clearly show that the proposed satellite configuration need to be equipped with an active TCS in order to keep the interface temperature within the desirable range, and the passive solutions described here are not efficient for such satellites with a cryogenic instrumentation.

5.3. Radiator Performance Evaluation Studies

As illustrated in Figures 22–25, the temperature of the body-mount and deployable radiators remains almost constant for different inclinations varying from 0 to 90 degrees for both worst orbital conditions. However, the temperature increases significantly with a reduction in radiator surface area. As described using Equation (23), β varies with varying orbital inclination (i). For every i, β changes due to Earth's seasonal variation. In this study, it is considered that the satellite is orbiting clockwise when observed from the sun or the sun is at north of the orbit plane and hence β assumed to vary from 0 to 90° [42]. As seen from Figure 13, there is a maximum β for every i and this maximum β us considered and plotted against varying inclination angles as shown in figures As illustrated in Figures 22–25. These figures illustrate the relationship between the temperature of radiator panels, i, β, and radiator surface area. For a constant radiator area, the temperature of the radiator panel will fluctuate while varying i and β. However, due to radiator's position in the satellite (placed just under the solar panels in this case), the temperature remains almost constant for all inclinations.

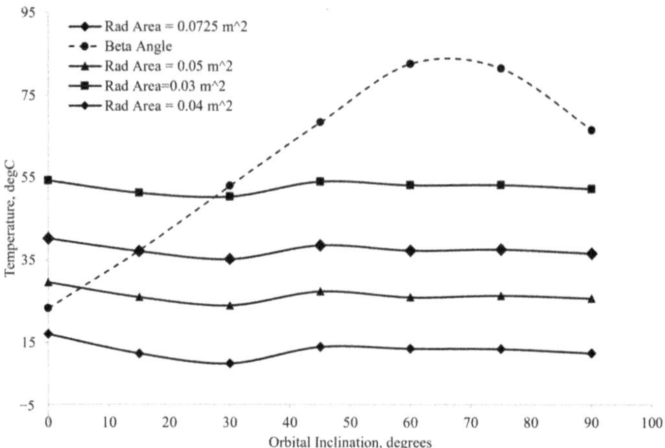

Figure 22. BMR temperature variation with inclination and beta angle—Worst hot condition.

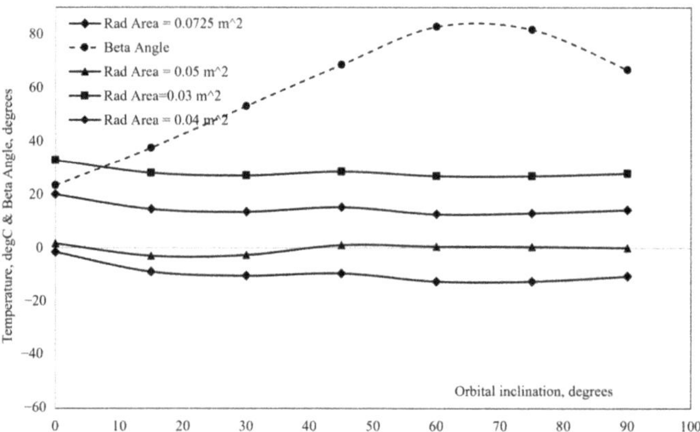

Figure 23. DR temperature variation with inclination and beta angle—Worst hot condition.

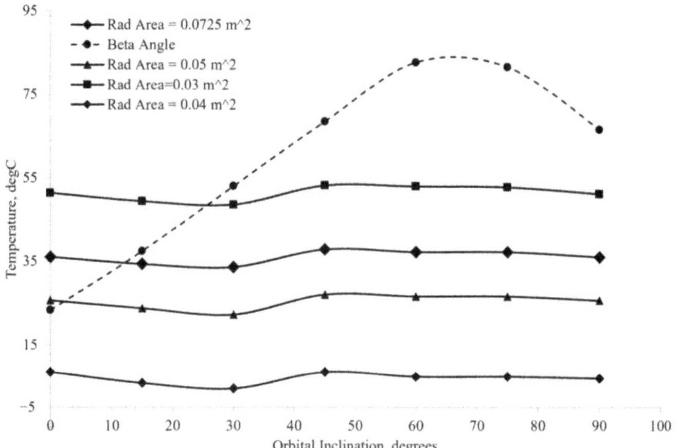

Figure 24. BMR temperature variation with inclination and beta angle—Worst cold condition.

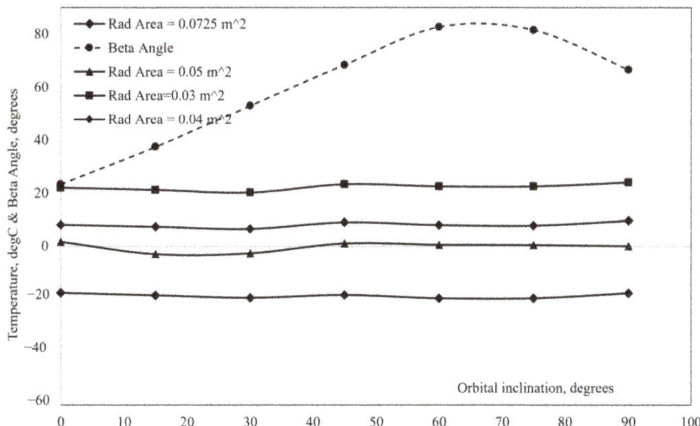

Figure 25. DR temperature variation with inclination and beta angle—Worst cold condition.

As with the body-mount radiator, the temperature of the deployable radiator for the orientation shown in Figure 6 also remains approximately constant for different inclination angles. Hence, the radiator designed for one orbit can also be used for other orbit conditions without altering too much and this is applicable only for the radiators that are not pointed to the sun at any instant in the orbit. If the mission requires specific pointing, additional environmental heat loads should be considered for the radiator design. It is observed from the plots that the effect of beta angle variation is not affecting the radiator temperature as the radiator is kept below the solar panels and the solar panels are assumed to have pointed to the sun.

5.4. Radiator Thickness and Thermal Coating Studies

The following set of simulations were carried out for sun-synchronous orbits. The sun synchronous orbit's ascending node moves in the same direction and at the same average rate as the sun's motion about the ecliptic plane by selecting the right combination of altitude and inclination [41]. Beta angle does not vary much for a satellite in a sun synchronous orbit and it helps to ignore environmental heat variation, but the seasonal variations still need to be considered. A special orbit, dawn-dusk, is studied and analyzed for the same thermal design, satellite attitude, orientation, parameters that were used for the equatorial orbit study.

From Figures 26 and 27, it can be proven that the radiator with higher thermal mass (higher thickness) exhibits better performance than lower thermal mass (lower thickness). Increasing the radiating panel thickness lowers the radiator maximum temperature in orbit while keeping the radiating surface area constant. However, the temperature range of the radiator panel is minimized with an increase in mass. The radiating capacity can be altered using specialized surface paints as discussed in Section 3.2. A few commercially available surface coatings are studied for the radiator performance, and the results are shown in Figures 28 and 29. The maximum radiator temperature can be lowered using low absorptivity paints and high emissivity paints which lowers the radiating temperature at hot and cold conditions. Both surfaces of the radiator are considered to have surface paints in the deployable radiator configuration as the panel radiates from both of its large surfaces, whereas in the body-mount configuration, only the radiating surface is applied with this coating and the other side is covered with a MLI. MLI in BMR helps to maintain the radiator at a steady state by avoiding radiation in and out from the radiator panel. Hence, body-mount radiator is considered to be a single-active face radiator, and deployable radiators are double-active face radiators. From the analysis, it can be said that the body-mount radiator panel can dissipate only 40–50% of the total heat that a full deployable radiator can dissipate. For the radiator mass of 0.4 kg, BMR's temperature range is 23.1 °C to −10.1 °C and DR's temperature range is 9.3 °C to −32.8 °C.

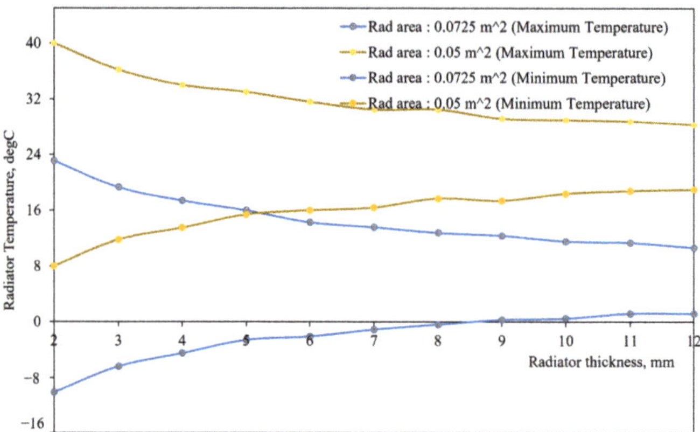

Figure 26. BMR —Thickness Study.

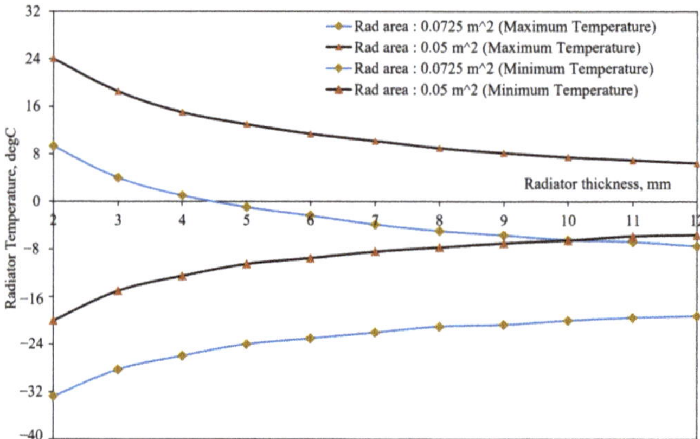

Figure 27. DR—Temperature variation with radiator thickness.

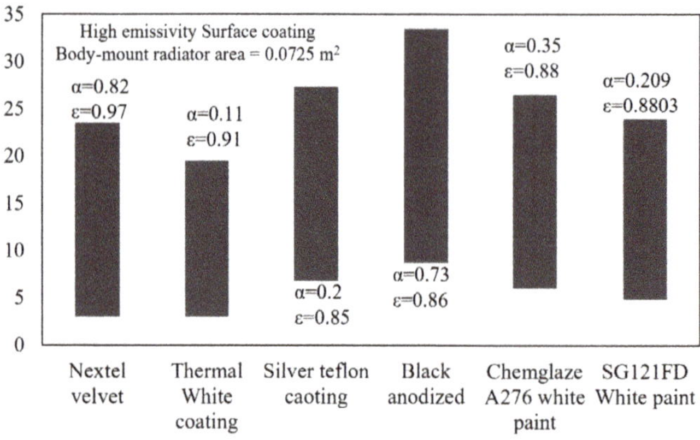

Figure 28. BMR Surface coating study.

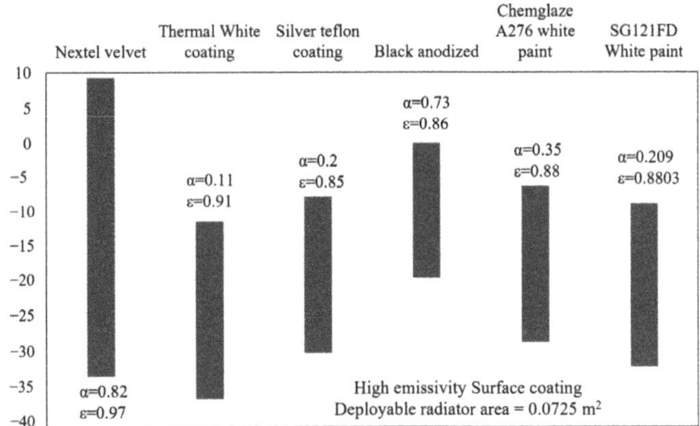

Figure 29. DR Surface coating study.

A constantly illuminated orbit with no eclipse phase is also simulated for the same thermal parameters and satellite configurations. This orbit provides constant power for the satellite components, but it increases the thermal load on different components exposed to the sun and this makes the thermal design more complex.

5.5. Discussion and Summary

Passive thermal control methods for cryogenic instrumentation, high power payloads and electronics on nano and micro satellite platforms may be a suitable solution, but in this design scenario, it fails to maintain the system interface temperature between 20 to 30 °C. However, there is a power threshold where a CubeSat will have adequate radiating capability. Hence, active thermal control techniques are recommended to be considered for such high power satellite missions. Active thermal control methods such as Loop Heat Pipes (LHP), mechanically pumped fluid loop systems, are being studied and considered suitable for solving thermal challenges in small satellite platforms. Radiator designs for different satellite configurations were analyzed and compared. A satellite radiator designed for a particular altitude, inclination, and specific orientation with respect to the sun can provide the same performance for other inclination angles only if the radiator panels are kept under the sun pointed solar panels. For this particular 6U design, as long as the BMR is not seeing the partial/full sun or exposed to direct solar flux, it is expected that the change in temperature of the radiator is minimal and within ±10 °C approximately. Radiator mass affects its performance to a larger extent. While keeping the available surface area constant, increasing the thickness increases the mass of the radiator panel fabricated from Al6061-T6. The higher the thickness or higher the mass of the panel, the lower the temperature difference and at the same time the overall radiator performance is increased.

Surface coatings were analyzed for a 6U satellite with body-mount and deployable radiator configurations. It is observed that the surface coatings greatly enhance the radiating capacity of the radiators for any satellite categories. Proper mounting methods and placement of the radiators are to be chosen very carefully in order to achieve the maximum heat loss into deep space. Most of the satellite radiators are designed to be pointed to deep space rather than at the Earth or the sun. Two 6U satellite configurations discussed in this study assumed that the radiators are pointed to deep space to minimize the environmental heat load. Some satellite missions or payload require specific pointing which could point the radiator to the sun for a few minutes. These special cases are carefully analyzed by calculating the total incoming environmental heat load to the radiator precisely. Radiator surfaces that faces the ram direction of the satellite need to be fabricated with high strength materials or should be thicker in order to deal with micrometeoroid impacts in low earth orbits and atomic oxygen effects which significantly erode the surface coatings before the end of the mission life.

6. Conclusions

Radiators are used on satellites to dump waste heat into deep space. Technological advancements and demand for high power payloads on small nanosatellite platforms increase the need for better thermal management systems. This study analyzed a few passive thermal control methods, such as multi-layer insulation, thermal straps, heat pipes, and passive radiators. Thermal straps are often used as high conductive flexible thermal links to transfer heat from inaccessible locations in the satellite to various other parts. An experimental investigation is carried out on a custom-designed thermal strap to characterize its conductance for different heat loads. The passive radiator is designed and studied in detail for many parameters, such as radiating surface area, radiator mounting type, surface coating, and radiator mass for a 6U satellite platform. The effect of different orbital inclinations on the passive radiator is studied for the same satellite geometric parameters.

Author Contributions: Design, S.S.; methodology, S.S., A.C.; Analysis, S.S.; Testing and validation, S.S.; investigation, S.S., A.C., D.V., B.L.; resources, S.S.; writing—original draft preparation, S.S.; writing—review and editing, S.S., D.V., B.L.; supervision, A.C., D.V., B.L.; project administration, S.S., A.C. All authors have read and agreed to the published version of the manuscript.

Funding: This work was completed from Amal Chandran's NTU startup grant "CubeSat for Earth Science observations".

Institutional Review Board Statement: Not applicable.

Informed Consent Statement: Not applicable.

Data Availability Statement: Data available on request due to restrictions on privacy.

Acknowledgments: The corresponding author would like to thank the Satellite Research Centre (SaRC) at Nanyang technological University (NTU) for providing sufficient support in the form of software, facility, and past satellite mission data to carry out this study. Also, the author would like to thank CR technologies for supporting with an academic Thermal Desktop® licence.

Conflicts of Interest: The authors declare no conflict of interest.

Abbreviations

The following abbreviations are used in this manuscript:

ATCS	Active Thermal Control System
PTCS	Passive Thermal Control System
MLI	Multi-layer Insulation
MWIR	Mid-Wave InfraRed
SWIR	Short-Wave InfraRed
TIR	Thermal InfraRed
DC	Dark Current
IR	InfraRed
EO	Earth Observation
ATOX	Atomic Oxygen
UV	Ultraviolet Radiation
EOL	End-of-Life
BOL	Beginning-of-Life
TCC	Thermal Contact Conductance
TVAC	Thermal Vacuum Chamber
LEO	Low Earth Orbit

GMM   Geometrical Mathematical Model
TMM   Thermal Mathematical Model
TD   Thermal Desktop
HX   Heat Exchanger
CCU   Cryo Cooler Unit
BMR   Body-Mount Radiator
DR   Deployable Radiator
TCS   Thermal Control System

References

1. Golkar, A.; Salado, A. Definition of New Space—Expert Survey Results and Key Technology Trends. *IEEE J. Miniaturizat. Air Space Syst.* **2021**, *2*, 2–9. [CrossRef]
2. Denis, G.; Alary, D.; Pasco, X.; Pisot, N.; Texier, D.; Toulza, S. From new space to big space: How commercial space dream is becoming a reality. *Acta Astronaut.* **2020**, *166*, 431–443. [CrossRef]
3. You, Z. (Ed.) Chapter 4—Micro/Nano Satellite Integrated Electronic System. In *Space Microsystems and Micro/nano Satellites*; Micro and Nano Technologies; Butterworth-Heinemann: Amsterdam, The Netherlands, 2018; pp. 115–146. [CrossRef]
4. Hevner, R.; Holemans, W.; Puig-Suari, J.; Twiggs, R. An Advanced Standard for CubeSat. In Proceedings of the 25th Annual AIAA/USU Conference on Small Satellites, Logan, UT, USA, 8 August 2011.
5. Rabionet, M.C. Study of Future Perspectives of Micro/Nanosatellites Constellations in the Earth Observation Market. 2019. Available online: https://upcommons.upc.edu/bitstream/handle/2117/174918/REPORT_fitxer%20de%20consulta.pdf?sequence=6 (accessed on 1 February 2022).
6. De Aragon, A.M. Future Applications of Micro/Nano-Technologies in Space Systems. 1996. Available online: https://www.esa.int/esapub/bulletin/bullet85/mart85.htm (accessed on 2 February 2022).
7. Inalhan, G.; Yillikci, K.; Ure, K.; Koyuncu, E. The Future of Micro/Nano-Satellite Based Earth Observation and Communication Systems. 2001. Available online: https://www.mcgill.ca/iasl/files/iasl/2015-lachs-panel3a-6-inalhan_et_al.pdf (accessed on 1 February 2022).
8. Han, Y.; Zhang, A. Cryogenic technology for infrared detection in space. *Sci. Rep.* **2022**, *12*, 2349. [CrossRef] [PubMed]
9. Grant, B.G.; Palmer, J.M. *The Art of Radiometry*; SPIE: Bellingham, WA, USA, 2009.
10. Thermal Control. 2021. Available online: https://www.nasa.gov/smallsat-institute/sst-soa/thermal-control (accessed on 6 March 2022).
11. David, G.G. (Ed.) *Spacecraft Thermal Control Handbook, Volume I: Fundamental Technologies*; AIAA Education Series; The Aerospace Press: El Segundo, CA, USA, 2002; pp. 19–36.
12. Anderson, L.; Davidson, R.L.; Swenson, C. SSC 18-WKVII-08 The Active CryoCubeSat Project: Testing and Preliminary Results. 2018. Available online: https://digitalcommons.usu.edu/cgi/viewcontent.cgi?article=4273&context=smallsat (accessed on 21 February 2022).
13. Xue, Y.; Li, Y.; Guang, J.; Zhang, X.; Guo, J. Small satellite remote sensing and applications—History, current and future. *Int. J. Remote Sens.* **2008**, *29*, 4339–4372. [CrossRef]
14. Boushon, K.E. Thermal Analysis and Control of Small Satellites in Low Earth Orbit. 2018. Available online: https://scholarsmine.mst.edu/cgi/viewcontent.cgi?article=8755&context=masters_theses (accessed on 16 January 2022).
15. Fick, W.; Gassmann, K.U.; Haas, L.D.; Haiml, M.; Hanna, S.; Hübner, D.; Höhnemann, H.; Nothaft, H.P., Thöl, R. Infrared detectors for space applications. *Adv. Opt. Technol.* **2013**, *2*, 407–421. [CrossRef]
16. Ross, R.G.J. Cryocoolers for Space Applications. 2015. Available online: https://www2.jpl.nasa.gov/adv_tech/coolers/Cool_ppr/CEC2015-Short%20Course%20Session%204.pdf (accessed on 8 February 2022).
17. Rando, N.; Lumb, D.; Bavdaz, M.; Martin, D.; Peacock, T. Space science applications of cryogenic detectors. *Nucl. Instrum. Methods Phys. Res. Sect. A Accel. Spectromet. Detect. Assoc. Equip.* **2004**, *522*, 62–68. [CrossRef]
18. Miyakita, T.; Hatakenaka, R.; Sugita, H.; Saitoh, M.; Hirai, T. Development of a new multi-layer insulation blanket with non-interlayer-contact spacer for space cryogenic mission. *Cryogenics* **2014**, *64*, 112–120. [CrossRef]
19. Miyakita, T.; Hatakenaka, R.; Sugita, H.; Saitoh, M.; Hirai, T. Evaluation of Thermal Insulation Performance of a New Multi-Layer Insulation with Non-Interlayer-Contact Spacer. In Proceedings of the 45th International Conference on Environmental Systems, Bellevue, WA, USA, 12–16 July 2015.
20. Corpino, S.; Caldera, M.; Nichele, F.; Masoero, M.; Viola, N. Thermal design and analysis of a nanosatellite in low earth orbit. *Acta Astronaut.* **2015**, *115*, 247–261. [CrossRef]
21. Ross, R. Quantifying MLI Thermal Conduction in Cryogenic Applications from Experimental Data. *IOP Conf. Ser. Mater. Sci. Eng.* **2015**, *101*, 012017. [CrossRef]
22. Kan, H.K.A. Space Environment Effects On Spacecraft Surface Materials. In *Radiation Effects on Optical Materials*; Levy, P.W., Ed.; International Society for Optics and Photonics, SPIE: Bellingham, WA, USA, 1985; Volume 0541, pp. 164–179. [CrossRef]
23. Anvari, A.; Farhani, F.; Niaki, K. Comparative Study on Space Qualified Paints Used for Thermal Control of a Small Satellite. *Iran. J. Chem. Eng.* **2009**, *6*, 50–62.

24. Roussel, J.F.; Alet, I.; Faye, D.; Pereira, A. Effect of Space Environment on Spacecraft Surfaces Control in Sun-Synchronous Orbits. *J. Spacecr. Rocket.* **2004**, *41*, 812–820. [CrossRef]

25. Liu, T.; Sun, Q.; Meng, J.; Pan, Z.; Tang, Y. Degradation modeling of satellite thermal control coatings in a low earth orbit environment. *Sol. Energy* **2016**, *139*, 467–474. [CrossRef]

26. Dhuley, R.; Ruschman, M.; Link, J.; Eyre, J. Thermal conductance characterization of a pressed copper rope strap between 0.13K and 10K. *Cryogenics* **2017**, *86*, 17–21. [CrossRef]

27. Henninger, J.H. Solar Absorptance and Thermal Emittance of Some Common Spacecraft Thermal Control Coatings. NASA. 2013. Available online: https://ntrs.nasa.gov/citations/19840015630 (accessed on 6 March 2022).

28. Lécossais, A.; Jacquemart, F.; Lefort, G.; Dehombreux, E.; Beck, F.; Frard, V. Deployable Panel Radiator. Available online: http://hdl.handle.net/2346/72955 (accessed on 16 July 2017).

29. Shanmugasundaram, S.; Chandran, A.; Joshi, S.C.; Edwin, T.H.T.; Valentini, D.; Olschewski, F. A Comparison Study on Thermal Control Techniques for a Nanosatellite Carrying Infrared Science Instrument. Available online: https://hdl.handle.net/2346/864 12 (accessed on 31 July 2020).

30. Hengeveld, D.; Wilson, M.; Moulton, J.; Taft, B.; Kwas, A. Thermal Design Considerations for Future High-Power Small Satellites. In Proceedings of the 48th International Conference on Environmental Systems, Albuquerque, NM, USA, 8–12 July 2018.

31. Karam, R.D. *Satellite Thermal Control for Systems Engineers*; American Institute of Aeronautics and Astronautics: Reston, VA, USA, 1998.

32. Sam, K.F.C.H.; Deng, Z. Optimization of a space based radiator. *Appl. Therm. Eng.* **2011**, *31*, 2312–2320.

33. Janzer, K.; Langer, M.; Killian, M.; Krejci, D.; Reissner, A. Thermal Control of High Power Applications on CubeSats. In Proceedings of the 69th International Astronautical Congress, Bremen, Germany, 1–5 October 2018. [CrossRef]

34. Sparrow, E. Radiation Heat Transfer between Surfaces. In *Advances in Heat Transfer*; Elsevier: Amsterdam, The Netherlands, 1965; Volume 2, pp. 399–452. [CrossRef]

35. Iwata, N.; Iwata, N.; Nakanoyaa, S.; Nakamura, N.; Takeda, N.; Tsutsui, F. Thermal Performance Evaluation of Space Radiator for Single-Phase Mechanically Pumped Fluid Loop. *J. Spacecr. Rocket.* **2021**, *59*, 225–235. [CrossRef]

36. Jentung, K. Introduction to Heat Pipes. Available online: https://tfaws.nasa.gov/wp-content/uploads/TFAWS2015-SC-Heat-Pipes.pdf (accessed on 21 January 2022).

37. Brouwer, H.; de Groot, Z.; Guo, J.; van Gerner, H.J. Solving the Thermal Challenge in Power-Dense CubeSats with Water Heat Pipes. In Proceedings of the 31st Annual AIAA/USU Conference on Small Satellites, Logan, UT, USA, 5–10 August 2017.

38. Meseguer, J.; Pérez-Grande, I.; Sanz-Andrés, A. 11—Heat pipes. In *Spacecraft Thermal Control*; Meseguer, J., Pérez-Grande, I., Sanz-Andrés, A., Eds.; Woodhead Publishing: Amsterdam, The Netherlands, 2012. [CrossRef]

39. Nemec, P.; Čaja, A.; Malcho, M. Mathematical model for heat transfer limitations of heat pipe. *Math. Comput. Model.* **2013**, *57*, 126–136. [CrossRef]

40. Weyant, J.; Garner, S.; Johnson, M.; Occhionero, M. Heat pipe embedded AlSiC plates for high conductivity—Low CTE heat spreaders. In Proceedings of the 2010 12th IEEE Intersociety Conference on Thermal and Thermomechanical Phenomena in Electronic Systems, Las Vegas, NV, USA, 2–5 June 2010; pp. 1–6. [CrossRef]

41. Rickman, S.L. *Introduction to Orbital Mechanics and Spacecraft Attitudes for Thermal Engineers*; NASA Tfaws; NASA: Washington, DC, USA, 2020; pp. 116–129.

42. Sumanth, R.M. Computation of Eclipse Time for Low-Earth Orbiting Small Satellites. *Int. J. Aviat. Aeronaut. Aerosp.* **2019**, *6*, 15. [CrossRef]

applied sciences

Article

Analysis on the Isostatic Bipod Mounts for the HERA Mission LIDAR

Nicole G. Dias [1], Paulo Gordo [1,2], Hugo Onderwater [2], Rui Melicio [3,4,*] and António Amorim [1]

[1] Centro de Astrofísica e Gravitação, Faculdade de Ciências, Universidade de Lisboa, 1749-016 Lisbon, Portugal; ngodias@fc.ul.pt (N.G.D.); prgordo@fc.ul.pt (P.G.); aamorim@sim.ul.pt (A.A.)
[2] Synopsis Planet—Advance Engineering Unipessoal LDA, 2810-174 Almada, Portugal; hugo.onderwater@synopsisplanet.com
[3] Instituto de Engenharia Mecânica, Instituto Superior Técnico, Universidade de Lisboa, 1049-001 Lisbon, Portugal
[4] Instituto de Ciências da Terra, Universidade de Évora, 7002-554 Évora, Portugal
* Correspondence: ruimelicio@gmail.com

Abstract: The Light Detection and Ranging (LIDAR) is a time-of-flight altimeter instrument being developed for the HERA mission, designated as Planetary ALTimeter (PALT). PALT is positioned in the center of the top face of the HERA probe, and therefore, it cannot use radiators to stabilize its internal temperature. The contribution of this paper is the design of isostatic bipod mounts for the LIDAR primary mirror. The performance of PALT must be maintained over a wide operational range, from $-60\,°C$ to $80\,°C$. These temperature requirements imply that a careful isostatic mount structure design is critical to maintaining performance in all operational scenarios. The purpose of the instrument is to perform range measurements from 500 m to 14 km. The instrument will contribute to the detailed characterization of the asteroid's topography, assist the probe navigation in operations such as fly-bys (including on the dark side of the asteroid) or landing. PALT has an emitter system that generates 2 ns, 100 μJ, 1535 nm laser pulses and a receiver system that collects the backscattered signal from the asteroid. The receiver system is composed of a 70 mm diameter Cassegrain telescope and a refractive system that focuses the signal on the sensor.

Keywords: isostatic bipod mounts; LIDAR; PALT; HERA mission; a-thermalization; optical performance

Citation: Dias, N.G.; Gordo, P.; Onderwater, H.; Melicio, R.; Amorim, A. Analysis on the Isostatic Bipod Mounts for the HERA Mission LIDAR. *Appl. Sci.* **2022**, *12*, 3497. https://doi.org/10.3390/app12073497

Academic Editors: Simone Battistini, Filippo Graziani and Mauro Pontani

Received: 28 January 2022
Accepted: 27 March 2022
Published: 30 March 2022

Publisher's Note: MDPI stays neutral with regard to jurisdictional claims in published maps and institutional affiliations.

1. Introduction

The HERA spacecraft includes several payload instruments, such as the Time-of-Flight (ToF) LIDAR that will measure the distances from the HERA spacecraft to the target. The measurement operations shall be performed at a distance from 500 m to 14 km, enabling operations such as fly-bys or landings. Previous space missions have deployed analogous instruments for specific requirements. One of the main challenges in those missions was the operational temperature range, since the LIDAR instruments were directly exposed to space and the required optical tolerances to maintain the instrument performance, i.e., internal alignment of optics and alignment between receiver and emitter. Each mission requires a specific LIDAR measurement range, operational temperature interval, radiation requirements and target objects, making the LIDAR design rather specific. The contribution of this paper is the design of isostatic bipod mounts for a small mirror of the LIDAR.

Within the mirror, the reflecting coating, the substrate and the supports have to be validated over the thermal range of the mission. When the mirror is subjected to a thermal condition, maintaining the surface shape is critical. "A temperature change induces a change in size which, for curving optical surfaces, produces a change in radius and hence a focus shift" [1]; therefore, a large change in temperature can cause a distortion.

Isostatic mounts, or flexures, are optomechanical components capable of compensating thermal and gravitational deformations in optical components. Isostatic bipod mounts are used to prevent deformations due to differences of thermal expansion coefficients, and to keep the mirror's optical axis in place [1]. Several design limitations occur due to the mission environmental requirements.

An a-thermal design must be taken into consideration since critical components have a different Coefficient of Thermal Expansion (CTE). Usually, optical components have a low CTE compared to the mount, thus the a-thermal design is a critical aspect to the mount design [2]. Pijnenburg [2] studied two different a-thermalization methods: one matching the bond thickness, and the second by applying elastic elements in the mount. The first approach is to adapt the thickness of the adhesive to compensate for the expansion of the optics and mounts. However, it resulted in the instability of the optical components, since it resulted in a thick bond line which does not work properly for strength and stiffness. The second approach included a tangential isostatic flexible element, a leaf spring. If well dimensioned, the mirror would be supported by three leaf springs, with a small bonding spot. The stability under inertial loads, such as gravity, is achieved by constraining with high stiffness the degree of freedom of the optical component, where "only residual local stresses in the glass arises around the bond spots" [2].

Following the second approach of Pijnenburg [2], several designs were presented for several space missions, especially for mirrors with large aperture. There are two types of conformities that should be followed: radial compliance for axisymmetric mirrors to compensate for thermal expansion mismatch, and tangential compliance to prevent assembly stress from navigating towards the mirror surface. Lateral mirror supports on the edge of the mirror help in fulfilling these compliances [3,4]. Isostatic mounts can be categorized according to the type of flexure element: simple blade flexures, that are used for tangential edge support for small axisymmetric mirrors, and a bipod flexure, which is a combination of two blade flexures forming a triangle [3,4]. Kihm [3] presented a design for a lightweight primary mirror, with pockets in the back surface, and with three square bosses extruded at the edge of the mirror for isostatic mounting. The isostatic mount presented by Kihm [3] had three different components: the flexure A, which is permanently bonded onto the mirror's boss; the flexure B, which is fastened to flexure A and is the support flexure; and shims that are placed between the flexures. In [5], an adjustable bipod flexure for a large aperture mirror is presented, formed by two flexure bars with tangential and radial blades. The connection between the flexure and the mirror is made by an invar connector since the invar's CTE matches with the material of the mirror, and it is glued to the mirror with epoxy adhesive.

In this paper, the isostatic bipod mounts design of PALT is presented. It follows a similar approach to the flexure presented by Kihm [3,4], which means that the apex of the triangle formed by the flexure should point to the center of mass of the mirror to minimize surface distortion [3,4]. Each bipod flexure has a symmetric combination of two tangential blades and the radial blade: tangential blades give tangential compliance, and radial blades give radial compliance to the mirror. Thus, when the mirror is accelerated in lateral directions or perpendicular to the mirror's optical axis, the tangential flexures compensate for this effect; however, radial expansion of the mirror due to thermal loads can be compensated for by the radial flexures [4]. The major differences from previous implementations lie in the mirror size and in the isostatic mount bipod mechanical design.

This paper is structured as follows: Section 2 presents the previous research. Section 3 presents the PALT design. Section 4 explains the assembly and bonding procedures. Section 5 reveals the thermoelastic simulations and respective results. Finally, in Section 6, the conclusions are outlined.

2. Previous Research

The Planetary ALTimeter (PALT) mechanical design has been optimized on different aspects to meet the requirements established for the HERA mission. In [6–8], the HELENA LIDAR's performance was evaluated regarding weaker requirement of shock, static simulation of 5G in two different directions, and thermal conductive simulations between −40 °C and +60 °C. The main objective was to study the secondary mirror deformation towards the base of the telescope, and maximum stresses. Two different materials (aluminum and titanium) were assessed for the upper part of the LIDAR telescope. The simulations revealed a higher factor of safety, and a minor displacement of the secondary mirror towards the base of the telescope for the titanium material. This modification was implemented on PALT.

In order to guarantee that the best optical performance is maintained, the primary mirror shall be a-thermalized, which is possible by the optimizing the isostatic bipod mounts (see Figure 1). Furthermore, the assembly and alignment procedures of the primary mirror should be as simple as possible to avoid damages on the mirror. This paper presents an optimization of the isostatic bipod mounts design, and details on the assembly procedure.

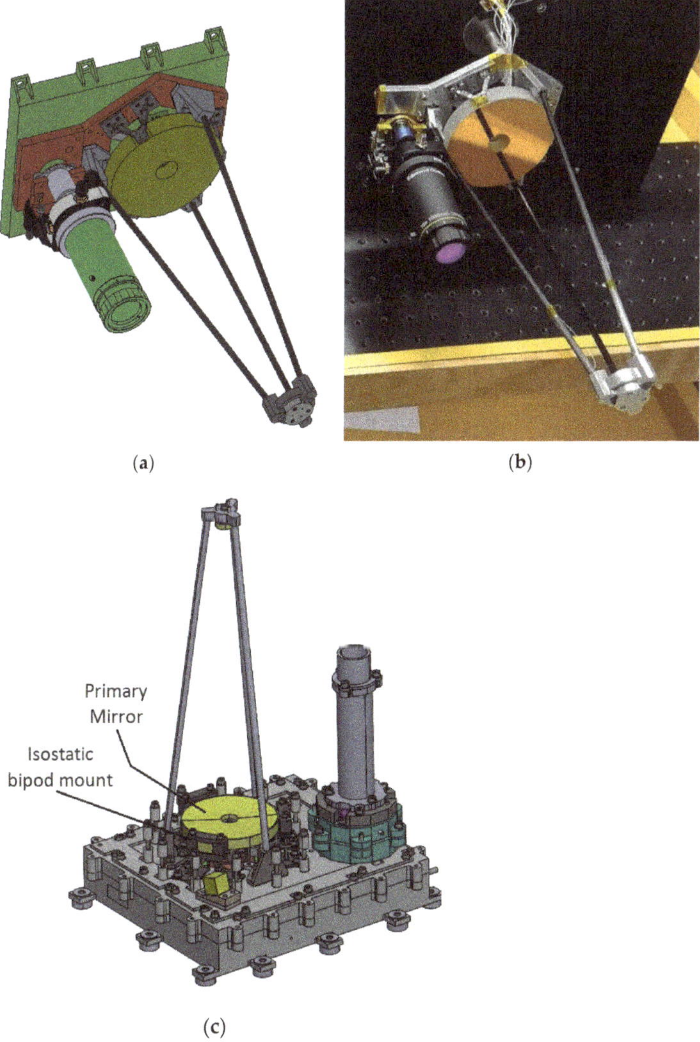

Figure 1. (**a**) HELENA design; (**b**) HELENA engineering model. (**c**) PALT design. Several transformations were made since HELENA design, with the isostatic bipod mounts being a critical design improvement.

3. Design

The design was iterated several times, with the focus being on the initial design and a final design. The first design of the isostatic bipod mounts was presented in HELENA [6,7], which was the starting point of the isostatic bipod mount for PALT. This first design is depicted in Figure 2.

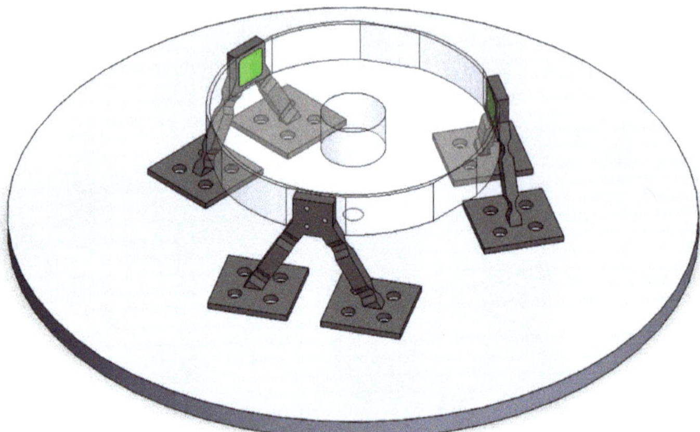

Figure 2. First design of isostatic bipod mount, which are positioned symmetrically around the mirror, and the representation of the adhesive pads area (highlighted in green).

This design followed the one implemented on HELENA LIDAR [6,7] (see Figure 3).

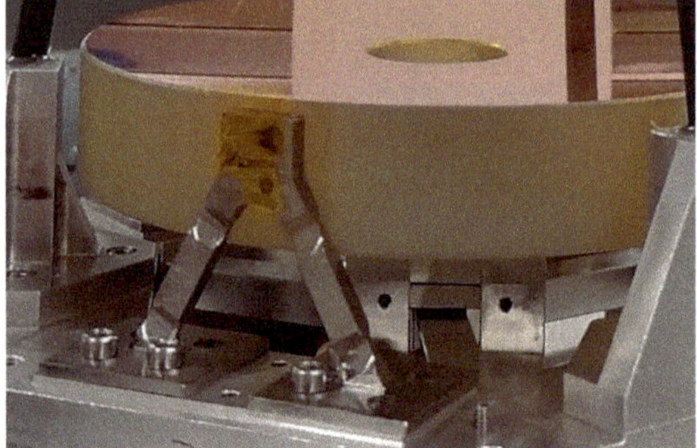

Figure 3. Isostatic bipod mount assembled on HELENA.

The primary mirror of HELENA had a 100 mm diameter and 11 mm thickness.

For PALT, a 70 mm diameter mirror was required. A first approach on the isostatic bipods mount design was based on the HELENA's optomechanics. The simulations of this design are presented in Section 4, and it was concluded that further optimization was desirable.

In a second version, the mirror's thickness was decreased to 8 mm and the three extrude bosses were equally distributed with an 8 mm length (where isostatic bipod mounts are assembled). Both isostatic bipod mounts combine tangential and radial blades in each support leg, arranged in a triangle. The proposed design includes two blade radial flexures and a tangential blade between the radial blades, whose purpose is to compensate for radial and tangential deformations, respectively. The material chosen for this isostatic mount is Titanium Grade 5 (Ti-6Al-4V) (see Table 1).

Table 1. Properties of the materials.

Material	Physical Property	Value
Zerodur	Density	2530 kg/m^3
	Coefficient of Thermal Expansion	1×10^{-7}/K
	Elastic modulus	84.7 GPa
	Poisson's ratio	0.25
	Thermal conductivity	1.46 W/m.K
	Ultimate strength	30 MPa
Aluminium 7075-T7351	Density	2800 kg/m^3
	Coefficient of Thermal Expansion	2.36×10^{-5}/K
	Elastic modulus	72 GPa
	Poisson's ratio	0.33
	Thermal conductivity	155 W/m.K
	Yield strength	435 MPa
Titanium Grade 5 (Ti-6Al-4V)	Density	4430 kg/m^3
	Coefficient of Thermal Expansion	8.6×10^{-6}/K
	Elastic modulus	113.8 GPa
	Poisson's ratio	0.33
	Thermal conductivity	6.7 W/m.K
	Yield strength	1100 MPa
Epoxy 2216B/A Gray	Density	1300 kg/m^3
	Coefficient of Thermal Expansion	102×10^{-6}/K
	Thermal conductivity	0.394 W/m.K
	Ultimate strength	10.2 MPa

The bipod isostatic mount presented in this article is shown in Figure 4, and it has dimensions of 33 mm height and a maximum length of 8 mm. Since the optimized implementation is a key element to achieve the required performance, the optimization of the design of the isostatic bipod mount was divided into four elements: a base flexure, two columns, and a top fitting. The separation of the bipod eases the bonding process between the mounts and the mirror extrude bosses.

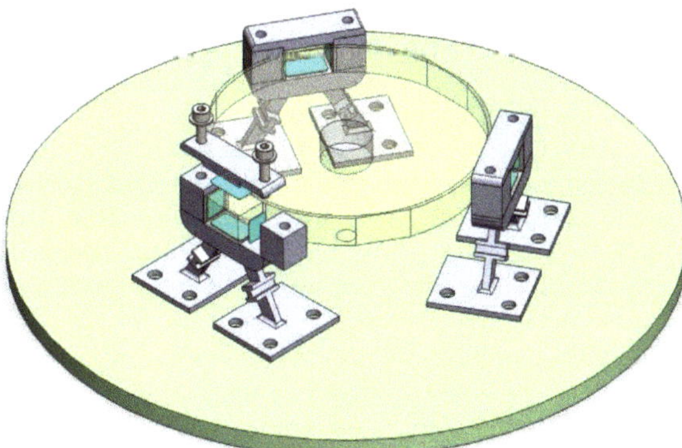

Figure 4. Isostatic bipod mount that will be assembled in PALT. They are positioned symmetrically around the mirror, and the adhesive pads area are displayed (highlighted in blue).

The extruded areas presented in the inner upper structure of the isostatic bipod mount are adhesive areas where the mirror is to be attached. The adhesive is presented as a thin pad with 0.1 mm thickness.

4. Assembly and Bonding Procedures

In HELENA, the isostatic bipod mount was bonded to the mirror with a 100 μm layer of adhesive. The bonding process was implemented by using a 100 μm diameter nylon wire, between the isostatic mount and the mirror, which were removed after the adhesive was fully cured. It was verified that the mount design blocked the application of the adhesive, which sprung the need for design changes. The isostatic bipod mount updated design is easily assembled and, since it is divided into four individual pieces, allow a better application of the epoxy without compromising the mirror. The bonding procedure is still to be qualified; however, the bonding procedure and the alignment procedure is defined. The first step fixates the mirror in space using three supports, assisted by Kapton pads and adjustment screws. A sheet made of Kapton, is to be placed on the mirror extrusions to ensure the adhesive is kept on the designated area and not touching the mirror optics surface (see Figure 5).

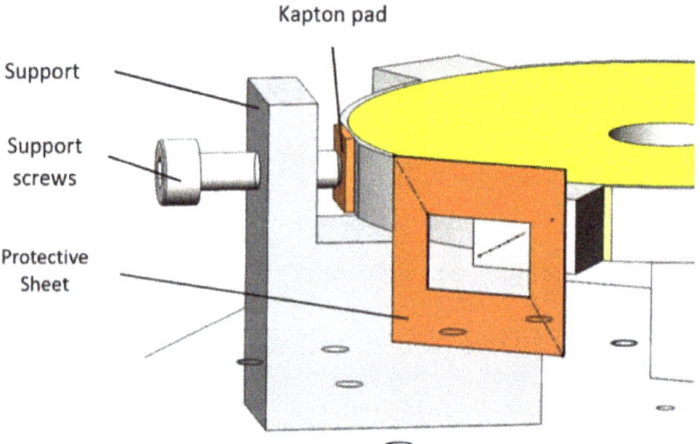

Figure 5. Mirror fixation and protection. The protective sheet is perforated to allow an easy removal after assembly and bonding of the flexures.

The adhesive thickness remains 100 μm apart from the flexures, on all four sides. To accomplish this, Kapton tape of a specific thickness is to be placed on the corners of the mirror extrusions. The tape serves, not only to guarantee that all four sides of the mirror extrusion will remain 100 μm apart from the flexures, but also to prevent any excess adhesive that may spread from the beyond the designated area (highlighted surfaces depicted in Figure 4). All the protective bonding components will be removed before the end of the work life of the adhesive.

After assuring that the mirror is not contaminated by adhesive, the flexure base is placed under the mirror extrusion, with the adhesive already applied. The columns of the flexure are placed on the flexure base, both with the adhesive already applied, finalizing with the top lid of the flexure, with the adhesive already applied, and screwed into place, fixating the upper part of the flexure (see Figure 6). The two screws guarantee the connection of the isostatic bipod mount: the screw is inserted on the top fitting, goes through the columns, and reaches the flexure base.

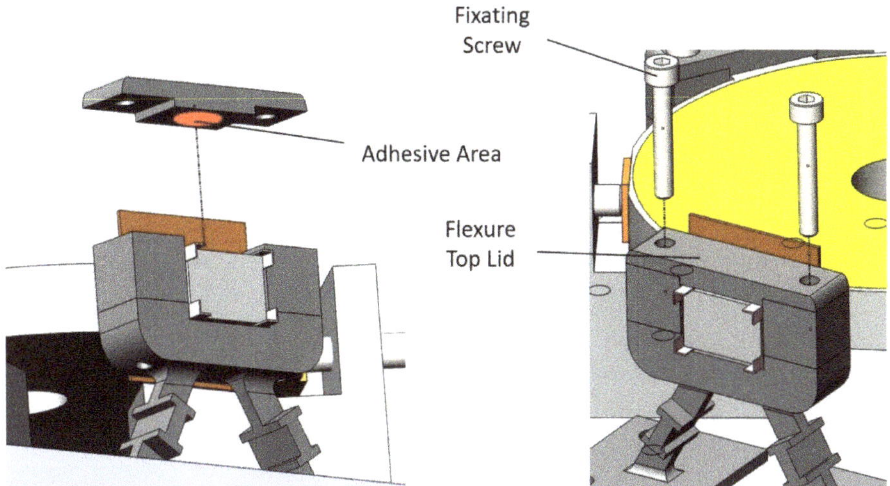

Figure 6. Final step on the bonding procedure. The protective sheets are removed before the end of the work life of the adhesive (schematic).

5. Thermoelastic Simulations

This section presents the thermoelastic simulations of both designs and the respective results.

5.1. Thermal Modeling and Materials

The isostatic bipod mounts were simulated in simple conditions to assess their behavior. The simulation only contemplated a circular base, to maintain the system symmetric, the three isostatic bipod mounts, and the mirror. In Figure 7, the simulation performed for the first design is depicted, and in Figure 8, the simulation performed for the optimized flexure design. The boundary conditions applied to the model were to prevent axial and tangential displacement, which means the base only has freedom radially, and a fixed-point support on the base center, which is coincident with the optical axis. The temperature was defined in the bottom surface of the base.

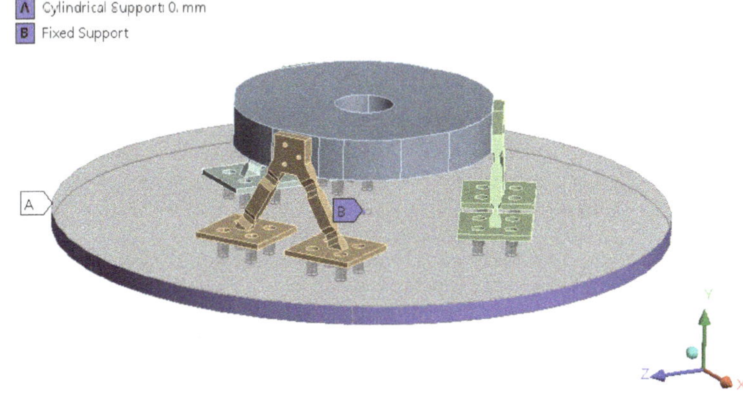

Figure 7. The boundary conditions of the first design.

Figure 8. The boundary conditions of the final design.

Both designs were subjected to the qualification non-operational temperature limits, i.e., +80 °C and −60 °C. The materials chosen were also the same: titanium grade 5 (Ti-6Al-4V) for the isostatic bipod mounts, aluminum 7075-T7351 for the base, Zerodur for the mirror, and Epoxy 2216B/A Gray for the adhesive pads. The properties of the chosen materials are presented in Table 1. The mechanical screw behavior was not assessed at this stage.

The Poisson's ratio and the Young's Modulus were given by the supplier and are confidential. The remaining properties are based on documentation [9,10]. The properties presented in Table 1 are the inputs needed for the thermo-elastic simulation.

The tolerance of the system is given by an optical study (i.e., tolerance analysis in ZEMAX). Regarding the telescope, it was identified that the worst tolerance offenders are the mirror radius and the mirror displacement. The tolerances that are assessed in this paper is the displacement on XZ plane of the primary mirror, the distance of the primary mirror to the base, and the curvature radius of the primary mirror (see Table 2).

Table 2. Thermal acceptable tolerances.

Component		Value [mm]	Comments
	XZ displacement	±0.01	
	Distance of Mirror to the Base	±0.09	
Primary Mirror	Curvature radius	±1	±1 mm radius variation is equivalent of a maximum surface displacement of ±2 µm

5.2. Thermoelastic Results

The aim of these simulations was to assess the isostatic bipod mount's behavior in the extreme temperatures. Two topics were studied, namely the survivability of the mirror (i.e., assessment of the stress in the mirror) and the analysis of the mirror displacement regarding its nominal position (to evaluate if the flexures were optimized and working correctly).

Figures 9 and 10 represent the XZ displacement results for the first isostatic bipod mount design for the hot and cold non-operational scenarios, respectively.

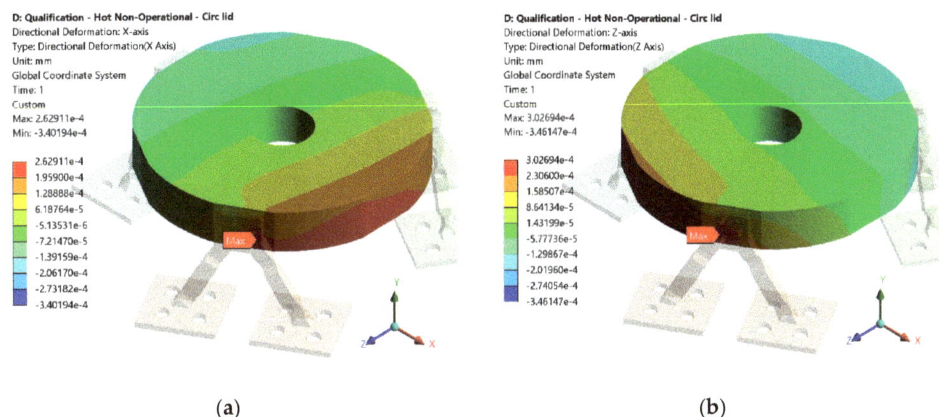

Figure 9. Result of hot (+80 °C) non-operational scenario—First design: (**a**) X-axis displacement, (**b**) Z-axis displacement.

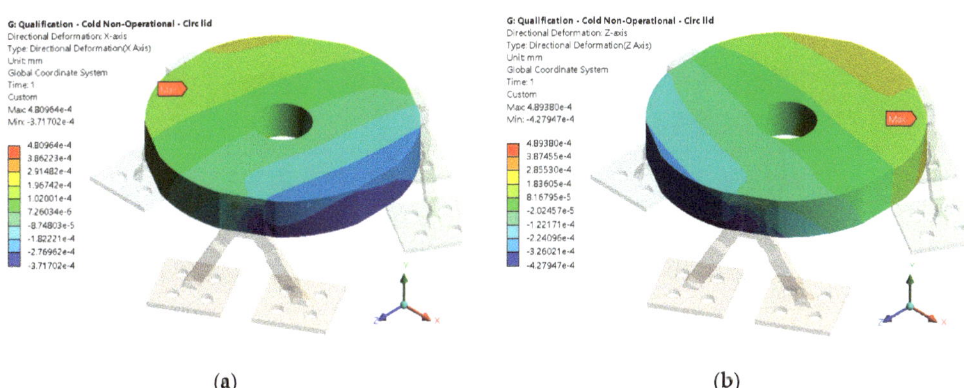

Figure 10. Result of cold (−60 °C) non-operational scenario—First design: (**a**) X-axis displacement, (**b**) Z-axis displacement.

From the results, it is evident that there is a displacement of the mirror relative to its nominal position. From the hot scenario, the extreme values of XZ displacement are in range of $\left[-3.461 \times 10^{-4};\, 3.027 \times 10^{-4}\right]$ mm, and for the cold scenario, the extreme values of XZ displacement are in range of $\left[-4.280 \times 10^{-4},\, 4.894 \times 10^{-4}\right]$ mm, which is fully within the range of acceptable tolerances from Table 2. An animation of the simulations can be found in (https://www.youtube.com/watch?v=M6bY21nINoE accessed on 23 March 2020)

Regarding the stress, the critical components were analyzed, namely the mirror and the isostatic bipod mounts. The thermal conductance along the mirror and the adhesives, and between the adhesives and the isostatic bipod mounts is established as 2500 W/(m²K). The contacts between the other materials (aluminum and titanium) are established as 150 W/(m²K) (as defined in [11]). The results of the von-Mises stress for the hot non-operational case and for the cold non-operational case are presented in Figures 11 and 12, respectively.

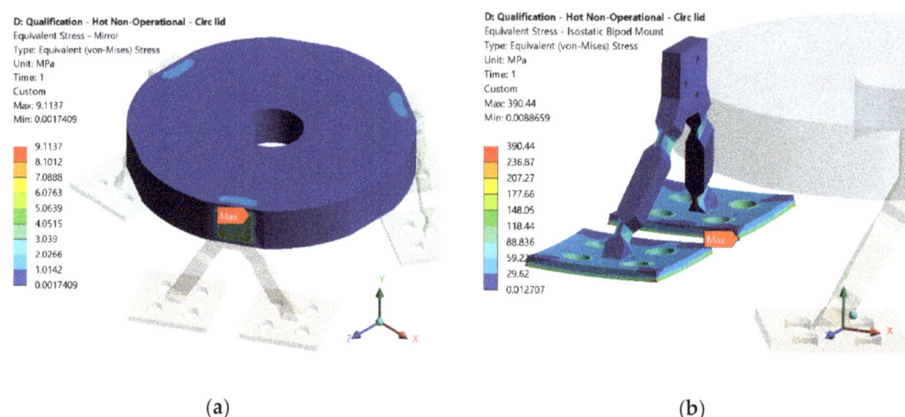

(a)   (b)

Figure 11. Result of hot (+80 °C) non-operational scenario—First design: (**a**) Mirror stress, (**b**) Isostatic bipod mount stress (deformation scale factor: 76).

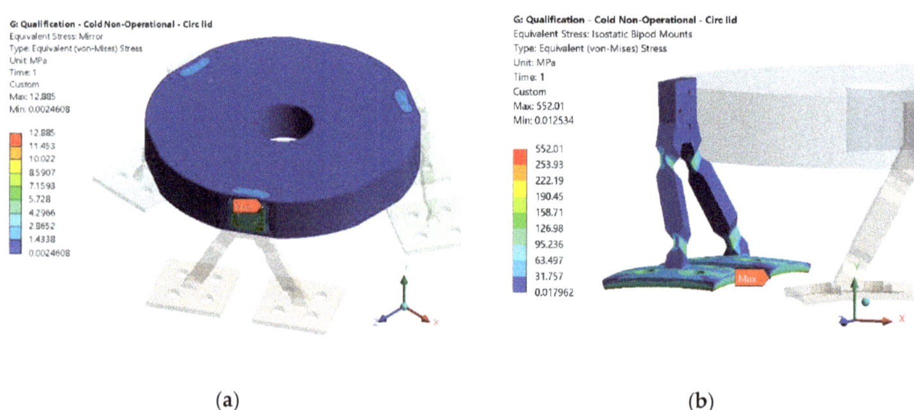

(a)   (b)

Figure 12. Result of cold (−60 °C) non-operational scenario—First design: (**a**) Mirror stress, (**b**) Isostatic bipod mount stress (deformation scale factor: 76).

The mirror maximum stress is sited on the bonding areas, with a safety factor of 3.29 for the hot scenario, and 2.33 for the cold scenario, which means that the mirror will withstand the non-operational temperatures.

The isostatic bipod mounts have a good overall performance, since their displacement is located on the radial blades. The tangential blade does not present a considerable stress since the displacement condition is essentially radial. The maximum stress is located on the base of the isostatic bipod mounts, but since the critical stress is on the blades and in the bonding area, that maximum is not considered. When the maximum stress range is decreased, the value of stress starts to appear on the radial blade, which is coincident with the radial deformation of the base (see Figures 11b and 12b). The safety factor for the isostatic bipod mount considers the maximum stress value, which converts into a safety factor of 2.82 for the hot scenario, and 1.99 for the cold scenario.

Making a similar study for the final design of the isostatic bipod mounts, the XZ displacement results are depicted in Figures 13 and 14 for the hot and cold non-operational scenarios, respectively.

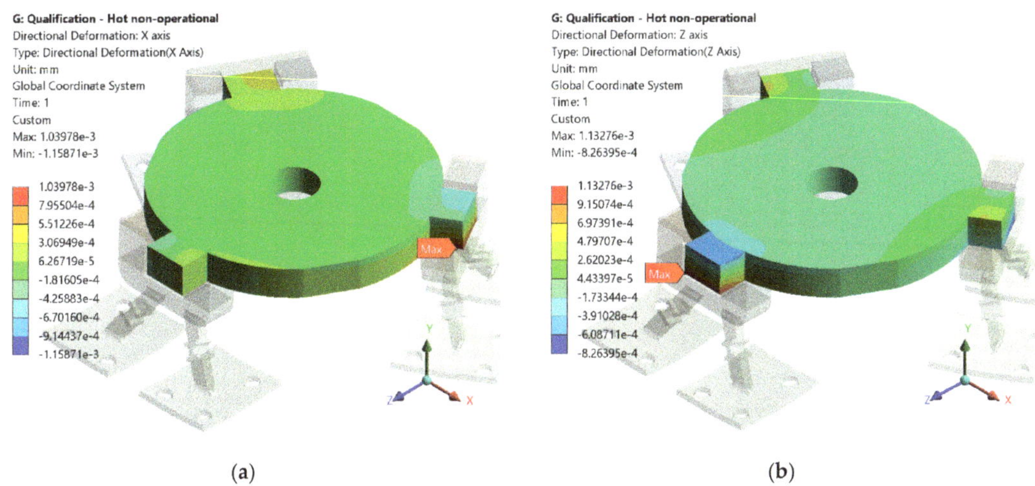

Figure 13. Result of hot (+80 °C) non-operational scenario—Final design: (**a**) X-axis displacement, (**b**) Z-axis displacement.

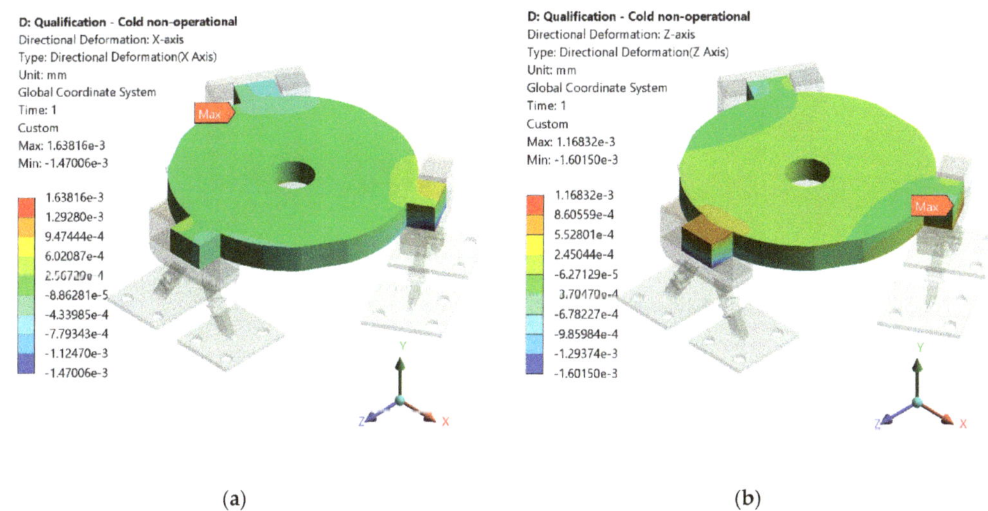

Figure 14. Result of cold (−60 °C) non-operational scenario—Final design: (**a**) X-axis displacement, (**b**) Z-axis displacement.

From the results, it is evident that there is a displacement of the mirror relative to its nominal position. Nevertheless, the results are within the established values for XZ displacement tolerances from Table 2. The extreme values of the hot non-operational temperatures for the final design of the isostatic bipod mounts are within the range of $\left[-1.159 \times 10^{-4}; 1.133 \times 10^{-4}\right]$ mm, and for the cold non-operational temperatures are in range of $\left[-1.602 \times 10^{-4}; 1.683 \times 10^{-3}\right]$ mm.

Regarding the stress in the mirror and the isostatic bipod mounts, the same assessment previously mentioned was made. The results are presented in Figures 15 and 16 for the hot and cold non-operational scenario, respectively.

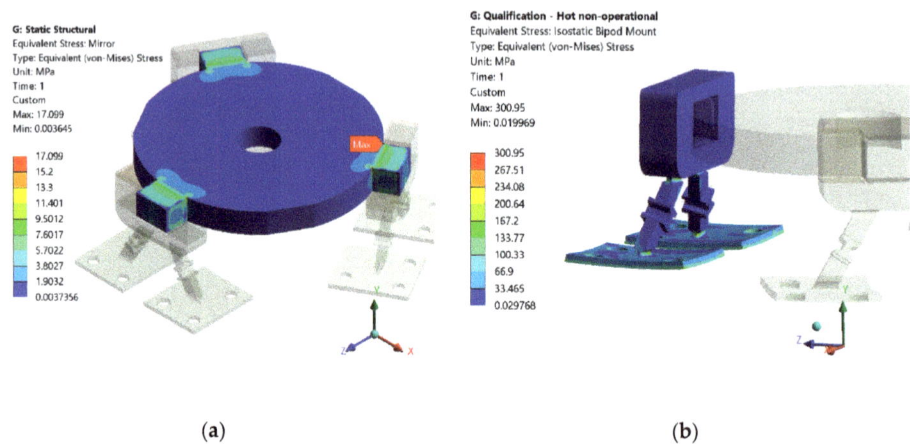

(a) (b)

Figure 15. Result of hot (+80 °C) non-operational scenario—Final design: (**a**) Mirror stress, (**b**) Isostatic bipod mount stress (deformation scale factor: 73).

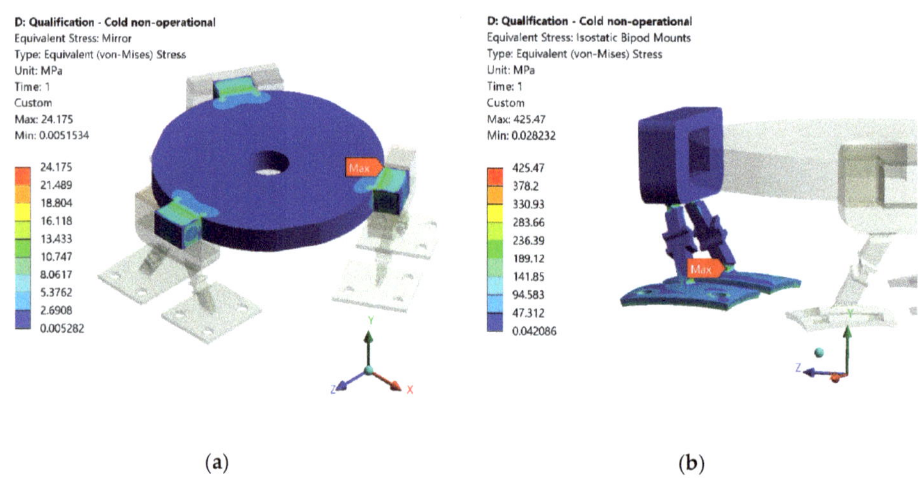

(a) (b)

Figure 16. Result of cold (−60 °C) non-operational scenario—Final design: (**a**) Mirror stress, (**b**) Isostatic bipod mount stress (deformation scale factor: 73).

The mirror's maximum stress is located on the bonding areas, with a safety factor of 1.75 for the hot scenario, and 1.24 for the cold scenario. Since the safety factor of the mirror is higher than unity, the mirror can withstand the hot and cold non-operational scenarios.

Regarding the isostatic bipod mounts stress distribution, it can be concluded that they have a good overall performance, since their displacement is located on the radial blades. The tangential blade does not present a considerable stress since the displacement condition is essentially radial.

The safety factor for the isostatic bipod mount converts into a safety factor of 3.66 for the hot scenario, and 2.59 for the cold scenario. Since the safety factor is higher than unity, the isostatic bipod mounts can withstand the hot and cold non-operational scenarios.

The maximum stress is found on the designated area of the radial blades (see Figures 15b and 16b), which is expected.

The distance between the primary mirror and the base is given in Figures 17 and 18.

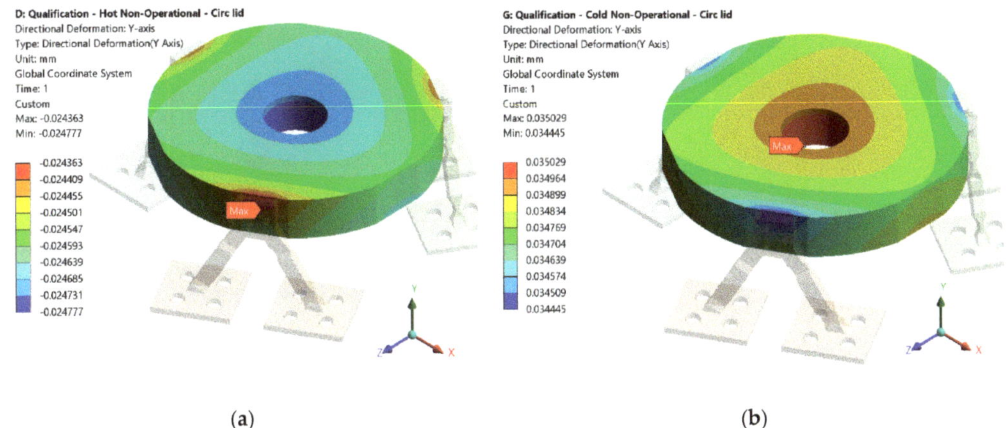

(a)   (b)

Figure 17. Result of Y-axis displacement for the first design: (**a**) hot (+80 °C) non-operational scenario, (**b**) cold (−60 °C) non-operational scenario.

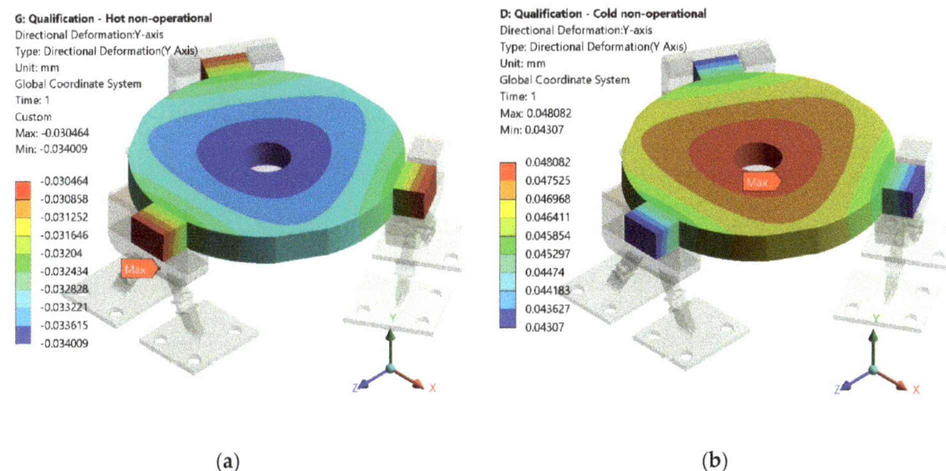

(a)   (b)

Figure 18. Result of Y-axis displacement for the final design: (**a**) hot (+80 °C) non-operational scenario, (**b**) cold (−60 °C) non-operational scenario.

The distance of the mirror towards the base for the first design is between [−0.0248, 0.035] mm, and for the final design is between [−0.0340, 0.0480] mm. These values are within the requirement of ±0.09 mm defined in Table 2.

For the analysis of curvature radius variation, the following result processing approach was implemented: (1) the displacement offset of the mirror was removed; (2) the difference between the mirror outer edge and mirror inner edge was calculated (see Figure 19). To this difference we named the radius curvature variation of the mirror, and its results are presented in Table 3.

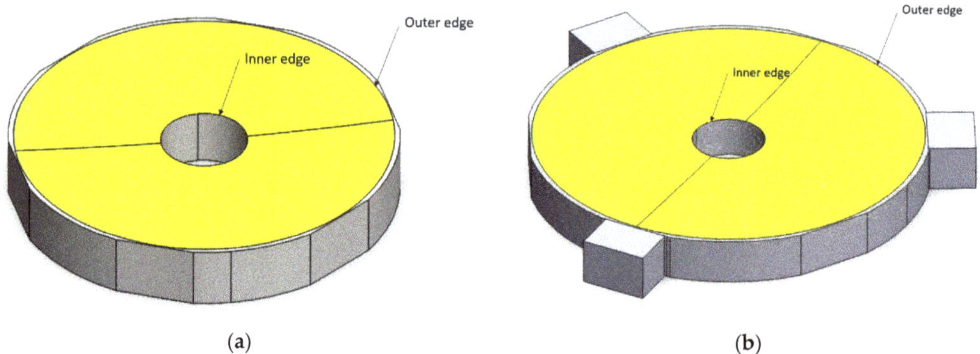

Figure 19. Representation of the inner and outer edge of the mirror: (**a**) first design, (**b**) final design.

Table 3. Curvature radius deformation.

	First Design		Final Design	
	Hot Scenario	Cold Scenario	Hot Scenario	Cold Scenario
Radius curvature variation [μm]	0.287	0.406	1.85	2.62

According to Table 2, a 1 mm curvature radius corresponds to a surface displacement of ± 2 μm. The cold scenario does not fulfil this requirement; and for this case, the LIDAR performance has a loss of energy.

6. Conclusions

The paper describes the development of an isostatic bipod mount design for small mirrors that withstand severe temperature requirements. The assessment of the design was presented considering a preliminary design, which was already built but not yet tested, and the updated design, which is going to be built and subjected to testing.

The first design presents several complicated processes for bonding and alignment, which prompt a design that would be easier to integrate. Both designs present a viable design for the LIDAR, having taken into account the XZ displacement, the distance of the mirror towards the base, the curvature radius, and the stress on the mirror. Both designs survive the non-operational scenarios; however, the final design provides an easier implementation and bonding procedure, since it is divided into four individual parts, in contrast to the first design. Additionally, it is expected that the final design has a higher resistance to shock, because the bonding of the mirror is implemented in four perpendicular areas per isostatic bipod mount instead of only one.

The future steps include the assessment of the performance of the isostatic bipod mounts assembled in the PALT, and consequently, the evaluation of the optical performance of PALT.

Author Contributions: Conceptualization, N.G.D., P.G., H.O., R.M. and A.A.; methodology, N.G.D., P.G., H.O., R.M. and A.A.; software, N.G.D.; investigation, N.G.D., P.G., H.O., R.M. and A.A.; writing—original draft preparation, N.G.D.; writing—review and editing, N.G.D., P.G., H.O., R.M. and A.A.; validation, N.G.D., P.G., H.O., R.M. and A.A.; supervision, N.G.D., P.G., H.O., R.M. and A.A. All authors have read and agreed to the published version of the manuscript.

Funding: This research received no external funding.

Institutional Review Board Statement: Not applicable.

Informed Consent Statement: Not applicable.

Data Availability Statement: Not applicable.

Acknowledgments: European Union's Horizon 2020 research and innovation programme under Grant Agreement No. 870377 (Project No. NEO-MAPP). FCT (Foundation for Science and Technology, I.P.), through: CENTRA, project UIDB/00099/2020; LAETA project UIDB/50022/2020; ICT project UIDP/04683/2020.

Conflicts of Interest: The authors declare no conflict of interest.

Nomenclature

CTE	Coefficient of Thermal Expansion
LIDAR	Light Detection and Ranging
PALT	Planetary ALTimeter
ToF	Time of Flight

References

1. Yoder, P.R., Jr. *Opto-Mechanical Systems Design*; CRC Press: Boca Raton, FL, USA, 2015.
2. Pijnenburg, J.A.; te Voert, M.J.; de Vreugd, J.; Vosteen, A.; van Werkhoven, W.; Mekking, J.; Nijland, B.A. Ultra-stable isostatic bonded optical mount design for harsh environments. In *Modern Technologies in Space-And Ground-Based Telescopes and Instrumentation II*; International Society for Optics and Photonics: Bellingham, DC, USA, 2012; Volume 8450, p. 845027.
3. Kihm, H.; Yang, H.S.; Moon, I.K.; Yeon, J.H.; Lee, S.H.; Lee, Y.W. Adjustable bipod flexures for mounting mirrors in a space telescope. *Appl. Opt.* **2012**, *51*, 7776–7783.
4. Kihm, H.; Yang, H.S.; Lee, Y.W. Bipod flexure for 1-m primary mirror system. *Rev. Sci. Instrum.* **2014**, *85*, 125101.
5. Liu, B.; Wang, W.; Qu, Y.J.; Li, X.P.; Wang, X.; Zhao, H. Design of an adjustable bipod flexure for a large-aperture mirror of a space camera. *Appl. Opt.* **2018**, *57*, 4048–4055.
6. Dias, N.G.; Arribas, B.N.; Gordo, P.; Sousa, T.; Marinho, J.; Melicio, R.; Amorim, A.; Michel, P. LIDAR altimeter conception for HERA spacecraft. *Aircr. Eng. Aerosp. Technol.* **2021**, *93*, 1018–1028.
7. Dias, N.G.; Arribas, B.N.; Gordo, P.; Sousa, T.; Marinho, J.; Melicio, R.; Amorim, A.; Michel, P. HERA mission LIDAR altimeter implementation. In *IOP Conference Series: Materials Science and Engineering*; IOP Publishing: Bristol, UK, 2021; Volume 1024, p. 012112.
8. Dias, N.G.; Arribas, B.N.; Gordo, P.; Sousa, T.; Marinho, J.; Melicio, R.; Amorim, A.; Livio, B.; Michel, P. HERA mission LIDAR mechanical and optical design. In *IOP Conference Series: Materials Science and Engineering*; IOP Publishing: Bristol, UK, 2022; Volume 1026, p. 012094.
9. Côté, P.; Desnoyers, N. Thermal stress failure criteria for a structural epoxy. *Optomech. Innov. Solut.* **2011**, *8125*, 81250K.
10. 3M Science, 3M™ Scotch-Weld™ Epoxy Adhesive 2216 B/A Gray. Available online: https://www.generaladhesivos.com/proveedor-pegamento/62hoja-tecnica-3M%20Scotch-Weld%20Epoxy%20Adhesive%202216%20B_A%20Gray%20(1).pdf (accessed on 1 December 2021).
11. Yovanovich, M.M. Thermal Interface (Joint) Conductance and Resistance, ECE 309 Course Notes. Available online: http://www.mhtl.uwaterloo.ca/courses_old/ece309/notes/conduction/cont.pdf (accessed on 1 December 2021).

Article

Analytic Derivation of Ascent Trajectories and Performance of Launch Vehicles

Paolo Teofilatto [1,*], Stefano Carletta [1] and Mauro Pontani [2]

[1] School of Aerospace Engineering, Sapienza University of Rome, Via Salaria 851, 00138 Rome, Italy; stefano.carletta@uniroma1.it

[2] Department of Astronautical, Electrical, and Energy Engineering, Sapienza University of Rome, Via Salaria 851, 00138 Rome, Italy; mauro.pontani@uniroma1.it

* Correspondence: paolo.teofilatto@uniroma1.it

Abstract: This papers introduces an analytic method to define multistage launcher trajectories to determine the payload mass that can be inserted in orbits of different semimajor axes and inclinations. This method can evaluate the gravity loss, which is the main term to be subtracted to the Tziolkowski evaluation of the velocity provided by the thrust of a launcher. In the method, the trajectories are dependent on two parameters only: the final flight-path angle γ_f at the end of the gravity-turn arc of the launcher trajectory and the duration t_c of the coasting arc following the gravity-turn phase. The analytic formulas for the gravity-turn phase, being solutions of differential equations with a singularity, allow us to identify the trajectory with a required final flight-path angle γ_f in infinite solutions with the same initial vertical launch condition. This can also drive the selection of the parameters of the pitch manoeuvre needed to turn the launcher from the initial vertical arc. For any pair γ_f and t_c, a launcher trajectory is determined. A numerical solver is used to identify the values γ_f and t_c, allowing for the insertion of the payload mass into the required orbit. The analytic method is compared with a numerical code including the drag effect, which is the only effect overlooked in the analytic formulas. The analytical method is proven to predict the payload mass with an error never exceeding the 10% of the actual payload mass, found through numerical propagation.

Keywords: multistage launch vehicles; ascent trajectory optimization; analytical performance evaluation; rocket staging

Citation: Teofilatto, P.; Carletta, S.; Pontani, M. Analytic Derivation of Ascent Trajectories and Performance of Launch Vehicles. *Appl. Sci.* 2022, *12*, 5685. https://doi.org/10.3390/app12115685

Academic Editor: Jérôme Morio

Received: 21 April 2022
Accepted: 30 May 2022
Published: 3 June 2022

Publisher's Note: MDPI stays neutral with regard to jurisdictional claims in published maps and institutional affiliations.

1. Introduction

The number of microsatellites to be orbited is increasing rapidly, in particular for the delivery and replacement of microsatellites as part of mega constellation programs. This is stimulating a new wave of dedicated launch vehicles capable of offering responsive, flexible and cost-effective services to this huge and new market. Some of these "micro-launchers" are listed in Table 1. The mass of payload m_{pay} to be inserted into LEO is below 1000 kg. In fact, m_{pay} is the driving element of a space rocket and it is important to define it at the first step of a launcher design. A very popular method of computation of m_{pay} is offered by the Tziolkowski formula. However, the results obtained using this method are not accurate since the space environment (gravity and aerdoynamic forces) and the guidance (direction of thrust) are not taken into account. On the other hand, the accurate computation of a launcher trajectory and the evaluation of launcher performances is a typical and complex problem in the optimization of aerospace trajectories. Launcher performances are evaluated maximizing the final mass that can be set into orbits with different parameters (semimajor axis, eccentricity, inclination). The problem has initial- and final-state variable constraints, path constraints and discontinuities in the mass variation, due to stage separation. This optimization problem can be approached in different ways: all the approaches require educated guesses to initialize some iterative algorithm able to refine the initial guesses and to achieve the optimal solution.

Table 1. Listof Micro-Launchers.

Launcher	Total Mass	Number of Stages	Max Payload to LEO	Maiden Flight
Simorgh, Iran	78,926 kg	3	350 kg	2021
Unha, North Korea	68,039 kg	3	200 kg	2012
LongMarc11, China	58,000 kg	4	700 kg	2015
Spectrum, Germany	50,000 kg	2	950 kg	2022 (planned)
Minotaur1, USA	36,200 kg	4	580 kg	2000
Zhu-Que1, China	24,600 kg	3	300 kg	2018
LauncherOne, USA	23,410 kg	2	500 kg	2000
Quased, Iran	22,700 kg	3	50 kg	2020
Kuoizhou1A, China	15,100 kg	3	300 kg	2017
Electron, USA	11,400 kg	2	225 kg	2017
Astra R3, USA	9980 kg	2	150 kg	2020

In the scientific literature, some papers [1–7] address the problem of optimizing the ascent trajectory of launch vehicles through indirect approaches. Examples are the multiple-subarc gradient restoration algorithm, proposed by Miele, and multiple shooting techniques [5–7]. The previously cited works [2–7] require a considerable deal of effort for deriving the analytical conditions for optimality and for the subsequent programming and debugging. Furthermore, these methods can suffer from a slow rate of convergence and an excessive dependence on the starting guess. Alternative approaches [8,9] are direct in nature, often are more robust, but require discretization of the problem (e.g., through collocation [8,9]) and the use of dedicated algorithms for nonlinear programming, without avoiding the need of a starting guess. Most recently, the indirect heuristic method [10] was proposed as an effective approach for trajectory optimization of ascent vehicles, with the remarkable feature of not requiring any starting guess. For the purpose of preliminary analysis of the performance of new ascent vehicles, fast, approximate approaches that are exempt from convergence issues and minimize the computational effort are desirable. With this intent, Pontani and Teofilatto [11] and Pallone et al. [12,13] propose two numerical approaches for performance evaluation of multistage launch vehicles that aim at obtaining accurate predictions of the performance attainable from a launch vehicle.

The aim of this paper is to derive formulas, possibly as simple as the Tziolkowski one, able to obtain in a few seconds the performance of a multistage space rockets and also able to offer trajectory parameters useful as input for the convergence of complex numerical programs with accurate representation of the rocket system and its space environment.

The formulas originate from previous work [14], which has been generalized to multistage rockets in [15]; all the rocket state variables are derived in [16]. In the present paper, the method is extended to include the optimal choice of a cost arc and the guidance of the last stage of the launcher till the injection into the selected orbit, so the performance of a space launcher is obtained. The method is based on simple mathematics and physics principles and only a few seconds are needed to determine the payload mass that can be set into orbits of different altitude and inclinations, starting from any location of the launch pad.

The major simplification made in order to obtain these results is that the aerodynamic drag is neglected, and this produces an error in the computation of the launcher trajectories. It is rather difficult to treat the aerodynamic effect analytically due to its highly non-linear dependence on velocity and altitude. Some analytic formulas including aerodynamics are derived in [17]; these however are restricted to the vertical arc of the trajectory. To evaluate the impact of drag, the method presented was compared to numerical results obtained by a high-fidelity level of representation of the launcher dynamics. It is shown that the error in the estimate of the optimal payload mass to be inserted into orbit is relevant only for orbits of altitude below 600 km, and in these cases the error does not exceed 10% Moreover, it is proved that the analytic approach proposed here can provide good guesses for numerical solvers of the optimal control problem.

The paper is organized as follows: in Section 2, a first sizing of a space launcher is given based on the Tziolkowski formula. Assuming a payload mass to be delivered in LEO orbit and assuming a fixed percentage of losses, a method is derived to size the staging of a multistage rocket in an optimal way. In Section 3, analytical formulas for the launcher gravity-turn phase are developed in detail for the case of multiple stage rockets. These formulas provide the launcher trajectory of the first stages and allow for the evaluation of the so-called gravity loss. Sections 4 and 5 describe the method of computation of the full launcher trajectory till the injection into orbit. Section 6 is dedicated to testing the method with respect to numerical computations.

2. First Sizing of a Launch Vehicle

Achieving a single stage to orbit is still an infeasible task and space launchers have a number N of stages ranging from 2 to 4. Each stage has a total mass m_i

$$m_i = m_{p_i} + m_{si}$$

where m_{p_i} is the propellant contained within the tank of each stage and the structural mass coefficient of each stage is defined as

$$\epsilon_i = \frac{m_{si}}{m_i} \tag{1}$$

The total mass at lift off is

$$M_1 = \Sigma_{i=1}^{N} m_i + \left(m_{hs} + m_{pay} \right)$$

with m_{hs} the mass of the heat shields (here released at the end of the gravity-turn phase) and m_{pay} is the satellite mass to be injected into orbit. At the time t_{b1} of the end of the first stage boost (burn time), the propellant mass m_{p_1} is consumed and the mass of the space rocket is

$$M_{1f} = M_1 - m_{p_1}$$

The velocity reached at burn time t_{b1} can be computed integrating the acceleration

$$\dot{V} = \frac{T}{m} \cos \alpha - \frac{\mu}{r^2} \sin \gamma - \frac{1}{2}\rho V^2 \frac{C_D S}{m} \tag{2}$$

here T is the thrust intensity with a direction determined by the angle α taken from the velocity $\vec{V}$. The velocity direction is determined by the flight-path angle γ taken from the local horizon direction $\hat{\theta}$, see Figure 1. The variable r is the radius distance from the centre of the Earth to the launcher centre of mass and μ is the Earth gravitational constant. The atmospheric density is denoted by ρ and C_D is the launcher drag coefficient computed with respect to the reference surface S.

To underline the thrust's contribution to the variation in velocity and consider all the other terms as "losses", let us write (2) as

$$\dot{V} = \frac{T}{m} - \frac{\mu}{r^2} \sin \gamma - \frac{1}{2}\rho V^2 \frac{C_D S}{m} - \frac{T}{m} \left(1 - \cos \alpha \right)$$

Integrating the above acceleration between time zero and t_{b1}, one obtains the velocity after the first-stage boost

$$\Delta V_1 = \int_{t_0}^{t_{b1}} \frac{T}{m} \, dt - \int_{t_0}^{t_{b1}} \frac{\mu}{r^2} \sin \gamma \, dt - \int_{t_0}^{t_{b1}} \frac{1}{2}\rho V^2 \frac{C_D S}{m} \, dt - \int_{t_0}^{t_{b1}} \frac{T}{m} \left(1 - \cos \alpha \right) dt \tag{3}$$

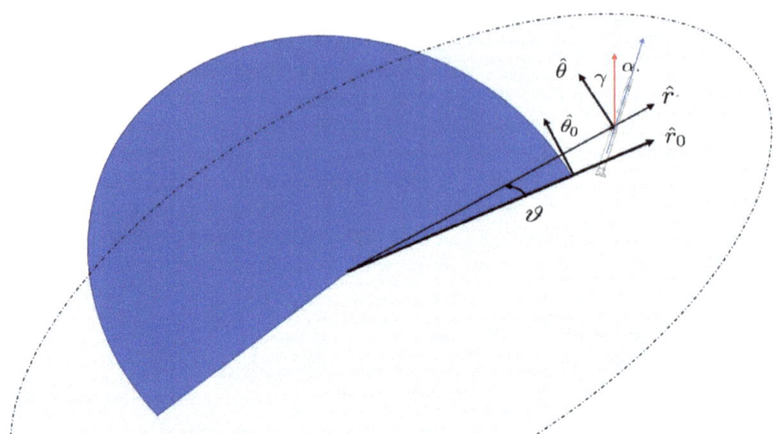

Figure 1. The state variable of the launch trajectory. The inertial frame $(\hat{r}_0, \hat{\theta}_0)$ has origin on the centre of the Earth (in dark blue). The orbit frame $(\hat{r}, \hat{\theta})$ has origin on the launcher centre of mass. The red arrow defines the velocity and the blue arrow the thrust direction

Equation (3) is called the "equation of losses"

$$\Delta V_1 = \Delta V_{Tz} - \Delta V_{grav} - \Delta V_{drag} - \Delta V_{mis}$$

where the first variation of velocity due to the engine boost is called the Tziolkowski variation of velocity and the other velocities subtracted to ΔV_{Tz} are called gravity, drag and misalignment losses, respectively. Average thrust is defined by $T = gIsp\,\dot{m}$, where $\dot{m}$ is equal to the mass rate $\dot{m} = -\frac{dm}{dt}$. Inserting $T = -gIsp\,\frac{dm}{dt}$ in (3), we obtain the Tziolkowski velocity variation provided by the first-stage engine

$$\Delta V_{Tz} = gIsp \, \log\left(\frac{M_1}{M_{1f}}\right)$$

Introducing the subrocket structural mass ratios

$$U_k = \frac{M_{kf}}{M_k}\,, \quad M_k = \Sigma_{i=k}^{N} m_i + \left(m_{pay} + m_{hs}\right)\,, \quad M_{kf} = M_k - m_{p_k}\,, \quad k = 1, N$$

the variation of the velocity provided by the engines of a multistage rocket is

$$\Delta V_{Tz} = \Sigma_{k=1}^{N} g \, Isp_k \, \log\left(\frac{1}{U_k}\right) \tag{4}$$

The values of the specific impulses Isp_k depend on the kind of propellant, for instance liquid or solid. Figure 2 shows the specific impulses for stages with solid propellant: these values are taken from the database [18] and we have average values

$$Isp_1 = 280\,s\,, \; Isp_2 = 290\,s\,, \; Isp_3 = 270\,s \tag{5}$$

Other parameters defined by technical constraints are the structural mass ratio of each stage (1): Figure 3 shows these values for a number of stages taken from the database [18], and we have average values

$$\epsilon_1 = 0.11\,, \; \epsilon_2 = 0.13\,, \; \epsilon_3 = 0.3 \tag{6}$$

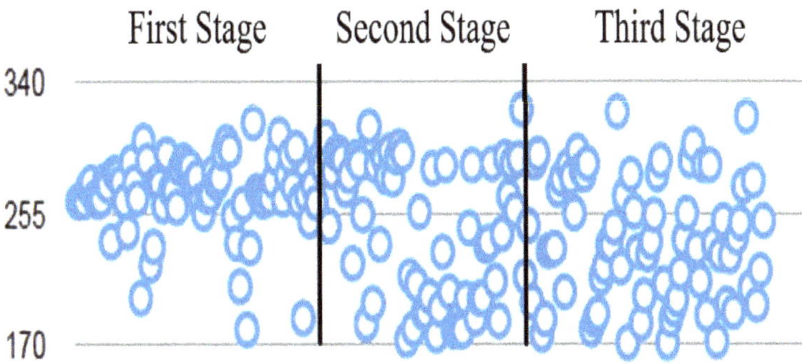

Figure 2. The specific impulse of different stages taken from the database ©Mark Wade [18].

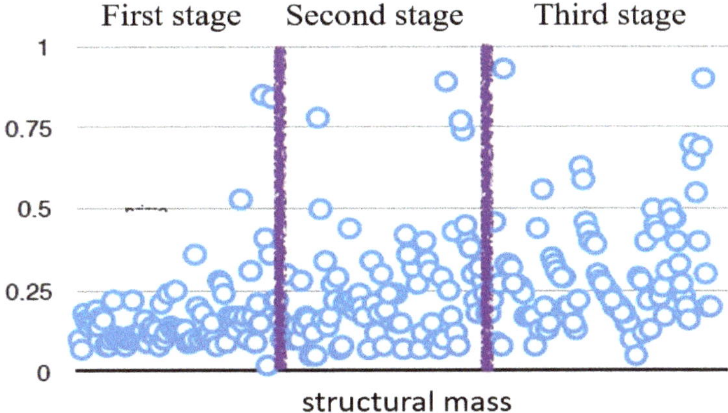

Figure 3. The structural mass ratio of different stages taken from the database ©Mark Wade [18].

It is important to know that if the total mass M_1, the payload mass m_{pay} and the values $\epsilon_k$ith the five unknow, Isp_k are given, there is an optimal selection of U_k, maximizing the Tziolkowski velocity (4). The procedure to identify the optimal subrocket mass ratios U_k^* is in Appendix A.

The parameters ϵ_k, Isp_k can be easily guessed at the first step of the rocket design, and the maximum payload mass m_{pay} to be inserted into a specific reference orbit is the performance characterizing a space launcher.

With ϵ_k, Isp_k, and m_{pay} fixed, the total mass M_1 determines the optimal mass distribution U_k^* (staging) and the maximum Tziolkowski velocity, according to Appendix A. If this velocity is equal to the velocity required to inject the payload mass m_{pay} into the reference orbit, M_1 is the launcher lift-off mass achieving this performance.

One point to underline is that the velocity to be provided by the thrust is not equal to the orbital velocity of a spacecraft on a circular orbit of altitude h: $V_a = \sqrt{\frac{\mu}{h+R_E}}$. In fact, we must consider the losses: these depend on each phase of the ascent trajectory. In the present Section it is assumed that the launcher velocity due to thrust must be overcome by 30% the required value for orbit injection. The orbit velocity relative to the Earth $\vec{V}_R$ depends on the latitude L of the launch site and on the azimuth ψ

$$V_R = \sqrt{V_a^2 + V_E^2 - 2V_a V_E \sin \psi}$$

where $V_E = \omega_E\,R_E\cos(L)$ is the velocity of the launch site and the azimuth angle ψ is related to the launch site latitude and orbit inclination by the equation: $\sin\psi = \frac{\cos i}{\cos L}$.

Then, the required velocity is here defined as

$$V_{req} = V_R(1 + 30\%)$$

Figure 4 shows the relative (in blue) and required (in red) velocities to achieve a reference circular orbit of altitude h = 650 km with different inclinations from a launch site at L = 30 deg N. Table 2 shows the minimum mass at lift off for a micro-launcher having parameters (5) and (6) for different payload masses (from 100 to 1000 kg) to be injected into the reference orbit ($V_{req} = 9805\,\text{m/s}$). The ratio $\frac{M_1}{m_{pay}}$ is constant.

Table 2. Minimum lift-off mass to inject the payload mass into circular polar orbit of altitude $h = 650\,\text{km}$ (average parameters (5) and (6), and launch site L = 30 deg).

Initial Mass M_1	Payload Mass m_{pay}
18,000 kg	100 kg
36,000 kg	200 kg
54,000 kg	300 kg
72,000 kg	400 kg
90,000 kg	500 kg

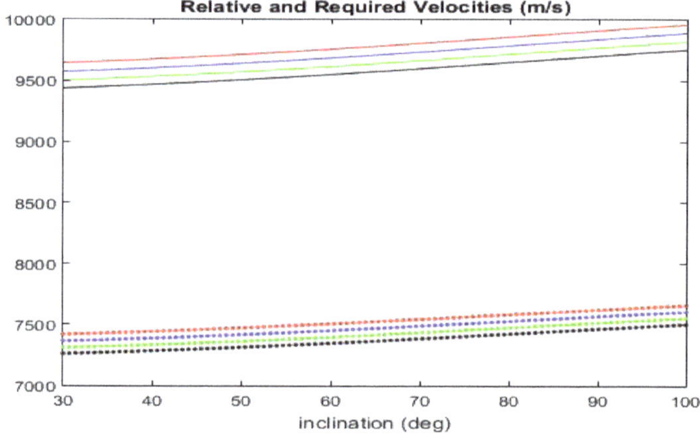

Figure 4. The relative (dotted) and required (solid line) velocities to achieve circular orbits of altitude h = 800 km (black), 700 km (green), 600 km (blue), 500 km (red) and different inclinations

3. A Second Step (Gravity Losses)

At lift off, the launcher performs a vertical trajectory followed by a pitch manoeuvre to select the required orbit plane. Then, to limit the aerodynamic loads on the structure, any launcher is forced to follow a zero angle of attack trajectory ($\alpha = 0$) during the atmospheric flight. This trajectory is called the gravity-turn trajectory, and it is performed by the first stages of space launchers. The trajectory is basically planar and the state variables can be the relative velocity, flight-path angle, altitude and range (V, γ, h, s). The equations of motion are

$$\begin{cases} \dot{V} = \frac{T}{m} - \frac{\mu}{r^2}\sin\gamma - \frac{1}{2}\rho V^2\frac{C_D\,S}{m} \\ \dot{\gamma} = \frac{\dot{s}}{R_E} - \frac{\mu}{r^2 V}\cos\gamma + \frac{1}{2}\rho V\frac{C_L\,S}{m} \\ \dot{h} = V\sin\gamma \\ \dot{s} = \frac{R_E}{r}V\cos\gamma \end{cases} \tag{7}$$

In fact, these equations are singular under the lift-off condition; $V = 0, \gamma_0 = \frac{pi}{2}$.

As observed in [14], this singularity generates an infinite number of solutions corresponding to the same lift-off condition: one can use this singularity to select among the different solutions the one with a prescribed final value, for instance a fixed flight-path angle γ_f [15].

An analytic formula for the velocity V, due to Culler and Fried, was obtained under the following hypotheses:

- Constant gravity;
- Negligible drag and lift effect;
- Constant thrust-over-mass ratio $n = \frac{T}{g\,m}$.

The last of these is the strongest assumption, and is handled considering an average value $\bar{n}$ of the thrust-to-weight ratio. This average value can be computed using the formulas for the variation of mass $m = m_0 - m_d\,t = m_0(1 - q\,t)$ and the definition of the mass ratio $U = \frac{m_s}{m_0}$, which verifies $U = 1 - q\,t_b$.

Then

$$
\begin{aligned}
\bar{n} &= \frac{1}{t_b}\int_0^{tb} \frac{T}{mg}(t)\,dt = \frac{T}{g\,t_b\,m_0}\int_0^{tb}\frac{dt}{1-qt} = \\
&= \frac{n_0}{t_b}\frac{1}{-q}\left[\ln(1-qt)\right]_0^{t_b} = \frac{n_0}{-q\,t_b}\ln(1-qt_b) = \\
&= -\frac{n_0}{1-U}\ln(U) = \frac{n_0}{1-U}\ln\left(\frac{1}{U}\right)
\end{aligned}
$$

where $n_0 = \frac{T}{gm_0}$ is the initial thrust-to-weight ratio.

Now two transformations are defined:

1. From the flight-path angle variable γ to the kick angle $\chi = \frac{\pi}{2} - \gamma$.
2. From the kick angle χ to the variable $z = \tan\frac{\chi}{2}$.

By the transformation in Equation (1) we have the following differential equations for the velocity and kick angle

$$
\begin{cases}
\dot{V} = -g\cos\chi + g\bar{n} \\
\dot{\chi} = \frac{g}{V}\sin\chi
\end{cases}
\tag{8}
$$

Now the following trigonometrical equations are recalled

$$
\cos\chi \quad - \quad \frac{1-\tan^2\frac{\chi}{2}}{1+\tan^2\frac{\chi}{2}} = \frac{1-z^2}{1+z^2}
$$

$$
\sin\chi = \sin\left(2\frac{\chi}{2}\right) = 2\sin\frac{\chi}{2}\cos\frac{\chi}{2} = 2\tan\frac{\chi}{2}\cos^2\frac{\chi}{2} = \frac{2z}{1+z^2}
$$

showing that the transformation to the z-variable allows us to transform the trigonometric function in rational functions. Then, Equation (8) becomes

$$
\begin{cases}
\dot{V} = -g\bar{n} - g\dfrac{1-z^2}{1+z^2} \\
\dot{z} = \dfrac{g}{V}z
\end{cases}
\tag{9}
$$

Dividing the first by the second equation of the system (9), one has

$$
\frac{dV}{dz} = \bar{n}\frac{V}{z} - \frac{V}{z}\frac{1-z^2}{1+z^2}
$$

that is

$$
\frac{dV}{V} = \left(\frac{\bar{n}}{z} - \frac{1}{z}\frac{1-z^2}{1+z^2}\right)dz
$$

By integration

$$V = A z^{\tilde{n}-1}(1+z^2) \tag{10}$$

The velocity in (10) is a function of the variable z. The time dependence can be obtained deriving the relation between time and z. Substitution of (10) in the second equation of the system (9) gives

$$\dot{z} = \frac{dz}{dt} = -\frac{g}{A}\frac{1}{z^{\tilde{n}-2}(1+z^2)}$$

by separation of variables and integrating from $t_0 = 0$ to t

$$t = \frac{A}{g}\left(\frac{z^{\tilde{n}-1}}{\tilde{n}-1} + \frac{z^{\tilde{n}+1}}{\tilde{n}+1}\right) \tag{11}$$

The launch initial conditions are in the given hypothesis

$$\begin{cases} V_0 = 0 \\ \gamma_0 = \frac{\pi}{2} \\ \chi_0 = 0 \Rightarrow z_0 = 0 \end{cases}$$

Imposing these initial conditions in the solution (10), one has

$$V_0 = A z_0^{\tilde{n}-1}(1+z_0^2) = 0 \quad \forall A$$

Then, the initial conditions are verified for any value of the constant A. That is, there are infinite (gravity-turn) solutions for the velocity, parameterized by A, corresponding to the same initial condition. This violation of the Chauchy theorem on uniqueness of the solution of a differential equation is due to the singularity of $\dot{z}$ for $V = 0$ in the system (9).

One can take advantage of this multiplicity of solutions to select for instance the solution that achieves a chosen flight-path angle γ_f at the burn-out time t_b. The chosen γ_f defines a value $z_f = \tan(\frac{\pi/2-\gamma_f}{2})$. By Equation (11), one obtains

$$A = g\, t_b \left(\frac{z_f^{\tilde{n}-1}}{\tilde{n}-1} + \frac{z_f^{\tilde{n}+1}}{\tilde{n}+1}\right)^{-1} \tag{12}$$

and the related velocity is obtained inserting this value of A into Equation (10).

In particular, the velocity at burn out is

$$V_f = A z_f^{\tilde{n}-1}(1+z_f^2) \tag{13}$$

In addition, the altitude h and range s can be easily derived [16]. For the altitude

$$\dot{h} = V \sin\gamma = V \cos\chi = V\frac{1-z^2}{1+z^2}$$

Then

$$\frac{dh}{dz} = \frac{\dot{h}}{\dot{z}} = \frac{V^2}{g\,z}\frac{1-z^2}{1+z^2}$$

By Equation (10), one has

$$\frac{dh}{dz} = \frac{A^2}{g}\left(z^{2\tilde{n}-3} - z^{2\tilde{n}+1}\right)$$

The elementary integration of the above equation between current z and initial z_0 gives the formula for the altitude

$$h(z) = h(z_0) + \frac{A^2}{g} \left[\frac{z^{2\bar{n}-2}}{2\bar{n}-2} - \frac{z^{2\bar{n}+2}}{\bar{n}+2} \right]_{z_0}^{z_f} \tag{14}$$

For the range s

$$\dot{s} = V \cos\gamma = V \sin\chi = 2\,V \frac{z}{1+z^2}$$

then

$$\frac{ds}{dz} = \frac{\dot{s}}{\dot{z}} = 2\frac{A^2}{g} \left(z^{2\bar{n}-1} + z^{2\bar{n}+1} \right)$$

The integration gives the formula for the range

$$s(z) = s(z_0) + 2\frac{A^2}{g} \left[\frac{z^{2\bar{n}}}{2\bar{n}} + \frac{z^{2\bar{n}+2}}{2\bar{n}+2} \right]_{z_0}^{z_f} \tag{15}$$

Equations (10), (12), (14) and (15), together with (11), provide the state variables of the gravity-turn trajectory of a single-stage rocket with flight-path angle γ_f at burn-out time t_b.

For instance, let us consider initial lift off condition $(V_0 = 0, \chi_0 = 0, h_0 = 0, s_0 = 0)$ and choose the final condition $\gamma_f = 0$ this condition corresponds to $z_f = 1$. Then Equation (12) gives

$$A = g\,t_b\,\frac{\bar{n}^2-1}{2\bar{n}}$$

By Equations (10), (14) and (15), the velocity, altitude and range at burn out are equal to

$$V_f = 2A = g\,t_b\,\frac{\bar{n}^2-1}{\bar{n}} \tag{16}$$

$$h_f = \frac{A^2}{g}\frac{1}{\bar{n}-1} = \frac{g t_b^2}{2}\frac{\bar{n}^2-1}{\bar{n}^2}$$

$$s_f = \frac{A^2}{g}\left(\frac{1}{\bar{n}} + \frac{1}{\bar{n}+1}\right) = \frac{g t_b^2}{2}\left(\frac{\bar{n}^2-1}{\bar{n}}\right)^2\left[\frac{1}{2\bar{n}} + \frac{1}{2\bar{n}+2}\right]$$

Note that the final velocity formula of a rocket is generally obtained by the Tziolkowski formula that takes into account only the thrust acceleration. In the present setting, the Tziolkowski formula gives the final velocity

$$V_{Tz} = g\,\bar{n}t_b$$

The final velocity obtained here, which includes gravity effect, is given by (16): the difference among the two is the gravity loss and it is equal to

$$\Delta V_g = V_{Tz} - V_f = \frac{g\,t_b}{\bar{n}} \tag{17}$$

The above results can be generalized to multistage launchers.

4. Analytic Derivation of Launcher Trajectories

The above analytic formulas can be generalized to multistage launchers. Let us start with two stages with the following average parameters (n_1, Isp_1, u_1) and (n_2, Isp_2, u_2).

The first stage has variables z_1, $V_1 = A_1 z_1^{\bar{n}_1-1}(1+z_1^2)$ in the time interval $[t_{10} = 0, t_{1f} = t_{b_1}]$. The second stage has variables z_2, $V_2 = A_2 z_2^{\bar{n}_2-1}(1+z_2^2)$ in the time interval $[t_{20} = t_{b_1}, t_{2f} = t_{b_2} + t_{b_1}]$ and the final value z_{2_f} is fixed (see Figure 5).

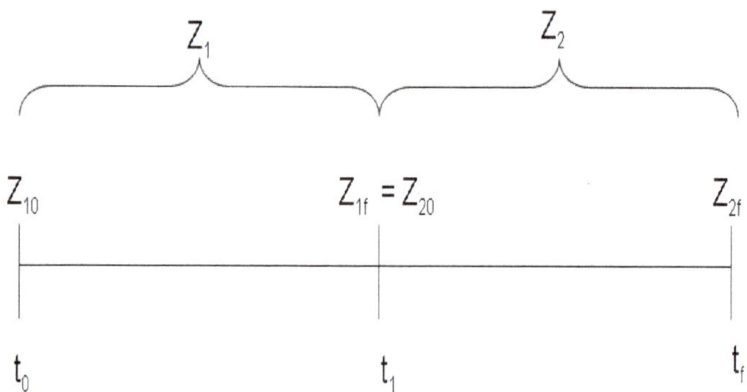

Figure 5. Two stages.

Of course the two solutions z_1, V_1 and z_2, V_2 must match at the boundary point

$$z_{1f} = z_{20}$$
$$V_{1f} = V_1(z_{1f}) = V_2(z_{20}) = V_{20}$$

The two matching conditions imply

$$\frac{A_1}{A_2} = z_{20}^{\bar{n}_2 - \bar{n}_1} \tag{18}$$

Equation (11) gives at first-stage burn out

$$t_{b1} = \frac{A_1}{g}\left(\frac{z_{20}^{\bar{n}_1 - 1}}{\bar{n}_1 - 1} + \frac{z_{20}^{\bar{n}_1 + 1}}{\bar{n}_1 + 1} \right) \overset{def}{=} f_1(z_{20}) \tag{19}$$

At the burn out of the second stage, one has z_{2f} and

$$t_{b2} = \frac{A_2}{g}\left(\frac{z_{2f}^{\bar{n}_2 - 1}}{\bar{n}_2 - 1} + \frac{z_{2f}^{\bar{n}_2 + 1}}{\bar{n}_1 + 1} - \frac{z_{20}^{\bar{n}_2 - 1}}{\bar{n}_2 - 1} + \frac{z_{20}^{\bar{n}_2 + 1}}{\bar{n}_1 + 1} \right) \overset{def}{=} f_2(z_{20}) \tag{20}$$

We have a system of three Equations (18)–(20) in the three unknowns A_1, A_2 and z_{20}. To solve it, let us consider the ratio

$$\frac{t_{b2}}{t_{b1}} = \frac{f_1}{f_2}$$

Because of (18), this ratio is a function of z_{20} only

$$f(z_{20}) = \frac{t_{b2}}{t_{b1}} - \frac{f_1}{f_2} = 0$$

The solution z_{20} of $f(z_{20}) = 0$ (in the range [0,1]) provides the value of the flight-path angle at the end of the first stage. Moreover, by Equation (20), one finds A_2 and by Equation (19) one finds A_1. Hence, the velocity profile of the two-stage launcher with a prescribed flight-path angle z_{2f} at second-stage burn out is equal to

$$\begin{aligned} V(z) &= A_1\, z_1^{\bar{n}_1 - 1}(1 + z_1^2) \ for \ z \in [0, z_{20}] \\ V(z) &= A_2\, z_2^{\bar{n}_2 - 1}(1 + z_2^2) \ for \ z \in [z_{20}, z_{2f}] \end{aligned} \tag{21}$$

With same arguments, we obtain the altitude and range

$$
\begin{aligned}
h(z) &= \frac{A_1^2}{g}\left[\frac{z_1^{2\bar{n}_1-2}}{2\bar{n}_1-2}-\frac{z_1^{2\bar{n}_1+2}}{\bar{n}_1+2}\right]\ ,\ for\ z\in[0,z_{20}] \\
h(z) &= h(z_{20})+\frac{A_2^2}{g}\left[\frac{z_2^{2\bar{n}_2-2}}{2\bar{n}_2-2}-\frac{z_2^{2\bar{n}_2+2}}{\bar{n}_2+2}\right]_{z_{20}}^{z}\ ,\ for\ z\in[z_{20},z_{2f}]
\end{aligned}
\tag{22}
$$

$$
\begin{aligned}
s(z) &= 2\frac{A_1^2}{g}\left[\frac{z^{2\bar{n}_1}}{2\bar{n}_1}+\frac{z^{2\bar{n}_1+2}}{2\bar{n}_1+2}\right]\ ,\ for\ z\in[0,z_{20}] \\
s(z) &= R(z_{20})+2\frac{A_2^2}{g}\left[\frac{z^{2\bar{n}_2}}{2\bar{n}_2}+\frac{z^{2\bar{n}_2+2}}{2\bar{n}_2+2}\right]_{z_{20}}^{z}\ ,\ for\ z\in[z_{20},z_{2f}]
\end{aligned}
\tag{23}
$$

Note that the gravity loss of the two-stage launcher is

$$
\Delta V_g = g Isp_1 log\left(\frac{1}{U_1}\right)+g Isp_2 log\left(\frac{1}{U_2}\right)-V(z_{2f})
$$

Any number of stage N can be handled: of course, more numerical work is needed to find out the unknown variables. The N-1-matching conditions on the flight-path angle generate the following N-1-matching conditions on the velocity

$$
\begin{aligned}
V_1(z_{20},A_1) &= V_2(z_{20},A_2) \\
V_2(z_{30},A_2) &= V_3(z_{30},A_3) \\
&\vdots \\
V_{N-2}(z_{N-2\,0},A_{N-2}) &= V_{N-1}(z_{N-1\,0},A_{N-1})
\end{aligned}
$$

These, together with the N relationships related to each burn time

$$
\begin{aligned}
t_{b_1} &= t_{b_1}(A_1,z_{20}) \\
t_{b_2} &= t_{b_2}(A_2,z_{20},z_{30}) \\
&\vdots \\
t_{b_N} &= t_{b_N}(A_N,z_{20},\ldots,z_{N\,0})
\end{aligned}
$$

generate $2N-1$ equations to be solved with respect to the $2N-1$ unknowns $z_{20},z_{30},\ldots z_{N\,0}$, $A_1,\ldots A_N$.

By these solutions, the velocity, altitude and range profiles can be computed using Equations (13)–(15) in the ranges $[0,z_{20}],\ldots[z_{N\,0},z_f]$.

For instance, for a three-stage launcher, one has

$$
t_{b1} = \frac{A_1}{g}\left(\frac{z_{20}^{n_1-1}}{n_1-1}+\frac{z_{20}^{n_1+1}}{n_1+1}\right)
\tag{24}
$$

$$
t_{b2} = \frac{A_2}{g}\left[\left(\frac{z_{30}^{n_2-1}}{n_2-1}+\frac{z_{30}^{n_2+1}}{n_2+1}\right)-\left(\frac{z_{20}^{n_2-1}}{n_2-1}+\frac{z_{20}^{n_2+1}}{n_2+1}\right)\right]
\tag{25}
$$

$$
t_{b3} = \frac{A_3}{g}\left[\left(\frac{z_3^{n_3-1}}{n_3-1}+\frac{z_3^{n_3+1}}{n_3+1}\right)\right]_{z_{30}}^{z_{3f}}
\tag{26}
$$

$$
A_1 = A_2 z_{20}^{\bar{n}_2-\bar{n}_1}
\tag{27}
$$

$$
A_2 = A_3\, z_{30}^{\bar{n}_3-\bar{n}_2}
\tag{28}
$$

(these five Equations (24)–(28) with the five unknowns $A1,A_2,A_3,z_{20},z_{30}$).

One solution is to write the variables A_1 and A_2 as function of A_3 using Equations (27) and (28). Then, the ratios $\frac{t_{b_2}}{t_{b_1}}$, $\frac{t_{b_3}}{t_{b_1}}$ are obtained dividing (25) and (26) by the Equation (24). The two ratios give two equations on the two unknown z_{20}, z_{30}. The solutions can be found numerically within the interval [0,1], and they determine the values of the flight-path angles at the end of the first and second stages. The values A_1, A_2, A_3 are derived by Equations (24) and (26) and and the velocities at the end of each stage are obtained.

For a four-stage launcher, there are seven equations in the seven unknowns $A1$, A_2, A_3, A_4, z_{20}, z_{30}, z_{40}

$$A_1 = A_2 z_{20}^{\bar{n}_2 - \bar{n}_1} \tag{29}$$

$$A_2 = A_3\, z_{30}^{\bar{n}_3 - \bar{n}_2} \tag{30}$$

$$A_3 = A_4\, z_{40}^{\bar{n}_4 - \bar{n}_3} \tag{31}$$

$$t_{b1} = \frac{A_1}{g}\left(\frac{z_{20}^{n_1 - 1}}{n_1 - 1} + \frac{z_{20}^{n_1 + 1}}{n_1 + 1} \right) \tag{32}$$

$$t_{b2} = \frac{A_2}{g}\left[\left(\frac{z_{30}^{n_2 - 1}}{n_2 - 1} + \frac{z_{30}^{n_2 + 1}}{n_2 + 1} \right) - \left(\frac{z_{20}^{n_2 - 1}}{n_2 - 1} + \frac{z_{20}^{n_2 + 1}}{n_2 + 1} \right) \right] \tag{33}$$

$$t_{b3} = \frac{A_3}{g}\left[\left(\frac{z_{40}^{n_3 - 1}}{n_3 - 1} + \frac{z_{40}^{n_3 + 1}}{n_3 + 1} \right) - \left(\frac{z_{30}^{n_3 - 1}}{n_3 - 1} + \frac{z_{30}^{n_3 + 1}}{n_3 + 1} \right) \right] \tag{34}$$

$$t_{b4} = \frac{A_4}{g}\left[\frac{2\, n_4}{n_4^2 - 1} - \left(\frac{z_{40}^{n_4 - 1}}{n_4 - 1} + \frac{z_{40}^{n_4 + 1}}{n_4 + 1} \right) \right] \tag{35}$$

The first three equations allow us to write A_1, A_2, A_3 as function of A_4 only. Then, the three ratios $\frac{t_{b_2}}{t_{b_1}}$, $\frac{t_{b_3}}{t_{b_1}}$, $\frac{t_{b_4}}{t_{b_1}}$ are functions of the three unknowns z_{20}, z_{30}, z_{40}. The solution is used then to derive $A_1 - A_4$.

Note that the above formulas for three- or four-stage launchers provide the values of the flight-path angles and the velocities at the end of each stage for the launcher trajectory with a fixed value of the final flight-path angle.

5. Guidance

The gravity-turn trajectory constrains the guidance and should be abandoned if possible; the relevant condition is the factor $q\,\alpha$, where $q = \frac{1}{2}\rho V^2$ is the dynamical pressure and α is the angle of attack. This factor must be below a limit $[q\,\alpha]_{max}$ which is peculiar to any launcher and can be satisfied by the gravity-turn ($\alpha = 0$) trajectory. As the altitude increases, the atmospheric density ρ decreases exponentially, so at a certain point of the flight the condition of $q\,\alpha$ disappears and a different guidance can be applied. The guidance applied here is based on: (A) a possible coasting arc, and (B) thrust along the horizontal direction.

Generally, a coasting arc is performed after the burn out of the stage $N - 1$ of an N stage launcher to increase the altitude at the cost of a moderate decrease in velocity. After the coasting arc, the launcher is close to the target altitude h_f, whereas the required velocity V_{req} must be gained. This velocity basically has the horizontal direction $\hat{\theta} = \cos\gamma\,\hat{V} + \sin\gamma\,\hat{I}$, then the thrust direction is along $\hat{\theta}$ during the N-stage boost.

With the above guidance scheme, the full launcher trajectory depends on two parameters: the flight-path angle γ_f at the end of the gravity-turn phase and the duration t_c of the coasting arc. The two parameters (γ_f, t_c) are introduced into an iterative routine (such as the Matlab routine fsolve.m) and updated to set the errors on the final altitude and velocity $(h(t_b) - h_f, V(t_b) - V_{req})$ to zero, where t_b is the N-stage burn-out time.

Let us consider a three-stage rocket as an example. The first two stages are in gravity turn with a prescribed final flight-path angle γ_{2f}, then a costing arc of duration t_c is performed. Finally the third-stage boost is applied with horizontal thrust direction.

The Matlab routine fsolve.m is used to update the values (γ_{2f}, t_c), for instance, to achieve the polar circular orbit of altitude $h_f = 650$ km from a launch site with a latitude of 30 deg N. The required velocity is then equal to $V_R = 7542$ m/s.

The launcher has the parameters reported in Table 3. The Tziolkowski velocity is equal to $\Delta V_{Tz} = 9743$ m/s. From the data of Table 3 it is possible to derive the mass parameters of each stage $m_{s_1}, m_{p_1}, m_{s_2}, m_{p_2}, m_{s_3}, m_{p_3}$ and the mass of each substage M_2, M_3, see Equations (A1) and (A2) in Appendix A. From the values $n_{0_1}, n_{0_2}, n_{0_3}$ and M_1, M_2, M_3, it is possible to derive the propellant mass rates $\dot{m}_1, \dot{m}_2, \dot{m}_3$, hence the burn times $t_{b_i} = \frac{m_{p_i}}{\dot{m}_i}, i = 1, 3$. For any fixed γ_{2f}, the Equations (21)–(23) give the state of the launcher at the end of the second stage where the gravity-turn phase ends. After a coasting of duration t_c, the third stages thrust the rocket up to the burn time t_{b_3}; the final altitude and velocity are compared with the required (h_f, V_R). The parameters γ_{2f}, t_c are updated in order to match the required final orbit conditions. After few iterations, the routine finds the following parameters to reach the required orbit: $\gamma_{2f} = 26.5$ deg, $t_c = 340$ s.

Table 3. Dataof the three-stage launcher.

Total mass $M1$	55,500 kg
Payload mass m_{pay}	405 kg
$\epsilon_1, \epsilon_2, \epsilon_3$	0.10, 0.11, 0.22
Isp_1, Isp_2, Isp_3	294 s, 295 s, 300 s
U_1, U_2, U_3	0.35, 0.32, 0.31
n_{01}, n_{02}, n_{03}	2.40, 2.06, 2.32
$\bar{n}_1, \bar{n}_2, \bar{n}_3$	3.85, 3.45, 3.95

6. Numerical Comparison

The results obtained by the analytic formulas are now compared with a numerical code simulating the launcher trajectory with better accuracy. The equations of motion are the three-dimensional equations of flight in the relative velocity frame with state variables $(r, \lambda, L, V_R, \gamma_R, \psi_R)$

$$
\begin{aligned}
\dot{r} &= V_R \sin \gamma_R \\
\dot{\lambda} &= V_R \cos \gamma_R \frac{\cos \psi_R}{r \cos L} \\
\dot{L} &= \frac{V_R \cos \gamma_R \sin \psi_R}{r} \\
\dot{V}_R &= r \omega_E^2 \cos L (\cos L \sin \gamma_R - \sin L \cos \gamma_R \sin \psi_R) + \frac{f_V}{m} \\
\dot{\gamma}_R &= \frac{V_R}{r} \cos \gamma_R + 2\omega_E \cos L \cos \psi_R + \\
&\quad + r \frac{\omega_E^2}{V_R} \cos L (\cos L \cos \gamma_R + \sin L \sin \gamma_R \sin \psi_R) + \frac{f_l}{m V} \\
\dot{\psi}_R &= -\frac{V_R}{r} \tan L \cos \gamma_R \cos \psi_R + 2\omega_E \cos L \tan \gamma_R \sin \psi_R + \\
&\quad - r \frac{\omega_E^2}{V_R \cos \gamma_R} \cos L \sin L \cos \psi_R - 2\omega_E \sin L
\end{aligned}
\tag{36}
$$

with

$$
\frac{f_V}{m} = \frac{T}{m} \cos \alpha - \frac{\mu}{r^2} \sin \gamma_R - \frac{1}{2}\rho\, C_D \frac{S}{m} V_R^2
\tag{37}
$$

$$
\frac{\dot{f_l}}{m} = \frac{T}{mV} \sin \alpha - \frac{\mu}{V_R\, r^2} \cos \gamma_R + \frac{1}{2}\rho\, C_L \frac{S}{m} V_R
$$

Equations (36) and (37) are used with $\alpha = 0$ for the gravity-turn arc, with $T = 0$ for the coasting arc and with $\alpha = \gamma$ for the third-stage boost. Lift off begins with an initial vertical arc of small duration t_V; this is computed in the Cartesian frame with non-singular coordinates. After the vertical arc, to turn the trajectory from the vertical direction, a pitch manoeuvre starts with a fixed direction of thrust $\alpha = \theta_k$ and ends when $\gamma_R = \theta_k$. At this time (called pitch over), the gravity-turn trajectory starts. The pitch direction of thrust θ_k is chosen so to have the prescribed value γ_{2_f} at the second-stage burn out. Then, the coasting arc duration is chosen to achieve the required orbit.

Figures 6–8 show the analytical and numerical graphs of the velocity, flight-path angle and altitude during gravity turn. Note that at the end of the second stage the discrepancy between analytic and numeric results is rather small for the velocity, whereas a reduction of 15% occurs in the altitude due to the numerical integration of the drag force. In the numerical calculation the atmospheric density is represented by the USATMO 76 model and the drag coefficient is kept constant at an average value $C_D = 0.4$. The reference surface is $S = \frac{\pi d^2}{4}$, with a diameter of $d = 2$ m. The drag effect reduces the payload mass that can be inserted into the reference orbit.

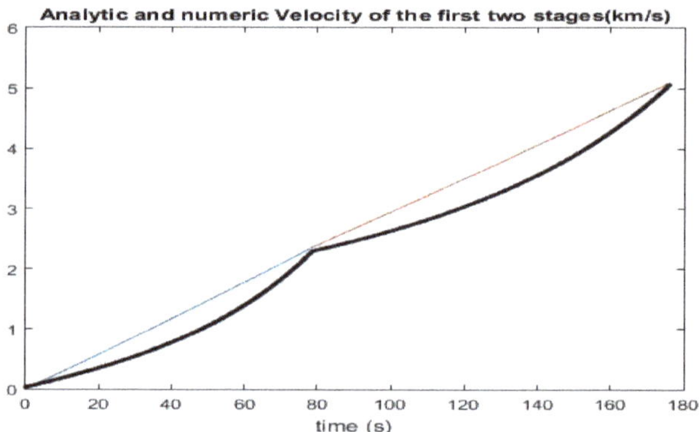

Figure 6. The velocity plot: analytic and numeric results during the gravity-turn arc. Numerical velocity of the first and second stage is in black, analytic velocity of the first stage is in blue and second stage in orange line.

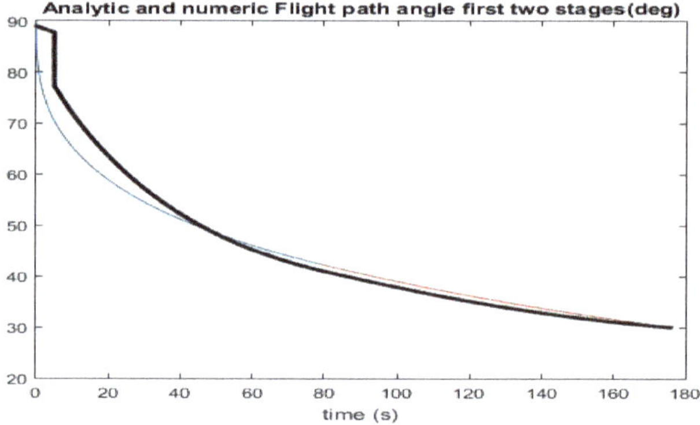

Figure 7. The flight-path angle plot: analytic and numeric results during gravity-turn arc. Numerical flight-path angle of the first and second stage is in black, analytic flight-path angle of the first stage is in blue and second stage in orange line.

In fact, the numerical results give insertion in the polar circular orbit of altitude 650 km of a payload of mass $m_{pay} = 390$ kg, vertical flight duration $t_v = 5$ s, and pitch angle $\theta_k = 76$ deg (generating $\gamma_{2_f} = 26.5$ deg), and coasting time $t_c = 355$ s. The payload mass reduction with respect to the analytical results is about 4%. The evolution of the orbit parameters during the launch is shown in Figures 9 and 10, and Figure 11 shows the launcher ground track.

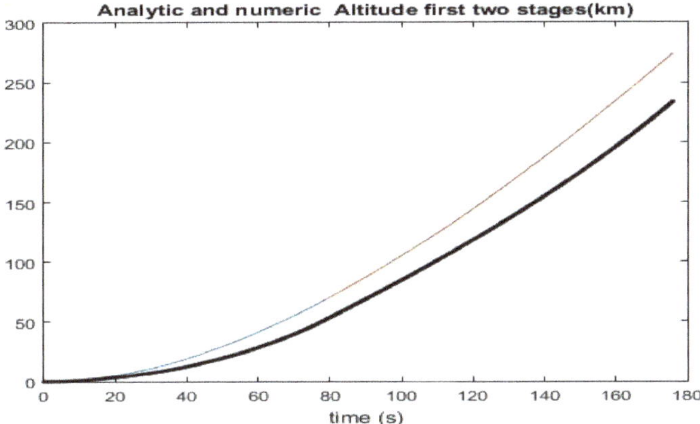

Figure 8. The altitude plot: analytical and numerical results during gravity-turn arc. Numerical altitude of the first and second stage is in black, analytical altitude of the first stage is in the blue line and the second stage in the orange line.

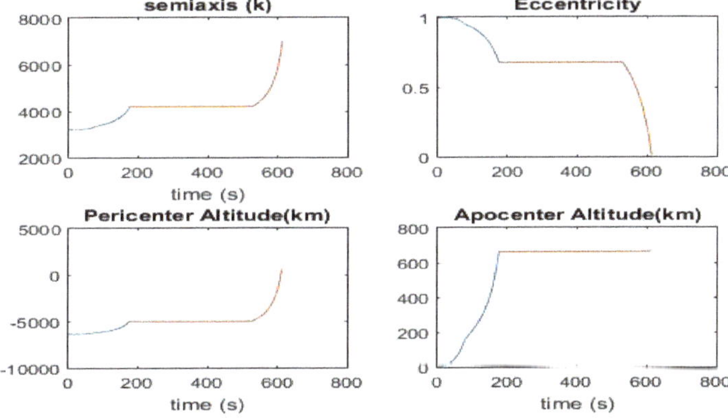

Figure 9. Thein-plane orbit parameters during the ascent. First-stage parameters are the blue line; coasting and second stage are the orange line.

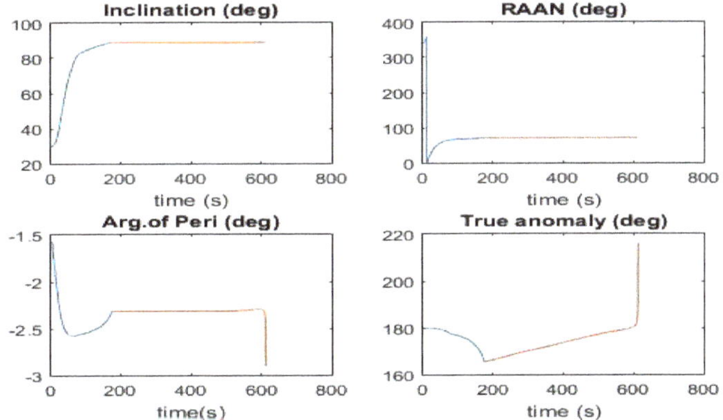

Figure 10. Out–of-plane orbit parameters during the ascent. First-stage parameters are in the blue line; coasting and second stage are in the orange line.

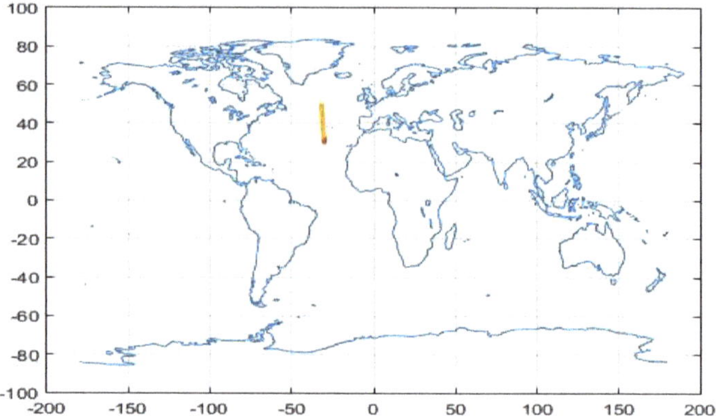

Figure 11. Launcher ground track.

Figure 12 shows the payload mass evaluation for polar circular orbits of different altitudes obtained by the analytic (blue curve) and numerical (orange curve) algorithms. Of course, the bigger differences are for orbits of lower altitude; however, the reduction in the payload mass does not exceed 10% of the value found analytically.

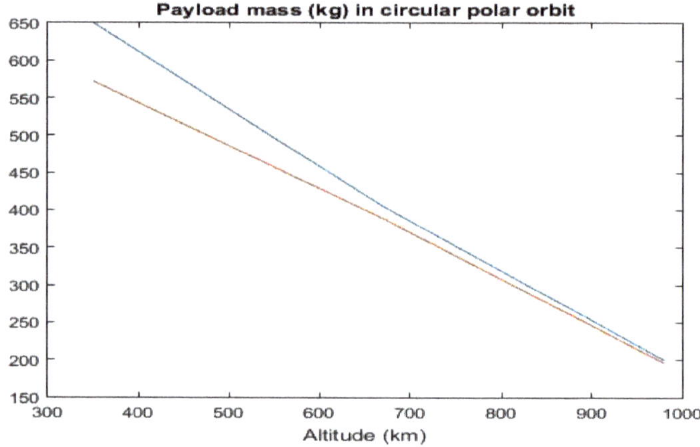

Figure 12. Payload mass of polar circular orbits of different altitudes, analytical (blue line) vs numeric results (orange line).

7. Concluding Remarks

This research extends the preceding generalization of the Culler and Fried formulation to ascent trajectories that include a coast arc and the guidance of the upper stage. The analytical developments related to the multistage formulation of the Culler and Fried analysis are contained in Section 4, dedicated to closed-form expressions for some fundamental parameters of multistage launch vehicles. Section 5 is focused on inclusion of the coast arc and guidance of the upper stage. These steps lead to the defining of an analytical methodology for the preliminary evaluation of the performance and trajectory of a launch vehicle, whose preliminary mass distribution is optimized through the solution of a polynomial equation of the same order as the number of stages (cf. Appendix A). The comparison of the analytical method at hand with the ascent path obtained through numerical integration testifies its accuracy, particularly in terms of prediction of the final payload mass, especially for medium- and high-altitude final circular orbits, with the error never exceeding the 10% of the actual payload mass.

Author Contributions: Conceptualization, P.T. and M.P.; methodology, P.T.; software, P.T.; validation, S.C. and M.P.; formal analysis, P.T.; investigation, P.T., S.C. and M.P.; resources, P.T., S.C. and M.P.; data curation, P.T., S.C. and M.P.; writing—original draft preparation, P.T.; writing—review and editing, S.C. and M.P.; supervision, P.T.; project administration, P.T. All authors have read and agreed to the published version of the manuscript.

Funding: This research received no external funding.

Data Availability Statement: Not applicable.

Conflicts of Interest: The authors declare no conflict of interest.

Appendix A

Let us recall the definitions of stage and subrocket structural mass ratios of an N-stage launcher:

$$\epsilon_k = \frac{m_{s_k}}{m_k} \ , \quad U_k = \frac{M_{k-m_{p_k}}}{M_k} \quad k = 1, \ldots N$$

where $m_k = m_{s_k} + m_{p_k}$ is the mass of each stage (structure plus propellant mass) and $M_k = \Sigma_{j=k}^{N} m_k + m_{pay}$ is the mass of each subrocket.

$$m_{s_k} = \frac{M_k(1 - U_k)}{1 - \epsilon_k} \ , \quad m_{p_k} = M_k(1 - U_k) \tag{A1}$$

$$M_k = M_{k-1} \frac{(U_{k-1} - \epsilon_{k-1})}{1 - \epsilon_{k-1}} \ , k = 2, N - 1 \ , \quad M_N = m_{pay} \tag{A2}$$

Let the values M_1 and m_{pay} be given, then, by

$$\frac{m_{pay}}{M_1} = \Pi_{i=2}^{N} \frac{M_i}{M_{i-1}} \tag{A3}$$

The Tziolkowski velocity is equal to

$$\Delta V_{Tz} = -g \sum_{i=1}^{N} Isp_i \log(U_i)$$

The problem is now: select the values U_k^* to minimize

$$J = g\Sigma_{i=1}^{N} Isp_i \log(U_i)$$

under the constraint (A3), which is equal to (see (A2)):

$$\Phi = \frac{m_{pay}}{M_1} - \Pi_{i=2}^{N} \frac{U_i - \epsilon_i}{1 - \epsilon_i} = 0 \tag{A4}$$

To solve the constrained minimum problem, the Hamilton function is introduced with Lagrangian multiplier λ:

$$H = J + \lambda \Phi$$

and the solution for the $N + 1$ unknowns U_i, λ derives from the $N + 1$ equations

$$\frac{\partial H}{\partial U_i} = 0$$

$$\frac{\partial H}{\partial \lambda} = 0 \tag{A5}$$

These equations correspond to

$$\frac{gIsp_i}{U_i} - \frac{\lambda}{1-\epsilon_i}\Pi_{k=1,k\neq i}^{N}\frac{U_k-\epsilon_k}{1-\epsilon_k} = 0 \tag{A6}$$

and to Equation (A4). From (A6), the Lagrangian multiplier λ is derived:

$$\lambda = \frac{gIsp_i(U_i-\epsilon_i)}{U_i}\Pi_{k=1}^{N}\frac{1-\epsilon_k}{U_k-\epsilon_k} \tag{A7}$$

Equation (A7) holds true for any $i = 1, N$, and this corresponds to

$$\frac{gIsp_1(U_1-\epsilon_1)}{U_1} = \cdots\cdots = \frac{gIsp_N(U_N-\epsilon_N)}{U_N}$$

that is

$$\frac{gIsp_i(U_i-\epsilon_i)}{U_i} = \frac{gIsp_1(U_1-\epsilon_1)}{U_1} \quad, i = 2, N \tag{A8}$$

Equation (A8) corresponds to

$$U_i = \frac{Isp_i\,\epsilon_i}{U_1(Isp_i - Isp_1) + Isp_1\,\epsilon_1}U_1 \quad, i = 2, N \tag{A9}$$

Replacing the Formulas (A9) into (A4), one has the following algebraic equation of order N in the unique unknown U_1:

$$\frac{m_{pay}}{M_1}\Pi_{i=1}^{N}[U_1(Isp_i - Isp_1) + Isp_1\,\epsilon_1] - \Pi_{i_1}^{N}\frac{\epsilon_i}{1-\epsilon_i}(Isp_1(U_1-\epsilon_1))^N = 0 \tag{A10}$$

The solution U_1^* of (A10) is used in (A9) to obtain all the subrocket structural mass ratios $\{U_i^*\}_{i=1,N}$, maximizing the Tziolkowski variation of velocity ΔV_{Tz}.

References

1. Miele, A. Multiple-subarc gradient-restoration algorithm, part 1: Algorithm structure. *J. Optim. Theor. Appl.* **2003**, *116*, 1–17. [CrossRef]
2. Miele, A. Multiple-subarc gradient-restoration algorithm, part 2: Application to a multistage launch vehicle design. *J. Optim. Theor. Appl.* **2003**, *116*, 19–39. [CrossRef]
3. Calise, A.J.; Tandon, S.; Young, D.H.; Kim, S. Further improvements to a hybrid method for launch vehicle ascent trajectory optimization. In Proceedings of the AIAA Guidance, Navigation, and Control Conference and Exhibit, Denver, CO, USA, 14–17 August 2000.
4. Gath, P.F.; Calise, A.J. Optimization of launch vehicle ascent trajectories with path constraints and coast arcs. *J. Guid. Contr. Dynam.* **2001**, *24*, 296–304. [CrossRef]
5. Lu, P.; Pan, B. Trajectory optimization and guidance for an advanced launch system. In Proceedings of the 30th Aerospace Sciences Meeting and Exhibit, Reno, NV, USA, 6–9 January 1992.
6. Lu, P.; Griffin, B.J.; Dukeman, G.A.; Chavez, F.R. Rapid optimal multiburn ascent planning and guidance. *J. Guid. Contr. Dynam.* **2008**, *31*, 45–52. [CrossRef]
7. Weigel, N.; Well, K.H. Dual payload ascent trajectory optimization with a splashdown constraint. *J. Guid. Contr. Dynam.* **2000**, *23*, 45–52. [CrossRef]
8. Jamilnia, R.; Naghash, A. Simultaneous optimization of staging and trajectory of launch vehicles using two different approaches. *Aero. Sci. Technol.* **2012**, *23*, 85–92. [CrossRef]
9. Roh, W.; Kim, Y. Trajectory optimization for a multi-stage launch vehicle using time finite element and direct collocation methods. *Eng. Optim.* **2002**, *34*, 15–32. [CrossRef]
10. Pontani, M. Particle swarm optimization of ascent trajectories of multistage launch vehicles. *Acta Astronaut.* **2014**, *94*, 852–864. [CrossRef]
11. Pontani, M.; Teofilatto, P. Simple method for performance evaluation of multistage rockets. *Acta Astronaut.* **2014**, *94*, 434–445. [CrossRef]
12. Pallone, M.; Pontani, M.; Teofilatto, P. Performance evaluation methodology for multistage launch vehicles with high-fidelity modeling. *Acta Astronaut.* **2018**, *151*, 522–531. [CrossRef]

13. Pallone, M.; Pontani, M.; Teofilatto, P. Modeling and performance evaluation of multistage launch vehicles by firework algorithm. In Proceedings of the 6th ICATT, Darmstad, Germany, 14–17 March 2016.
14. Culler, G.; Fried, D. Universal gravity turn trajectories. *J. Appl. Phys.* **1957**, *28*, 672. [CrossRef]
15. Sotto, E.D.; Teofilatto, P. Semi-analytic formulas for launcher performances evaluation. *J. Guid. Control.* **2002**, *25*, 538–545. [CrossRef]
16. Teofilatto, P. A paper and pencil method of evaluating launcher performances. In Proceedings of the International Astronautical Conference, Adelaide, Australia, 25–29 September 2017.
17. Campos, L.; Gil, R. On four new methods of analytic calculation of rocket trajectories. *Aerospace* **2018**, *5*, 88. [CrossRef]
18. Encyclopedia Astronautica. Available online: www.astronautix.com (accessed on 21 March 2022).

applied sciences

Article

Assessing the Capacity and Coverage of Satellite IoT for Developing Countries Using a CubeSat

Pooja Lepcha [1,*], Tharindu Dayarathna Malmadayalage [2], Necmi Cihan Örger [1], Mark Angelo Purio [1], Fatima Duran [1], Makiko Kishimoto [1], Hoda Awny El-Megharbel [1] and Mengu Cho [1]

[1] Laboratory of Lean Satellite Enterprises and In-Orbit Experiments, Kyushu Institute of Technology, Kitakyushu 804-8550, Japan

[2] Arthur C. Clarke Institute for Modern Technologies, Moratuwa 10400, Sri Lanka

* Correspondence: lepcha.pooja586@mail.kyutech.jp

Abstract: Many regions in developing countries do not have any access to communication networks even though the number of devices connected through the Internet of Things (IoT) is increasing significantly. A small satellite platform could provide global network coverage in low Earth orbit to these remote locations at a low cost. This paper describes the overall mission architecture and the implementation of remote IoT using a 1U volume in 6U CubeSat platform named KITSUNE. In KITSUNE, one of the missions is to leverage IoT for building a network of remote ground sensor terminals (GST) in 11 mostly developing countries. This paper evaluates the capacity and coverage of a satellite-based IoT network for providing remote data-collection services to these countries. The amount of data that could be collected from the GSTs and forwarded accurately to the users determines the actual capacity of the Store and Forward (S&F) mission. Therefore, there are several proposed parameters to estimate this capacity in this study. In addition, these parameters are retrieved from the simulations, ground test results, and on-orbit observations with the KITSUNE satellite. The proposed IoT system, which is composed of the GSTs and IoT subsystem onboard KITSUNE satellite, is determined to be capable of providing valuable information from remote locations. In addition, the collected data are achieved and analyzed to monitor sensory data specific to each country, and it could help to generate prediction profiles as well.

Keywords: small satellite; LoRa; ground sensor terminal; store and forward; satellite IoT; capacity building

Citation: Lepcha, P.; Malmadayalage, T.D.; Örger, N.C.; Purio, M.A.; Duran, F.; Kishimoto, M.; El-Megharbel, H.A.; Cho, M. Assessing the Capacity and Coverage of Satellite IoT for Developing Countries Using a CubeSat. *Appl. Sci.* **2022**, *12*, 8623. https://doi.org/10.3390/app12178623

Academic Editors: Filippo Graziani, Simone Battistini and Mauro Pontani

Received: 27 July 2022
Accepted: 25 August 2022
Published: 28 August 2022

Publisher's Note: MDPI stays neutral with regard to jurisdictional claims in published maps and institutional affiliations.

1. Introduction

Small satellites, which are mostly made of CubeSats in the 1–14 kg category, have seen an exponential growth since 2012 [1], and it could be explained by their compact size, shorter development time, and significantly lower development and launch costs compared with conventional large satellites while achieving similar missions to a certain degree. Even though this category of small satellites is attractive to test payloads/instruments or subsystems without sufficient flight heritage to evaluate on-orbit performance, there could be technology demonstration missions that are not particularly brand-new for the conventional large satellites such as IoT missions. However, the development and testing of any payload for a CubeSat mission introduces additional challenges compared to the traditional satellites considering the limited resources available for a single mission. As a result, a CubeSat mission could be evaluated through available mass/volume budget, communication capability, power generation, storage, and revisit time considering a ground system, and so on.

Recent developments in non-space-grade and commercial-off-the-shelf (COTS) components have led to satellite system development being confined to not only the most developed countries, but also the proliferation of university experiments, commercial startup spin-offs and developing countries with highly constrained space programs using small satellites. Kyushu Institute of Technology (Kyutech) has been maintaining a database of small satellites launched since 2013 that considers satellites with fueled mass less than or equal to 500 kg. The database is built on open-source information available on web pages, papers, conference papers, etc. In 2020, 1156 small satellites were launched as in Table 1, and 86% of these satellites were used for commercial communication missions to provide global Internet services and leveraging the Internet of Things (IoT), machine-to-machine (M2M) communications, and store and forward (S&F) missions.

Table 1. Distribution of small satellite missions in 2020.

Missions	Number of Satellites	Percentage of Satellites
Military	1	0.09
Astronomy	3	0.26
Others	4	0.35
Technology demonstration	68	5.88
Earth observation and geosciences	84	7.27
Communication	996	86.16

In addition, recent applications of satellite-based S&F systems have enabled achieving IoT/M2M, which are also referred to as satellite IoT (SIoT), and automatic identification system (AIS) applications for tracking ships. At least 18 space startups entered the IoT market in 2018, and an exponential growth in the number of satellites employing S&F for IoT/M2M could be observed from this period in Figure 1. Most startups plan to launch dozens to hundreds of CubeSats or satellites even smaller than the 1U CubeSat form factor in low Earth orbit (LEO) to provide global coverage and lower latency time in data retrieval [2].

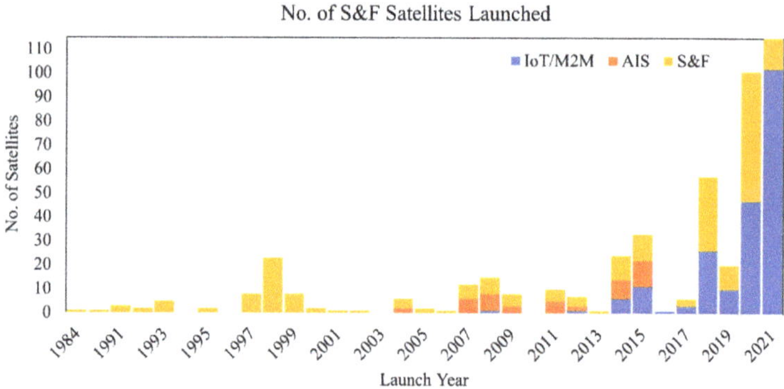

Figure 1. Timeline of the number of S&F satellites launched over the years.

A satellite-based S&F system is a communication technique in which a satellite is used as a gateway for data collection and dissemination. Therefore, the basic principle of operation requires establishing communication between a terrestrial station and a satellite to uplink successfully. The next step could be described as storing data in the satellite through that uplink. Finally, the stored data could be downloaded to another ground station to be distributed or forwarded to its destination [3]. The time difference between the first and last step indicates the latency time through the operation, and it could be a critical parameter with regards to the application [4,5]. Figure 2 shows the illustration for the basic principle of operation of a satellite-based S&F system.

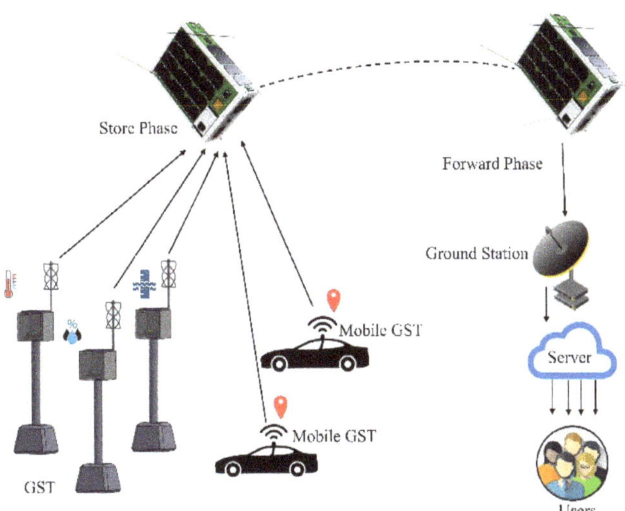

Figure 2. Illustration for a satellite-based S&F system.

The first proof of concept for the digital S&F communications experiment was carried out in the UoSAT-2 satellite as one of the missions by the University of Surrey in 1984 [6]. Some notable investments for global coverage using S&F satellites include service providers such as Iridium constellations, Argos Data collection system, and Orbcomm constellations, and they have been operating since the mid-nineties. Although it has been predicted that there will be more than 83 billion IoT connections by 2024 [7], there are still inaccessible regions by terrestrial networks in most developing countries. According to a report by International Telecommunication Union (ITU) for 2021 [8], 37% of the total population do not have access to the Internet, and 98% of this population reside in developing countries. Furthermore, it was reported that 84 out of 160 weather stations are still manually operated in Bhutan in South Asia [9]. The lack of automatic weather stations is observed in other developing countries such as [10], where the near-surface temperatures and rain levels are expected to significantly alter the water resources, agriculture, energy resources, infrastructure and so on [11]. As a result, a CubeSat-based SIoT in LEO could offer a solution to several existing and approaching needs [12], and it could be considerably valuable to retrieve sensory measurements of the critical parameters related to the weather, soil, and water quality from the most inaccessible rural regions. The advantages of leveraging SIoT in LEO, which are not only limited to CubeSats, could be listed below:

- Satellites could provide global coverage and provide access to any remote topography where terrestrial networks are not viable.
- Satellite communications are offered in a variety of frequencies, orbits, and data rates. In addition, they are resilient to natural disasters compared to terrestrial communication, which could be overloaded in developing countries during disasters such as earthquakes, tsunamis, and so on [13].
- Utilization of small satellite communications in the LEO enables the use of low power terminals in the remote stations due to lesser signal loss compared to satellites in higher orbits such as sun-synchronous orbits, medium earth orbit (MEO) or geostationary earth orbit (GEO) [14].

Implementation of SIoT applications is promising using CubeSats; however, it is also challenging due to limitations in power budget, available space for payloads and requirements for fulfilling adequate link margin between the satellite and the GSTs. In addition to these constraints, evolving wireless technologies with power efficient communication protocols enabling Low Power Wide Area (LPWA) coverage such as Sigfox [15],

NB-IoT [16,17], and LoRa have been studied for their adaptability to be employed for communication between the GST and the onboard payload. LoRa has been implemented as the communication protocol in this study for several reasons, which are explained in the following sections.

LoRa stands for Long Range, and it is a proprietary modulation scheme based on chirp spread spectrum (CSS) technique. Since it can establish communication over long distances, communication protocols for satellites with ground LoRa terminals have been of great interest. In addition, it has low transmission power requirements while being robust against multipath fading and the Doppler effect. Many studies and experiments have been performed for assessing LoRa adaptability in the LEO SIoT. While LoRa performance over long-distance links up to 250 km was tested in [18], a satellite communication scenario with a Doppler residual frequency deviation of 22.3 Hz was tested in [19]. Various analyses regarding the performance of LoRa in LEO have also been performed in [20–22], and constellation designs using LoRa-based SIoT have been proposed in [23] for achieving low-latency IoT. Although there are many works in the literature that studied and predicted the expected performance of LoRa transceivers in LEO, many of these studies have not been validated through flight demonstrations. A small number of satellites launched with LoRa mission payloads, including a university satellite project TRICOM-1R that was successfully able to confirm the reception of data in the 920 MHz band over 300 km distance [24]. Following the design of the TRICOM-2 satellite, JPRWASAT was launched by Rwanda with a LoRa receiver, and the result of this mission has not been published thus far [25]. Sapienza University of Rome launched WildTrackCube—SIMBA [26] with LoRa receivers for monitoring the behavior of animals in Kenya. Commercial companies such as Lacuna Space have launched 5 out of a planned constellation of 240 satellites, and Swarm Technologies launched 93 out of a planned constellation of 150 satellites for achieving SIoT using LoRa modulation in the Very High Frequency (VHF) band [7]. Fossa Systems launched FOSSASAT-1 with a LoRa transceiver that could communicate with their ground module in amateur UHF band, and the on-orbit results are unavailable due to antenna deployment failure. In addition, FOSSASAT-1B had a launch failure; however, it is still planned to launch a constellation of 80 satellites for providing IoT connectivity [27].

Although many of these technology demonstrations of satellites with LoRa transceivers have been launched, a limited number of publications discuss details of LoRa payload design, on-orbit operation results, capacity, experimental outcomes, and the performance of the LoRa physical layer in LEO. This paper describes how a small satellite platform named KITSUNE has the capability to provide services to multiple developing countries for remote data collection. This paper discusses the implementation of LoRa modulation on a CubeSat and GSTs developed in domestic environments for achieving SIoT. LoRa modulation had been used for communication as it has been studied extensively, and the technology is more viable for ground-to-satellite links. The unique aspects of this study are listed below:

- The system model is described with comprehensive details on the design of SIoT payload and the GSTs.
- The payload design could be accommodated in a 1-unit CubeSat, and it could be directly plugged into the open-source BIRDS bus developed at Kyutech [28].
- The payload design is software-configurable in terms of data rates and sensitivity of the receiver on orbit.
- The capacity of the proposed network is evaluated through several parameters defined through ground testing results, simulations, and on-orbit results.
- It fosters the involvement of 11 mostly developing countries while promoting capacity-building in space activities within these countries.
- The GST design is open source for any country that aspires to be part of this SIoT mission.

This paper is made of five sections. Section 2 outlines the satellite mission concept and the definition of capacity in terms of the satellite design. It also includes details of the hardware design for both satellite payload and ground segment including design considerations that were made to be hosted in a CubeSat platform. Section 3 defines the parameters that are used to assess the capacity of the mission through various tests such as receiver-sensitivity tests and antenna-radiation pattern tests, in addition to estimations for parameters such as communication time between the satellite and GST. The parameters derived in Section 3 are used to quantify the capacity and coverage of the proposed SIoT system in Section 4, and the overall study is concluded in Section 5.

2. Mission Concept

Considering the KITSUNE mission in LEO, the communication time between the GSTs and the satellite could be estimated for a single satellite in orbit. However, the potential capacity of the SIoT mission without limiting to a single satellite could be evaluated based on the constellation configuration, distribution of Ground Stations (GSs) and GSTs, and the communication capability as listed below:

- The number of satellites in orbit, number of orbital planes, elevation angle, orbital plane spacing and orbital eccentricity for continuous data reception from GSTs and relaying to multiple ground stations if a constellation is employed considering global coverage and target application scenarios [29,30].
- The number of GSTs in optimized locations for maximizing ground coverage of each GST [31,32].
- Number of geographically distributed ground stations capable of downloading the data from the satellites and forwarding it to a database accessible by the users [33].
- Interlink capability of satellites in a constellation to relay the data to ground stations for lower latency in data retrieval from the GSTs [34].
- The uplink and downlink capability of the satellites and the GSTs, which depend on the communication frequency, data rates, transmit power and receiver sensitivities.

The full capacity of SIoT that provides continuous coverage almost in real time could be realized by using multiple GSTs that transmit the data to a constellation of satellites communicating with each other and distributed ground stations, in order to minimize the latency time. Furthermore, the ground stations forward data to a common database that is accessible to the users as illustrated in Figure 3a.

In this paper, an SIoT mission considering a single satellite, KITSUNE, with 11 geographically distributed GSTs and one ground station capable of downloading the data is proposed as shown in Figure 3b. A single satellite is incapable of providing continuous coverage, but it could be employed for delay-tolerant applications (DTA).

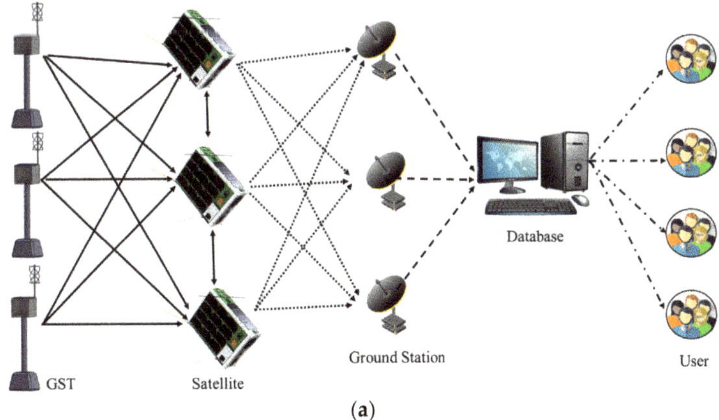

Figure 3. *Cont.*

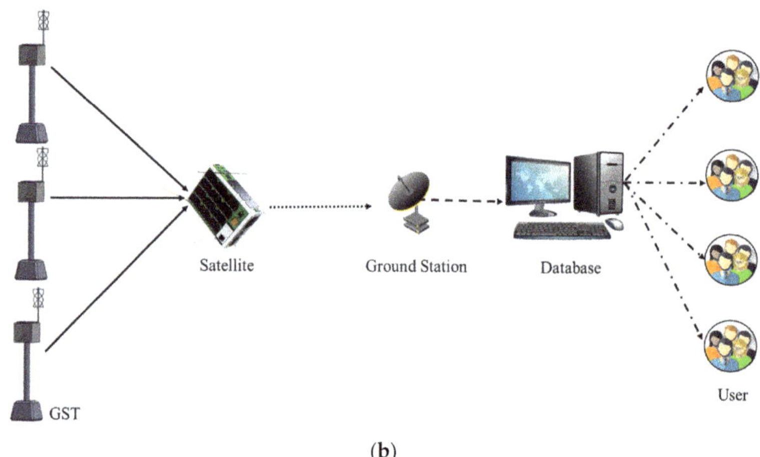

Figure 3. Implementation architecture of (**a**) SIoT for achieving full potential and (**b**) the proposed SIoT.

The proposed method is targeted towards offering a low-cost solution to developing countries and collaborative capacity-building programs such as the utilization of KiboCUBE, which considers a piggyback launch to the International Space Station (ISS) [35]. Therefore, the methods for optimizing the coverage performance by using a minimum number of satellites in a LEO constellation as in [23,36,37] are not applicable to this study since it only uses a single satellite at ISS orbit. Furthermore, the variation of the coverage and capacity in different latitudes have been shown through a metric of coverage degree and IoT device density [36]. Therefore, there are multiple aspects to accomplish the full capacity of an IoT mission considering the space-segment and the ground-segment simultaneously. In this study, the choice of orbital parameters is limited to orbital inclination of 51.6° and altitude of approximately 400 km as in KITSUNE satellite. The idea of the proposed method is not to launch a constellation of satellites to achieve continuous coverage. Instead, it is critical to prove that a CubeSat mission could help to collect data from remote locations in developing countries, which could be a significant step towards capacity building and space utilization as a demonstration mission.

2.1. KITSUNE Mission

The KITSUNE satellite has dimensions 340.5 mm × 226.3 mm × 100.0 mm and a weight of 7.544 kg, and the internal structure of the satellite is shown in Figure 4. KITSUNE was launched to ISS on 19 February and deployed on 24 2022. KITSUNE was developed as a dual satellite system; separating the components based on utilization of amateur and non-amateur frequencies. The amateur radio frequencies are selected for 2-unit main bus (2UMB) and 3-unit camera payload, whereas 1-unit SPATIUM-II (space precision atomic-clock timing utility mission) hosts the total electron content (TEC) and SIoT payload while using non-amateur frequencies for communications.

The 2UMB controls the main payload of the Earth-observation mission to capture 5-m class images through amateur radio frequencies. It also provides direct power from its battery to SPATIUM-II, which has its own bus system. In addition, there is no control command from the 2UMB to SPATIUM-II; therefore, they operate as a dual-satellite system housed in the same structure with a common battery. SPATIUM-II is controlled independently by a separate GS in Kyutech through non-amateur radio frequencies, and its SIoT mission is the described with space-segment and the GSTs in this paper.

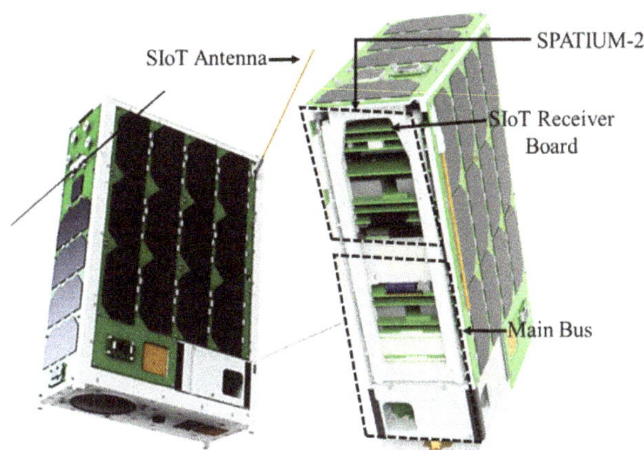

Figure 4. KITSUNE satellite configuration and internal arrangement.

SIoT Mission

The main aim of the SIoT mission can be described as a demonstration of a CubeSat-based SIoT system for achieving remote data collection and IoT service from GSTs in multiple countries. The mission objectives considering the satellite payload and the ground segment are listed below:

- Development of SIoT payload using COTS components.
- Development of low-cost and lower power GSTs, which could be fixed or mobile stations that could collect a wide range of sensor data.
- Collaborative development of GSTs to employ the most suitable sensor configurations in developing countries and deploy GSTs in remote locations around the world.
- Remote data collection of sensor data using KITSUNE and distribution of data using an online database management system developed at Kyutech.
- Human capacity building, knowledge transfer and promotion of international collaboration in developing countries for satellite-based missions.

In order to achieve the mission objectives, the technical mission requirements for the satellite payload segment were determined as:

- The payload shall have multiple channels to receive data from various GSTs simultaneously.
- The SIoT receiver shall have sensitivity close to ideal sensitivity, the permissible sensitivity deviation should be less than 10 dB.
- The receivers shall be able to operate in two frequency bands (400 MHz and 433 MHz).
- The receiver modules shall be reconfigurable on-orbit in terms of spreading factor (SF), coding rate (CR), bandwidth (BW) and center frequency (CF).
- The payload shall be accommodated in a 1-unit CubeSat with current consumption of the receiver less than 250 mA to meet the available power budget of a 1-unit CubeSat.
- The payload shall be able to operate in the space environment.

2.2. Payload Design

SIoT mission payload has eight LoRa receivers to ensure data reception from at least eight GSTs simultaneously. Each pair of these receivers is controlled and configured by one receiver microcontroller unit (MCU) using serial peripheral interface (SPI) communication protocol. Four of these receiver MCUs are in turn controlled by the main mission MCU, and the block diagram of the payload is shown in Figure 5.

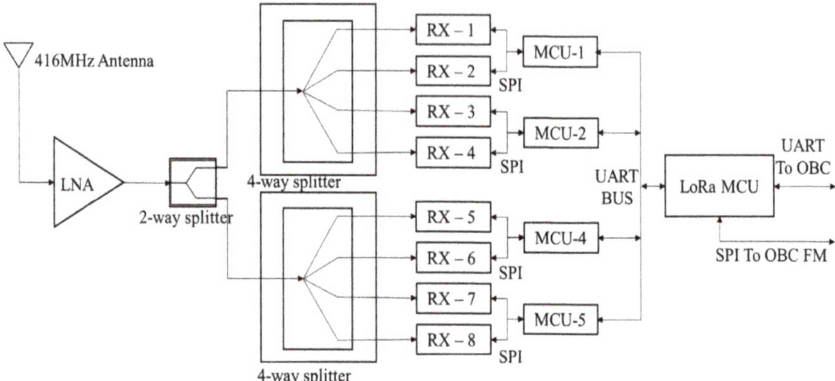

Figure 5. SIoT payload block diagram.

The receivers could be remotely configured even in orbit, allowing upgrades and optimizations for the execution of the SIoT mission. In addition, the data rates and center frequencies of each of these receivers with a resolution of 1 kHz can be configured remotely by sending an uplink command from the ground station, maximizing the capabilities of in-orbit demonstrations. All the receivers share the same monopole antenna with a wide bandwidth of 50 MHz tuned at a center frequency of 416 MHz.

The signals received from the antenna are firstly amplified using a low noise amplifier (LNA) with a gain of 22 dB. The signal passes through a two-way splitter and further through a four-way splitter until it reaches individual receivers. The design is optimized to compensate for the 10 dB losses in the splitters by the gain of the LNA. The receivers can detect signals as weak as −135 dBm from the ground. In addition, the PCB design of the payload is compatible with the open-source BIRDS Bus [28] for 1-unit CubeSat developed at Kyutech following the backplane type interface, which uses a 50-pin connector as shown in Figure 6.

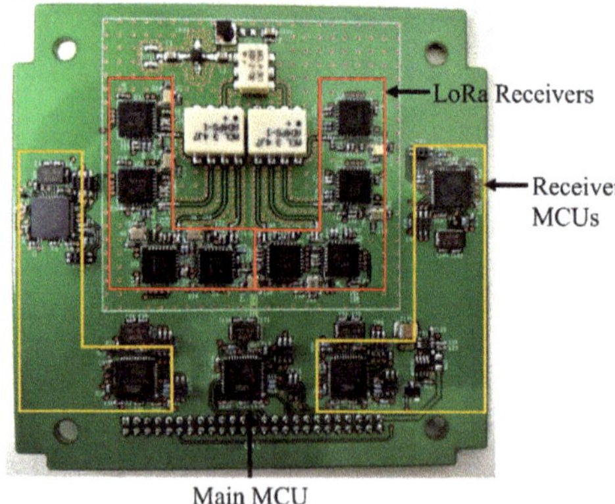

Figure 6. S&F receiver payload design.

SX1278 LoRa transceiver chips developed by Semtech have been used as onboard receivers, and these chips have flight heritage from the BIRDS-3 LoRa demonstration mission (LDM), where the basic operation was confirmed for more than 2 years in orbit. In the LDM mission, two RFM98 modules with SX1278 chips, one as a transmitter and another as a receiver, were connected in a PCB with a 20 dB attenuator between them to avoid emission outside of the satellite as mandated in the frequency license application process. When the mission was executed, a 10-byte data packet was sent 20 times from the transmitter to the receiver, and during this transmission the current consumption and temperature of both the transmitter and receiver were recorded to detect anomalies. The received data packets were checked using CRC bytes, and the recorded data were downloaded to perform analysis on the ground [28].

The patented LoRa modulation technique using the LoRa transceivers can achieve a sensitivity of over -148 dBm according to the datasheet [38]. The transmission range and the data rates are determined by bandwidth and the SF. A combination of lower SF and higher BW allows higher data rates with shorter transmission range while higher SF and lower BW provide a longer range with reduced data rate. In addition, these transceivers could be a suitable option for CubeSat-based SIoT data collection systems due to high interference immunity and low power consumption. Several commercially available complete modules such as RFM98 have been designed by companies such as HopeRF using SX1278 transceiver chips, and these modules have a predesigned matching circuit for both input and output. In KITSUNE, the matching circuit was entirely designed by the team to optimize the PCB design while improving the RF performance. Since the payload board requires us to employ "receive-only" capability from the GSTs, the input matching circuit was optimized and designed according to the datasheet recommendation for reception only. All the components used for the payload design are COTS components and were chosen for their ease of availability, affordable costs, and their performance. The power consumption of the receiver payload is 0.84 W and 1.26 Wh per orbit. In addition, the specifications of the receiver payload board are detailed in Table 2 below.

Table 2. Specifications of SIoT receiver payload board.

Parameters	Values
Dimensions	90 mm × 86 mm
Operating Voltage	3.5 V
Current consumption	240 mA
Configurable Frequency range	399.900 MHz–401.948 MHz 433.000 MHz–435.048 MHz with a resolution of 1 kHz
Measured Sensitivities	-116 dBm~-135 dBm
Tested Data rates	46 bps–488 bps
Reconfigurable SF	6–12
Reconfigurable BW	7.8 kHz, 10.4 kHz, 156 kHz, 20.8 kHz, 31.25 kHz, 41.7 kHz, 62.5 kHz, 125 kHz
Reconfigurable Coding Rate	4/5–4/8
No. of receivers	8

The satellite has three modes of operation for SIoT mission in orbit as listed below:

- **Instant mode:** The mission payload is turned on instantly over a ground station using an uplink command. The mission duration is specified in the command, and the receivers are turned on for that specific period.
- **Target Mode:** The reservation commands are uplinkeded to the satellite to turn on the mission payload over targeted locations as the GST network countries. These reservation commands can be decided based on satellite time and TLE, and a maximum of 10 reservation commands can be stored in the satellite.
- **24-h Mode:** The mission payload can be turned on for 24 h continuously, and the receivers are always in the active listening mode.

The operation and performance of the integrated flight model (FM) were tested in various emulated space environment conditions in order to demonstrate the satellite is able to operate properly in LEO. Fault mode analysis of each subsystem and SIoT mission were also performed during the satellite development process to identify possible causes of failures and contingency plans were prepared to make the system tolerable to these faults.

2.3. Ground Segment

The requirements for the GSTs are stringent as they must be placed in remote locations inaccessible by terrestrial communications. Therefore, the GSTs are required to be simple, robust, and standalone devices that can be powered by solar panels and batteries. The requirements for GST are listed below:

- The design shall be capable of incorporating most kinds of commercially available sensors.
- The structure shall be able to encompass antenna and solar panels externally, protect internal elements and withstand harsh temperature ranges.
- It shall be placed at an adequate height of approximately 1.5 m above the ground for additional protection.
- It shall be able to operate continuously and have a low power consumption (<5 Wh/day). GSTs should have standalone operating power using low-efficiency solar cells and commercially available rechargeable batteries.
- It shall transmit sensor data to the satellite with an adequate link margin.

On the ground segment, there are two types of GST designs such as fixed GST that could be placed in any remote place and mobile GST that could be placed on moving vehicles such as public transport. The system design of the fixed GST is guided by the block diagram shown in Figure 7.

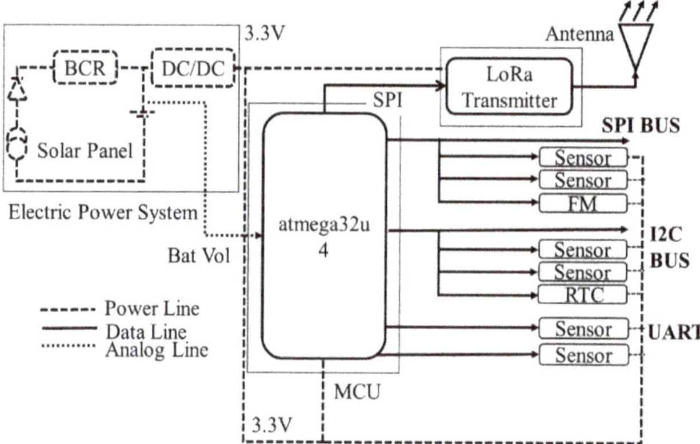

Figure 7. Block diagram of GST for implementation.

As illustrated in Figure 8, a wide range of sensors could be integrated with the MCU using any of the three-serial data transmission formats. Inter-integrated circuit (I2C), UART, SPI, and analog pins produce a universal sensor port. In addition, the sensors are directly plugged into ports that connect to the MCU, and the collected sensor data is time stamped using a real time clock (RTC). Furthermore, they are transformed into data packets before being transmitted through an antenna via LoRa transmitter. The data packets are also stored in the flash memory (FM). The electrical power system (EPS) of the conventional sensor nodes is typically substantial, owing to their complexity and considerable power requirement. While the EPS in the proposed method is simple and consumes less power to operate, it also comprises low-efficiency, outdoor-grade solar panels, a battery charging circuit (BCR) and a Li-Ion rechargeable battery. The detailed GST design has been explained in [39].

The fundamental advantage of this system is design simplicity, compact size, robustness, and the ability to operate in harsh environments. Furthermore, another additional advantage is the ability to be installed in any required location without relying on on-ground infrastructure and terrestrial communication networks. The PCB design of the GST is shown in Figure 8.

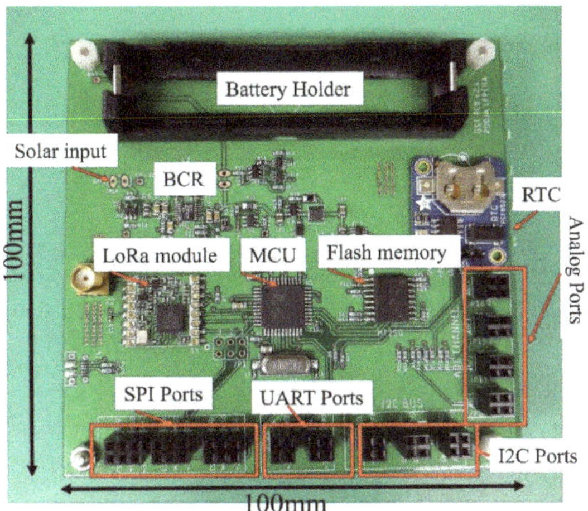

Figure 8. PCB design of GST.

The antenna used for the GST is a right hand circularly polarized quadrifilar helix (QFH) antenna. The antenna was designed using the guide mentioned in [39] for 433 MHz using COTS materials. The standards for electrical enclosures are classified and rated based on the degree of protection provided against dust, water, and other intrusions by the International Protection (IP Code) Marking also known as IEC standard 60,529 [40]. The MCUs and sensors are enclosed within an IP66 housing that provides complete protection against dust and water while allowing the GST to be placed in remote terrains vulnerable to varying climatic conditions throughout the year. In addition, the specifications of the GST boards are detailed in Table 3.

Table 3. Specifications of the fixed GST board.

Parameters	Values
Dimensions	100 mm × 100 mm
Operating Voltage	3.3 V
Current consumption	35 mA standby, 90 mA transmit
Frequency range	400 MHz–440 MHz
Transmit power	Up to 100 mW (20 dBm)
Data rate	46 bps–781 bps
Reconfigurable SF	6–12
Bandwidth	7.8 kHz to 125 kHz
Communication ports	I2C, SPI, UART, analog

The basic mode of operation for GSTs is to sample and transmit sensor data in real time. In addition, the power consumption of the GST is minimized to achieve the power budget for continuous operation with low-efficiency solar panels and rechargeable batteries.

The data packets collected by the satellite from GSTs around the world are downloaded and parsed based on GST ID through a common database. To achieve this, the database is designed to accept the data in standard file formats—either tab separate (TSV), comma separated (CSV) or text (TXT). From the parsed information, data will be encoded in the database through an online webpage by the ground stations, which receive the data. A dedicated user account is provided for each country for data encoding, access, and download. While all data could be accessed by an admin user account, the dedicated user account is restricted to access only for the GSTs of the relative countries. The data collected from the GSTs of other countries are not accessible to anyone else except the country that transmitted the data. To facilitate data sharing, the country could access their own data in the database, and reports could be generated based on their selected data in TSV, CSV or TXT format.

3. Coverage and Capacity Evaluation

To derive the coverage and capacity of SIoT using the payload design and the GST, Equation (1) is used to estimate the data that can be collected per day by KITSUNE.

$$Data\ per\ day\ (in\ bytes) = \sum_{i=1}^{n_{GST}} \tau(i) * B * \beta * \acute{\eta} \tag{1}$$

Data per day is estimated with the communication time between satellite and GST in minutes in a pass $\tau(i)$, the number of GSTs visible per day n_{GST}, the number of bytes received per minute B, the ratio of meaningful bytes to total bytes in a packet β, the efficiency factor $\acute{\eta}$ that incorporates the overall efficiency reduction due to the factors such as performance dependence on temperature and the Doppler effect. These parameters are described in the following subsections that includes overall implementation plan for the mission, the design of the satellite and the GSTs, and the communication protocols between them.

3.1. Communication Time between the Satellite and GST($\tau(i)$)

The communication time $\tau(i)$ between the satellite and the GST depends on the link budget, and the link budget depends on the sensitivity of the receiver, elevation angles, antenna gains and losses such as free space path loss (FSPL), polarization loss, and ionospheric loss. These parameters are first determined through the ground tests using the designed payload and GST, and the communication time is derived using the parameters.

3.1.1. Antenna Radiation Pattern

The radiation pattern of the receiver antenna was measured in the anechoic chamber tests. The antenna radiation pattern was measured for three planes such as X-Y plane, Y-Z plane, and X-Z planes, and in each plane for vertical and horizontal orientation of the reference antenna. Figure 9 shows the rotation axis of the antenna for determining the antenna radiation pattern.

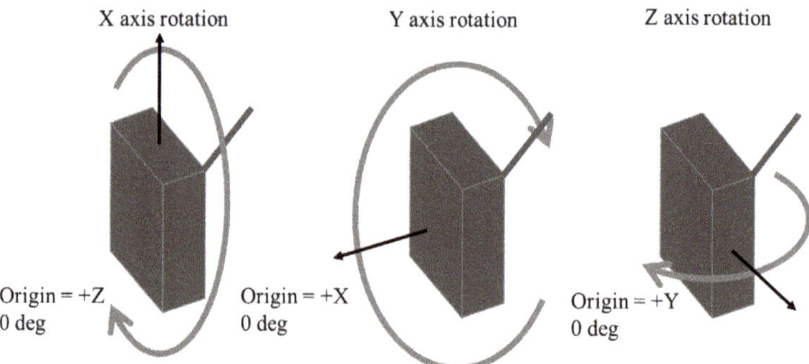

Figure 9. Satellite rotation configuration for three axes radiation pattern measurement.

The maximum gain of the receiver antenna integrated together with the satellite body was 0.27 dBi for 400 MHz and 0.36 dBi for 433 MHz as in Figure 10.

3.1.2. Receiver Sensitivity

Theoretical receiver sensitivities up to -148 dBm can be achieved with SX1278 modules according to the datasheet; however, the actual receiver sensitivity could be less than the initial design parameters. To determine the functional receiver sensitivity of the produced subsystem, communication tests between the integrated satellite operating at nominal mode and the GST were performed in the anechoic chamber, and the test setup is shown in Figure 11.

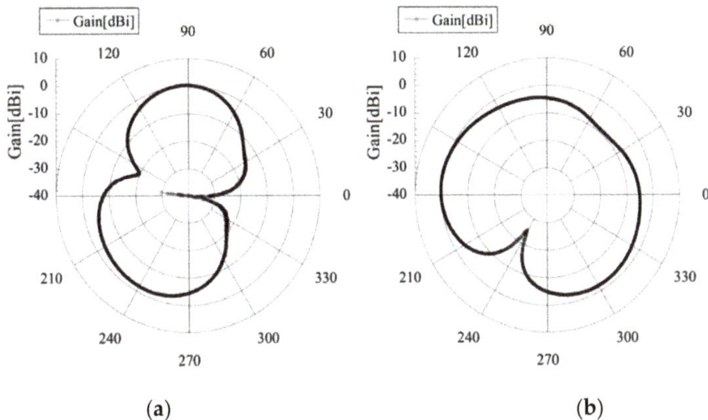

Figure 10. Radiation pattern of the designed monopole antenna for (**a**) 400 MHz and (**b**) 433 MHz.

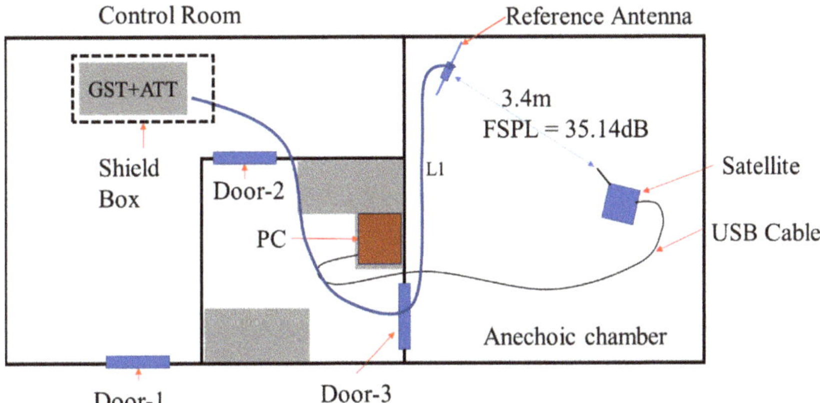

Figure 11. Test configuration for receiver sensitivity tests in an anechoic chamber.

The GST was placed inside the shield box outside the control room to reduce the effect of possible leakage from the GST LoRa transceiver to the payload receiver. A reference antenna with a known gain (G1) was placed 3.40 m away from the satellite, which was connected to the GST using a cable with a cable loss of L1. The GST transmitted data packets to the satellite continuously, and the data reception was recorded on the satellite memory while monitoring through the serial monitor connected to the receiver payload. In addition, it was routed to the control room using a USB cable and attenuators (ATT) were added to the GST side until the receiver payload could no longer receive the data packets. The success rate was calculated by checking whether correct data packets without byte errors were received at the receiver side. The sensitivity was taken for the attenuation for which at least 60% of the overall packets were received correctly.

The GST transmitted an 18 dBm signal using a reference antenna with a maximum gain of 2.15 dBi. The cable loss L1 is determined to be 22.04 dB, and the FSPL was calculated as 35.14 dB. The Equation (2) was used to measure the sensitivity of the receiver, and all terms are given in decibel (dB) scale in the equation below.

$$P_{RX} = P_{TX} - ATT - L_1 - L_{cable} + G_1 + G_2 \tag{2}$$

Various combinations of SF and BW were tested for achieving different data rates and receiver sensitivities by the same setup. As a result, the optimal settings obtained balancing the date rate and the sensitivity are determined to be with SF 12, BW 31.25 kHz, and CR 4/8. The resulting data rate is 46 bps. Additionally, a receiver with SF 8, BW 31.25 kHz and CR 4/8 resulting in a data rate of 488 bps was also tested to understand the relationship between data rate and sensitivity. The results of the tests are shown in Figure 12.

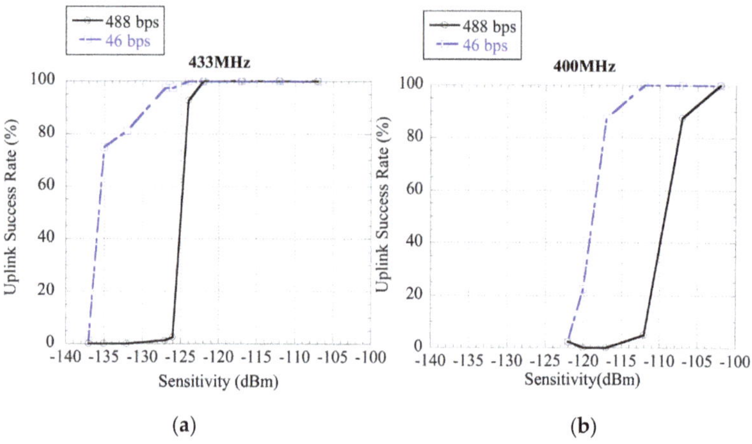

Figure 12. Uplink success rate for receiver sensitivity of (**a**) 400 MHz and (**b**) 433 MHz.

The minimum measured sensitivity considering 60% successful data reception is -118 dBm for 400 MHz, and -135 dBm for a data rate of 46 bps. The measured sensitivity is 26 dBm less for 400 MHz and 9 dBm less for 433 MHz compared to the ideal sensitivity that could be achieved at these settings, which is -144 dBm. According to these results, the reduced sensitivity could be attributed to deviation in performance of the electronic components from designed parameters due to mismatch in circuit blocks and non-ideal components and internal EMI of the satellite. Furthermore, the additional deviation for 400 MHz could be attributed to environmental noise. Although the satellite was placed inside the anechoic chamber that can provide attenuation of 100 dB in the 400 MHz range according to MIL STD 285 standards, noise peaks from outside the chamber were observed in the spectrum analyzer captured by the dipole antenna. These noise peaks were not observed in the 433 MHz range, and the receiver sensitivity at 400 MHz is expected to be similar to 433 MHz when the satellite is in orbit.

The measured sensitivity for 433 MHz at 46 bps is considered adequate for communication with the satellite in orbit. Higher sensitivities can be achieved using lower data rates, but the data transmission time increases as well. For instance, decreasing data rate from 46 bps to 11 bps, the transmission time increases from 3 s to 14 s. As a result, data transmission for longer periods would result in higher power consumption for the GSTs, and the data packets could be affected by doppler shift rate and the monopole antenna rotation with tumbling satellite.

3.1.3. Link Budget Calculation

The measured sensitivity is accounted to calculate the link budget in orbit, considering the satellite orbit as 400 km and FSPL at elevation angles of 10°, 20°, 30°, 40°, 60°, 80° and 90°. The transmit power of 20 mW is used since in most countries with this power, a license is not required to transmit with 433 MHz frequency an Industrial, Scientific, and Medical (ISM) band. The maximum permissible transmit power is determined by the radio communications regulation of each country. Using this sensitivity, the link budget is calculated considering the losses detailed in Table 4, both on the satellite side and the GSTs.

Table 4. The parameters used for the link budget calculation.

The parameters from GST to the satellite	
Transmission power	20 mW (13 dBm)
Maximum antenna gain	6 dBi
GST transmission line loss	1 dB
GST Antenna pointing loss	3 dB *
Polarization loss	3 dB *
Atmosphere + ionosphere loss	1.5 dB *
The parameters for the satellite receiver	
Antenna pointing loss	3 dB *
Antenna gain	0 dBi
Receiver sensitivity	−135 dBm for 433 MHz
Transmission power	20 mW (13 dBm)
Maximum antenna gain	0 dBi
Receiver sensitivity	−135 dBm for 433 MHz

* Estimated values.

The link margins for varying elevation angles in the 400 km orbit are detailed in Table 5. The antenna gain variation of the GST for each elevation angle of the satellite is considered for the link margin calculations.

Table 5. Link budget calculations for different elevations.

Satellite Elevations	Path Loss	GST Antenna Gain	Link Budget
0 degrees	152.39 dB	−3.50 dBi	−19.38 dB
10 degrees	148.35 dB	2.10 dBi	−9.84 dB
20 degrees	145.04 dB	3.20 dBi	−5.53 dB
30 degrees	142.56 dB	5.50 dBi	−0.55 dB
40 degrees	140.72 dB	6.00 dBi	1.79 dB
60 degrees	138.39 dB	6.10 dBi	4.22 dB
90 degrees	137.22 dB	5.90 dBi	5.29 dB

According to the link margin calculations, a successful uplink of sensor data from the GST to the satellite can be achieved for elevation angles greater than 33 degrees. The communication time between the satellite and the GST can hence be determined for satellite passes with elevations greater than 33 degrees with Equation (3).

$$\tau(i) \text{ in minutes} = 10 - \tau < 33 \tag{3}$$

$\tau < 33$ is the time in minutes when the elevations are less than 33 degrees. Satellites in LEO have an orbital period of approximately 90 min with an average of 10 min ground pass for at least four times in a day. To derive actual communication time between KITSUNE and the GSTs, the simulation for satellite orbital path is performed in Section 3.2.

3.2. Number of GSTs Visible per Day (n_{GST})

The GSTs for the KITSUNE SIoT mission are simultaneously being developed in 11 different agencies/institutions predominantly from the developing countries. The participating countries are part of the BIRDS Ground Station Network (GSN) that was initially formed to support the missions of BIRDS constellation of satellites [41]. The countries that are a part of the GST Network along with the application are explained in Table 6. Most of these participating countries have significantly limited access to space technologies with highly constrained space programs.

Table 6. The countries participating in S&F IoT mission in alphabetical order.

Sl No.	Countries	Target Application
1	Bhutan	Weather parameters
2	Costa Rica	Weather Parameters
3	Malaysia	Air quality, temperature, and humidity
4	Mongolia	Temperature, air pressure and humidity
5	Nepal	Water level of glacier lakes and weather parameters
6	Paraguay *	Environmental parameters (AEP) Monitoring of transmission lines (ANDE)
7	Philippines	Gas, humidity, pressure, temperature
8	Sri Lanka	Water level in reservoirs, weather data
9	Taiwan	Earthquake detection using GNSS sensors
10	Uganda	Air quality measurement and atmospheric humidity
11	Zimbabwe	Soil moisture for landslide prediction

* Two GSTs.

Currently, one GST is being developed in each of these countries (except Paraguay with two GSTs) but prospects for multiple GSTs for providing more coverage are underway. The GSTs developed in some of these developing countries are shown as an example in Figure 13.

Figure 13. The GSTs developed in various countries as Paraguay, Malaysia, Taiwan, and Mongolia from left to right.

The orbital path of the KITSUNE is simulated for 24 h in STK software, and all possible passes for GST network countries participating in the SIoT mission are generated. On average, each country has five passes per day while each pass varies between four minutes to ten minutes. The path of the passes is shown in Figure 14.

In Section 3.1 a positive link margin is derived for elevations greater than 33 degrees. According to the simulation, there is at least one pass per day with elevation greater than 30 degrees in each country. For delay-tolerant applications, collection of data once per day is tolerable for these kinds of applications.

For in-depth analysis of number of visible GSTs for a longer period, the orbital path was simulated for one month (31 days) for elevation angles greater than 33 degrees. The result of the simulation is summarized in Figure 15. In summary, the number of GSTs that are visible for elevation greater than 33 degrees ranges between a minimum of 12 passes to a maximum of 22 passes in a day and an average of 16 passes for the entire period. The average elevation angles per day varies between 47 degrees to 64.50 degrees and an average of 56 degrees for the whole period.

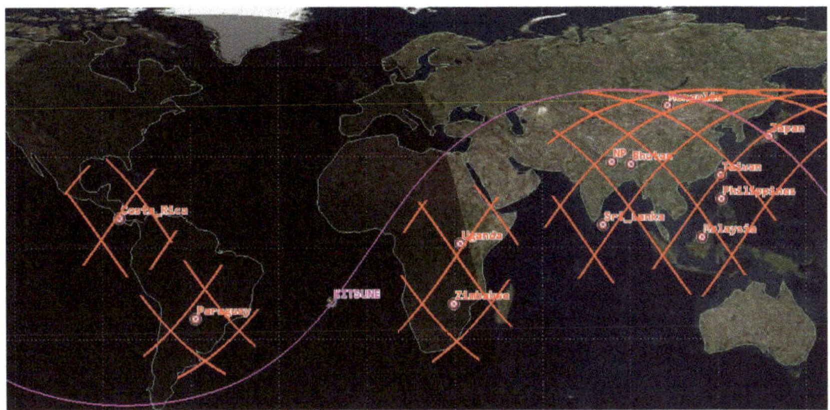

Figure 14. Satellite passes for KITSUNE in GST countries for 24 h.

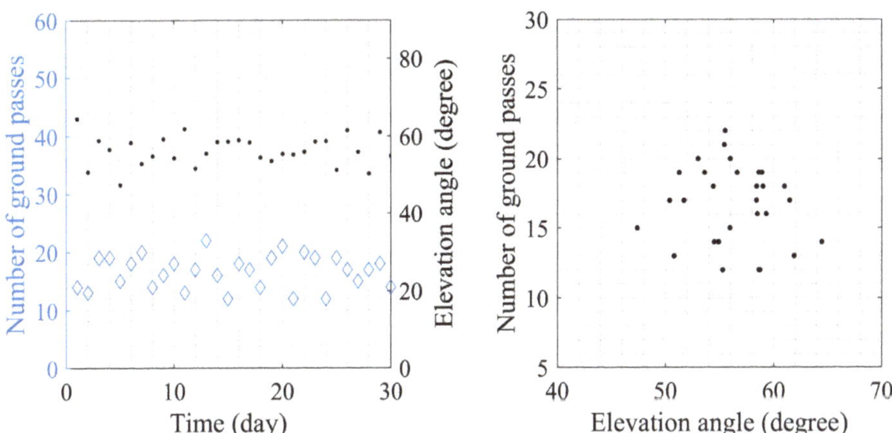

Figure 15. Number of ground passes and their average elevations for KITSUNE in GST countries for one month.

The number of visible GSTs in a day, n_{GST}, derived from the simulation results is taken as 16 and the average elevation is taken as 56 degrees. Assuming a proportional elevation increment during a satellite pass of 10 min of 56-degree elevation, using Equation (4), the communication time $\tau(i)$ between satellite and GST is calculated as approximately 4 min.

3.3. Number of Bytes per Minute (B)

The number of bytes B that the SIoT receiver payload can receive depends on the programmed algorithm for data reception. The receivers are programmed to receive data packets with a maximum packet length of 32 bytes. The receiver MCUs store five packets of 32 bytes of data temporarily from each receiver before forwarding it to the main mission LoRa MCU, using universal asynchronous receiver-transmitter (UART) line every one minute. The number of bytes the satellite can receive per minute is given by Equation (4).

$$B = n_p \times n_B \times n_{MCU} \tag{4}$$

The number of bytes that the satellite can receive depends on the number of packets the receiver can transfer to MCU per minute, n_p, the number of bytes in a packet, n_B and the number of MCUs, n_{MCU}. Therefore, the number of bytes B that can be received in one minute is approximately 1280 bytes.

The LoRa MCU stores the received data in a 1 Gbit flash memory in the OBC/EPS board based on the day number, and each day 64 kB of data storage is assigned in the flash memory. This design enables ease of data download after the satellite is deployed in orbit. This MCU also communicates with the onboard computer (OBC) using the UART protocol. The collected data will be downlinked using another transceiver in the licensed UHF band with a data rate of 4800 bps.

3.4. Ratio of Meaningful Bytes to Total Bytes in a Packet (β)

Since many GSTs are expected to be deployed in various parts of the world, it is imperative to identify the data source as the origin country and the location of the GST. As a result, a data packet format has been standardized for all GSTs, which includes GST ID, data type, time of data collection and sensor data. Two bytes are assigned for GST ID, and the first byte is for country identification. The second byte is used when there is more than one GST in a country, and it would be identified by using the second byte. The data type specifies what kind of data and how many kinds of data are uplinkeded in the packet. Four bytes are assigned for timestamp, which is the time at which the sensor data was recorded, and it should include seconds, minutes, hour, day, month, and year. Each country can choose whether they want to send data with either short or long data packet format. The optimal data packet length is set to 32 bytes since longer data transmission time consumes more power, and the data packets can have errors due to doppler shift rate and unstable attitude of the satellite. A checksum byte is assigned at the end of the packet to ensure the satellite received the correct packet. The packet formats for GSTs are shown in Figure 16.

Byte 0	Byte 1	Byte 2	Byte 3	Byte 4	Byte 5	Byte 6	Byte 7	Byte 8	Byte 9	Byte 10	Byte 11	Byte 12	Byte 13	Byte 14	Byte 15
0×00	0×00	0×00	0×00	0×00	0×00	0×00	0×00	0×00	0×00	0×00	0×00	0×00	0×00	0×00	0×00
GST ID		Data Type	Time				Sensor Data (8 bytes)								Check Sum

Byte 0	Byte 1	Byte 2	Byte 3	Byte 4	Byte 5	Byte 6	Byte 7	Byte 8	Byte 9	Byte 10	Byte 11	...	Byte 29	Byte 30	Byte 31
0×00	0×00	0×00	0×00	0×00	0×00	0×00	0×00	0×00	0×00	0×00	0×00		0×00	0×00	0×00
GST ID		Data Type	Time				Sensor Data (24 bytes)								Check Sum

Figure 16. Short data packet format (16 bytes) (**top**) and long data packet format (32 bytes) (**bottom**).

The ratio of meaningful bytes to the total number of bytes can be deduced from Equation (5) that considers only the bytes that contain the time and sensor data.

$$\beta = \begin{cases} 0.75, & \text{for short data packet format} \\ 0.88, & \text{for long data packet format} \end{cases} \tag{5}$$

3.5. Efficiency Factor ($\acute{\eta}$)

The efficiency factor is introduced in the equation to include the impacts of temperature and the Doppler effect on the reception capability of the designed system. The efficiency factor is derived using Equation (6).

$$\acute{\eta} = \eta_T \times \eta_D \tag{6}$$

The efficiency factor, $\acute{\eta}$ depends on the efficiency factor due to temperature impact, η_T and the efficiency factor due to the Doppler-shift-impact mitigation method, η_D. The approach of determining the efficiency factors for the proposed system are described in the following subsections.

3.5.1. Efficiency Factor due to Temperature (η_T)

The effects of temperature variations on LoRa modulation have been tested and found to reduce the received signal strength when the ambient temperature increased in [42,43]. Inverse correlation between low temperatures and signal-to-noise ratio (SNR) was observed in [43], but extensive tests for low temperatures have not been performed, especially in the space environment. To deduce the impact of temperature on the data reception quality, an experiment was performed to derive the efficiency factor.

A test with SIoT receiver board and backplane was conducted in Peltier-based Thermal Testing (PeTT) Chamber located at Kyutech. The chamber pressure was maintained below 1×10^{-3} Pa throughout the test and the temperature was varied between $-22\,°C$ to $+72\,°C$. The center frequencies of the eight receivers were set 5 kHz apart from 433.060 MHz to 433.085 MHz, and data from the GST were transmitted at 433.080 MHz.

The reception of the data was evaluated on the weighted success that is a measure of number of packets received for transmitted packets and the success rate of the bytes in the data. The test results are summarized in Figure 17 for the three receivers; RX7 is the tuned frequency, RX6 is receiver with 5 kHz lower CF and, RX8 with 5 kHz higher CF.

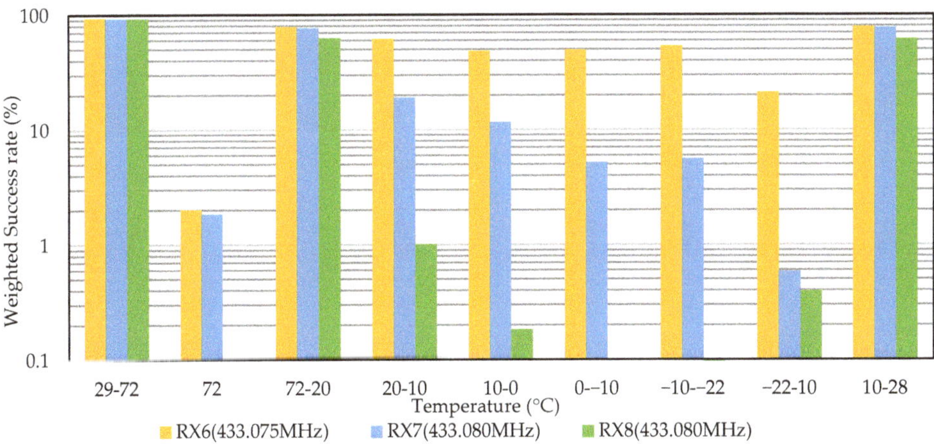

Figure 17. Summary of temperature dependence test of receivers in PeTT chamber.

It was observed that the receiver performance was not affected by increase in temperature up to 72 °C; however, only 3 packets out of 105 packets were received by RX6 and RX7 with a byte success rate of 64% at 72 °C. The performance of the receivers deteriorated when the temperature was reduced below 20 °C, an apparent change in the center frequency was noted since the receiver RX6 received more packets with correct bytes than RX7 and RX8. At cold temperatures, the center frequency of the receiver increased. The performance of the RX-7 further reduced at lower temperatures while RX6 continued to receive packets with 50%-byte success rate. At the minimum of $-22\,°C$, the success rate was low for all the receivers, nevertheless RX6 continued to perform better than the other two. The temperature was increased to set the receivers to ambient temperature and just over 10 °C, and all receivers received the correct data packets.

This temperature dependence could be correlated to the temperature profile in-orbit for deriving the efficiency factor for temperature. The on-orbit temperature of the backplane, which is similar to the temperature of the SIoT board, was below 10 °C when the beta angle was below 25 degrees and increased up to +42 °C at maximum beta angle. Figure 18 shows the temperature of SIoT board obtained from on-orbit housekeeping data with error bars for varying beta angles.

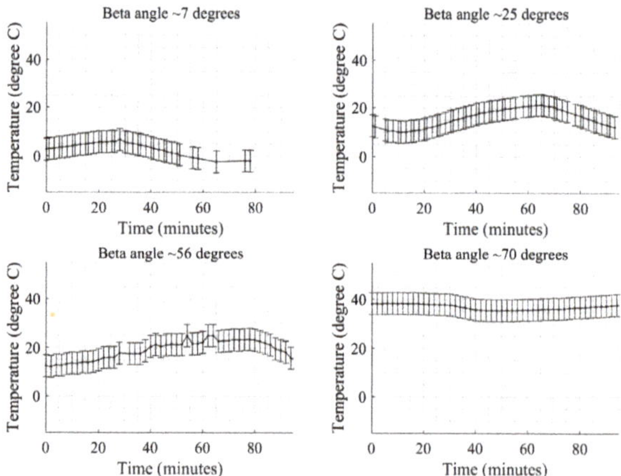

Figure 18. On-orbit temperature variation of SIoT board with respect to the beta angle.

The efficiency factor for temperature can hence be derived from the performance of the receivers with a threshold of 10 °C, which is true when the beta angle is above or below 25 degrees, as shown in Equation (7).

$$\eta T = \begin{cases} 0.85 \text{ for beta angles} > 25 \text{ degrees} \\ 0.50 \text{ for beta angles} < 25 \text{ degrees} \end{cases} \tag{7}$$

The absolute beta angles in the LEO orbit of 400 km vary between 0–74 degrees, and the beta angle simulation is performed in STK software for the period of one year, as shown in Figure 19. The beta angle is above 25 degrees for 59% of the total simulated duration. Using Equation (7) for the total amount of data that can be collected inclusive of temperature impact, about 70% from the total estimated data for one year could be retrieved.

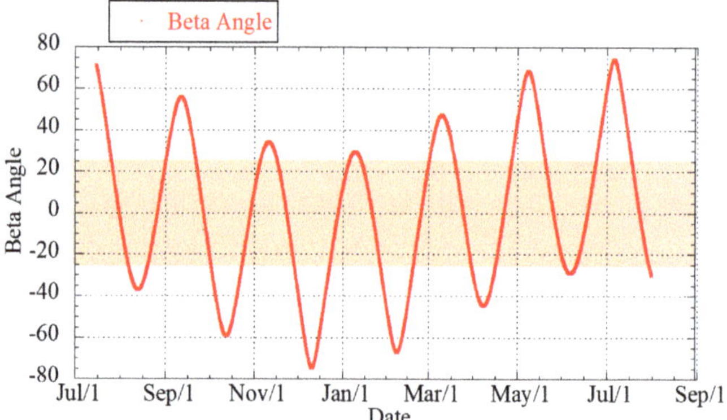

Figure 19. Beta angle variation of KITSUNE for one year from July 2022 to September 2023.

3.5.2. Efficiency Factor due to Doppler Effect Mitigation (ηD)

In LEO satellite at 400 km, the Doppler effect can cause a frequency offset of up to ±10 kHz to the data signal arriving at the receiver. Many studies on the ground for LoRa modulation immunity to the Doppler effect were performed. To confirm the immunity of the designed receivers against the static Doppler effect, the designed receivers were configured to have center frequency deviation in steps of 1 kHz up to ±15 kHz and 100 data packets were transmitted at each deviation. The success rate was calculated based on the number of packets received and the correctness of the bytes. The tests confirmed that the receivers could receive data packets with 95% accuracy up to −14 kHz deviation, but positive deviations were not tolerated. The summary of the results is depicted in Figure 20. This means the receivers have a higher probability of receiving the packets if they were transmitted at higher frequencies than the tuned center frequency. This shall be verified on-orbit further.

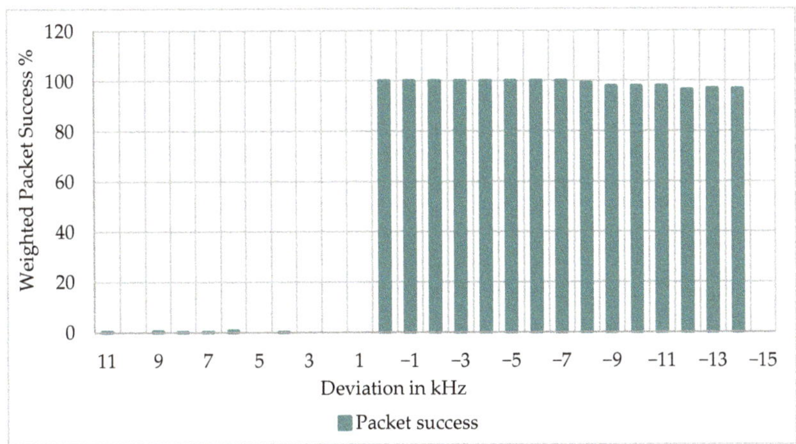

Figure 20. Weighted packet success rate for frequency deviation in steps of 1 kHz.

Since there are eight receivers in the SIoT board, the change in center frequency due to the Doppler effect can be accommodated by overlapping the center frequencies of the receiver so that other receivers tuned at slightly higher or lower frequencies can still receive the data packets. The configuration of eight receivers for passive Doppler shift mitigation is shown in Figure 21, with the channel of each receiver overlapping each another. This makes the system mitigate the impact of center frequency offsets due to the Doppler effect.

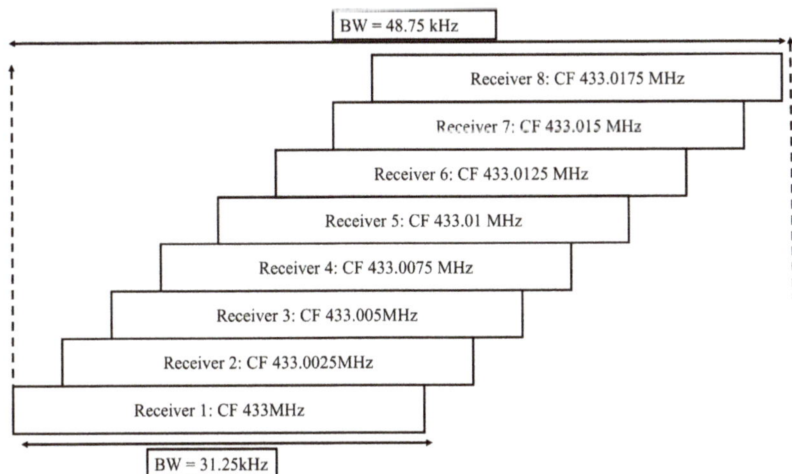

Figure 21. Passive Doppler correction method on-orbit.

The overlap of center frequencies is subjected to reduction in the overall system efficiency by at least 25% considering the receivers in the extreme ends may not receive the data packets correctly. The efficiency factor due to Doppler-effect mitigation is defined in Equation (8).

$$\eta D = 0.75. \tag{8}$$

The efficiency factor can hence be calculated using Equation (6) using the efficiencies obtained for temperature and Doppler-impact mitigation as follows:

$$\acute{\eta} = \begin{cases} 0.64 \text{ for beta angles} > 25 \text{ degrees} \\ 0.37 \text{ for beta angles} < 25 \text{ degrees} \end{cases}$$

4. Results and Discussion

For estimating the coverage and capacity of a satellite in the ISS orbit to provide services to 11 mostly developing countries, Equation (1) was used, the parameters for which were derived in Section 3. In summary, the parameters that were obtained for the estimate are detailed in Table 7.

Table 7. Parameters for estimation of average SIoT capacity.

Parameter	Value
Satellite orbit altitude (km)	400
Orbital inclination (degree)	51
Number of satellites in orbit	1
Number of GSTs on the ground	11
Average satellite elevation per day (deg)	56
Average satellite pass time (min)	4
Average number of GSTs visible per day	16

The parameters are based on the ground test results, simulations and on-orbit results for the designed SIoT payload board and the GST design. The parameters include the antenna radiation pattern for both the satellite monopole antenna and the GST antenna, the sensitivity tests that were performed for the integrated satellite inclusive of all the EMI noises generated by the satellite subsystems, and the link budget calculation inclusive of the losses.

The capacity of the system per day is summarized in Table 8 for scenarios considering the average of all satellite elevations simulated for a month and the duration of satellite to GST communication during this period. For the maximum capacity estimation, the simulation yielded a total of 22 satellite passes in a day with average elevation of 56 degrees and for minimum, a total of 12 passes with average elevation of 55 degrees were derived. The minimum amount of data per GST that can be received by the satellite per day is 45.50 kB.

Table 8. Summary of SIoT capacity estimation for one year.

Communication Time	Short Format		Long Format	
	Per Day	Per Year	Per Day	Per Year
Minimum	45.50 kB	16.04 MB	52.50 kB	18.71 MB
Average	60.50 kB	21.39 MB	70.00 kB	24.95 MB
Maximum	82.50 kB	29.41 MB	96.25 kB	34.31 MB

Although the estimation is based on the known system inference, it is still optimistic speculating other factors that could affect the performance of the receiver, such as impact of temperature on the receiver capability and the Doppler effect. Therefore, the efficiency factor is introduced in the equation that incorporates the efficiency due to temperature impact and the Doppler-effect mitigation method. Since the temperature on orbit depends on the beta angle of the satellite that varies between 0–74 degrees absolute values, the simulated beta angles with a threshold of 25 degrees for a period of one year are used to estimate the average capacity of the SIoT mission for a year in Table 9.

Table 9. Summary of SIoT data capacity estimation for one year including efficiency factor.

Parameters	Short Format		Long Format	
	Per Day	**Per Year**	**Per Day**	**Per Year**
Beta angle < 25 degrees	9.23 kB	3.29 MB	10.76 kB	3.84 MB
Beta angle > 25 degrees	22.57 kB	8.04 MB	26.33 kB	9.38 MB
Total	31.79 kB	11.33 MB	37.09 kB	13.22 MB

The measured environmental parameters in most participating GST Network countries include temperature, humidity, and pressure. Considering at least two decimal digit resolutions for each sensor and assigning 4 bytes for each sensor, 31.79 kB of data in a day and 1.99 kB of data per pass, which is approximately 169 samples, can be collected from each country per day. This data capacity can be divided between multiple GSTs on the ground. In typical applications, the parameters for LoRa modulation such as the SF, CR and bandwidth can affect the amount of data that can be collected from the sensor terminals. Since the receivers are set to receive only five packets of data per minute, reconfiguration of these parameters allows for improving the quality of data packets that shall be received and not increase the upper limit of the capacity.

The received data packets are downlinked at Kyutech ground station, the configuration of which is shown in Figure 22. From the operation experience of previous satellites from the same ground station, approximately 7.9 kB of data can be downlinked per ground pass using 400 MHz UHF frequency. Integrating the data downlinked for four passes a day, all the data collected by the payload receiver can be downlinked and uploaded to the database within a day. The maximum latency between data transmission from the GST to the data reception from the database is 24 h. This latency is acceptable for delay-tolerant applications.

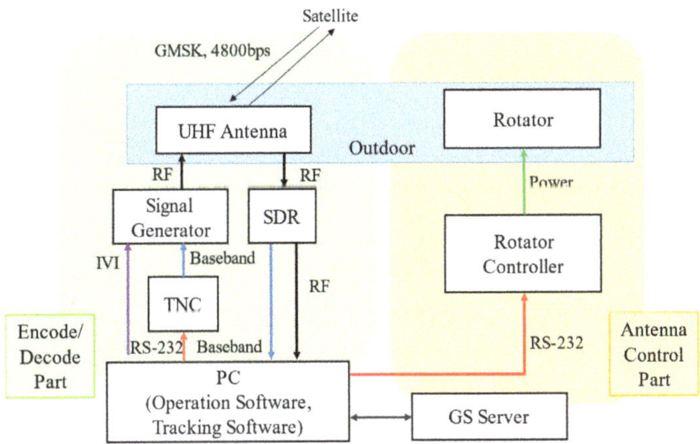

Figure 22. Configuration of ground station at Kyutech.

The estimation for the capacity of the mission is made as realistic as possible, inclusive of known parameters for the designed system. The unknown parameters are the on-orbit dynamics, which may change the performance and the capacity of the receivers. The capacity shall be re-evaluated once the mission is made fully functional on-orbit.

In addition to the technical edge of the designed system for the SIoT mission, the involvement of developing countries for the GST network provided a platform to experience 'hands-on' while building their own GSTs. Each of these participating countries customized the GST design based on material substitutes available locally, optimizing the design for that country. The overall development cost was targeted to be achieved under $150 based on a market survey of components in Japan to achieve low-cost requirements [39] but the actual development cost in the GST countries ranged from $200 to $300 due to variation in market prices. The GST community from each country comprises individuals actively involved in developing the space sector, including students from universities, officials in space agencies, and space enthusiasts, who would like to improve space technologies in their country. Most of these countries are participating in demonstrating a proof of concept so that these technologies could be applied in their future projects to solve local issues.

Since there are no publications in the literature that discuss the capacity of a SIoT based on the actual design of a CubeSat and the limitations due to the design, the proposed system cannot be compared quantitively for the performance evaluation. Instead, this study could serve as a base for quantitative comparison for satellites that will be developed in the future for S&F remote data-collection missions. There are areas of improvement that are currently being implemented in the Micro Orbiter-1 (MO-1) mission, which is a 1-unit CubeSat being developed at Kyutech. MO-1 will have two S&F payloads, which will have eight receivers each onboard, for the frequencies of 400 MHz and 920 MHz, and the 400 MHz board is a replica of the KITSUNE payload board; however, the 920 MHz is an improvised version of the 400 MHz board with 70 mA lesser power consumption. Communication tests are underway, and the preliminary test results showed similar receiver sensitivity to the KITSUNE S&F receiver. As a result, it is under study to improve the receiver sensitivity further.

5. Conclusions

Small satellites in LEO can provide global coverage and access to any remote topography. The utilization of CubeSat offers a cost-effective solution in developing countries where data collection from remote areas using terrestrial networks is impractical. This paper explored the coverage and capacity of a single satellite with a SIoT mission capability and multiple low-cost GSTs developed locally to provide services to developing countries. The SIoT satellite payload for KITSUNE was designed and developed to be accommodated within the dimensions and power budget of a 1-unit volume of the 6U CubeSat. The GSTs that can measure any kind of sensor data and transmit the data directly to KITSUNE were developed for the ground segment. The GSTs are simultaneously being developed in 11 mostly developing countries, and they will be placed in remote locations to measure environmental parameters. The involvement in the GST network provided the participating countries with access to space technologies without launching their own satellites.

The capacity of the SIoT mission for providing services to 11 mostly developing countries was evaluated based on the actual performance of the designed receiver and the ground tests results, simulations and on-orbit results that were performed to determine the parameters that affect the performance of the designed system. The tests included antenna-radiation pattern test for both the payload antenna and the GST, receiver-sensitivity tests for fully integrated satellites, and link budget calculations based on the on-orbit parameters.

KITSUNE was launched to ISS in February 2022 and deployed into orbit in March 2022. The functionality of the payload receiver was tested but the on-orbit results are not in the scope of this paper since the GSTs are being prepared to be deployed in participating countries. On-orbit data collection from the GSTs, validation of the ground testing results with the on-orbit performance and further analysis will be performed. Future work will involve comprehensive testing of the GSTs made by 11 mostly developing countries with KITSUNE. The on-orbit results will be beneficial for developing countries as a first step towards space utilization for solving domestic issues.

Author Contributions: Conceptualization, P.L., T.D.M., N.C.Ö., M.A.P., F.D., M.K., H.A.E.-M. and M.C.; methodology, P.L., T.D.M. and N.C.Ö.; software, P.L. and T.D.M.; validation, P.L., T.D.M., N.C.Ö. and M.C.; formal analysis, P.L., T.D.M. and N.C.Ö.; investigation, P.L., T.D.M. and N.C.Ö.; resources, M.C.; data curation, P.L.; writing—original draft preparation, P.L.; writing—review and editing, P.L., N.C.Ö. and M.C.; visualization, P.L. and N.C.Ö.; supervision, M.C.; project administration, M.C.; funding acquisition, M.C. All authors have read and agreed to the published version of the manuscript.

Funding: This research was partially funded by JSPS Core-to-Core Program B: Asia-Africa Science Platforms (JPJSCCB20200005).

Institutional Review Board Statement: Not applicable.

Informed Consent Statement: Not applicable.

Data Availability Statement: Not applicable.

Acknowledgments: The authors would like to express their appreciation to all KITSUNE team members for their contribution in satellite development namely; Jose Rodrigo Cordova Alarcon, Victor Hugo Schulz, Abhas Maskey, Adolfo Javier Jara Cespedes, Anibal Antonio Mendoza Ruiz, Cosmas Kiruki, Daisuke Nakayama, Dmytro Faizullin, Dulani Chamika Withanage, Ei Phyu Phyu, Eyoas Ergetu Areda, Hari Ram Shrestha, Ibukun Oluwatobi Adebolu, Kateryna Aheieva, Kentaro Kitamura, Mariko Teramoto, Mazaru Ariel Manabe Safi, Muhammad Hasif Bin Azami, Takashi Oshiro, Victor Mukungunugwa, Yuma Nozaki, and Yuta Kakimoto. The authors also expresses gratitude to staffs Sangkyun Kim, Hirokazu Masui, Takashi Yamauchi, and Ofosu Joseph Ampadu for providing unrelenting support during the development and testing of KITSUNE.

Conflicts of Interest: The authors declare no conflict of interest.

References

1. Pradhan, K.K.; Cho, M. Shortening of Delivery Time for University-Class Lean Satellites. *J. Small Satell.* **2020**, *9*, 881.
2. Jenkinsa, M.G.; Alvarado, J.C.; Calvo, A.J.; Jiménez, A.C.; Godínez, J.C.; Salazar, A.V.; Molina, J.R.; Rosales, L.C.; Martinez, E.; Jiménez-Salazar, V.; et al. Testing and Operations of a Store and Forward CubeSat for Environmental Monitoring of Costa Rica. In Proceedings of the International Astronautical Congress, IAC, Bremen, Germany, 1–5 October 2018; International Astronautical Federation, IAF: Bremen, Germany, 2018; Volume 2018-October.
3. Havlicek, J.P.; McKeeman, J.C.; Remaklus, P.W. Networks of Low-Earth Orbit Store-and-Forward Satellites. *IEEE Trans. Aerosp. Electron. Syst.* **1995**, *31*, 543–554. [CrossRef]
4. Perumal, S.; Raman, V.; Samy, G.N.; Shanmugam, B.; Kisenasamy, K.; Ponnan, S. Comprehensive Literature Review on Delay Tolerant Network (DTN) Framework for Improving the Efficiency of Internet Connection in Rural Regions of Malaysia. *Int. J. Syst. Assur. Eng. Manag.* **2022**, *13*, 764–777. [CrossRef]
5. Tariq, N.; Asim, M.; Al-Obeidat, F.; Farooqi, M.Z.; Baker, T.; Hammoudeh, M.; Ghafir, I. The Security of Big Data in Fog-Enabled Iot Applications Including Blockchain: A Survey. *Sensors* **2019**, *19*, 1788. [CrossRef]
6. Sweeting, M.N. The University of Surrey UoSAT-2 Spacecraft Mission. *J. Inst. Electron. Radio Eng.* **1987**, *57*, S99. [CrossRef]
7. Centenaro, M.; Costa, C.E.; Granelli, F.; Sacchi, C.; Vangelista, L. A Survey on Technologies, Standards and Open Challenges in Satellite IoT. *IEEE Commun. Surv. Tutor.* **2021**, *23*, 1693–1720. [CrossRef]
8. ITU. *ITU 2021 Measuring Digital Development: Facts and Figures*; ITU: Geneva, Switzerland, 2021; ISBN 978-92-61-35401-5.
9. World Bank. *World Bank Modernizing Weather, Water, and Climate Services: A Road Map for Bhutan*; World Bank: Washington, DC, USA, 2018; pp. 1 88.
10. Nsabagwa, M.; Byamukama, M.; Kondela, E.; Otim, J.S. Towards a Robust and Affordable Automatic Weather Station. *Dev. Eng.* **2019**, *4*, 100040. [CrossRef]
11. Nsubuga, F.W.; Rautenbach, H. Climate Change and Variability: A Review of What Is Known and Ought to Be Known for Uganda. *Int. J. Clim. Chang. Strateg. Manag.* **2018**, *10*, 752–771. [CrossRef]
12. Chippalkatti, V.S.; Biradar, R.C. Review of Satellite Based Internet of Things and Applications. *Turk. J. Comput. Math. Educ.* **2021**, *12*, 758–766.
13. Suematsu, N.; Kameda, S.; Oguma, H.; Sasanuma, M.; Eguchi, S.; Kuroda, K. Multi-Mode Portable VSAT for Disaster-Resilient Wireless Networks. In Proceedings of the Asia-Pacific Microwave Conference, Sendai, Japan, 4–7 November 2014; IEEE: Sendai, Japan, 2014; pp. 549–551.
14. Wang, A.; Wang, P.; Miao, X.; Li, X.; Ye, N.; Liu, Y. A Review on Non-Terrestrial Wireless Technologies for Smart City Internet of Things. *Int. J. Distrib. Sens. Networks* **2020**, *16*, 155014772093682. [CrossRef]
15. Raza, U.; Kulkarni, P.; Sooriyabandara, M. Low Power Wide Area Networks: An Overview. *IEEE Commun. Surv. Tutor.* **2017**, *19*, 855–873. [CrossRef]
16. Kodheli, O.; Maturo, N.; Andrenacci, S.; Chatzinotas, S.; Zimmer, F. Link Budget Analysis for Satellite-Based Narrowband IoT Systems. In Proceedings of the International Conference on Ad-Hoc Networks and Wireless, Luxembourg, 1–3 October 2019; Springer: Cham, Switzerland, 2019; pp. 259–271.
17. Kodheli, O.; Andrenacci, S.; Maturo, N.; Chatzinotas, S.; Zimmer, F. An Uplink UE Group-Based Scheduling Technique for 5G MMTC Systems over LEO Satellite. *IEEE Access* **2019**, *7*, 67413–67427. [CrossRef]
18. Demetri, S.; Zúñiga, M.; Picco, G.P.; Kuipers, F.; Bruzzone, L.; Telkamp, T. Automated Estimation of Link Quality for Lora: A Remote Sensing Approach. In Proceedings of the 18th ACM/IEEE International Conference on Information Processing in Sensor Networks (IPSN), Montreal, QC, Canada, 16 18 April 2019; pp. 145–156.
19. Doroshkin, A.A.; Zadorozhny, A.M.; Kus, O.N.; Prokopyev, V.Y.; Prokopyev, Y.M. Experimental Study of LoRa Modulation Immunity to Doppler Effect in CubeSat Radio Communications. *IEEE Access* **2019**, *7*, 75721–75731. [CrossRef]

20. Fernandez, L.; Ruiz-De-Azua, J.A.; Calveras, A.; Camps, A. Assessing LoRa for Satellite-to-Earth Communications Considering the Impact of Ionospheric Scintillation. *IEEE Access* **2020**, *8*, 165570–165582. [CrossRef]

21. Wu, T.; Qu, D.; Zhang, G. Research on LoRa Adaptability in the LEO Satellites Internet of Things. In Proceedings of the 15th International Wireless Communications and Mobile Computing Conference, IWCMC, Tangier, Morocco, 24–28 June 2019; IEEE: Piscataway, NJ, USA, 2019; pp. 131–135.

22. Anantachaisilp, P.; Muangkham, M.; Punpigul, N.; Thammawichai, M. Store and Forward CubeSat Using LoRa Technology and Private LoRaWAN-Server. In Proceedings of the 34th Annual Small Satellite Conference, Online, 1–6 August 2020; Utah State University: Logan, UT, USA, 2020.

23. Fraire, J.A.; Henn, S.; Dovis, F.; Garello, R.; Taricco, G. Sparse Satellite Constellation Design for LoRa-Based Direct-to-Satellite Internet of Things. In Proceedings of the 2020 IEEE Global Communications Conference, GLOBECOM 2020, Taipei, Taiwan, 8–10 December 2020; pp. 1–6.

24. Verspieren, Q.; Obata, T.; Nakasuka, S. Innovative Approach to Data Gathering in Remote Areas Using Constellations of Store & Forward Communication Cubesats. In Proceedings of the 31st International Symposium on Space Technology and Science (ISTS), Matsuyama, Japan, 3–9 June 2017; pp. 1–6.

25. Verspieren, Q.; Matsumoto, T.; Aoyanagi, Y.; Fukuyo, T.; Obata, T.; Nakasuka, S.; Kwizera, G.; Abakunda, J. Store-and-Forward 3U Cube Sat Project Tricom and Its Utilization for Development and Education: The Cases of TRICOM-1R and JPRWASAT. *Trans. Jpn. Soc. Aeronaut. Space Sci.* **2020**, *63*, 206–211. [CrossRef]

26. Marziolia, P.; Frezzaa, L.; Curianòa, F.; Amadioa, D.; Mwanikib, C.; Makindib, S.; Okellob, C.; Jahjahc, M.; Toninellic, M.; Santoni, F. The WildTrackCube-SIMBA CubeSat: Italian-Kenyan Mission for Wildlife Monitoring. In Proceedings of the 72nd International Astronautical Congress (IAC), Dubai, United Arab Emirates, 25–29 October 2021; International Astronautical Federation (IAF): Dubai, United Arab Emirates, 2021.

27. Fossa Systems Global IoT Connectivity. Available online: https://fossa.systems/es/home-spanish/ (accessed on 25 October 2021).

28. Kim, S.; Yamauchi, T.; Masui, H.; Cho, M. BIRDS BUS: A Standard CubeSat BUS for an Annual Educational Satellite Project. *J. Small Satell.* **2021**, *10*, 1015–1034.

29. Van't Hof, J.; Karunanithi, V.; Speretta, S.; Verhoeven, C.; McCune, E. Low Latency IoT/M2M Using Nano-Satellites. In Proceedings of the 70th International Astronautical Congress (IAC), Washington DC, USA, 21–25 October 2019; International Astronautical Federation (IAF): Washington, DC, USA, 2019.

30. Qu, Z.; Zhang, G.; Cao, H.; Xie, J. LEO Satellite Constellation for Internet of Things. *IEEE Access* **2017**, *5*, 18391–18401. [CrossRef]

31. Zhao, M.C.; Lei, J.; Wu, M.Y.; Liu, Y.; Shu, W. Surface Coverage in Wireless Sensor Networks. In Proceedings of the IEEE INFOCOM, Rio de Janeiro, Brazil, 19–25 April 2009; IEEE: Piscataway, NJ, USA, 2009; pp. 109–117, ISBN 9781424435135.

32. Wang, B.; Lim, H.B.; Ma, D. A Survey of Movement Strategies for Improving Network Coverage in Wireless Sensor Networks. *Comput. Commun.* **2009**, *32*, 1427–1436. [CrossRef]

33. Vasisht, D.; Chandra, R. A Distributed and Hybrid Ground Station Network for Low Earth Orbit Satellites. In HotNets 2020—Proceedings of the 19th ACM Workshop on Hot Topics in Networks, Chicago, IL, USA, 4–6 November 2020; pp. 190–196. [CrossRef]

34. Leyva-Mayorga, I.; Soret, B.; Popovski, P. Inter-Plane Inter-Satellite Connectivity in Dense LEO Constellations. *IEEE Trans. Wirel. Commun.* **2021**, *20*, 3430–3443. [CrossRef]

35. Kojima, A.; Yárnoz, D.G.; Di Pippo, S. Access to Space: A New Approach by the United Nations Office for Outer Space Affairs. *Acta Astronaut.* **2018**, *152*, 201–207. [CrossRef]

36. Tondo, F.A.; Montejo-Sánchez, S.; Pellenz, M.E.; Céspedes, S.; Souza, R.D. Direct-to-Satellite IoT Slotted Aloha Systems with Multiple Satellites and Unequal Erasure Probabilities. *Sensors* **2021**, *21*, 7099. [CrossRef]

37. Zhou, H.; Liu, L.; Ma, H. Coverage and Capacity Analysis of LEO Satellite Network Supporting Internet of Things. In Proceedings of the IEEE International Conference on Communications, Shanghai, China, 20–24 May 2019; Volume 2019–May.

38. Semtech Corporation. *Low Power Long Range Transceiver Datasheet SX1276/77/78/79—137 MHz to 1020 MHz*; Semtech Corporation: Camarillo, CA, USA, 2016.

39. Lepcha, P.; Kim, S.; Masui, H.; Cho, M. Application of Small Satellites for Low-Cost Remote Data Collection Using LoRa Transmitters. *Trans. Jpn. Soceity Aeronaut. Sp. Sci. Aerosp. Technol.* **2021**, *19*, 224–230. [CrossRef]

40. *International Electrotechnical Commission Degrees of Protection Provided by Enclosures (IP Code) (IEC 60529: 1989+ A1: 1999+ A2: 2013)*; International Electrotechnical Commission: Geneva, Switzerland, 2013.

41. Jirawattanaphol, A.; Kurahara, N.; Cho, M. Design and Development of Ground Station Network for CubeSats Constellation, Joint Global Multi-Nation BIRDS. In Proceedings of the 60th Joint Conference on Space Science and Technology, Hakodate, Japan, 6–9 September 2016; p. 6.

42. Boano, C.A.; Cattani, M.; Römer, K. Impact of Temperature Variations on the Reliability of LoRa an Experimental Evaluation. In Proceedings of the SENSORNETS 2018—Proceedings of the 7th International Conference on Sensor Networks, Funchal, Portugal, 22–24 January 2018; Volume 2018-January.

43. Souza Bezerra, N.; Åhlund, C.; Saguna, S.; de Sousa, V.A. Temperature Impact in LoRaWAN—A Case Study in Northern Sweden. *Sensors* **2019**, *19*, 4414. [CrossRef]

Article

Performance Evaluation of Machine Learning Methods for Anomaly Detection in CubeSat Solar Panels

Adolfo Javier Jara Cespedes *, Bramandika Holy Bagas Pangestu, Akitoshi Hanazawa and Mengu Cho

Laboratory of Lean Satellite Enterprises and In-Orbit Experiments (LaSEINE), Kyushu Institute of Technology, Kitakyushu 8048550, Japan
* Correspondence: javier.jara-cespedes758@mail.kyutech.jp

Abstract: CubeSat requirements in terms of size, weight, and power restrict the possibility of having redundant systems. Consequently, telemetry data are the primary way to verify the status of the satellites in operation. The monitoring and interpretation of telemetry parameters relies on the operator's experience. Therefore, telemetry data analysis is less reliable, considering the data's complexity. This paper presents a Machine Learning (ML) approach to detecting anomalies in solar panel systems. The main challenge inherited from CubeSat is its capability to perform onboard inference of the ML model. Nowadays, several simple yet powerful ML algorithms for performing anomaly detection are available. This study investigates five ML algorithm candidates, considering classification score, execution time, model size, and power consumption in a constrained computational environment. The pre-processing stage introduces the windowed averaging technique besides standardization and principal component analysis. Furthermore, the paper features the background, bus system, and initial operational data of BIRDS-4, a constellation made of three 1U CubeSats released from the International Space Station in March 2021, with a ML model proposal for future satellite missions.

Keywords: anomaly detection; BIRDS project; CubeSat; machine learning; electrical power system

Citation: Cespedes, A.J.J.; Pangestu, B.H.B.; Hanazawa, A.; Cho, M. Performance Evaluation of Machine Learning Methods for Anomaly Detection in CubeSat Solar Panels. *Appl. Sci.* **2022**, *12*, 8634. https://doi.org/10.3390/app12178634

Academic Editors: Simone Battistini, Filippo Graziani and Mauro Pontani

Received: 19 July 2022
Accepted: 25 August 2022
Published: 29 August 2022

Publisher's Note: MDPI stays neutral with regard to jurisdictional claims in published maps and institutional affiliations.

1. Introduction

The development of small satellites by non-space-faring nations is significantly driven by the availability of low-cost launch and commercial off-the-shelf (COTS) components. A clear example is the Joint Global Multi-National Birds or BIRDS program, a multinational small satellite research and educational program led by the Kyushu Institute of Technology (Kyutech) in Japan [1]. The BIRDS program allows non-space-faring nations to design, integrate, build, test, launch, and operate their respective first satellites.

BIRDS-4 is the fourth iteration of the BIRDS program (previously BIRDS-1, 2, and 3). It is a constellation of three CubeSats: GuaraniSat-1, Maya-2, and Tsuru, deployed into orbit on 14 March 2021 [2]. The satellites use the BIRDS standardized bus, an open-source initiative by Kyutech designed for educational CubeSat projects inherited from the previous generation [3].

The BIRDS program follows the approach of a Lean Satellite concept [3]. The concept relies on utilizing commercially available yet non-space-proven components to obtain effective and efficient development [4]. On the other hand, the mass and size limitations usually indicate only a few or no redundant systems available [5]. The intermittent and short-term communication window limits the data transmission capability, potentially affecting the housekeeping-data-monitoring analysis. Furthermore, the CubeSat system's limitations generally include power generation, telemetry bandwidth, computational power, and memory [6].

The BIRDS team uses conventional threshold values to perform telemetry analysis for the satellite's health monitoring. However, considering the available bandwidth and large volumes of data collected onboard the satellite, it is impossible to download all the data during the operational phase. In addition, data monitoring requires a team with sufficient knowledge and experience due to the complex variations in telemetry patterns.

According to the BIRDS-3 telemetry data, two satellites revealed power generation loss in one of their panels. The team could not immediately detect the symptoms within 29 (on Raavana) and 72 (on Uguisu) days after being deployed into orbit. Recently, the BIRDS-4 team discovered a critical issue related to the electrical power subsystem of GuaraniSat-1 [2]. The satellite stopped transmitting after three days of operation. Subsequent analysis of the continuous wave (CW) beacon data showed a lower charging capacity than its sister satellite, Tsuru, confirming no power generation on two solar panels. Therefore, there is an urgent need to develop solutions to mitigate the issue from any possible technical approach. One of the possible approaches would be utilizing Machine Learning (ML), considering the substantial amount of data collected by the BIRDS team.

The analysis for related research comes from these three main aspects: pre-processing techniques, proximity-based algorithms, and linear models. As a representative case, the work presented by Wu J. et al. [7] used Long Short-Term Memory (LSTM) and Ensembled One-Class Support Vector Machines to perform anomaly detection. Multiple One-Class Support Vector Machines were used to obtain high-precision and high-recall outputs. The method was evaluated using the telemetry data of the SMAP (Soil Moisture Active Passive) satellite and MSL (Mars Science Laboratory). Jin W. et al. [8] proposed a cluster-based anomaly threshold determination method in another paper. Experiments were conducted on satellite telemetry data, showing that the proposed method outperformed the autoencoders.

Probabilistic clustering approaches, such as the work presented by Yairi T. et al. [9], are also intensely interesting. In this work, the authors proposed a data-driven health monitoring and anomaly detection method for artificial satellites based on probabilistic dimensionality reduction and clustering. The proposed method was experimentally applied to Japan Aerospace Exploration Agency (JAXA) housekeeping data of Small Demonstration Satellite 4 (SDS-4) and validated over two years.

Finally, different ML-based approaches have also received considerable attention for anomaly detection. For example, techniques based on Principal Component Analysis (PCA), such as the technique presented by Zamry N. et al. [10], have been used to improve efficiency by reducing the computational complexity and improving memory utilization overhead while maintaining high accuracy. The works presented by Pan D. et al. [11], Peng Y. et al. [12], and Li J. et al. [13] also used PCA for feature extraction and fault detection in telemetry data. The k-nearest neighbor (kNN), Recurrent Neural Networks (RNN), autoencoder, and One-Class Support Vector Machine (OC-SVM) have also been studied with spacecraft (Suzaku) electrical power system data [14].

The term "anomaly" refers to any point or value within the dataset that produces patterns of oddity, novelty, fault, deviation, or exceptions. Hence, the activity of identifying anomalies from normal data is known as anomaly detection.

This paper presents a comparative study of different ML techniques to detect anomalies in CubeSat telemetry data, considering system limitations such as computational power, memory footprint, and communication window. The approach considers both pre-processing methods and ML models. We applied the process to a solar panel's dataset generated by BIRDS-3 (NepaliSat, Raavana, Uguisu) and BIRDS-4 (Tsuru) satellites. It is important to note that this is the first utilization of such data. In summary, the contributions of this work are as follows:

- Analysis of the novel solar panel dataset collected from four CubeSats.
- Analysis of five different ML models based on their classification scores, execution times, model sizes, and power consumption.
- The proposal of ML model candidates for solar panel anomaly detection on CubeSat systems.

The contributions of this work to the research community are the findings showing that specific ML models, i.e., linear models, are most suitable for solar panel anomaly detection onboard CubeSats, given the lack of attitude control and constrained computational capability of future satellite projects. The rest of the paper is organized in the following way: Section 2 introduces the satellite system and the dataset. Section 3 describes the methods utilized. Section 4 provides the results obtained. Section 5 presents analysis and discussion regarding the results of the experiments on ML modelling and future works. Finally, Section 6 summarizes the outcome of the research and proposes the best model to detect the anomalies within the dataset.

2. Materials

2.1. BIRDS Satellite System

The satellites of the BIRDS-3 and BIRDS-4 constellations are based on the 1U CubeSat standard. The external dimensions are (114.5 × 107.8 × 104) mm, and they have an average mass of 1.3 kg. The bus system consists of the front access board (FAB), the electrical power system (EPS), the onboard computer (OBC), the communication board (COM), and the rear access board (RAB). All printed circuit boards (PCBs) are connected to a backplane board via 50-pin connectors. Figure 1 shows the bus system configuration.

Figure 1. Bus system configuration.

The RAB is used for programming and monitoring each mission's microcontrollers. The FAB is used for programming and monitoring the Main PIC, FAB PIC, RESET PIC, and COM PIC microcontrollers. The details of the satellite system can be found in the BIRDS Open-Source Repository [15].

In BIRDS-3, two available mission boards are utilized to space-qualify a low-powered long range (LoRa) modulation module, which has massive potential for future S&F missions. This mission is called a LoRa Demonstration Mission (LDM) [16].

On the other hand, the BIRDS-4 constellation uses the mission boards for the nine satellite missions. The satellite structure was renewed by taking the BIRDS structure as a base to satisfy the requirements of the HNT mission [2,17]. The downlink frequency is 437.375 MHz, whereas APRS-DP and store-and-forward missions use the VHF band.

Through the uplink signal sent by the ground station, the satellite receives, processes, and downlinks telemetry and mission data back to the ground station. The satellites use dipole antennas for communication. The antennas are deployed 30 min after satellite deployment. The uplink is performed through the BIRDS Ground Station Network, which has 13 participating countries worldwide, except in Europe and North America.

Both constellations share the same EPS design and components. Solar panels placed on five sides of the CubeSat (+X, +Y, +Z, −X, −Z) are responsible for power generation. The solar panel consisting of two cells connected in series is shown in Figure 2.

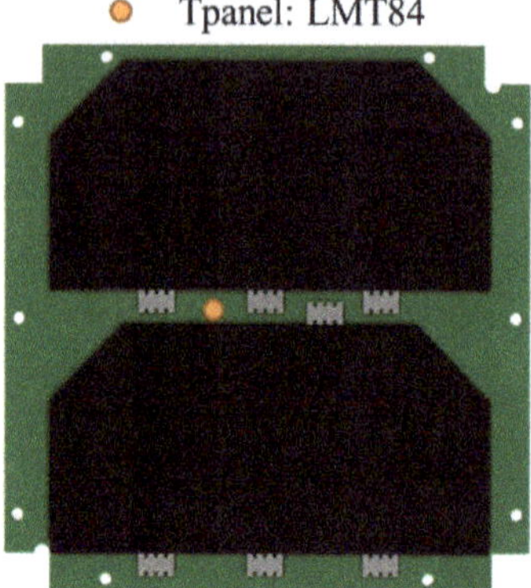

Figure 2. Solar panel with two cells connected in series.

The specific solar cell (AZURESPACE 3G30A) has an ideal maximum power generation of 1.2 W. The BIRDS BUS uses Linear Technology Corporation's LTC3119 as the Battery Charge Regulator (BCR), due to its ability to control the maximum power point of solar cells. The energy generated is stored in six rechargeable Eneloop NiMH batteries with a capacity of 1900 mAh per battery arranged in a 3-series 2-parallel configuration. The total power generation per orbit is 2600 mWh, and the total power available to the satellite load is 1600 mWh. The power system has two 3.3 V lines, one 5 V line, and two unregulated lines. All of the EPS's vital components are illustrated in Figure 3.

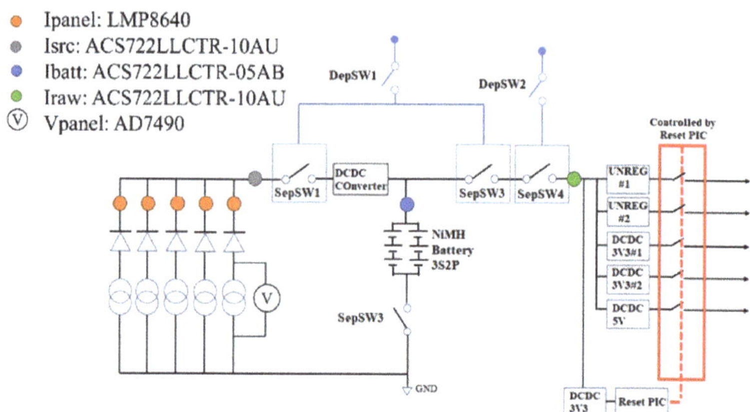

Figure 3. EPS block diagram with power distribution.

2.2. Dataset Overview

The dataset contains solar panel data of Tsuru satellite (from March 2021 to November 2021). It includes 16,704 temperature, current, and voltage data points from the satellite panels. Additionally, the dataset consists of 8439 samples from Nepalisat, Raavana, and Uguisu Satellites (from June 2019 to October 2020). Please refer to TSURU.xlsx, NEPAL-ISAT.xlsx, RAAVANA.xlsx and UGUISU.xlsx in Supplementary Materials section. The satellites use the sensors described in Table 1.

Table 1. Descriptions of solar panel sensors.

Sensor	Variable	Unit	Data Size
LMP8640	Current	mA	8 bits
AD7490	Voltage	mV	12 bits
LMT84	Temperature	°C	12 bits

The BIRDS team operators periodically execute the High Sampling Sensors Data Collection Mission (HSSC) to obtain the data during one orbit. Subsequently, the team downloads housekeeping data to the ground station. The sampling rate is 10 s for the Tsuru satellite and 5 s for Nepalisat, Raavana, and Uguisu Satellites.

2.3. Data Exploration and Pre-Processing

Before detecting outliers in the data, it is essential to understand how the satellite's orbit and attitude affect the panel's voltage, current, and temperature. The BIRDS-4 constellation was deployed into orbit from the ISS (altitude 400 km, inclination: 51.6°, duration: 92.6 min). Since no attitude control is applied, the satellites are in free rotation at approximately 3 °/s on each axis. Therefore, the panel condition is affected by two factors: the beta angle that determines the eclipse and sunlight periods and the attitude perturbation of the satellite. In Figure 4a, at a low beta angle we can observe significant temperature changes (~48 °C) due to the transition of the satellite from eclipse to sunlight, and a slight temperature variance (~4 °C) due to the satellite's rotation on its axes. In Figure 4b, at a high beta angle, we observe temperature variation (~30 °C) only due to the satellite's rotation.

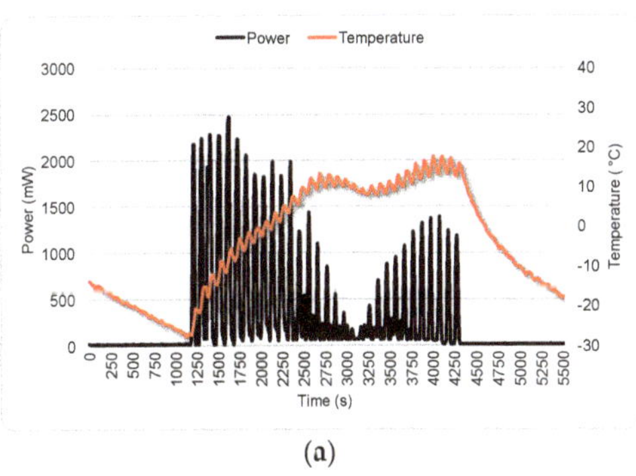

(a)

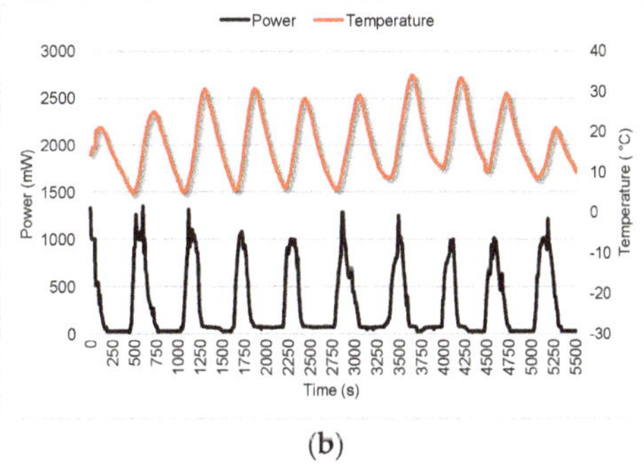

(b)

Figure 4. One orbit +Y panel power generation and temperature: (**a**) beta angle 15°; (**b**) beta angle 70°.

In addition, at a low beta angle, the power requirement from the satellite is high right after the eclipse period, so the power generation is high. However, at a high beta angle, since the satellite does not experience an eclipse throughout the orbit, the power requirement is less and so the generation is also less. The generated power of each panel shows large fluctuations due to the satellite's attitude (i.e., rotation), which causes different sides of the satellite to face the sun at different instants.

2.4. Dataset Correlation

We can find a correlation among the solar panel's parameters by analyzing the telemetry data. Figure 5a shows the correlation between the voltage and the temperature of the +Y panel when it faces the sun for approximately 90 s at varying angles while rotating. Figure 5b shows the correlation between the current and the temperature of the same panel for the same time window. A slight delay can be observed in the current graph compared to the voltage graph. Significant current only flows for approximately 40 s because of a specific potential difference required in the circuit between the panel output and the input of the DC/DC converter for the current to start flowing.

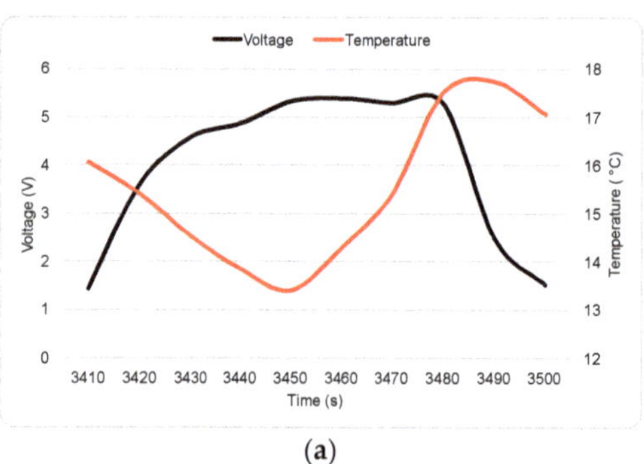

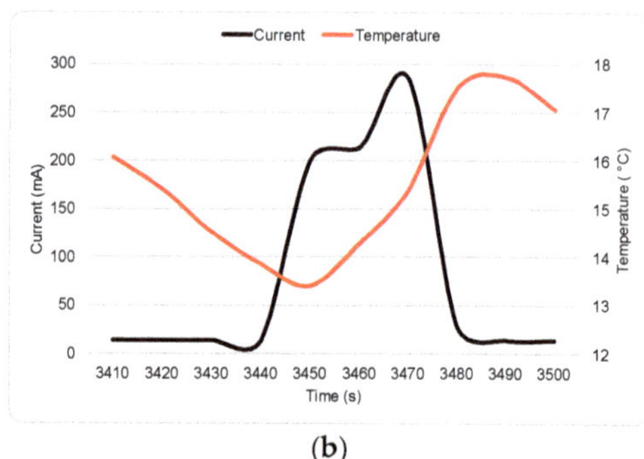

(a)   (b)

Figure 5. +Y panel parameters correlation: (**a**) voltage vs. temperature, (**b**) current vs. temperature.

Due to the photovoltaic characteristic of the solar cell, the voltage and current readings are more sensitive to the change in light intensity. However, for the temperature reading, due to the heat transfer from the solar cell to the temperature sensor located behind the panel, a delay is observed for the change in temperature as compared to instant change in voltage and current during the transition from eclipse to sunlight. Hence, the satellite's temperature readings were not utilized in this study due to the low correlations of the trend with the current and voltage samples. From here on, the dataset comprised ten dimensions of current and voltage telemetry data from each panel.

3. Methods

3.1. Anomaly Definition

This section presents is a critical step in the dataset's preparation to define anomalies. The process determines the scope of anomalies investigated in the study. The definition of anomalous and normal data is based on the analysis results of the operational satellite team. The analysis generally depends on the satellite engineering model, on-orbit condition, and several previous tests before launch. Moreover, the definition is a reference for the subsequent process of dataset labeling. Labeling is a manual process of classifying the data points as normal (label: 0) or anomalous (label: 1). According to the telemetry data analysis, two types of anomalies are happening in the solar panel system of BIRDS satellites: solar panel failure (more than one cell) and solar cell failure, which are described in detail in the following subsections.

The anomalous points are marked in red, and the regular points are in black. The data point is marked as an anomaly if it relates to the conditions of a solar panel or solar cell failure.

3.1.1. Type 1: Solar Panel Failure

Solar panel failure (or type 1) specifies an anomalous condition wherein the voltage sensor reads a value below an expected threshold of 1200 mV on one or more panels at any given observation time. The expected lowest values of the sensor readings for a non-illuminated solar panel must be within 1200 to 1500 mV. The numbers refer to the residual voltage generated by the panel system's Analog to Digital Converter (ADC) circuit. Any data below the threshold indicate a bad voltage condition for a panel of two solar cells connected in series. The team specified those expected values from the operational telemetry data of the BIRDS-3 satellite (Raavana). Thus, the dataset is labeled with the following condition:

$$V_i(t) < 1200 \text{ mV}, \tag{1}$$

with $V_i(t)$ being the values of voltage sensor readings at samples of t and for i panel. Correspondingly, Figure 6 presents the voltage and current sensor readings in both normal (black) and anomalous (red) sections for type 1 anomalies by applying (1). It covers approximately 2.5 orbital periods. Type 1 anomalies in the dataset contaminate 8.59% of the total sample.

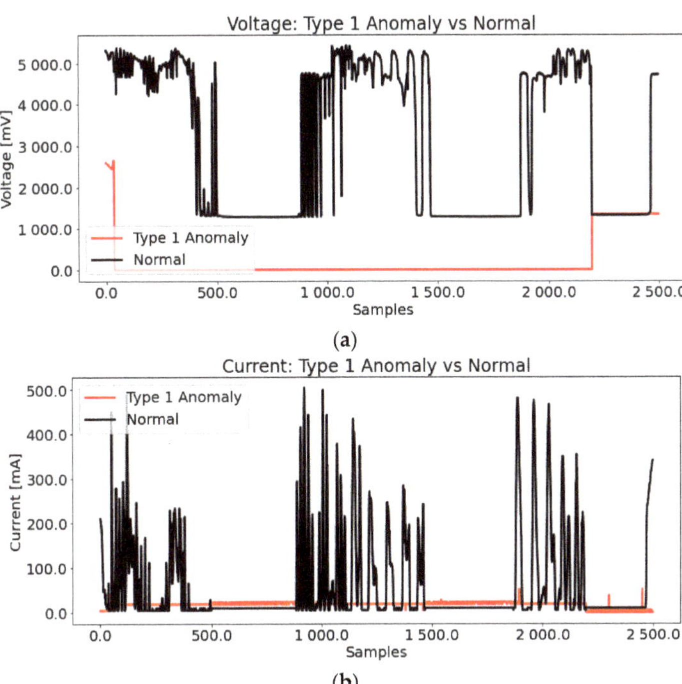

Figure 6. Type 1 anomalies in (**a**) voltage and (**b**) current data.

3.1.2. Type 2: Solar Cell Failure

Solar cell failure (or type 2) characterizes the anomaly of a compromised cell in a panel (of any panels). This specific failure is only noticeable during the illumination period. Based on the actual telemetry data of the BIRDS-3 satellite (Uguisu), the voltage sensor reading shows consecutive values (at least three samples) within 2000 to 3000 mV with no expected electrical current reading (<50 mA). Analytically, one solar cell in the Uguisu satellite is subject to an anomaly condition. However, this particular condition can happen instantly in the transition period of entering and exiting the eclipse. Therefore, the labeling method considers the previous and the following data points in a sequence. We define such anomaly as

$$2000\,\text{mV} \leq V_i(t-1,\, t,\, t+1) \leq 3000\,\text{mV} \wedge I_i(t) < 50\,\text{mA}, \tag{2}$$

with $V_i(t)$ and $I_i(t)$ represent the value of voltage and current sensor readings in a sample of t and for panel i, respectively. Accordingly, Figure 7 depicts the voltage and current sensor readings in both normal (black) and anomalous (red) sections for type 2 anomaly by applying (2). It covers approximately 2.5 orbital periods. Type 2 anomalies in the dataset contaminate 7.22% of the samples within the dataset.

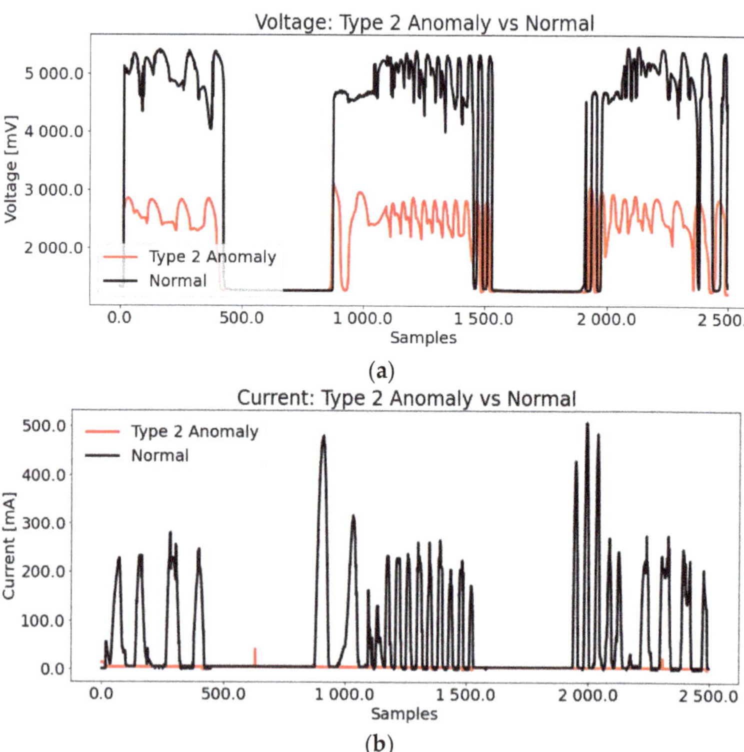

Figure 7. Type 2 anomalies in (**a**) voltage and (**b**) current data.

3.2. Pre-Processing Techniques

Pre-processing techniques attempt to extract and enhance specific features (e.g., patterns, structure, trends) within the dataset sample. In the non-sequential analysis, the model expects this stage to increase the variance between classes in the dataset. Hence, the output forms distinctive features among classes. Implicitly, it allows the ML models to distinguish the samples. The process begins with data pre-processing using windowed averaging, standardization, and Principal Component Analysis. All three techniques were applied to the input samples that were the output of the dataset preparation phase. The analysis used two telemetry data, namely, current and voltage. Figure 8 shows the initial data distribution patterns for the voltage parameter before applying the pre-processing techniques. The figure indicates the different sample distributions of the anomalous and normal data classes. Through visual observation, we can clearly distinguish the two categories. However, it is difficult to identify the anomaly class if the analysis is based on simple thresholding. It can be seen that the anomalous patterns are located in the center of the vertical axis. However, the problem with simple thresholding is that the process may incorrectly categorize a sample in the transition state as an anomaly, especially a type 2 anomaly. Thus, a data pre-processing step is crucial to extract the anomalous patterns so that the ML model has distinguishable input samples.

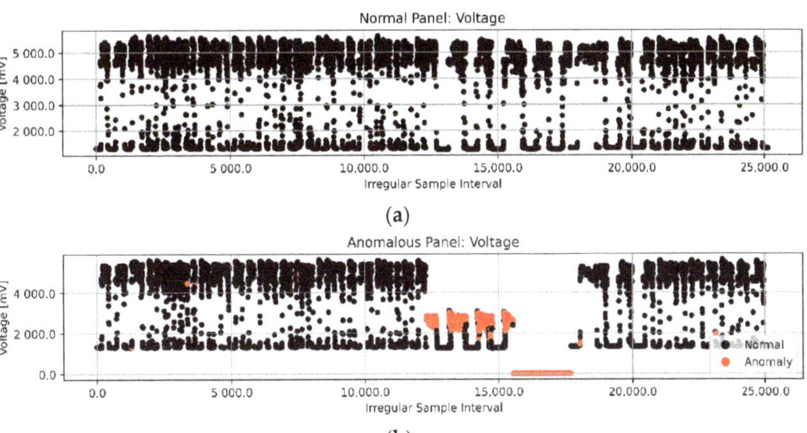

(a)

(b)

Figure 8. Initial voltage data plots: (**a**) normal and (**b**) anomalous.

3.2.1. Windowed Averaging

We propose a unique statistical method to enhance local or temporary (specific interval) trends within the dataset's parameters. Despite its utilization for non-sequential analysis, the technique introduces sequential information from the local trend intervals. It calculates an average value (average window, $\overline{X_{a,k}}$) for a particular interval (window size, a) in data points. Then, it replaces all values within the specific estimated window. Applying those processes for the rest of the data points creates isolated window size groups for the entire dataset. We define the required iteration (b) for any given number (n) of data points as

$$b = \left\lceil \frac{n}{a} \right\rceil ,\tag{3}$$

and formulate the average window calculation as

$$\overline{X_{a,k}} = \begin{cases} \dfrac{\sum_i^{ak} X_i}{a}, & k < b \\[2mm] \dfrac{\sum_i^{ak} X_i}{n - a(b-1)}, & k = b \end{cases},\tag{4}$$

with k representing the iteration index of a specific data point and X a variable of interest within the dataset. For each iteration, we define the beginning of variable index i as

$$i = a(k-1) + 1 .\tag{5}$$

The results of feature extraction by windowed averaging cause rapid fluctuations in data changes to be isolated for a specific interval. Figure 9 illustrates how the windowed averaging process results in a voltage dataset from an anomaly-contaminated panel by applying (3)–(5). We decided on the value of 50 as an optimal value of the window interval for the windowed averaging technique.

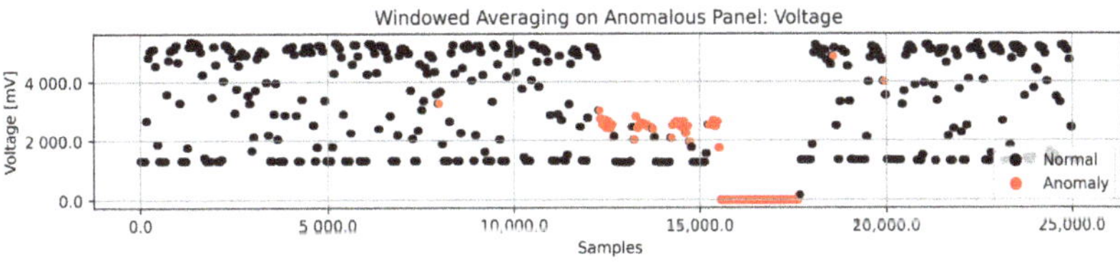

Figure 9. Initial voltage data plot after windowed averaging.

3.2.2. Standardization

Z-score standardization or standardization is a method to standardize a set of data based on its global mean distance relative to all samples to balance the analysis. It begins with computing the global mean value for each parameter in the dataset. Then, it replaces all initial values with the differences between the global mean and the initial values. In this study, we simplified the dataset distribution according to Z-score to balance the analysis based on the global mean value of the dataset. It was applied directly after the windowed averaging technique. We used the standardization implemented by the sklearn library [18]. The method was implemented based on the z-score calculation [19].

3.2.3. Principal Component Analysis

Principal Component Analysis, or PCA, is the most common dimensionality reduction technique [20]. PCA discovers the directions of the highest variance from higher-dimensional data and projects them onto a lower-dimensional subspace without losing much information. In practice, this method can reduce the number of parameters fed into the ML model without sacrificing necessary details. The lowest-dimensional projection of the PCA is known as the Principal Component (PC). Therefore, the first PC, or PC1, represents the axis consisting of the highest-variance projection. In contrast, the following PCs (PC2, PC3, and so forth) are the axis directions perpendicular to all previous PCs. We implemented PCA using the sklearn library, following the procedure in previous research [18,21].

We applied the PCA technique to the whole dataset after windowed averaging and standardization. It reduced the ten dimensions of the original dataset into two dimensions. Subsequently, the PCA-transformed data were fed into the ML modelling process.

3.3. Model Candidates

We utilized ML algorithms or models capable of classifying values in an instance without considering trends or values from the previous measurement. The implemented code for all ML models is publicly available from sklearn and PyOD library [18,22]. In this work, the models originated from two categories: proximity-based algorithms and linear models. Typically, proximity-based algorithms classify data by computing the distances between samples to compare each sample's relative distance to its surroundings (e.g., local outlier factor, cluster-based local outlier factor, and k-nearest neighbor). On the other hand, the linear models learn to formulate a mathematical function to create class-based boundaries within the dataset (e.g., Linear Discriminant Analysis and One-Class Support Vector Machine). The section focuses on brief descriptions of the investigated models without going in-depth into the models' technical details. Moreover, we intended to discover the best of the two ML categories for future CubeSat system implementation.

3.3.1. Proximity-Based Algorithms

Proximity-based algorithms use proximity or distance in hyperspace to classify each data point. It estimates the distance of one data point relative to its surroundings. Thus, the proximate sample number of surroundings is crucial for the result.

A.   Local Outlier Factor

Local Outlier Factor (LOF) is one of the first famous local anomaly detection algorithms [23]. The LOF is essentially a ratio of local densities. It examines the local density of any sample in relation to its neighbor. Local means it depends on the object's isolation in its surroundings. It considers a significantly lower density than others as an outlier [24].

B.   Cluster-Based Local Outlier Factor

Cluster-Based Local Outlier Factor (CBLOF) adopts a clustering algorithm (e.g., k-Means) to determine the area density in the dataset. Afterwards, it performs a density estimation for each cluster. The size of the cluster and the distance to the nearest large cluster define the anomaly score [25].

C. K-Nearest Neighbor

K-nearest neighbor (kNN), in the context of anomaly detection, is an algorithm that measures the distance of a data point to its kth nearest neighbor as the anomaly score [26]. We used this approach to measure the density of a dataset probability distribution. The value of the anomaly score depends on the dataset, the number of dimensions, and its normalization. As a result, it is challenging to select an appropriate value for the nearest neighbor.

3.3.2. Linear Model

The ensuing algorithm learns a decision function from the dataset to create a boundary for each class. It generates a formula to create a function to predict unknown values. The linear model offers a relatively quick training phase and is straightforward to interpret [20].

A. Linear Discriminant Analysis

Linear Discriminant Analysis (LDA) is a classifier with a linear decision function to set the boundary for each class [27]. It creates a function from the conditional densities of the labelled dataset using Bayes' rule. LDA assumes the dataset has a Gaussian distribution and each parameter has the same variance. The algorithm estimates the mean and variance of the dataset for each class.

B. One-Class Support Vector Machine

The One-Class Support Vector Machine (OC-SVM) is an ML algorithm intended to detect a novelty, i.e., a rare event. OC-SVM does not model any probability distribution from the dataset. It learns to find a function for the high- and low-density regions in the dataset based on max-margin methods. A lower density region implies a rare event, or a novelty, within the dataset [28].

3.4. Experimental Setup

We performed experiments based on a computer simulation to investigate the performances of the ML algorithm candidates. The simulation followed the methodology illustrated in Figure 10.

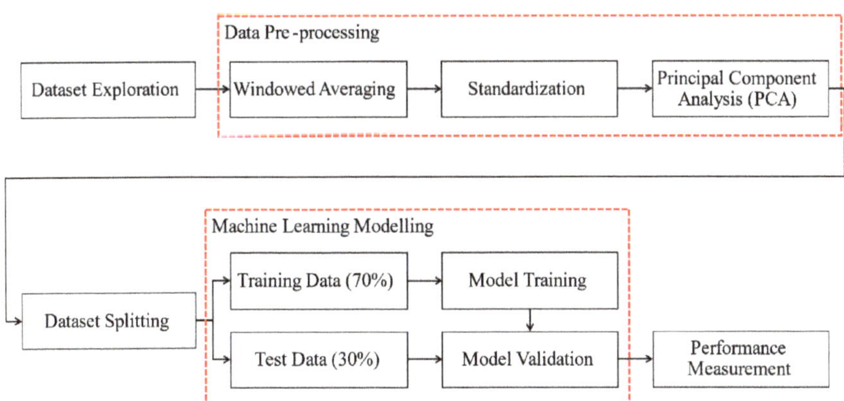

Figure 10. Experiment methodology.

First, we explored variables within the dataset for their correlations, characteristics, and anomaly types. Then, we manually categorized each observation point according to the anomaly definition for evaluation purposes. We applied some pre-processing techniques to the entire dataset using the following steps. By this point, we expected the data formed a distinct pattern or region density for each class. Afterwards, we randomly split the dataset into a training set (70%) and a test set (30%) to prepare for the ML modelling stage. The ML model learned from the training set to generate a decision function. The training stage also introduced optimization, or tuning, for some variables within the ML model. Finally, we validated the optimum ML model with the test set and evaluated its performance.

3.4.1. Environment and System

We performed simulations for its experimental investigation by utilizing a Jupyter Notebook (6.3.0) environment running Python (3.8.8) on Windows 11 [29,30]. A personal computer used to run the simulation was equipped with an Intel Core i7 8565U processor and 16 GB of memory. In addition, we also investigated the models' performances in the constrained computational environment, i.e., Raspberry Pi 2 Model B. The Raspberry Pi runs Raspbian OS with Python (3.9.2) on a Broadcom BCM2863 SoC (900-MHz quad-core ARM Cortex-A7 CPU 1 GB RAM). We selected this particular device as it has less computational capabilities than the planned future system implementing a CubeSat (Raspberry Pi CM3+) onboard. Raspberry Pi CM3+'s in-orbit performance has been proven in a recent operational satellite project (KITSUNE). It is an acceptable system representation to demonstrate the feasibility of model deployment. Finally, the simulation libraries were utilized as follows: pre-processing, ML modelling, evaluation, and plotting (sklearn, PyOD, pickle, NumPy, pandas, SciPy, matplotlib, scikitplot, and seaborn) [18,23,30–36].

3.4.2. Simulation Parameters

According to the anomaly definition, the dataset has 8.59% type 1 and 7.22% type 2 anomalies. In total, the anomalous data across the dataset is 15.78%. Figure 11 depicts the labelled data according to the definition. Hence, it implies that the available dataset is imbalanced, with the majority consisting of normal class conditions.

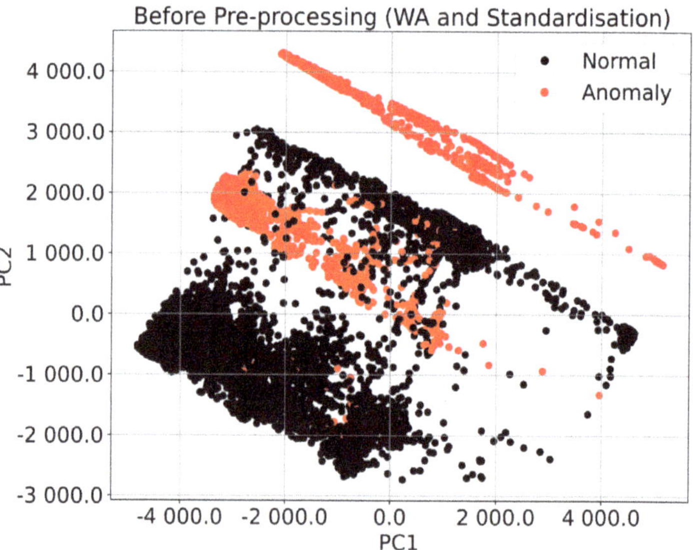

Figure 11. The first two principal component plots of the labelled dataset.

We implemented WA techniques in the pre-processing stage using the rolling and replace function in the pandas DataFrame library. The standardization was performed via the StandardScaler() function, available from the sklearn library. Next, the simulation implemented principal component analysis through a function provided by sklearn. Table 2 describes all necessary values for the simulation. Although we did not systematically tune the parameters, we determined the best values from an informal search for each algorithm.

Table 2. Simulation parameters.

Algorithm	Optimized Parameters	Trials	Best Value
Windowed Averaging	window size	10–300	50
PyOD Models	outlier fraction	0.05–0.2	0.16
kNN	neighbors	10–300	255
LOF	neighbors	10–350	280

4. Results

This section presents the evaluation and results regarding the experimental methodology. The pre-processing techniques were applied to emphasize any distinct pattern from each class. Figure 12 shows the labelled dataset after the pre-processing stage.

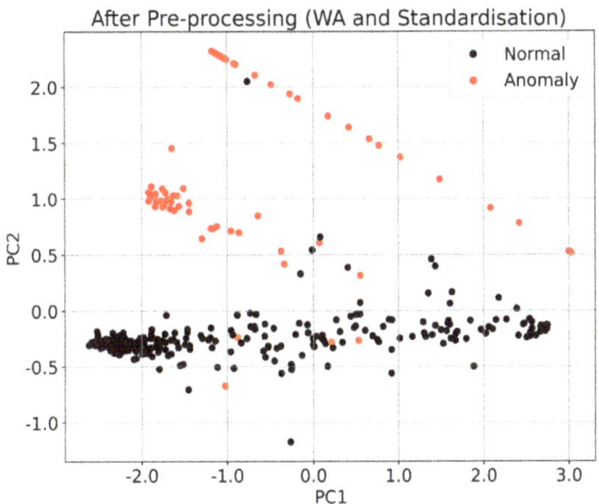

Figure 12. The first two principal component plots after pre-processing.

It provides a much more separable pattern between classes than the dataset plotted in Figure 11. The dataset forms a unique but linearly distinguishable pattern favoring the linear models over proximity-based models. Subsequently, the ML data were split and fed into the ML modelling phase. The rest of the section explains the evaluation of the ML models' performances.

4.1. Performance Evaluation

The performance evaluation of the ML model candidates was based on classification score, inference time, model size, and power consumption. The measurement for each performance parameter was taken for an average of 10 iterations. This approach was performed to ensure that any external factors would not influence the consistency of the results.

This work evaluated a performance using F-score metrics from the classification report. F-score or F-measure represents the harmonic mean of the precision and recall scores from the predicted label to the actual label [37]. Precision is the number of true positives (tp) divided by the total number of positive classifications, including false positives (fp). Recall is the number of true positives divided by the total number of positives, including false negatives (fn). In binary classification, recall is usually referred to as sensitivity. The F-score (F_β) is formulated as follows:

$$F_\beta = \left(1 + \beta^2\right) \times \frac{Precision \times Recall}{(\beta^2 \times Precision) + Recall} \, , \tag{6}$$

or

$$F_\beta = \frac{(1+\beta^2)tp}{(1+\beta^2)tp + \beta^2 fn + fp} \ . \tag{7}$$

The F1-score ($\beta = 1$) is selected as the minimum number of false positives, and false negatives are paramount for future onboard implementation. Further, the evaluation used F-score metrics considering the binary classification and imbalance class problem on the available dataset [38].

The inference time evaluation was measured directly by the standard package provided by the Jupyter Notebook. It measures the processing time of the CPU for specific instructions. The pickle library estimated the model size from a converted python object into a byte stream. Then, the power consumption was measured for the Raspberry Pi experiment, which considered the pre-processing stage and model inference.

4.2. Analysis

Accordingly, we compare the five selected ML algorithms based on their classification performances described in Table 3.

Table 3. F-score for each algorithm regarding the anomaly class.

Models	Precision *	Sensitivity *	F1-Score *
LOF	0.26 ± 0.02	0.26 ± 0.03	0.26 ± 0.02
CB-LOF	0.64 ± 0.02	0.65 ± 0.02	0.64 ± 0.01
kNN	0.34 ± 0.01	0.34 ± 0.01	0.34 ± 0.01
LDA	**0.97 ± 0.01**	0.75 ± 0.01	**0.85 ± 0.01**
OC-SVM	0.83 ± 0.01	**0.83 ± 0.02**	0.83 ± 0.01

* Mean ± standard deviation. Maximum scores are bolded. The metrics range from 0 to 1.

The results suggest that the LDA algorithm performed best for performance evaluation. It achieved the highest capability for anomaly detection concerning its F1-score. It is also important to acknowledge that this algorithm is the simplest to implement without advanced libraries. Although it showed the overall highest F1-score, there is a significant disparity between the precision and the sensitivity. A higher precision score means the model has low false-positive rates, whereas a higher sensitivity score means the model has a low false-negative rates. Table 3 suggests that the precision score of LDA is proportionally better than its sensitivity score, which means that the model might introduce more false negatives compared to false positives. The OC-SVM reached second in anomaly detection. It was slightly behind the LDA in the overall score but offered a better balance of the precision and sensitivity scores.

Furthermore, this study investigated both linear models in a Raspberry Pi. It was essential to confirm the model's technical viability in this study for future implementation. Table 4 shows the experimental result on the Raspberry Pi.

Table 4. The experimental results on Raspberry Pi.

Models	Execution Time [s] *	Model Size [kB]	Power Consumption [mWh]
LDA	**3.00 ± 0.17**	40.17	**4.288**
OC-SVM	9.45 ± 0.39	**15.25**	9.342

* Mean ± standard deviation. Maximum scores are bolded.

The LDA performed faster than OC-SVM in overall execution time, caused by it being a relatively more straightforward algorithm. This would influence the total power consumption in general. Both models result in a nominal size for the Raspberry Pi in terms of memory footprint. The OC-SVM offers a smaller size because it learns only from the original data (unsupervised) instead of finding correlations from the actual labels (supervised).

5. Discussion

In this work, both linear models indicated better performance than proximity-based algorithms, as expected. As for the proximity-based algorithms, the CB-LOF attained the best results. It had a relatively good F1-score compared to the rest of the proximity group. The performance differences can be explained by the characteristics of linear models that create a boundary function between classes. In contrast, the proximity-based algorithms use the mean value as a center point for a clustered class. In the dataset, the features within each class prefer a specific boundary function rather than clustered classes. Therefore, our dataset favors linear models.

The F1-score represents the ability of a specific algorithm to classify data points. The score is influenced by how each model approaches the problem. On the other side of the coin, the inference time is directly associated with the computational effort required by the model. Algorithm improvement (e.g., simplified code) can accelerate the model inference regarding the target system constraint or limitation. This improvement also directly impacts power consumption, as the faster the inference, the less power consumed. Further, the model size correlates with the number of training data used in the ML modelling. It varies between the models, but the proximity-based algorithms require more memory to calculate a specific number of neighboring or surrounding data.

In the future, we plan to extend the results of this study in two aspects: First is the battery system analysis, another critical system included in the EPS. The scope of this preliminary work was limited to the solar panel system only. Second, this study examined the ML approach algorithm for semi-sequential types of data points. Therefore, it will be essential to investigate another approach: full-time series analysis. Finally, from the results of this study, we have favorable model candidates to be developed for analytical tools of ground station telemetry data and onboard system model development.

Accordingly, the future system implementation will be developed simultaneously with the CubeSat system. The model has to be adjusted with every satellite sensor calibration process to determine any hyperparameter within the ML models or pre-processing techniques which significantly influences the model's sensitivity to the anomalies. Therefore, the final embedded ML-based anomaly detection has the expected nominal operation in the ground-tested flight model. Ultimately, it could provide critical information immediately after the satellite's deployment and early operation. For example, by having an onboard autonomous anomaly detector, the CubeSat might be able to broadcast its initial condition by CW beacon without the need for the operational team to perform a long traditional analysis of the telemetry data, which potentially takes weeks after the deployment.

6. Conclusions

In this work, we analyzed a novel dataset of BIRDS CubeSats. We investigated anomalies in the CubeSat solar panel system through an experimental ML study to compare proximity-based algorithms and linear model algorithms using on-orbit fault data obtained from four different CubeSats. Further, we introduced the windowed averaging method, standardization, and PCA in the pre-processing stage. The evaluation results on F-score metrics indicate that LDA and OC-SVM are the best models for detecting these particular anomalies within the dataset. LDA and OC-SVM models were tested on Raspberry Pi to confirm their technical viability for future onboard implementation. The results confirm their feasibility for onboard applications and prove that LDA outperforms OC-SVM in terms of execution time and power consumption.

Supplementary Materials: The following supporting information can be downloaded at https://www.mdpi.com/article/10.3390/app12178634/s1. TSURU.xlsx: TSURU satellite dataset; NEPALISAT.xlsx: NEPALISAT satellite dataset; RAAVANA.xlsx: RAAVANA satellite dataset; UGUISU.xlsx: UGUISU satellite dataset.

Author Contributions: Conceptualization, A.J.J.C. and B.H.B.P.; methodology, A.J.J.C.; software, A.J.J.C. and B.H.B.P.; validation, A.J.J.C., B.H.B.P., A.H. and M.C.; formal analysis, A.J.J.C. and B.H.B.P.; investigation, A.J.J.C. and B.H.B.P.; resources, M.C.; data curation, A.J.J.C.; writing—original draft preparation, A.J.J.C.; writing—review and editing, B.H.B.P., A.H. and M.C.; visualization, A.J.J.C. and B.H.B.P.; supervision, M.C.; project administration, M.C.; funding acquisition, M.C. All authors have read and agreed to the published version of the manuscript.

Funding: This research was partially funded by JSPS Core-to-Core Program B: Asia–Africa Science Platforms (JPJSCCB20200005).

Institutional Review Board Statement: Not applicable.

Informed Consent Statement: Not applicable.

Data Availability Statement: Not applicable.

Acknowledgments: The authors would like to express their appreciation for the valuable comments of the associate editors and anonymous reviewers. This work was supported by BIRDS-3 and BIRDS-4 engineering team members. In addition, the authors would like to acknowledge the support provided by George Maeda, Sankyun Kim, Hirokazu Masui, and Takashi Yamauchi.

Conflicts of Interest: The authors declare no conflict of interest.

References

1. BIRDS Program Digital Textbook CubeSTD-2019-001G. Available online: https://www.birds-project.com (accessed on 10 August 2021).
2. Jara, A.; Bautista, I.; Maeda, G.; Kim, S.; Masui, H.; Yamauchi, T.; Cho, M. An Overview of the BIRDS-4 Satellite Project and the First Satellite of Paraguay. In Proceedings of the 35th Conference on Small Satellites, Logan, UT, USA, 7–12 August 2021.
3. Cho, M.; Graziani, F. *Definition and Requirements of Small Satellites Seeking Low-Cost and Fast-Delivery*; IAA SG-2007-4.18; IAA: Paris, France, 2017.
4. Wertz, J.R.; Everett, D.F.; Puschell, J.J. *Space Mission Engineering: The New SMAD*, 1st ed.; Microcosm Press: Hawthorne, CA, USA, 2011; pp. 45–59.
5. Langer, M.; Bouwmeester, J. Reliability of CubeSats—Statistical Data, Developers' Beliefs and the Way Forward. In Proceedings of the 30th Conference on Small Satellites, Logan, UT, USA, 6–11 August 2016.
6. Maskey, A.; Cho, M. CubeSatNet: Ultralight Convolutional Neural Network designed for on-orbit binary image classification on a 1U CubeSat. *Eng. Appl. Artif. Intell.* **2020**, *96*, 103952. [CrossRef]
7. Wu, J.; Yao, L.; Liu, B.; Ding, Z.; Zhang, L. Combining OC-SVMs With LSTM for Detecting Anomalies in Telemetry Data with Irregular Intervals. *IEEE Access* **2020**, *8*, 106648–106659. [CrossRef]
8. Jin, W.; Sun, B.; Li, Z.; Zhang, S.; Chen, Z. Detecting Anomalies of Satellite Power Subsystem via Stage-Training Denoising Autoencoders. *Sensors* **2019**, *19*, 3216. [CrossRef]
9. Yairi, T.; Takeishi, N.; Oda, T.; Nakajima, Y.; Nishimura, N.; Takata, N. A Data-Driven Health Monitoring Method for Satellite Housekeeping Data Based on Probabilistic Clustering and Dimensionality Reduction. *IEEE Trans. Aerosp. Electr. Syst.* **2017**, *53*, 1384–1401. [CrossRef]
10. Zamry, N.M.; Zainal, A.; Rassam, M.A.; Alkhammash, E.H.; Ghaleb, F.A.; Saeed, F. Lightweight Anomaly Detection Scheme Using Incremental Principal Component Analysis and Support Vector Machine. *Sensors* **2021**, *21*, 8017. [CrossRef] [PubMed]
11. Pan, D.; Liu, D.; Zhou, J.; Zhang, G. Anomaly detection for satellite power subsystem with associated rules based on Kernel Principal Component Analysis. *Microelect. Reliab.* **2015**, *55*, 2082–2086. [CrossRef]
12. Peng, Y.; Jia, S.; Feng, X.; Su, F. Telemetry fault detection for meteorological satellite based on PCA. In Proceedings of the 16th International Symposium on Communications and Information Technologies, Qingdao, China, 8–10 December 2016.
13. Li, J.; Yang, W.; Liu, D.; Liu, J. Kernel Self-Adaptive Learning-Based Satellite Telemetry Data Classification. In Proceedings of the 3rd International Conference on Computing Measurement Control and Sensor Network, Matsue, Japan, 20–22 May 2016.
14. Umezu, R.; Sugie, T.; Nagase, M.; Kokai, R.; Takeshima, T.; Ebisawa, K.; Mitsuda, K.; Yamamoto, Y. Detection of Failure Sign of Spacecraft using Machine Learning. *Space Sci. Inf. Japan* **2019**, *8*, 11–20.
15. BIRDS Open-Source Repository. Available online: https://github.com/BIRDSOpenSource (accessed on 13 March 2022).
16. Chamika, D.; Cho, M.; Maeda, G.; Kim, S.; Masui, H.; Yamauchi, T.; Panawennage, S.; Shrestha, B. BIRDS-3 Satellite Project Including the First Satellites of Sri Lanka and Nepal. In Proceedings of the 70th International Astronautical Congress, Washington, DC, USA, 21 October 2019.
17. Nakayama, D.; Yamauchi, T.; Masui, H.; Kim, S.; Toyoda, K.; Malmadayalage, T.L.D.; Cho, M.; the BIRDS-4 Project Team. On-Orbit Experimental Result of a Non-Deployable 430-MHz-Band Antenna Using a 1U CubeSat Structure. *Electronics* **2022**, *11*, 1163. [CrossRef]

18. Pedregosa, F.; Varoquaux, G.; Gramfort, A.; Michel, V.; Thirion, B. Scikit-learn: Machine Learning in Python. *J. Mach. Learn. Res.* **2011**, *12*, 2825–2830.

19. Kreyszig, E. *Advanced Engineering Mathematics*, 10th ed.; Wiley: Hoboken, NJ, USA, 2011; p. 880.

20. Murphy, K. *Machine Learning: A Probabilistic Perspective*, 1st ed.; MIT Press: Cambridge, UK, 2012.

21. Minka, T.P. Automatic choice of dimensionality for PCA. In Proceedings of the 2000 Neural Information Processing Systems (NIPS) Conference, Denver, CO, USA, 27 November–2 December 2000; Volume 13, pp. 598–604.

22. Zhao, Y.; Nasrullah, Z.; Li, Z. PyOD: A Python Toolbox for Scalable Outlier Detection. *J. Mach. Learn. Res.* **2019**, *20*, 1–7.

23. Goldstein, M.; Uchida, S. A Comparative Evaluation of Unsupervised Anomaly Detection Algorithms for Multivariate Data. *PLoS ONE* **2016**, *11*, e0152173. [CrossRef] [PubMed]

24. Breunig, M.M.; Kriegel, H.P.; Raymond, T.; Sander, J. LOF: Identifying density-based local outliers. *ACM Sig. Rec.* **2000**, *29*, 93–104. [CrossRef]

25. He, Z.; Xu, X.; Deng, S. Discovering cluster-based local outliers. *Patt. Recog. Lett.* **2003**, *24*, 1641–1650. [CrossRef]

26. Ramaswamy, S.; Rastogi, R.; Shim, K. Efficient algorithms for mining outliers from large data sets. *ACM Sig. Rec.* **2000**, *29*, 427–438. [CrossRef]

27. Duda, R.; Hart, P.; Stork, D. *Pattern Classification*, 2nd ed.; Wiley: Hoboken, NJ, USA, 2000; pp. 215–264.

28. Scholkopf, B.; Platt, J.C.; Shawe-Taylor, J.; Smola, A.J.; Williamson, R.C. Estimating the Support of a High-dimensional Distribution. *Neural Comp.* **2001**, *13*, 1443–1471. [CrossRef] [PubMed]

29. Kluyver, T.; Ragan-Kelley, B.; Pérez, F.; Granger, B.E.; Bussonnier, M.; Frederic, J.; Kelley, K.; Hamrick, J.B.; Grout, J.; Corlay, S.; et al. Jupyter Notebooks—A publishing format for reproducible computational workflows. In *Positioning and Power in Academic, Players, Agents and Agendas*; IOS Press: Amsterdam, The Netherlands, 2016; pp. 87–90.

30. The Python Library Reference, Release 3.8.8, Python Software Foundation. Available online: https://www.python.org/downloads/release/python-388/ (accessed on 13 March 2021).

31. Harris, C.R.; Millman, K.J.; van der Walt, S.J. Array programming with NumPy. *Nature* **2020**, *585*, 357–362. [CrossRef] [PubMed]

32. McKinney, W. Data Structures for Statistical Computing in Python. In Proceedings of the 9th Python in Science Conference, Austin, TX, USA, 28 June 2010.

33. Virtanen, P.; Gommers, R.; Oliphant, T.E. SciPy 1.0: Fundamental Algorithms for Scientific Computing in Python. *Nat. Methods* **2020**, *17*, 261–272. [CrossRef]

34. Hunter, J.D. Matplotlib: A 2D Graphics Environment. *IEEE Comput. Sci. Eng.* **2007**, *9*, 90–95. [CrossRef]

35. Reiinakano/Scikit-Plot: V0.3.5. Available online: https://zenodo.org/record/1245853#.Yogp4ujMJPY (accessed on 13 March 2021).

36. Waskom, M.L. Seaborn: Statistical Data Visualization. *J. Open Source Softw.* **2021**, *6*, 3021. [CrossRef]

37. Manning, C.D.; Raghavan, P.; Schütze, H. *Introduction to Information Retrieval*, 1st ed.; Cambridge University Press: Cambridge, UK, 2008; pp. 151–175.

38. Sokolova, M.; Lapalme, G. A Systematic Analysis of Performance Measures for Classification Tasks. *Inf. Process. Manag.* **2009**, *45*, 427–437. [CrossRef]

Article

Scalable and Configurable Electrical Interface Board for Bus System Development of Different CubeSat Platforms

Marloun Sejera [1,2,*], Takashi Yamauchi [1], Necmi Cihan Orger [1], Yukihisa Otani [1] and Mengu Cho [1]

[1] Laboratory of Lean Satellite Enterprises and In-Orbit Experiments (LaSEINE), Department of Electrical and Space Systems Engineering, Kyushu Institute of Technology, Kitakyushu 804-8550, Japan

[2] School of Electrical, Electronics, and Computer Engineering, Mapua University, Manila 1002, Philippines

* Correspondence: sejera.marloun-pelayo148@mail.kyutech.jp or mpsejera@mapua.edu.ph

Abstract: A flight-proven electrical bus system for the 1U CubeSat platform was designed in the BIRDS satellite program at the Kyushu Institute of Technology. The bus utilizes a backplane board as the mechanical and electrical interface between the subsystems and the payloads. The electrical routes on the backplane are configured by software using a complex programmable logic device (CPLD). It allows for reusability in multiple CubeSat projects while lowering costs and development time; as a result, resources can be directed toward developing the mission payloads. Lastly, it provides more time for integration and system-level verification, which are critical for a reliable and successful mission. The current trend of CubeSat launches is focused on 3U and 6U platforms due to their capability to accommodate multiple and complex payloads. Hence, a demonstration of the electrical bus system to adapt to larger platforms is necessary. This study demonstrates the configurable electrical interface board's scalability in two cases: the capability to accommodate (1) multiple missions and (2) complex payload requirements. In the first case, a 3U-size configurable backplane prototype was designed to handle 13 mission payloads. Four CPLDs were used to manage the limited number of digital interfaces between the existing bus system and the mission payloads. The measured transmission delay was up to 20 ns, which is acceptable for simple serial communications such as UART and SPI. Furthermore, the measured energy consumption of the backplane per ISS orbit was only 28 mWh. Lastly, the designed backplane was proven to be highly reliable as no bit errors were detected throughout the functionality tests. In the second case, a configurable backplane was implemented in a 6U CubeSat with complex payload requirements compared to the 1U CubeSat platform. The CubeSat was deployed in ISS orbit, and the initial on-orbit results indicated that the designed backplane supported missions without issues.

Keywords: CubeSat; electrical interface; scalability; bus system development

Citation: Sejera, M.; Yamauchi, T.; Orger, N.C.; Otani, Y.; Cho, M. Scalable and Configurable Electrical Interface Board for Bus System Development of Different CubeSat Platforms. *Appl. Sci.* **2022**, *12*, 8964. https://doi.org/10.3390/app12188964

Academic Editors: Filippo Graziani, Simone Battistini and Mauro Pontani

Received: 27 July 2022
Accepted: 5 September 2022
Published: 6 September 2022

Publisher's Note: MDPI stays neutral with regard to jurisdictional claims in published maps and institutional affiliations.

1. Introduction

A CubeSat is a class of satellites with a defined size and form factor. A 1U CubeSat, for example, has dimensions of $10 \times 10 \times 11.35$ cm^3 and a mass of up to 2 kg, as defined in the CubeSat Design Specifications from California Polytechnic State University (CalPoly) [1]. The document describes the mechanical, electrical, and operational specifications of CubeSats from 1U to 12U. However, it does not cover how the components in a CubeSat, i.e., both the bus and the payload, are interfaced. This lack of such a definition allows CubeSat developers the freedom to choose which interface method to use. More importantly, this aspect could cause incompatibility issues between components, and solving these issues could considerably consume time that could be used for other verification activities to ensure mission success [2]. Furthermore, CubeSat projects can be developed by multiple collaborators, and clearly defining an interface standard between developers during the initial phase could prevent project delays, compatibility issues, and increased costs while improving overall mission success.

The first CubeSats developed used the stacking approach, where components were placed on top of each other using a connector. This interfacing method follows the PC/104 specifications [3] employed in embedded computers. The adoption of this specification to CubeSats defines the wiring harness, the printed circuit board (PCB) footprint, and the mechanical mounting of the boards, while the boards are stacked using 104-pin connectors. One of the first developers to employ the PC/104 specification in CubeSat applications is Pumpkin, Inc., who introduced the CubeSat Kit Bus (CSKB) [4]. The CSKB has become the de facto standard in CubeSat design and has been adopted by many commercial CubeSat developers.

However, using the PC/104 specification in CubeSats has several issues. According to a survey in [5], 51% of 36 respondents agreed that the size of the connector is too big. Ref. [6] confirmed that the connector occupies up to 20% of the PCB space. This limits the board designers in placing components on the board, resulting in low PCB utilization. The stacking height of the connector is also considerable, making the spacing between boards particularly wide. Another issue is that the number of pins interconnecting the board is often not utilized to its fullest extent. This may increase the risk of human error when assigning pins and mapping during the development and integration phases. Even though there could be many unused pins, a harness is extensively used in stacked systems. Lastly, top boards need to be disassembled if there is a need to take out a middle board, especially during troubleshooting. This leads to additional development and integration time.

There have been multiple efforts to resolve issues in using PC/104, specifically in terms of its connector. ISIS, for example, started to consider alternatives to CSKB connectors [7]. The company introduced CSKB Lite, two 28-pin connectors with just enough pins for full utilization. It is also backward-compatible with the standard CSKB connector. In the case of Nagoya University's NUcube satellites, where high-density interfaces are necessary [8], 144-pin connectors were used but with only a 9 mm stack-up height to reduce the volume occupied by the PCBs and a pitch of 0.50 mm to allow more space for components on the boards. This, in turn, made the connector incompatible with CSKB. Lastly, Korea Space University's KAUSAT-5 CubeSat used a flexible flat cable (FFC) connector instead of PC/104 to save volume and mass [6]. Despite these efforts, issues such as extensive use of harnesses and difficulty when assembling and disassembling were not addressed.

Another interface method uses a dedicated PCB called a backplane board that provides mechanical and electrical connections between the bus and the payloads. UWE-3, a 1U CubeSat from the University of Würzburg in Germany, first carried out the use of a backplane and became the reference for UNISEC Europe's CubeSat specification interface document (CSID) interface [9,10]. The BIRDS satellite program at the Kyushu Institute of Technology (Kyutech) in Japan also adopted the backplane board approach to its 1U CubeSats as an interface to the bus and the payload [11]. The bus and payload boards have 50-pin male connectors connected to the 50-pin female connectors of the backplane board. These connectors have a smaller form factor, which provides more space for electronic components on the bus and payload boards. The spacing between boards can also be adjusted by moving the female connectors on the backplane board. This provides efficient utilization of the limited volume of a CubeSat. In addition, power lines, as well as analog and digital signals, are routed through the PCB. This greatly reduces the use of the wiring harness, which is considered one of the fundamental reasons for satellite failure. Lastly, the backplane approach makes satellite assembly and disassembly significantly more straightforward. Therefore, it allows integration and troubleshooting to be performed in a shorter time.

A survey was conducted to determine the interface method used in CubeSats launched from 2003 to 2019 [12]. Of the 397 CubeSats surveyed, only 170 CubeSats had an identified interface since the sources of information (e.g., web pages, papers, conference papers, etc.) did not provide the details. A total of 137 CubeSats used the PC/104 interface, which is about 80% of the total identified CubeSats. However, the use of backplanes on CubeSats started gaining favor in 2013, with 24 satellites launched from then until 2019. For example,

CalPoly's Aerospace Engineering Department developed its own kit as an educational platform for satellite development known as CalPoly CubeSat Kit MK1. Its internal configuration uses a backplane that can connect five boards through its 48-pin female connectors [13]. However, this backplane's design lacks flexibility because of changes in interface definition from one CubeSat project to the other. These changes lead to the complete reproduction of the backplane, adding cost and development time [14].

The third generation of the BIRDS project introduced a standard software-configurable backplane board as one of its technology demonstrations [14]. A complex programmable logic device (CPLD) was placed on the backplane, and digital signals between the bus and the payload were routed through the CPLD. By reprogramming the chip, rerouting can be performed without changes in the hardware design. This makes the backplane flexible and reusable by future satellites with minimal modification while saving time and cost. The configurable backplane was proven to work in space during satellite deployment at the International Space Station (ISS) in June 2019 until it was deorbited in October 2021. The software-configurable backplane has become an integral part of the BIRDS 1U bus architecture. The standard bus, however, has limited digital interfaces that constrain the number of payloads a CubeSat can carry out.

The demand for launching CubeSats is on a continuous uptrend, and most CubeSats being launched are 3U and 6U platforms mainly because they can accommodate multiple and complex payloads. Implementing a configurable interface board on larger CubeSat platforms could provide benefits similar to those in the 1U platform, such as a reduction in development throughput and cost.

This paper aims to demonstrate how the configurable interface board can be scaled up and adapted to different CubeSat sizes using the BIRDS 1U standard bus system. The novelty of the present work is described as a confirmation of the configurable interface board's flexibility in absorbing challenges encountered when scaling up to larger CubeSat platforms. Since implementing a configurable backplane to larger platforms such as 3U or 6U CubeSats has not been achieved before, several challenges such as managing communication between the existing standard bus system and multiple mission payloads, as well as meeting the mission requirements of complex payloads, are extensively covered in this study. In addition, the contribution of this paper is that it presents a 3U configurable backplane designed to manage several missions on the limited number of available electrical interfaces of a standard bus. The design concept could benefit satellite developers who provide hosted payload services where bus resources are maximized to accommodate as many payloads as possible. This paper also demonstrates the modifications in a bus system necessary to scale up and handle complex missions in a W6U CubeSat. The CubeSat was deployed from the ISS in March 2022, and it has been successfully supporting the execution of the missions.

This paper is composed of five sections. Section 2 discusses the standard 1U bus system. It also discusses two backplane designs in different CubeSat platforms—a 3U backplane prototype for multiple payloads and a backplane for a W6U CubeSat with complex mission requirements. Section 3 discusses the tests conducted and the results of the two backplane designs. Lastly, Sections 4 and 5 present the Discussion and Conclusion.

2. BIRDS 1U Standard Bus System

The BIRDS Program of Kyutech is an educational, capacity-building satellite program that aims to empower participants from non-space-faring countries to lead or start satellite projects in their home countries [15]. In the program, the graduate students from the participating countries gain hands-on experience on how to design, develop, test, and operate CubeSats. A total of thirteen countries have participated in the program, nine of which built the first satellite in their countries. The program has deployed a total of seventeen satellites in its five generations of constellations from 2015 to 2022—five in BIRDS-1 and three each in BIRD-2, BIRDS-3, BIRDS-4, and BIRDS-5. A total of sixteen satellites are 1U CubeSats, while one of the three satellites in BIRDS-5 is a 2U CubeSat.

Kyutech has made the BIRDS 1U bus system available as open-source information [16,17]. CubeSat developers can gain full access to all information necessary to build a 1U satellite, and these include technical drawings, source code, PCB design, assembly and testing procedures, test reports, and interface control documents (ICDs). This effort allows more people to develop their satellites in an easier, faster, and cheaper way. At present, there are two universities, two high school projects, and a company in Japan that are benefiting from this initiative. In addition, BIRDS members from Malaysia, Mongolia, the Philippines, and Sri Lanka are making satellites in their countries using the BIRDS 1U bus.

Figure 1 shows the internal boards of a 1U BIRDS satellite. The components are described below.

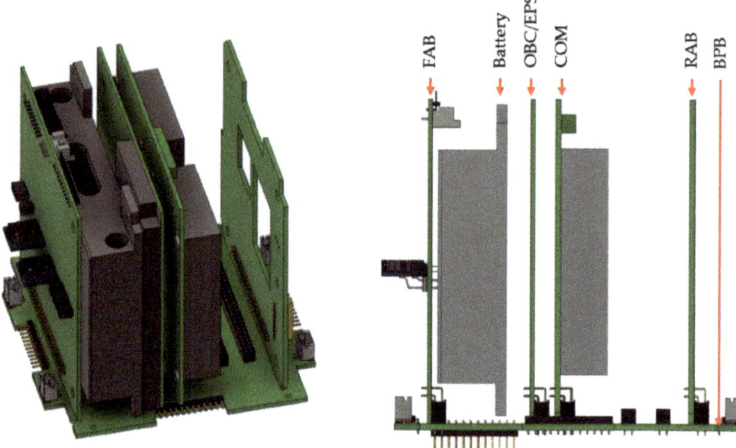

Figure 1. BIRDS 1U satellite internal boards.

- *Front access board (FAB).* This board offers external interfaces or umbilical connections. These interfaces or connectors give users access to subsystem microcontrollers for programming and debugging, monitoring voltage levels, and battery charging. The FAB is also in charge of managing the batteries, collecting the power produced by the solar panels, and controlling electrical power safety. The FAB microcontroller is a Microchip PIC 16F1789 that manages the housekeeping data, including the battery health and solar panel power generation.
- *Battery.* The bus uses six commercial nickel–metal hydride (NiMH) batteries in a three-series, two-parallel (3S2P) configuration. The batteries are housed in a battery box and are connected to the FAB via a 10-pin connector.
- *On-board computer/electrical power system (OBC/EPS).* The board serves three primary functions. A dedicated Microchip PIC microcontroller handles each process. The main PIC (PIC 18F67J94) oversees command and data handling. The uplink command is received and executed by the main PIC. It also obtains the FAB PIC housekeeping data and mission data for downlink. The COM PIC (PIC 16F1789) manages communication between the radio transceiver and the main PIC. It sends commands from the radio transceiver to the main PIC. It also forwards data from the main PIC to the radio transceiver for downlink. The reset PIC (PIC 16F1789) carries out the electrical power subsystem, and it is in charge of power distribution to other subsystems and the payload. The radio transceiver and antenna deployment system use two unregulated power lines, and there are also two 3.3 V power lines and one 5 V line that the mission payload can use. All three microcontrollers are connected via universal asynchronous receiver/transmitter (UART) serial interfaces to form a ring network.
- *Communication (COM) board.* This board is a dedicated radio transceiver that handles both uplink and downlink communication from and to the ground. The transceiver

employs Gaussian minimum-shift keying (GMSK) modulation with a baud rate of 4800 bps in both directions and an AX.25 protocol for data format. The frequency of operation is in the ultra-high frequency (UHF) amateur band.

- *Rear access board (RAB)*. This board provides external access to the mission payload for programming and debugging.
- *Backplane board (BPB)*. This backplane connects all internal and external boards electrically and mechanically. Figure 2 shows the top side of a backplane board. The internal boards are connected to the backplane via 50-pin female connectors with a 2 mm pitch (C101 to C106). The pins are assigned as power lines, digital lines (for programming, debugging, and data communication), and analog lines. The connector has two pins assigned to each power line and the system ground. The same pins are set to all the connectors. Since the maximum current for each pin is rated as 1 A, the power line current is limited to 2 A. The remaining pins of the connectors are for digital lines and analog lines.

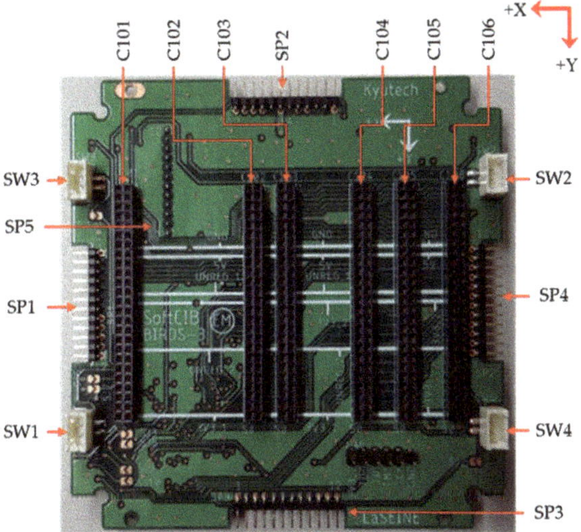

Figure 2. Backplane board (top side).

The antenna panel and four solar panels are connected to the backplane via 12 pin male connectors with a 2.54 mm pitch (SP1 to SP5). The fifth solar panel is connected directly to the FAB. The connectors in the solar panels route the generated power and temperature readings to the FAB, and the connector on the antenna panel provides an unregulated power line to the antenna deployment system. In addition, the backplane also includes two-pin male connectors (SW1 to SW4) for the deployment switches connected to the satellite structure.

The space between the COM (C103) and RAB (C104) connectors is allocated for the mission payload. The maximum board thickness of the payload that can be accommodated is 22.35 mm. Up to two mission boards (C104, C105) can be placed in the given space. It is possible to customize the backplane board to reduce the number of 50-pin female connectors to one and shift its position.

A satellite ICD contains the overall information on the mechanical and electrical interfaces between the components. The connector pin assignment is an example of an electrical interface specification found in this document, and the satellite developer can use the pin assignment to determine how components are connected. Table 1 shows the pin assignment of the 50-pin connector on the OBC/EPS board. In addition, it specifies the name of each pin, the connected microcontroller, the pin route, the rated voltage and current, and the protocol and baud rate for digital pins. The pins are named according to their function or connection. For example, PROG_GIO_1 denotes the first general I/O pin for programming. Another example is OBC-COM_1, which denotes the first pin connecting the OBC/EPS board and COM board. Pins 9–12, 21, 22, 27, 28, 37, 38, and 42 are linked to the main PIC and routed to the mission board. These 11 pins serve as digital interfaces for the bus and the payload, whereas pins 31–34 are serial peripheral interface (SPI) connections to the flash memory (FM) that the main PIC and mission payload share.

Table 1. OBC/EPS board pin assignment.

Pin No.	Pin Name	MCU	Destination	Voltage	Current	Baud Rate	Protocol
1	PROG_GIO_1	All	FAB	3.3 V	<30 mA	-	PIC
2	PROG_GIO_2	All	FAB	3.3 V	<30 mA	-	PIC
3	No connection	-	-	-	-	-	-
4	PROG_GIO_4	Reset	FAB	3.3 V	<30 mA	-	PIC
5	PROG_GIO_5	COM	FAB	3.3 V	<30 mA	-	PIC
6	PROG_GIO_6	Main	FAB	3.3 V	<30 mA	-	PIC
7	OBC-COM_1	COM	COM board	3.3 V	<30 mA	115,200	RS232
8	OBC-COM_2	COM	COM board	3.3 V	<30 mA	115,200	RS232
9	FAB_to_RAB_GIO_3	Main	Mission	3.3 V	<30 mA	-	DIO
10	FAB_to_RAB_GIO_4	Main	Mission board	3.3 V	<30 mA	-	DIO
11	FAB_to_RAB_GIO_5	Main	Mission board	3.3 V	<30 mA	115,200	RS232
12	FAB_to_RAB_GIO_6	Main	Mission board	3.3 V	<30 mA	115,200	RS232
13	GND_SYS	-	All boards	GND	1 A	-	Power
14	GND_SYS	-	All boards	GND	1 A	-	Power
15	SUP_5 V0	-	All boards	5 V	1 A	-	Power
16	SUP_5 V0	-	All boards	5 V	1 A	-	Power
17	FAB_to_OBC_GIO_1	Main	FAB	3.3 V	<30 mA	9600	RS232
18	FAB_to_OBC_GIO_2	Main	FAB	3.3 V	<30 mA	9600	RS232
19	FAB_to_OBC_GIO_3	Main	FAB	3.3 V	<30 mA	9600	RS232
20	FAB_to_OBC_GIO_4	Main	FAB	3.3 V	<30 mA	9600	RS232
21	CPLD_8	Main	Mission board	3.3 V	<30 mA	9600	RS232
22	CPLD_9	Main	Mission board	3.3 V	<30 mA	9600	RS232
23	SUP_UNREG_1	-	All boards	Unreg	1 A	-	Power
24	SUP_UNREG_1	-	All boards	Unreg	1 A	-	Power
25	SUP_3 V3_2	-	All boards	3.3 V	1 A	-	Power
26	SUP_3 V3_2	-	All boards	3.3 V	1 A	-	Power
27	CPLD_10	Main	Mission board	3.3 V	<30 mA	9600	RS232
28	CPLD_11	Main	Mission board	3.3 V	<30 mA	9600	RS232
29	RAW_POWER	-	FAB	Raw power	1 A	-	Power
30	RAW_POWER	-	FAB	Raw power	1 A	-	Power
31	CPLD_12	Memory	Mission board	3.3 V	<30 mA	1,000,000	SPI
32	CPLD_13	Memory	Mission board	3.3 V	<30 mA	1,000,000	SPI
33	CPLD_14	Memory	Mission board	3.3 V	<30 mA	1,000,000	SPI
34	CPLD_15	Memory	Mission board	3.3 V	<30 mA	1,000,000	SPI
35	SUP_UNREG_2	-	All boards	Unreg	1 A	-	Power
36	SUP_UNREG_2	-	All boards	Unreg	1 A	-	Power
37	CPLD_16	Main	Mission board	3.3 V	<30 mA	9600	RS232
38	CPLD_17	Main	Mission board	3.3 V	<30 mA	9600	RS232
39	Kill_SW	Main	FAB	3.3 V	<30 mA	-	DIO
40	No connection	-	-	-	-	-	-
41	No connection	-	-	-	-	-	-
42	CPLD_18	Main	Mission	3.3 V	<30 mA	-	DIO
43	OBC-COM_3	COM	No connection	3.3 V	<30 mA	-	DIO
44	OBC-COM_4	COM	COM board	3.3 V	<30 mA	-	DIO
45	OBC-COM_5	COM	COM board	3.3 V	<30 mA	-	DIO
46	OBC-COM_6	COM	COM board	3.3 V	<30 mA	-	DIO
47	OBC-COM_7	COM	COM board	Analog	<30 mA	-	Analog
48	OBC-COM_8	COM	COM board	Analog	<30 mA	-	Analog
49	SUP_3 V3_1	-	All boards	3.3 V	1 A	-	Power
50	SUP_3 V3_1	-	All boards	3.3 V	1 A	-	Power

Figure 3 shows the bottom side of the backplane, where all active devices are placed. This is to avoid interference with the components at the top. One of the active components is a CPLD that can be programmed to perform specific logical functions. The CPLD is a lattice semiconductor ispMACH LC4256ZE, an ultralow power device that uses 1.8 V LVCMOS (low-voltage CMOS) technology with a standby current of 13 uA. A voltage regulator that uses 3.3 V input from one of the power lines supplies 1.8 V to the integrated circuit. In addition, a joint test action group (JTAG) connection is used to program the chip.

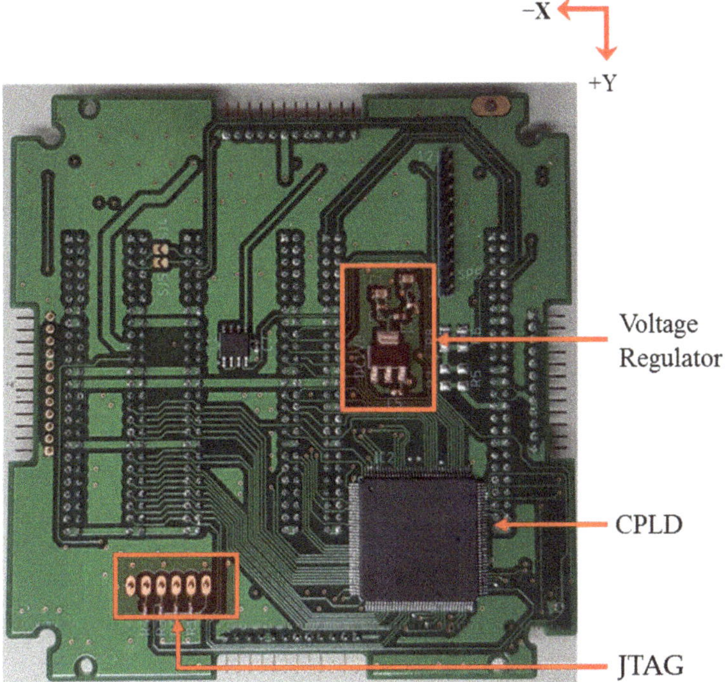

Figure 3. Backplane board (**bottom** side).

Table 2 represents the list of CubeSats with their CPLD families and corresponding functionalities. The CPLDs are most commonly implemented within the telemetry command and data-handling subsystem. On the other hand, the CPLD in BIRDS-3 was implemented on the backplane board as an interface between the PIC and the mission payloads. The table also shows the propagation delay and internal voltage supply of the CPLD according to the datasheet. The propagation delay varies depending on the chip package, and the data are based on the 144-pin quad-flat package (QFP) chip if not indicated by the resources. The ispMach4000 ZE CPLD was chosen for its optimal performance in terms of propagation delay and power consumption in addition to ease of availability, implementation, and cost.

Figure 4 illustrates how data communication is performed between the bus system and mission boards. The 11 digital interfaces from the bus are routed to the mission boards through the CPLD. Depending on the implementation, these interfaces can be UART, SPI, digital input/output (DIO), or their combination. There is also an SPI line for transferring mission data to the shared FM in the OBC/EPS board, and the CPLD manages these digital interfaces to allow mission payload communication to the bus system. In addition, Figure 5 shows how a CPLD on the backplane operates where the CPLD is programmed as a voltage follower, with the output pin logic levels matching the paired input pin. Digital interfaces can be rerouted without requiring hardware changes by reprogramming the CPLD. This saves both cost and time when redesigning the board. This also makes the backplane significantly more adaptable, especially during the initial development phase when routing changes are expected.

Table 2. Comparison of CPLD family used in CubeSats.

Satellite	CPLD Function	CPLD Family	Propagation Delay (Max)	Voltage Supply (Internal)
AAReST MirrorSat [18]	On-board Computer (OBC); Support for switching between the two Raspberry Pi compute modules and power sequencing	Xilinx XC9500 XL	6 ns	3~3.6 V
OPTOS [19]	Distributed On-board Data-Handling Terminal (DOT); Interface between the CPU and mission experiments	Xilinx Cool Runner II	7 ns	1.7~1.9 V
ARISSat-1 [20]	Integrated Housekeeping Unit (IHU); Glue logic between the video input processor, the SDRAM, and MCU	Intel Altera MAX II	5.4 ns	2.5 V, 3.3 V
BIRDS-3 [21]	Backplane; Interface between the main PIC and payloads	Lattice ispMach4000ZE	5.8 ns	1.7~1.9 V

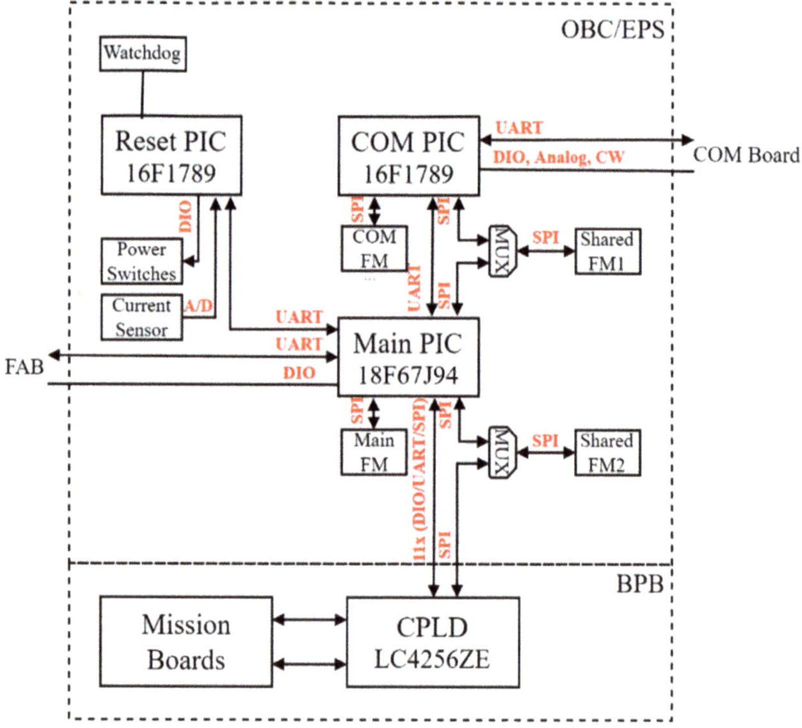

Figure 4. Block diagram of data handling between the bus and the payload.

Figure 5. Digital line routing using a CPLD.

Using this standard bus system, the configurable interface board can be scaled up and adapted to different CubeSat sizes. As a result, resources can be more focused on developing the mission payload and instruments. Lastly, it allows more time for integration and system-level verification, which are critical for a reliable and successful mission [22,23]. Since implementing a configurable backplane to larger platforms has not been achieved before, challenges such as managing communication between the existing standard bus system and multiple mission payloads, as well as meeting the mission requirements of complex payloads, may occur. To prove that the standard bus system can be scaled and address the challenges encountered when scaling up to larger CubeSat platforms, two cases have been studied. In Section 2.1, a backplane prototype was developed to handle several missions based on the limited number of available electrical interfaces of a standard bus. In Section 2.2, an actual implementation of a backplane that can handle missions with complex requirements was demonstrated.

2.1. 3U-Size Configurable Backplane

The BIRDS bus system is not only intended for 1U CubeSats but is designed to scale up to a 3U platform with minor modifications [24]. One modification is in the design of the backplane, where a larger CubeSat provides more space for mission boards than in a 1U. A 3U configurable backplane prototype was developed, as shown in Figure 6. The backplane is a six-layer PCB and measures 320 mm × 90 mm × 1.6 mm, and all internal boards and deployment switch connectors are placed on the top side of the board, which is the same as the 1U standard bus. In addition, the bus system components (FAB, OBC/EPS, COM, and RAB) and their arrangement were kept unchanged. Lastly, the space between the COM board and the RAB was allocated for the 13 mission boards. Since the 3U platform would require more power for the mission payloads, the battery capacity would also increase, leading to a bigger battery box than in the 1U. Therefore, the space for the battery box, which is between the FAB and OBC/EPS board, is wider than in the 1U backplane.

A standard pin assignment for all mission boards is described in detail in Table 3. While 12 power pins are pre-assigned based on the power distribution from the OBC/EPS board, the developers have the flexibility to assign the remaining pins to CPLD, miscellaneous, and umbilical connections. In addition, the 20 pins assigned to CPLD connections are configurable even after the backplane is fabricated. This number of pins allows the payload developers to assign any kind of serial digital interface to link over the bus.

Table 3. Standard 50-pin assignment for mission boards.

50-Pin Mission Board Allocation	No. of Pins
CPLD connections	20
Miscellaneous (e.g., analog, direct connections)	8
Power	12
Umbilical (programming and debugging)	10
Total	**50**

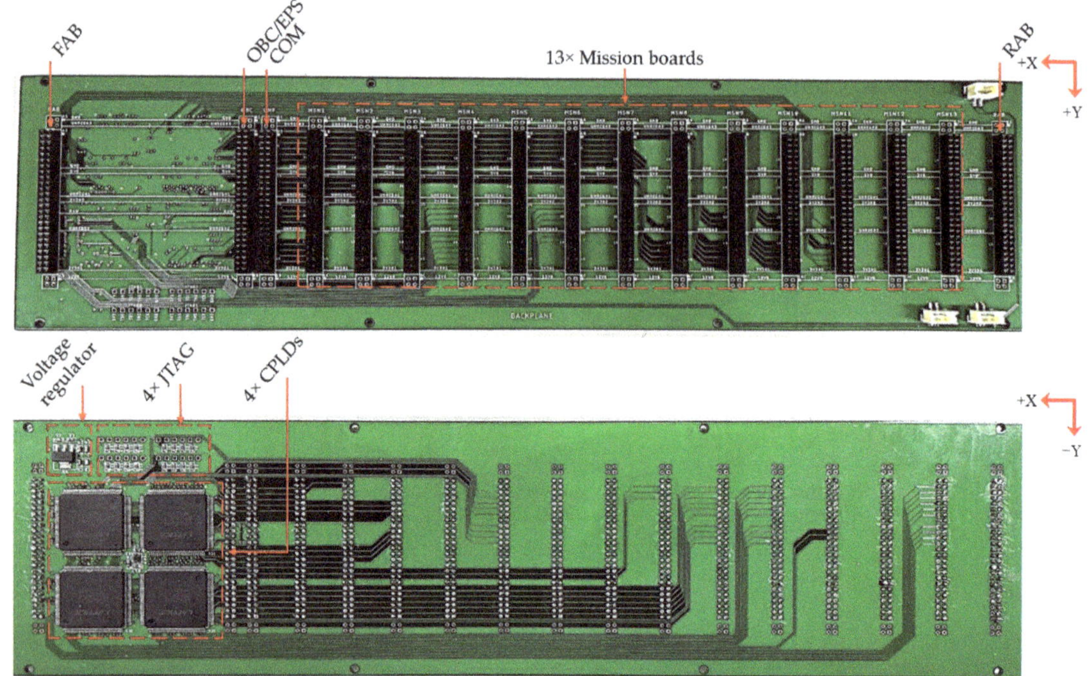

Figure 6. The 3U configurable backplane (**top** and **bottom** sides).

Four CPLDs are laid on the bottom side of the backplane, while the battery is placed on the other side. Cumulatively, the CPLDs provide the interface between the bus system and mission payloads. A voltage regulator supplies 1.8 V to all CPLDs, and each device has its JTAG pins for programming. Figure 7 shows the logical connections between the bus and the mission payloads. The SPI (from the shared FM) and 11 digital interfaces (from the main PIC) in the bus system are directly connected to CPLD1. The CPLDs are cascaded to each other through the 15 I/O pins. These are later configured to correspond to the bus system SPI and 11 digital interfaces. The remaining I/O pins of the CPLD are distributed to the mission (MSN) boards. CPLD1 to CPLD3 manage three mission boards, while CPLD4 manages four. To manage the bus system digital interfaces, the CPLDs are programmed to function as four-to-one multiplexers with four select (SEL) lines.

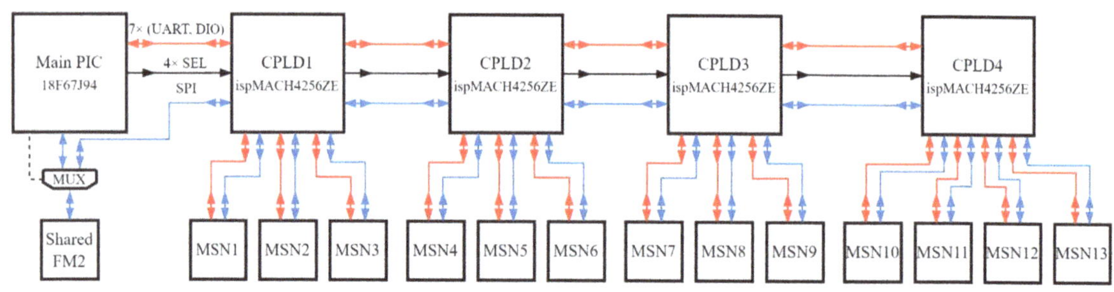

Figure 7. Logical connection of bus system and payload through CPLDs.

Table 4 shows the truth table of how the CPLDs function. A total of 4 of the 11 digital interfaces (SEL0 to SEL3) in the bus are used as select pins of the multiplexer function of the CPLDs. A specific logic state of the select pins allows a mission payload to access the seven remaining digital interfaces and the SPI in the bus system. For example, if the select pin logic values are 0001, CPLD1 allows MSN1 to access the bus system. Conversely, if the select pin logic values are 0110, CPLD1 routes all 15 digital lines to CPLD2. CPLD2 then routes the digital lines to MSN6. There are 13 combinations of select pin logic states corresponding to the 13 mission boards.

Table 4. The 3U backplane truth table.

Input				Output			
SEL3	SEL2	SEL1	SEL0	CPLD1	CPLD2	CPLD3	CPLD4
0	0	0	1	MSN1	-	-	-
0	0	1	0	MSN2	-	-	-
0	0	1	1	MSN3	-	-	-
0	1	0	0	CPLD2	MSN4	-	-
0	1	0	1	CPLD2	MSN5	-	-
0	1	1	0	CPLD2	MSN6	-	-
0	1	1	1	CPLD2	CPLD3	MSN7	-
1	0	0	0	CPLD2	CPLD3	MSN8	-
1	0	0	1	CPLD2	CPLD3	MSN9	-
1	0	1	0	CPLD2	CPLD3	CPLD4	MSN10
1	0	1	1	CPLD2	CPLD3	CPLD4	MSN11
1	1	0	0	CPLD2	CPLD3	CPLD4	MSN12
1	1	0	1	CPLD2	CPLD3	CPLP4	MSN13

Verifications were conducted on the backplane to test the performance. First, a functional test was performed to check whether the multiplexing function worked. Signal propagation delay and overall power consumption were also measured. Lastly, a bit error check was performed. The results and discussions of the test are given in Section 3.1.

2.2. KITSUNE W6U CubeSat

The previous section explained how a configurable backplane is designed to handle multiple payloads with relatively basic requirements that are related to power and data communication to the bus system. As a result, the bus system is not modified for integration. In addition, the payload operation does not require control of the satellite attitude. Therefore, the attitude determination and control subsystem (ADCS) was not included in the bus system. To implement payloads with advanced requirements, this section describes how a configurable backplane was modified, including additional subsystems.

KITSUNE satellite is a W6U CubeSat platform designed and developed in Japan as a collaboration project by the Kyushu Institute of Technology (Kitakyushu, Japan), Harada Seiki Co., Ltd. (Hamamatsu, Japan), and Addnics Corp. (Tokyo, Japan) [25]. The satellite project kicked off in September 2019, and KITSUNE was delivered to the Japan Aerospace Exploration Agency (JAXA) in November 2021. The satellite was deployed from the ISS on 24 March 2022 and is now in operation.

Figure 8 shows the 3D model of the W6U CubeSat. The CubeSat is divided into three sections: a 3U section for the camera payload that can capture 5 m class resolution images, a 1U section with technology demonstration and scientific experiment missions, and a 2U section for the main bus system. The 1U section is known as SPATIUM-II. It is basically a 1U satellite with its own bus that manages the missions. In addition to drawing power from the main bus, SPATIUM-II can work independently. A configurable backplane, developed to serve as the interface to all three sections, was placed in the middle of the satellite.

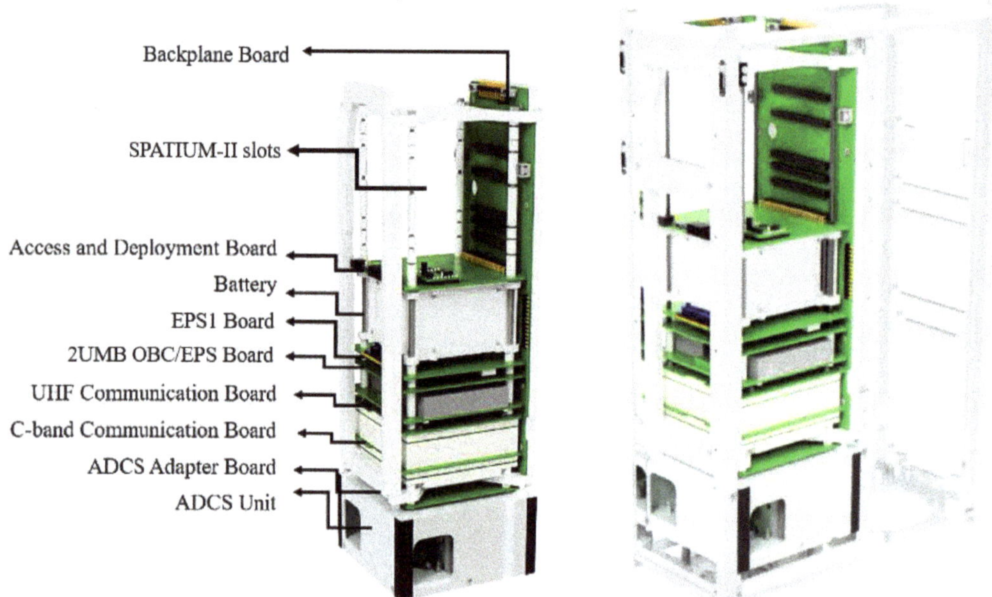

Figure 8. A 3D model of the W6U CubeSat bus system rack assembly (**left**) and the integrated rack assembly in the structure (**right**).

The main bus system was modified according to the mission requirement. The modifications were as follows:

- *FAB.* The functions of the FAB were separated into two boards. The umbilical to the bus system was assigned to the access and deployment board (ADB). The ADB provides external access to the microcontrollers (the main PIC, reset PIC, and COM PIC) for programming and debugging, battery charging, and voltage monitoring. The ADB has a microcontroller for antenna deployment. Power-related functions such as power generation, battery management, and power safety were assigned to the EPS1 board, as shown in Figure 8.
- *Battery.* The bus uses six commercial lithium-ion (Li-ion) batteries in a two-series, three-parallel (2S3P) configuration. The batteries are housed in a battery box and are connected to the FAB via a 12-pin connector.
- *OBC/EPS.* The three unused pins in the 50-pin connector were utilized—pin 3 was connected to a DIO of the main PIC, while pins 40 and 41 were connected to two DIOs of the COM PIC. The additional DIOs of the main PIC make the digital interfaces between the bus and payload available to 12 pins. Another radio transceiver can use the two DIOs of the COMP PIC. Lastly, the reset PIC provides an additional unregulated, 12 V power line for the camera payload [26].
- *Communication.* A C-band transceiver was utilized in addition to the UHF transceiver. The C-band transceiver is used for the high-speed download of data necessary in high-resolution images. It can also receive uplink commands as a backup to the UHF transceiver. The UHF (COM1) and C-band (COM2) transceivers use amateur radio bands.
- *ADCS.* The ADCS is necessary to obtain quality Earth images. A 0.5U ADCS from Adcole Maryland Aerospace was selected. This plug-and-play module has a standalone computer that manages its three reaction wheels, a three-axis magnetometer, two Earth

sensors, and three electromagnets [27]. An adapter board was used to connect the module to the backplane.

On the SPATIUM-II side, the components are as follows:

- *Access board (AB).* This board is where the umbilical for SPATIUM-II was placed. It has external access to the OBC/EPS board and payload microcontrollers for programming and debugging. It is similar to the main bus ADB, except that it does not have a microcontroller for deployment.
- *OBC/EPS.* The three unused pins (pins 3, 40, and 41) in the 50-pin connector were connected to the DIO pins of the main PIC. The additional DIOs of the main PIC make the digital interfaces between the bus and the payload available to 14 pins. Lastly, the two 3.3 V lines distributed by the reset PIC were changed to 3.5 V and 4.5 V, as required by the payloads.
- *COM.* This is similar to the UHF transceiver in the BIRDS 1U bus system, except that it operates in the non-amateur UHF band.
- *Payloads.* SPATIUM-II has two missions, store-and-forward and total electron content (TEC). Store-and-forward uses an on-board LoRa payload that collects sensory data from ground sensor terminals (GST) and downlinks the data to the Kyutech non-amateur ground station. On the other hand, the TEC mission measures the total electron content in the ionosphere. The mission requires two connectors to the backplane for its payloads. The first payload is for the chip-scale atomic clock (CSAC) module [28]. The second payload has the Raspberry Pi module, software-defined radio (SDR), and radiofrequency (RF) switch. Lastly, the mission's global positioning system (GPS) receiver was placed in one of the solar panels.

The backplane board is a six-layer PCB with dimensions of 250.5 mm × 90 mm × 1.6 mm, as shown in Figure 9. The top side has connectors to the internal components of the main bus and the SPATIUM-II. All are 50-pin male connectors with a 2 mm pitch. The OBC/EPS, EPS1, and ADB in the main bus have additional 4-pin female connectors on each end of the 50-pin connector, which are allocated for the additional power lines, system ground, and battery power. The additional pins make the total pin count 58 for the three boards.

The two biggest solar panels (+X and −X) are connected to the backplane via 13-pin male connectors (A), while the −Y and +Z panels are connected via a 26-pin male connector (B). All connectors have a 2.54 mm pitch between pins. The fourth solar panel (+Y) is connected directly to the EPS1 board. In addition to routing the generated power and temperature readings to the EPS1 board, the connectors connect to the sun sensors and antenna deployment circuits. Additionally, B has a route to the two GPS modules. Lastly, C denotes two-pin male connectors for the deployment switches connected to the satellite structure.

At the bottom side of the backplane, there is a two-pin connector (D) that connects the sun sensor in the -Z panel and a 30-pin connector (E) that connects the camera controller to the backplane. There are two CPLDs for the main bus and SPATIUM-II working independently. A voltage regulator in the main bus converts 3.3 V from the power line to 1.8 V. There are two voltage regulators on the SPATIUM-II side. The first regulator converts 5 V from the power line to 3.3 V. Then, the second regulator converts 3.3 V to 1.8 V. The CPLD uses 3.3 V as the output supply voltage, whereas 1.8 V is the LVCMOS supply voltage.

Both CPLDs in the KITSUNE satellite function as voltage followers. The available interfaces on the bus are enough for the payloads to use. Thus, the multiplexing function is not necessary. The main bus CPLD has 43% utilization, while the SPATIUM-II CPLD has 35% utilization. This means that, out of the 96 I/O pins in a CPLD, the main bus and SPATIUM-II utilized 42 and 34 pins, respectively. These digital connections are combinations of DIO, SPI, and UART interfaces. Ground verification and on-orbit results are discussed in Section 3.2.

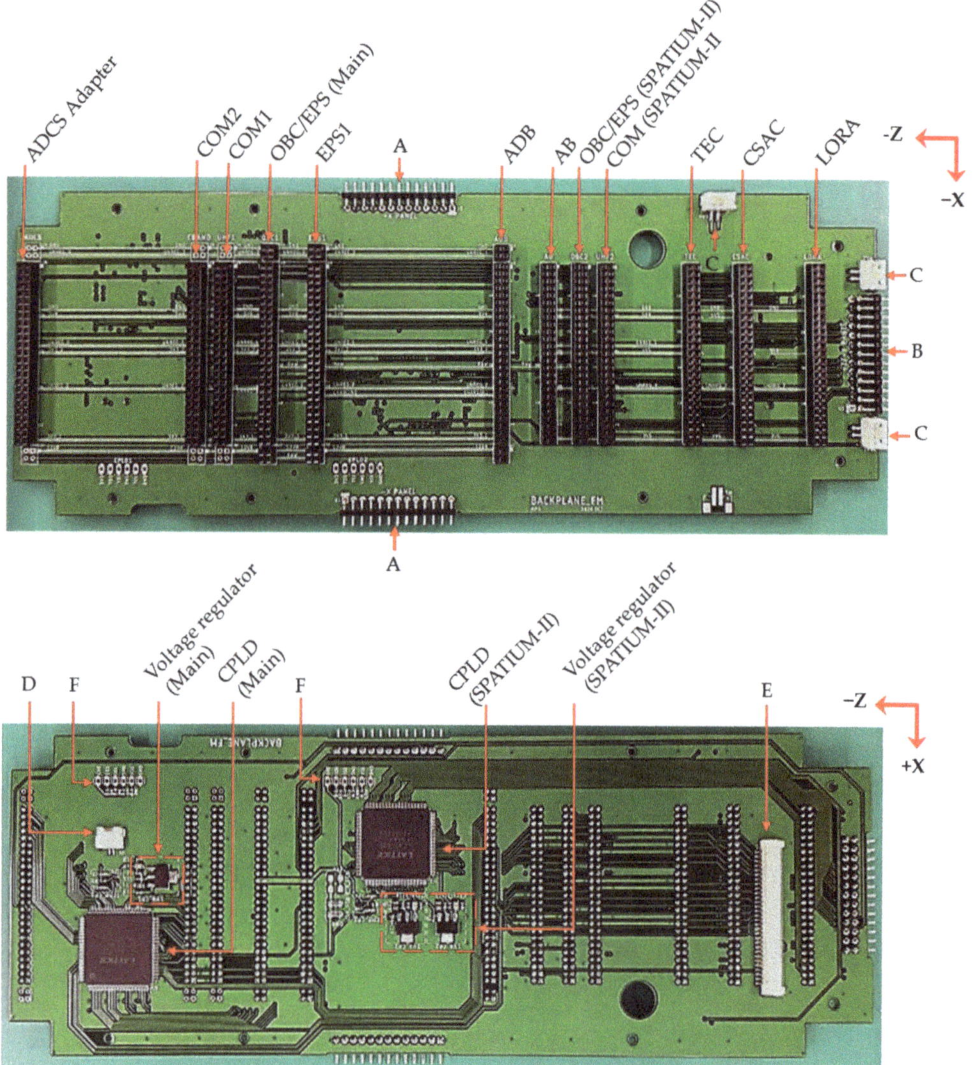

Figure 9. KITSUNE configurable backplane (**top** and **bottom** sides).

3. Tests and Results

This section discusses the tests that were conducted and the results. The first subsection covers the 3U backplane. The second subsection covers the KITSUNE backplane ground tests and on-orbit results.

3.1. 3U Backplane Verification

The four CPLDs in the backplane served as multiplexers, allowing the 13 mission boards to access the bus system's digital interfaces. Each CPLD's code was generated using very high-speed integrated circuit description language (VHDL). Lattice ispLEVER Classic was the design environment tool used to complete device design, including concept, synthesis, and simulation, as well as to generate the device joint electron device engineering council (JEDEC) programming file. Lastly, the JEDEC file was loaded into Lattice Diamond Programmer to program the CPLD via its JTAG pins.

The backplane board's functionality was validated by comparing the input and output signals of the OBC/EPS and mission board interfaces. To pass the functionality test, the two signal waveforms must be identical. The combination of logic levels on the four select pins determined which mission board had access to the bus interfaces. For example, in Figure 10, the select pins (9–12) were set to 0001. This combination allows Mission 1 access to the bus. The Digilent Digital Discovery instrument was used in the test as both a pattern generator and a logic analyzer. The instrument generated a 1 MHz clock as input to the OBC/EPS board, and the output signal from Mission 1 was compared to the input clock using the logic analyzer function. According to the waveforms, the output signal followed the logic values of the input signal. The same test was run on each of the 13 mission boards, and no differences were found.

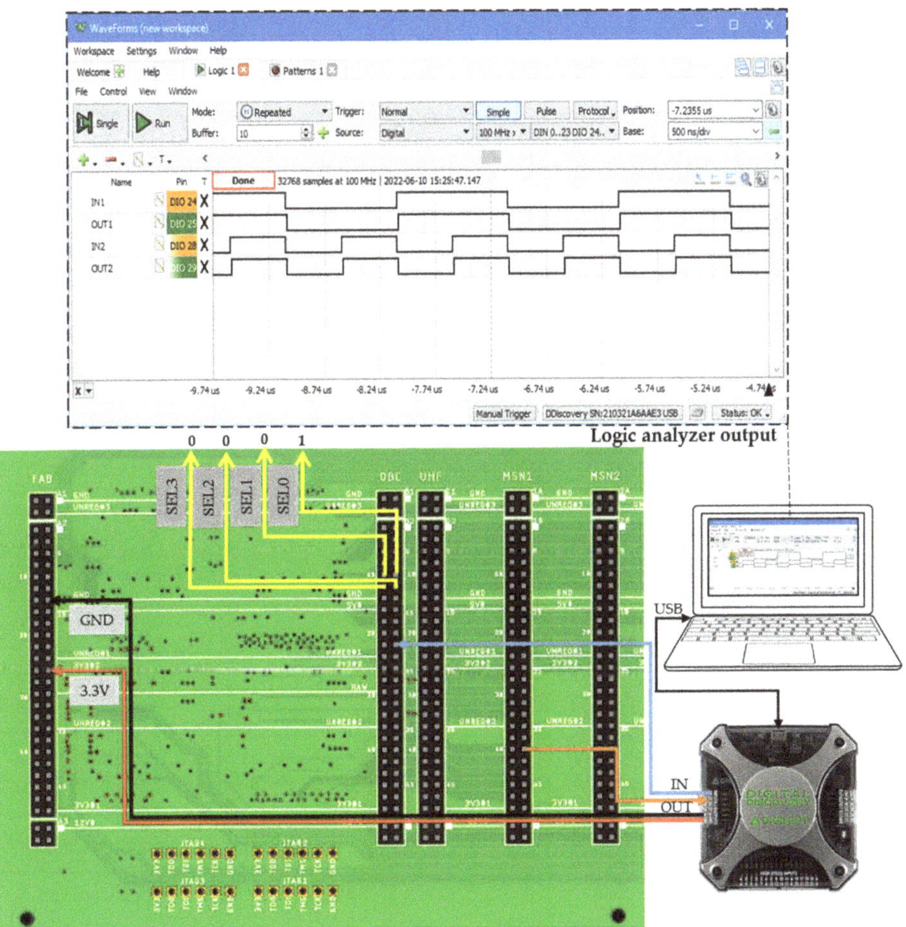

Figure 10. Functional test setup.

A bit error test was also performed to further validate the functionality of the backplane. For this test, a data stream was sent to the OBC/EPS digital interface and received by a mission through a CPLD. The data received from the mission were then compared to the data transmitted to check for possible bit differences. As illustrated in Figure 11, the Raspberry Pi module transmitted a bit array every 300 ms, and it reported the number of bit errors when it detected differences between transmitted and received data. This method was repeated for three different baud rates, such as 1 Mbps, 2 Mbps, and 4 Mbps, on each mission board. While all mission boards had no bit differences recorded, the findings demonstrated the reliability and integrity of the backplane. One important note from this test is that the time difference between the transmission and reception of the data stream does not represent the transmission delay, which is explained next.

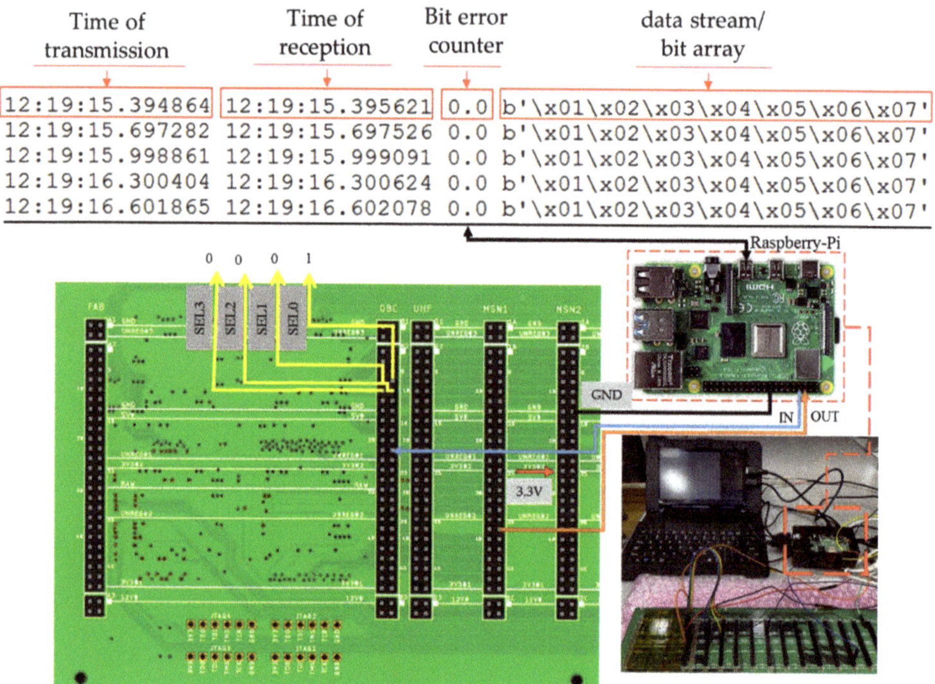

Time of transmission	Time of reception	Bit error counter	data stream/ bit array
12:19:15.394864	12:19:15.395621	0.0	b'\x01\x02\x03\x04\x05\x06\x07'
12:19:15.697282	12:19:15.697526	0.0	b'\x01\x02\x03\x04\x05\x06\x07'
12:19:15.998861	12:19:15.999091	0.0	b'\x01\x02\x03\x04\x05\x06\x07'
12:19:16.300404	12:19:16.300624	0.0	b'\x01\x02\x03\x04\x05\x06\x07'
12:19:16.601865	12:19:16.602078	0.0	b'\x01\x02\x03\x04\x05\x06\x07'

Figure 11. Bit error check test setup.

The transmission delay, defined as the time it takes for the data packet to arrive at the mission board, is another validated parameter. A significant delay may result in errors in the received data. Figure 12 shows the input and output signals as observed in the oscilloscope. The right photo shows that the two signals were identical. The waveforms were zoomed in for accurate delay measurements. All mission boards had their measurements taken, and Table 5 summarizes the measured transmission delay. The measured values were grouped based on the number of CPLDs through which the signal passed. In addition, the delay was measured twice for each mission board using different signal input sources—a Digilent Analog Discovery pattern generator (Test 1) and a function generator (Test 2). According to the results, the transmission delay of a single CPLD was approximately 5.0 ns. When a signal was transmitted through all four CPLDs, the transmission delay was measured as approximately 20.0 ns. Lastly, it is concluded that the measured transmission delay could not produce bit errors in data arrays.

Lastly, the power consumption of the backplane board was investigated since the available power for a small satellite platform such as CubeSats with limited resources determines survivability in orbit as well as the ability to support multiple payloads. The current consumption was measured at the 3.3 V input to the voltage regulator under two conditions as idle mode and active CPLDs. When all four CPLDs were active, the measured current increased from 4.3 mA to 5.6 mA. As a result, the maximum power consumption was determined as approximately 18.5 mW. This result confirmed two important points. First, the power drawn by the backplane would have negligible impact on the overall power consumption of a satellite. For instance, the backplane would only need 28.0 mWh of energy per cycle in ISS orbit. Second, the four CPLDs consumed significantly low power compared to the voltage regulator. According to datasheets, the voltage regulator quiescent current was 4.0 mA, whereas the CPLD quiescent current was only 13 uA. Therefore, if it is necessary to reduce the power consumption of the backplane even further, the focus should be on the voltage regulator rather than the CPLD.

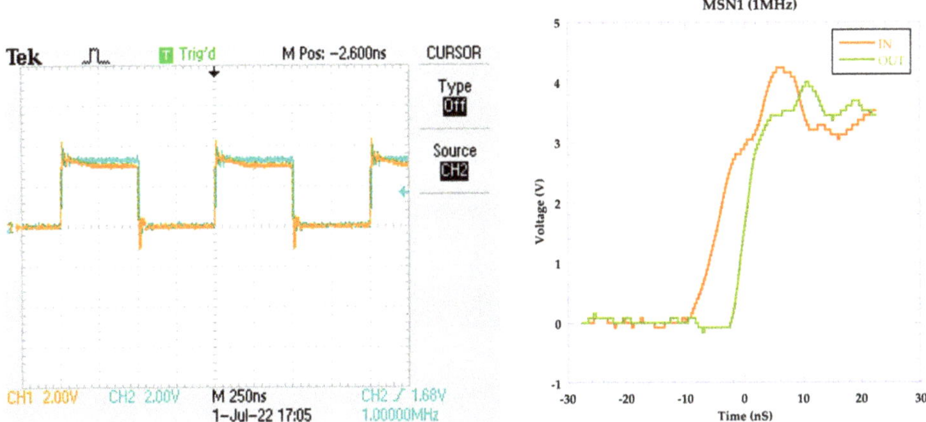

Figure 12. Input and output signals on a CPLD.

Table 5. Summary of the measured transmission delay.

Mission Board	No. of CPLDs	Transmission Delay	
		Test 1	Test 2
Mission 1–3	1	5.2 ns	6.0 ns
Mission 4–6	2	9.4 ns	10.2 ns
Mission 7–9	3	13.3 ns	14.4 ns
Mission 10–13	4	18.5 ns	16.8 ns

3.2. KITSUNE Backplane Verification

During the satellite's development, board- and system-level verifications were performed. In addition, satellite on-orbit data were available. The subsections that follow discuss both ground and on-orbit results.

3.2.1. Ground Tests

The main bus, as well as the SPATIUM-II CPLDs in the backplane, served as voltage followers. The input and output signals were compared during the board-level verification. The same test method as was used on the 3U backplane prototype was used. Table 6 lists the digital interfaces that were routed to the CPLD. The main bus CPLD routed four pairs of UART, two sets of SPI, and five DIO lines. The SPATIUM-II CPLD routed three pairs of UART, one SPI, and seven DIO lines. The data confirmed that the CPLD could route serial interfaces such as UART and SPI.

Table 6. Summary of digital interfaces routed through CPLD.

Interfaces	No. of CPLD Pins	
	Main Bus	SPATIUM-II
UART	16	12
SPI	16	8
DIO	10	14
Total	**42**	**34**

A functional test to check for bit errors was performed in the KITSUNE backplane, which is similar to the 3U backplane prototype. In the test, the Raspberry Pi module transmitted a data stream to the OBC/EPS every 300 ms, and the device then compared the signal received from the ADCS adapter board. When a difference between the two datasets was detected, the Raspberry Pi module displayed the number of bit errors. The test was conducted at temperatures ranging from $-10\,^{\circ}\text{C}$ to $+70\,^{\circ}\text{C}$ in a vacuum chamber. There was no recorded bit error during the entire 12 h test.

Two backplane-related failures were observed during system integration. During the engineering model (EM) development, a line connecting an umbilical in the ADB to the EPS1 was accidentally routed through the main bus CPLD. It was found that the line was used to monitor the battery voltage. The CPLD was severely damaged due to the error, so the ICD was thoroughly reviewed before finalizing the design to prevent the incident from reoccurring. Another recorded failure was in one of the main bus SPI interfaces. The ADCS adapter board's microcontroller could not obtain data from the ADB's magnetometer placed on the ADB, and it was found that the memory input slave output (MISO) line's signal direction was incorrect. This issue was quickly resolved by updating the VHDL code and reprogramming the CPLD.

The operation and performance of the flight model (FM) satellite were tested in various space environment conditions to demonstrate that the satellite could operate properly in low Earth orbit. A summary of the thermal vacuum and vibration tests is shown in Table 7.

Table 7. Space environment test parameters.

Test	Information
Thermal Vacuum Test	
Temperature range	$-15\,^{\circ}\text{C}$ to $+60\,^{\circ}\text{C}$
Number of cycles	2
Vibration Test	
Acceleration level	5.77 Grms
Duration	2 min

The telemetry reading from the thermal vacuum test (TVT) in Figure 13 shows that the main bus CPLD temperature was about $2\,^{\circ}\text{C}$ higher than that of the SPATIUM-II CPLD. This is because the main bus CPLD was placed right below the battery, which is a heat source, for the entire test duration. The TVT ran for two cycles, and the recorded minimum and maximum CPLD temperatures were $-3\,^{\circ}\text{C}$ and $+50\,^{\circ}\text{C}$. There was no recorded anomaly in the performance of the backplane for the entire 87 h duration.

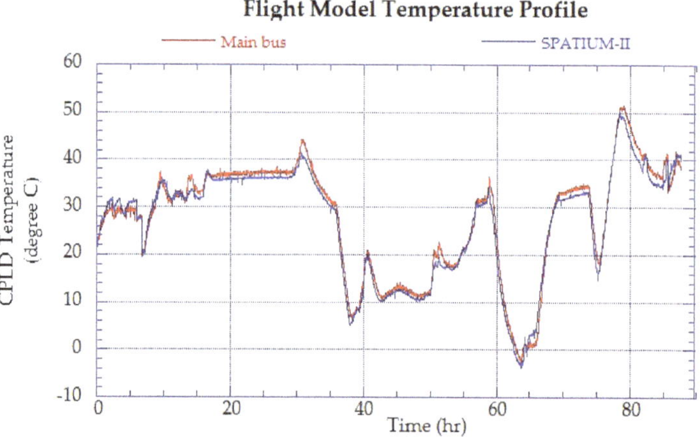

Figure 13. TVT CPLD temperature profile.

3.2.2. On-Orbit Results

Since the deployment of KITSUNE in March 2022, data communications between components in the main bus have been confirmed. For example, a ground command to activate the ADCS and set a specific ADCS mode was sent to the satellite. When the command was received, the main PIC sent a trigger through a DIO interface to activate the overcurrent protection (OCP) circuit that powered up the ADCS PIC. The main PIC then used UART lines to communicate with the ADCS PIC. The satellite log confirmed this set of actions. When a ground command to download ADCS telemetry data was sent to the satellite, the data stored in the ADCS FM were transferred to the bus system's shared FM. The data from the shared FM were then accessed by the COM PIC (via SPI) and downloaded to the ground. The downloaded satellite log and ADCS telemetry data confirmed that the routes between the bus and the ADCS were operational. DIO, UART, and SPI interfaces were connected to the CPLD. As a result, the CPLD was carrying out its function. Table 8 summarizes the on-orbit data communications via the main bus CPLD; the circles (○) in the last column indicate that all 21 digital lines were verified to be working.

Table 8. Summary of on-orbit data communications in the main bus through CPLD.

Digital Lines	Baud Rate	On-Orbit Result
2× UART (Main PIC–ADB PIC)	9600	○
2× UART (Main PIC–ADCS PIC)	9600	○
2× UART (Main PIC–CBAND)	115,200	○
2× UART (COM PIC–CBAND)	115,200	○
4× SPI (ADCS PIC–Magnetometer)	1,000,000	○
4× SPI (ADCS FM–Shared FM)	1,000,000	○
DIO (Main PIC–ADB OCP)	-	○
DIO (ADCS PIC–Magnetometer reset)	-	○
DIO (ADCS PIC–Magnetometer DRDY)	-	○
DIO (Main PIC–ADCS OCP)	-	○
DIO (COM PIC to CBAND CW)	-	○

The same verification was performed in the SPATIUM-II section. For example, a ground command to activate the LoRa payload was sent to the satellite. When the command was received, the main PIC sent a trigger through a DIO interface to activate the overcurrent protection (OCP) circuit that powered up the LoRa MCU. The main PIC then used UART lines to communicate with the LoRa MCU. The data from the LoRa payload were directly stored in the bus system's shared flash memory. The data from the shared flash memory were then accessed by the COM PIC (via SPI) and downloaded to the ground. The downloaded satellite log and LoRa data confirmed that the routes between the bus and the LoRa payload were operational. DIO, UART, and SPI interfaces were connected to the CPLD. As a result, the CPLD was carrying out its function. Table 9 summarizes the on-orbit data communications via the main bus CPLD. The circles in the last column indicate that all 15 digital lines were verified to be working.

Table 9. Summary of on-orbit data communications in the SPATIUM-II bus through CPLD.

Digital Lines	Baud Rate	On-Orbit Result
2× UART (Main PIC–LoRa MCU)	19,200	○
2× UART (Main PIC–RPI)	19,200	○
2× UART (Main PIC–CSAC)	19,200	○
4× SPI (LoRa /MCU–Shared FM)	1,000,000	○
3× DIO (Main PIC–BC OCP)		○
DIO (Main PIC–LoRa OCP)	-	○
DIO (Main PIC–RPI OCP)	-	○
DIO (Main PIC–GPS)	-	○
DIO (Main PIC to SDR)	-	○

Figure 14 shows that the on-orbit power consumption of the main bus and the SPATIUM-II was comparable to the ground data. The average power consumption per orbit of SPATIUM-II was 65 mW, while that of the main bus was 16 mW. The number of voltage regulators explains the difference in power consumption between the main bus and SPATIUM-II. The main bus has a 1.8 V regulator, whereas SPATIUM-II has 3.3 V and 1.8 V regulators. We can recall that CPLD uses 1.8 V as the supply for its LVCMOS and 3.3 V as the supply for the output logic voltage. Since SPATIUM-II does not have a 3.3 V power line, it uses the 5 V power line and first converts it to 3.3 V, then 1.8 V. The 1.8 V and 3.3 V regulators have standby currents of 4 mA and 8 mA, respectively.

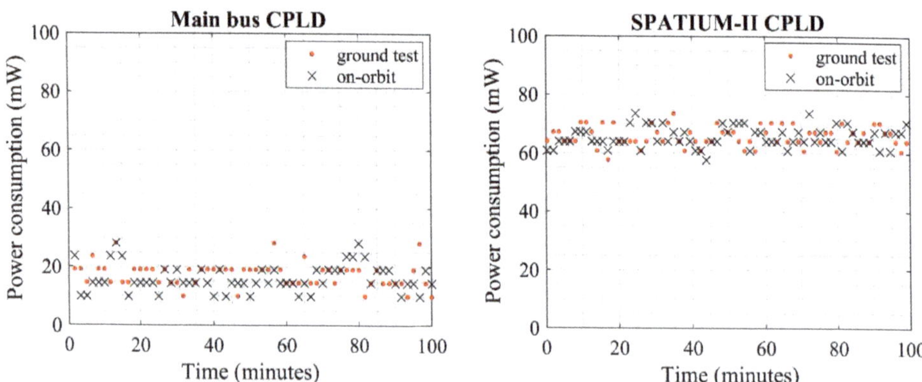

Figure 14. Ground data vs. on-orbit data power consumption for 2U main bus (**left**) and SPATIUM-II (**right**).

In Figure 15, four temperature profiles of the CPLDs per one orbit in different Sun beta angles were plotted. All graphs show that the main bus CPLD (CPLD1) temperature was at least 3 °C higher than that of the SPATIUM-II CPLD. This observation was the same as the TVT result, where the temperature difference was attributed to the main bus CPLD being placed close to the battery. The direct relation of the Sun beta angle to the device temperature was also evident in all four graphs.

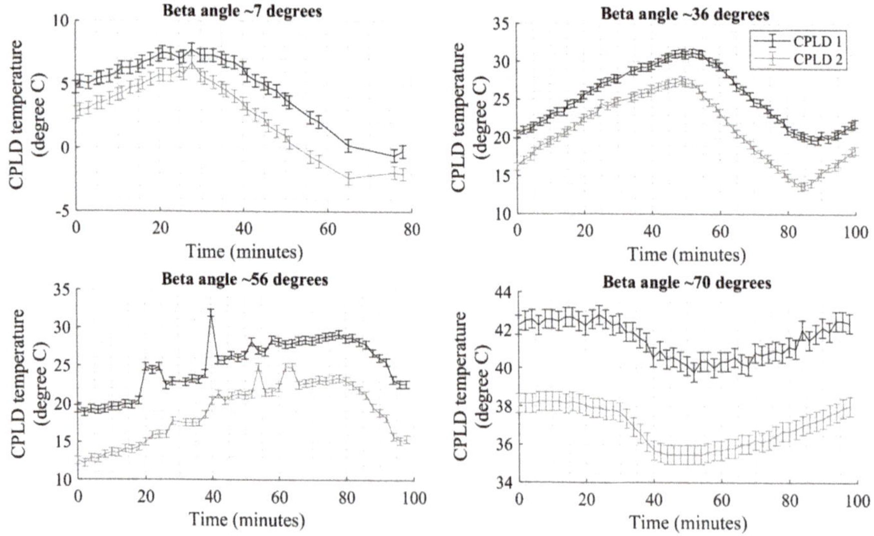

Figure 15. On-orbit CPLD temperature profile at different Sun beta angles.

3.3. Summary of Tests and Results

From the two cases above, the configurability of the backplane was demonstrated, and the following parameters were confirmed:

- *Number of configurable pins.* A maximum of 20 out of 50 pins on each mission board were made configurable. This number of pins allows the payload developers to assign any kind of serial digital interface to link over the bus.
- *Transmission bit errors.* The integrity of the data received over the configurable backplane was confirmed in two different instances. First, data were transmitted at baud rates up to 4 Mbps at room temperature. Second, the test was conducted in a vacuum environment at temperatures ranging from $-10\,^{\circ}$C to $+70\,^{\circ}$C for 12 h at a 1 Mbps baud rate. No bit errors were detected during either test, and the findings demonstrated the reliability and integrity of the backplane.
- *Transmission delay.* A maximum transmission delay of 20 ns over four CPLDs in the backplane was observed, which is acceptable for simple serial communications such as UART and SPI.
- *Power consumption.* The maximum power consumed by the backplane with four CPLDs was 18.5 mW or 28.0 mWh of energy per cycle in ISS orbit. The power drawn by the backplane is insignificant to the overall power consumption of a satellite.
- *Development time.* From the experience of satellite development, a simple routing error in a hardwired backplane would take at least three weeks and additional costs to fix. However, in the case of a configurable backplane, a simple change in the CPLD code could resolve the error within hours without incurring costs.

4. Discussions

The present paper studied the scalability of configurable electrical interfaces for two cases. In the first case, a 3U-size configurable backplane prototype was designed to support 13 mission payloads. The backplane contained four CPLDs that served as multiplexers, allowing the bus system to control all payloads with a limited number of digital interfaces. Several tests were conducted to verify its functionality and performance. The results showed that the backplane would only consume 28 mWh of energy per cycle in ISS orbit. This is considerably low and does not affect the overall power consumption of a CubeSat. A total transmission delay of up to 20 ns was measured, which is acceptable for serial communications such as UART and SPI. The short transmission delay also ensured data integrity after it was transmitted through the CPLDs. This was validated further in the bit error check, where the transmitted signal (of up to 4 Mbps) was compared to the received signal, and no bit errors were detected throughout the test.

In the second case, a configurable backplane was used in a W6U satellite that carried out complex missions. The electrical bus system was modified as needed to meet the mission requirements. Additional power lines, integration of commercial ADCS and high-speed transceivers, and an autonomous bus system for the 1U SPATIUM-II section were among the modifications. According to the on-orbit results, no anomalies were detected on any of the 21 digital connections in the main bus or on any of the 17 digital connections in SPATIUM-II, which passed through the CPLDs. Furthermore, the power consumption of the satellite's main bus and SPATIUM-II sections in orbit was comparable to the ground results.

The advantage of this study is that the scalable standard bus allows more time for integration and system-level verification, which is critical for a reliable and successful mission. The design concept could also benefit satellite developers who provide hosted payload services where bus resources are maximized to accommodate as many payloads as possible. UART and SPI communications were extensively used in the study since the bus system's command and data handling are based on these protocols. However, other protocols have not been supported by the configurable backplane. According to Cho et al. [2], other protocols, such as I2C, CAN, USB, and Ethernet, can be used and are expected to be used by CubeSat developers and vendors. A programmable backplane that supports these protocols should be developed in future work.

Turmenjargal et al. [14] recommended that the backplane is reconfigurable after the satellite has been fully assembled or even after the satellite has been launched into orbit. The former is easily implemented by inserting the JTAG pins into the umbilical. The 3U backplane prototype partially meets the latter recommendation. The preconfigured multiplexing function of the CPLDs in the backplane allows the selected payload(s) access to the bus system by sending ground commands to the satellite. Full on-orbit reconfiguration of the backplane CPLD as a contingency is still under investigation.

Kim et al. [24] stated that the flexibility of the BIRDS 1U electrical bus system is one of its key features. The BIRDS bus system was designed to be compatible with up to 3U platforms, with minimum modifications. The BIRDS bus command and data-handling architecture were retained in the 3U backplane prototype and the KITSUNE backplane. UART and SPI are the interfaces between the bus system and the payload. However, other protocols may be required by mission payloads. A bridge circuit capable of translating different protocols between the bus and payload is needed to keep the existing architecture. The bridge circuit can be implemented in the configurable backplane.

5. Conclusions

A backplane board provides an electrical interface among CubeSat components. It has additional advantages, such as ease of assembly/disassembly and fewer harnesses than a de facto standard PC/104 style interface. To confer more flexibility to the interface, a software-configurable backplane was developed. The backplane with an ispMACH LC4256ZE CPLD was demonstrated onboard a 1U CubeSat for more than 2 years in orbit. Through a CPLD, digital interfaces can be rerouted without requiring hardware changes by reprogramming the CPLD. This saves both money and time when redesigning the board.

The designed configurable interface board was verified to be scalable and adaptable to different CubeSat sizes while absorbing the challenges in the process. While a hardwired backplane is applied to a specific satellite, the scalable standard bus in this study exhibited its reusability in multiple satellite projects. Hence, it has the advantage of providing additional time for system-level integration and verification, which are essential for a reliable and successful mission. The design concept could also be advantageous to satellite manufacturers that offer hosted payload services, where bus resources are utilized to support many payloads.

Several future studies have been identified to fully realize the configurable backplane's scalability. The first is to verify that it can support communication protocols other than UART and SPI. Protocols such as I2C, CAN, and Ethernet are currently being used and are desired to be utilized in future projects by CubeSat developers and vendors. Secondly, a bridge circuit that translates different protocols can be incorporated into the configurable backplane. The bridge circuit will allow the use of the BIRDS electrical bus architecture on missions that require protocols other than UART and SPI. Lastly, on-orbit reconfiguration of the CPLD as a contingency could be explored.

Author Contributions: Conceptualization, M.S.; methodology, M.S.; software, M.S. and Y.O.; validation, M.S., Y.O., and N.C.O.; formal analysis, M.S.; investigation, M.S. and N.C.O.; resources, M.C.; data curation, M.S.; writing—original draft preparation, M.S.; writing—review and editing, N.C.O., T.Y., Y.O., and M.C.; visualization, M.S. and N.C.O.; supervision, M.C.; project administration, M.C.; funding acquisition, M.C. All authors have read and agreed to the published version of the manuscript.

Funding: This research was partially funded by JSPS Core-to-Core Program B: Asia–Africa Science Platforms (JPJSC2020005) and the "Acquisition and dissemination promotion project consignment fee for international standards related to energy conservation" by the Ministry of Economy, Trade and Industry, Japan.

Institutional Review Board Statement: Not applicable.

Informed Consent Statement: Not applicable.

Data Availability Statement: Not applicable.

Acknowledgments: The authors would like to express gratitude to the KITSUNE team members, particularly Jose Rodrigo Cordova Alarcon, Victor Hugo Schulz, Pooja Lepcha, Tharindu Lakmal Dayarathna Malmadayalage, Abhas Maskey, Adolfo Javier Jara Cespedes, Anibal Antonio Mendoza Ruiz, Cosmas Kiruki, Daisuke Nakayama, Dmytro Faizullin, Dulani Chamika Withanage, Ei Phyu Phyu, Eyoas Ergetu Areda, Fatima Gabriela Duran Dominguez, Hari Ram Shrestha, Hoda Awny A. A. Elmegharbel, Ibukun Oluwatobi Adebolu, Kateryna Aheiev, Kentaro Kitamura, Makiko Kishimoto, Mariko Teramoto, Mark Angelo Cabrera Purio, Masui Hirokazu, Mazaru Ariel Manabe Safi, Muhammad Hasif Bin Azami, Ofosu Joseph Ampadu, Sangkyun Kim, Takashi Oshiro, Victor Mukungunugwa, Yuma Nozaki, and Yuta Kakimoto. Their contributions to the development and operation of the satellite are highly appreciated.

Conflicts of Interest: The authors declare no conflict of interest.

References

1. Johnstone, A. CubeSat Design Specification Rev. 14.1, 2022. Available online: https://www.cubesat.org/cubesatinfo (accessed on 30 June 2022).
2. Cho, M.; Yamauchi, T.; Sejera, M.; Ohtani, Y.; Kim, S.; Masui, H. CubeSat Electrical Interface Standardization for Faster Delivery and More Mission Success. In Proceedings of the AIAA/USU Conference on Small Satellites, Virtual, UT, USA, 1–6 August 2020. SSC20-WKI-03.
3. PC/104 Embedded Consortium. PCI-104 Specification Version 2.6, 2008. Available online: https://pc104.org/hardware-specifications/pc104/ (accessed on 30 June 2022).
4. Pumpkin, Inc. CubeSat Kit Bus. Available online: http://www.cubesatkit.com/ (accessed on 26 July 2022).
5. Bouwmeester, J.; Langer, M.; Gill, E. Survey on the Implementation and Reliability of CubeSat Electrical Bus Interfaces. *CEAS Aeronaut. J.* **2017**, *9*, 163–173. [CrossRef]
6. Song, S.; Kim, H.; Chang, Y.-K. Design and Implementation of 3U CubeSat Platform Architecture. *J. Aerosp. Eng.* **2018**, *2018*, 2079219. [CrossRef]
7. ISISPACE ISIS. CubeSat Platform Optimization. In Proceedings of the International Workshop on Lean Satellites, International Workshop on Lean Satellites, Tokyo, Japan, 4–5 December 2019; Available online: https://lean-sat.org/proceedings_2019.html (accessed on 26 July 2022).
8. Park, J.H.; Yamaoka, K.; Tajima, H.; Inamori, T.; Miyata, K. Challenges in Common Connector Modifications: Towards a Reusable Satellite Bus. In Proceedings of the International Workshop on Lean Satellites, Virtual, 27 November–1 December 2020; Available online: https://lean-sat.org/proceedings_2020.html (accessed on 26 July 2022).
9. Busch, S.; Reichel, F.; Bangert, P.; Schilling, K. The UWE Satellite Bus, a Modular and Flexible Architecture for Future Picosatellite Formations. In Proceedings of the International Astronautical Congress, Beijing, China, 23–27 September 2013. IAC-13.B4.6 B.3.
10. UNISEC Europe. *CubeSat Subsystem Interface Definition (Proposal) Version 1.0*, 2017. Available online: http://unisec-europe.eu/wordpress/wp-content/uploads/CubeSat-Subsystem-Interface-Standard-V2.0.pdf (accessed on 26 July 2022).
11. Cho, M.; Teramoto, M.; Yamauchi, T.; Maeda, G.; Kim, S.; Masui, H. Program Management for Sustainable University CubeSat Programs Based on the Experience of Five Generations of CubeSat Projects, BIRDS Program. In Proceedings of the AIAA/USU Conference on Small Satellites, Logan, UT, USA, 6–11 August 2022. SSC22-WKV-05.
12. Cho, M. Introduction to Lean Satellites. In Proceedings of the International Workshop on Lean Satellites, Tokyo, Japan, 4–5 December 2019; Available online: https://lean-sat.org/proceedings_2019.html (accessed on 26 July 2022).
13. Faure, P.; Guerrero, G.; Diamantopoulos, S.; Pignatelli, D.; Poly, C.; Impact, C. Cal Poly CubeSat Kit—A Technical Introduction to Mk I. In Proceedings of the AIAA/USU Conference on Small Satellites, Logan, UT, USA, 7–12 August 2021. SSC21-WKI-07.
14. Tumenjargal, T.; Kim, S.; Masui, H.; Cho, M. CubeSat Bus Interface with Complex Programmable Logic Device. *Acta Astronaut.* **2019**, *160*, 331–342. [CrossRef]
15. Polansky, J.L.; Cho, M. A University-Based Model for Space-Related Capacity Building in Emerging Countries. *Space Policy* **2016**, *36*, 19–23. [CrossRef]
16. Cho, M. Open-Sourcing 1U CubeSat Platform for Education and Capacity Building. In Proceedings of the CubeSat Developers Workshop, San Luis Obispo, CA, USA, 26–28 April 2022; Available online: https://www.youtube.com/watch?v=I4FetOoJwhM (accessed on 15 July 2022).
17. Kyushu Institute of Technology. BIRDS Bus Open Source. Available online: https://birds-project.com/open-source/ (accessed on 1 July 2022).
18. Ramaprakash, P. AAReST MirrorSat Payload Interface Computer Version II. Master's Thesis, University of Surrey, Surrey, UK, 2017.
19. Cabo, P.; Lora, I. OPTOS: A pocket-size giant (MISION, OPERATION & EVOLUTION). In Proceedings of the AIAA/USU Conference on Small Satellites, Logan, UT, USA, 10–13 August 2009. SSC09-X-11.

20. EE Times | Bible, S. Chips in Space: Let's Look inside ARISSat-1 (Part 1). Available online: https://www.eetimes.com/author.asp?doc_id=1285317# (accessed on 15 July 2022).
21. Maskey, A.; Lepcha, P.; Shrestha, H.R.; Chamika, W.D.; Malmadayalage, T.L.D.; Kishimoto, M.; Kakimoto, Y.; Sasaki, Y.; Tumenjargal, T.; Maeda, G.; et al. One Year On-Orbit Results for Improved Bus, LoRa Demonstration and Novel Backplane Mission of a 1U CubeSat Constellation. *Trans. Jpn. Soc. Aeronaut Space Sci.* **2022**. to be published. [CrossRef]
22. Venturini, C.; Braun, B.; Hinkley, D.; Berg, G. Improving Mission Success of CubeSats. In Proceedings of the AIAA/USU Conference on Small Satellites, Logan, UT, USA, 12–15 March 2018. SSC18-IV-02.
23. Doyle, M.; Dunwoody, R.; Finneran, G.; Murphy, D.; Reilly, J.; Thompson, J.; Walsh, S.; Erkal, J.; Fontanesi, G.; Mangan, J.; et al. Mission Testing for Improved Reliability of CubeSats. In Proceedings of the International Conference on Space Optics 2020, Virtual, 30 March 2021; Volume 11852, pp. 2699–2718.
24. Kim, S.; Cho, M.; Masui, H.; Yamauchi, T. BIRDS BUS: A Standard CubeSat BUS for an Annual Educational Satellite Project. *JoSS* **2021**, *10*, 1015–1034.
25. Orger, N.C.; Cordova-Alarcon, J.R.; Schulz, V.H.; Dayarathna, T.; Cho, M.; Yamauchi, T.; Masui, H.; Ampadu, O.J.; Kim, S.; Lepcha, P.; et al. KITSUNE: A 6U CubeSat for 5-m Class Imaging, C-Band Radio. In Proceedings of the 33rd International Symposium on Space Technology and Science, Oita (Virtual), Japan, 28 February–4 March 2022. 2022-f-21.
26. Lepcha, P.; Dayarathna, T.; Schulz, V.H.; Orger, N.C.; Cordova-Alarcon, J.R.; Shrestha, H.R.; Kim, S.; Yamauchi, T.; Cho, M. Scalability of Kyutech Standardized Electrical Power System for Lean Satellites. In Proceedings of the 33rd International Symposium on Space Technology and Science, Oita (Virtual), Japan, 28 February–4 March 2022. 2022-f-53.
27. Cordova-Alarcon, J.R.; Jara-Cespedes, A.J.; Withanage, D.C.; Schulz, H.; Orger, N.C.; Kim, S.; Cho, M. Attitude Determination and Control System for the 6U CubeSat KITSUNE. In Proceedings of the 33rd International Symposium on Space Technology and Science, Oita (Virtual), Japan, 28 February–4 March 2022. 2022-f-40.
28. Aheieva, K.; Rahmatillah, R.; Ninagawa, R.; Adebolu, I.O.; Kim, S.; Kakimoto, Y.; Yamauchi, T.; Masui, H.; Cho, M.; Lap, C.C.; et al. Project Overview of SPATIUM-I: A Technology Demonstration Mission Toward Global Three-Dimensional Ionosphere Mapping via CubeSat Constellation Equipped with an Atomic Clock. In Proceedings of the 69th International Astronautical Congress, Bremen, Germany, 1–5 October 2018. IAC-18-B4.7.

Article

Development of Innovative CubeSat Platform for Mass Production

Eyoas Ergetu Areda *, Jose Rodrigo Cordova-Alarcon, Hirokazu Masui and Mengu Cho

Department of Electrical and Space System Engineering, Kyushu Institute of Technology, Kitakyushu 804-8550, Japan
* Correspondence: areda.eyoas-ergetu811@mail.kyutech.jp

Featured Application: The presented innovative design concept significantly impacts the development of nanosatellites such as CubeSats, particularly for mass production missions that demand high efficiency and fast delivery.

Abstract: With the recent increase in CubeSats' ability to undertake complex and advanced missions, they are being considered for missions such as constellations, which demand high development efficiency. From a satellite interface perspective, productivity can be maximized by implementing a flexible modular structural platform that promotes easy reconfigurability during the integration and testing phase. Thus, the structural design of a CubeSat plays a crucial role in facilitating the satellite integration process. In most cases, the mechanical interface implemented between the primary load-supporting structure and internal satellite subassemblies affects the speed and efficiency of satellite integration by adding or reducing complexity. Most CubeSat structural designs use stacking techniques to mount PCBs onto the primary structure using stacking rods/screws. As a result, the internal subsystems are interconnected. This conventional interface method is observed to increase the number of structural parts, while increasing complexity during integration. In this study, flexible 3U and 1U CubeSat platforms are developed, based on the slot concept. This innovative mounting design provides a simple method of mounting PCBs into the slots. The concept is evaluated and verified for its feasibility for mass production applications. Count and complexity analysis is carried to evaluate the proposed design against the conventional type of structural interface methods. The assessment reveals that this new concept demonstrates a significant improvement in the efficiency of the mass production process.

Keywords: flexible integration; CubeSat structure; slot; efficient; mass production

Citation: Areda, E.E.; Cordova-Alarcon, J.R.; Masui, H.; Cho, M. Development of Innovative CubeSat Platform for Mass Production. *Appl. Sci.* **2022**, *12*, 9087. https://doi.org/10.3390/app12189087

Academic Editors: Simone Battistini, Filippo Graziani and Mauro Pontani

Received: 28 July 2022
Accepted: 5 September 2022
Published: 9 September 2022

Publisher's Note: MDPI stays neutral with regard to jurisdictional claims in published maps and institutional affiliations.

1. Introduction

A CubeSat is a standardized, modular nanosatellite class satellite with a basic dimension of a cubic decimeter [1]. Over the last decade, CubeSats have been used as a fundamental tool for academic institutions to conduct innovative in-orbit technology demonstration missions [1] (p. 59) within a limited budget and human resources. Recently, small space businesses started looking at the business potential in the area by engaging in the development of CubeSats' subsystems. The continuous innovation in small satellite technology has been the driving force behind an increase in the capability of these classes of satellites in recent years. These tremendous advancements are mainly due to the continual miniaturization of microprocessors [1,2]. As a result, various missions are being executed on a single CubeSat platform, with increasing mission sophistication. For these reasons, subsystems are designed to have multi-functionality requirements within stringent mass and volume constraints.

As a result of this increased capability, CubeSats' applications have gradually expanded from simple technology demonstration missions and remote sensing mission platforms to more advanced and sophisticated missions. Mega-constellations using distributed space systems (DSS) [1] and deep space exploration [3] are some examples of current possibilities. These mega-constellation missions require the deployment of a massive number of CubeSats for the intended aims. Constellations using the CubeSat platform guarantee more ground coverage, greater built-in redundancy, and shorter revisit times, available for the same price as conventional big satellites [4].

Table 1 lists commercial CubeSat constellations with more than two functional spacecraft in orbit as of July 2022. This shows the gradual increase in CubeSat constellation missions in different application areas. The stakeholders are predominantly private space companies with commercial purposes to provide space-based satellite solutions by utilizing state-of-the-art technologies in the field. These services aim to provide affordable space-based solutions, while maintaining a profit margin [1] (p. 15).

The cost of mega-constellation development depends on several cost driver factors with different levels of effect on the total project cost. The supply chain is a crucial factor which should be managed strictly to avoid any delay in the schedule. However, efficiency during the design and development phase also plays a decisive part in determining the project time and cost. Inefficiency in the development process leads to unnecessary costs that lead to potential bankruptcy [5]. In addition, it causes unwanted workmanship errors. Incorporating the design for assembly and manufacturing (DFAM) principle during the conceptual design phase of the satellite development leads to an increased efficiency and/or productivity [6] (p. 19). By observing further down to the level of the structure of the subsystems in satellite development, improvement in efficiency can be attained by reducing the assembly steps and simplifying the effort required during the integration, assembly, and testing processes. This is done by modularizing and standardizing the design process, while keeping a certain level of flexibility [6] (p. 74–82). Therefore, it is vital to critically evaluate all design processes and identify inefficiency in the process as much as possible.

Design modularity is an essential element in satellite development which significantly affects the level of effort in the project team and its effect on development costs can be quantified [7]. A modularized satellite bus has standardized and reconfigurable components [7]. The reduction in design effort allows for shorter development time and this in turn results in an improvement in productivity and efficiency, particularly for mass production applications [7].

Mass production, by nature, requires frequent design iteration and continuous testing during the beginning phase, when a new batch of satellites is produced due to the required tests and verifications to verify the design [1]. These necessitate the frequent assembly and disassembly of satellite subsystems. Here, easy reconfigurability plays a key factor in determining the speed of the integration. The arrangement of components and the nature of interconnectedness inside the satellite affects the speed of the development when it comes to easy access. However, in the subsequent development, as the learning curve increases, the ease and small number of assembly steps become significant.

Table 1. Commercial CubeSat constellations with more than two functional spacecraft in orbit as of July 2022 [8].

Organization	Number Launched	Target Number	First Launch	Form Factor
Planet Labs	519	>150	2013	3U
Spire	160	>150	2013	3U
GeoOptics	9	50	2017	6U, 12U
Helios Wire	5	30	2017	16U
Swarm Technologies	177	150	2018	1U/4U
Kepler Communications	19	140	2018	3U, 6U
Fleet Space	7	140	2018	1.5U, 3U, 12U
Astrocast	12	80	2018	3U
Aistech	4	20	2018	6U, 2U
Guodian	15	38	2018	6U
Kuva Space	3	100	2023	6U, 2U
Lacuna Space	6	240	2019	3U, 6U
SatRevolution	7	1024	2019	6U, 2U
UnseenLabs	7	50	2019	6U
Kleos Space	12	40	2020	CUBESAT

On the other hand, an increase in the number of structural parts increases the inter-connectivity of parts, which also increases interdependency. Any problem with one of the components may affect the adjacent connected parts. Hence, the implemented mechanical interface between the internal subsystems and structure plays a crucial role in speeding up the integration and delivery of the final products. In the present paper, a unique interface design is introduced to tackle the issues associated with the most common mechanical interface.

Literature Review

The primary function of a CubeSat's structure is to provide support and protection to the internal and external satellite bus system and payload components throughout the satellite's lifetime, during development on the ground, during launch, and in the orbit environment [9] (p. 133). Besides this basic requirement, the CubeSat structure is also required to provide an easy and flexible platform for quick integration during frequent testing phases [1] (p. 322), especially for missions involving mass production. Satellite assembly and integration in a mass production environment is unique from mass production of products for ground use, as it is often limited by various tight constraints, such as time due to the fixed launch and specific market windows to meet the targeted customer demand. In constellation missions, however, as reliability gradually increases, generation by generation, the first batch of the satellite often requires extensive functional and environmental testing [1] (p. 324).

CubeSat structures can be developed from a custom design or procured from the CubeSat market as commercially off-the-shelf (COTS) designs. Several companies offer a set of ready-mades and verified CubeSat subsystems, including a fully assembled structural platform. This approach reduces the development time, since testing and verification requirements can be skipped in most cases, when the subsystems have already been verified and demonstrated in an orbit environment. Verification, however, is a mandatory requirement for custom structural designs to conduct rigorous screening and reliability tests/inspections [10]. The ready-made CubeSat platforms, to some extent, help to shorten the development time [9]. However, their fixed designs impose limitations on the flexibility in defining the placement of mission payloads that are unique in size. CubeSat vendors such as GOMSPACE [11], PUMPKIN [12], ISIS [13], and Complex system and small satellite (C3S) [14] provide several standard design options for CubeSat structures.

These COTS structural designs have several predefined attachment points, providing freedom when mounting the internal subassembly. Some of these COTS structures are made of several modular frames or plates which can easily be expanded to other CubeSat form factors. Examples can be found in references [12,13]. Most of these structural designs accommodate satellite subsystems that were developed on their own to satisfy their specific interface requirements. Custom structural design addresses these interface limitations by developing a structure that is tailored to specific satellite mission requirements. Despite its flexibility, the development and design verification steps take a comparatively long time. The decision to invest is mostly made based on the available budget and time.

Even if these design solutions are aimed at solving a specific problem, they still utilize many structural components/parts. The number of structural parts is one of the critical issues that should be addressed, particularly for an efficiency-demanding mass production application. Therefore, a standard, flexible platform with a minimum number of parts and joints is very important for mass-producible, fast-delivery applications.

The internal configuration of subsystems mostly depends on the type of electrical interface implemented. Commonly, there are two main electrical interface methods implemented in CubeSats, such as PC/104 and the backplane board interface. The first is the PC/104 interface, where PCBs are stacked one on top of the other through an extended "stack-through" connectors module. Images of this interface can be found in [15]. In this case, standoffs between the subsystems are used to provide mechanical support and transfer the load to the main structural frames. Many commercial companies make this PC/104 interface.

On the other hand, the backplane board (BPB) approach uses a common interface board, where all the other internal PCBs are connected to a motherboard, as shown in Figure 1.

Figure 1. BIRDS satellite BPB interface [16].

In terms of mechanical support, similar to the PC/104, four long stacking rods connect all of the individual PCB boards once electrical connection has been established with the connectors on the BPB. Cylindrical spacers are inserted into the rods between each consecutive PCB stack to constrain their position, as shown in Figure 2. Recently, several universities and some commercial vendors have increasingly adopted this interface approach, as it provides better flexibility in electrical connections. For instance, the University of Würzburg in Germany developed a UWE-4 satellite in collaboration with other institutions that implements a backplane board electrical interface [17]. The backplane board type of interface is also the currently preferred approach used on most of the satellite missions at the Kyushu Institute of Technology (Kyutech) due to its high modularity [16]. So far, 19 CubeSats have been built based on the backplane style and 16 satellites have already been launched.

Figure 2. Stacking of PCBs with long rods.

However, from a fast delivery and quick assembly point of view, both electrical interface approaches with a stacking-type mechanical interface still pose some fundamental problems, mostly due to the mounting method of internal subsystems using long stacking rods. The method of inserting these long screws through the PCBs and the structure generally requires some level of effort for maintaining the alignment in this conventional type mechanical interface. The placement of the spacers between the PCBs particularly is often a challenging task during assembly, due to the required simultaneous task of inserting the stacking rods and placing the spacers, aligned with the holes in the PCBs, not to mention the substantial number of structural parts utilized, which consumes time during assembly. This potentially introduce workmanship errors to the assembly process, damaging the connector pins, etc. In addition, the additional frames used to connect these stacking rods to the structural frames, as shown in Figure 3, increase the total part count of the structure.

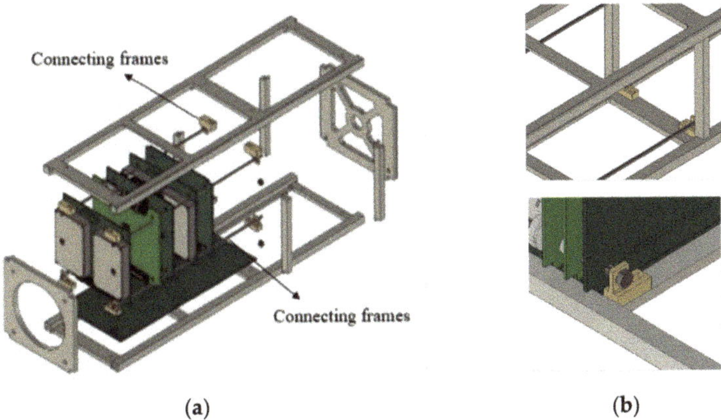

Figure 3. (**a**) Exploded 3U model based on conventional interface methods and (**b**) connecting rods for mounting internal subassemblies.

Both the above electrical interface approaches characterize the most conventional CubeSat interfaces. Several design solutions have been proposed to address the issues of flexibility and modularity. This study investigates similar design concepts available and tries to analyze their feasibility for the assumed applications.

A commercial space company developed a card-slot-type structural design in which individual subsystems are inserted into defined slots using special spacers holding the internal subsystems. An example can be seen on the company's website [14]. This design concept can solve the above issue of interdependency between subsystems during integration. However, the retractable PCB holder/spacer used to prevent the direct contact of the subsystems with the main structural frames increases the total structural part count and possibly the complexity during assembly.

Another study conducted by Istanbul Technical University for the PSAT-II mission [18] considered a modular 3U structure design with evenly spaced slit features on the interior faces of the four main frames to mount the internal subsystems. The PCBs were inserted into these tacks. The design in the study aimed to increase flexibility in rearranging internal subsystems, without the need to change the design of the structure, and to demonstrate it on a standardized bus in orbit. However, despite its concept of modularity, the study did not clearly show the electrical interface methods, nor the mechanical interface used to mount the internal PCB into the slots. insertion of PCBs directly into the given slits could potentially damage the PCB as well as the sensitive components, especially in the launch environment.

In general, the most important efficiency parameters, such as structural part count and number of assembly steps, which directly influence the level of complexity of the integration, have not been adequately addressed in either design. In fact, the complexity and the number of assembly steps has a direct relationship with the number of parts that exist. In addition, complexity can be a result of the mechanical interface method implemented. Therefore, it is important to have a platform where few parts and joints are needed for integration, with an assembly procedure easy to comprehend for someone who is not familiar with the tasks.

Once a design is optimized to the point where it has few structural parts and subsystems with little interdependence, it is important to standardize the platform so that few design changes are required in subsequent developments. This can be achieved by defining the interface between the satellite subsystems and the structure. At the same time, it is also important to allow some level of flexibility for subsystems or payloads with unique dimensions to fit into the structure without difficulty.

The goal of this research is to develop a flexible standard efficient 3U STM (structure and thermal model) which is suitable for mass production applications. The design concept is developed as an STM for an ongoing 3U CubeSat project.

Recalling the common conventional type of structure, which is made up of several structural frames, rods, and plates, it is important to address the associated challenges. Therefore, a slot-based structural design is proposed with a unique interface between the internal PCB and the structure. The following are design parameters defined to evaluate the design objective.

The design goals:

1. A reduction in the number of parts;
2. A reduction in complexity;
3. A reduction in the number of assembly steps;
4. To show the scalability of the design to the 1U CubeSat form factor;
5. To comply with all interface and launch requirements;
6. An adequate volume for the communication payload.

These design goals are used as evaluation parameters to compare the new design concept against the existing traditional, conventional CubeSat structural designs. After this initial phase of design, an EM and FM models are currently being developed with further design optimization. In this study, the initial STM development, from concept design to manufacturing and testing, is explained.

The purpose of the present paper is to develop a flexible and modular standard 3U CubeSat structure that is suitable for highly efficiency-demanding mass production applications. The novelty of this paper is the interface method used to mount the internal subsystems onto the slots, which provides a much lower part count with an easy assembly technique compared to the existing slot-type structural platforms. In addition, the slots are standardized to reduce the need for a change in the interface design of internal components, while at the same time facilitating the easy relocation of subsystems within the platform during the configuration definition phase.

This paper consists of seven sections. The second section describes the conceptual development of the structural design. The suitability of the proposed design is evaluated against the existing structural platform in the third section. The assessment results are analyzed in the fourth section. The fifth and the final sections provide the conclusion and directions for future work.

2. Conceptual Development of Structural Design

To come up with a design that solves the aforementioned issues, several design concepts are considered at the beginning of development based on the defined design goals. The CubeSat standard interface developed by California Polytechnic State University is used as a design reference [19]. An example of a design specification document is the JAXA's Japanese Experimental Module (JEM) Payload Accommodation Handbook, for those CubeSats that are released from the JEM Remote Manipulator System using the JEM Small Satellite Orbital Deployer J-SSOD installed in [20]. In addition, constraints from previous heritage subsystems are taken into account during the development of the design concept. Satellite bus subsystems, such as the electrical and power subsystem (EPS), onboard computer (OBC), and communication (COM) boards, have been used for several previous projects. Since design modification of these subsystems costs a significant amount of money, the basic shape and dimensions of these internal PCBs are kept and used as additional design constraints.

During the development of this new design, Fusion 360 software is used to model the structure. After a series of design iterations and evaluations, a slot-based design concept is chosen for a 3U STM using a total of eight aluminum structural components, as shown in Figure 4. Two parallel mirror slot plates are designed to mount the satellite's internal bus and payload. They are attached to the two bottom and top plates. Additionally, for this STM, two side frames on the Plus and Minus X axis are used to connect the top and bottom plates to give the structure extra stiffness and provide additional support for PC/104 standard subsystems such as the ADCS units.

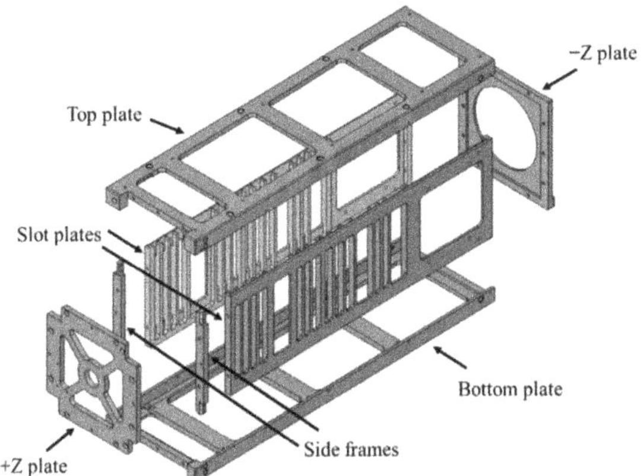

Figure 4. Exploded view of the Slot-based 3U structural model.

2.1. Standardizing the Slots

Greater attention is given to standardizing the width of the slot rails to reduce the variety of spacers used. To allow easy interchangeability between subsystems, three standard slot widths are defined, as shown in Figure 5: 6 mm, 9 mm, and 18 mm. Besides these standard slot widths, free-sized slots are also considered to accommodate a few oversized components, depending on the type of mission. In this particular design, the battery and the communication payload are assumed to have relatively large sizes; therefore, slots of 27 mm and 85 mm in width were adopted.

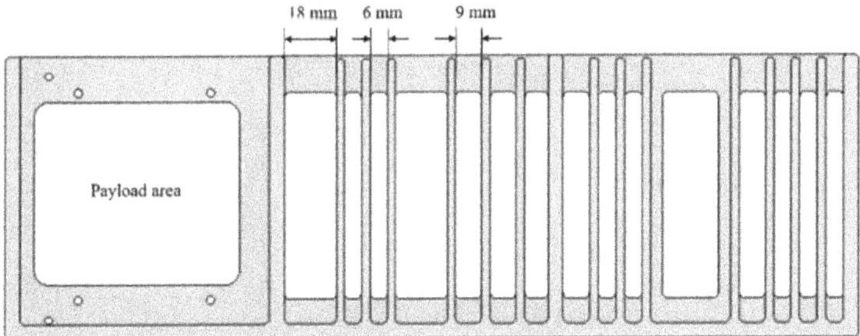

Figure 5. Slot plate with standardized slot width for 3U model.

2.2. PCB Holder (Spacers)

Internal subsystem components, mostly PCB boards, are inserted into these standard slots using a set of special PCB holders (spacers) made of softer plastic material to protect the sensitive surface of the PCB. The spacer removes the need for long screw rods to mount the internal subsystems, as in the case of conventional CubeSat design. The spacers' width is derived from the defined slot width. The spacers also have a cutout slit according to the thickness of each PCB or component, as shown in Figure 6. They are attached to the PCBs at four corners by just applying a small push force for a snap-fit. The tolerance of the cutout on the spacers should correspond to that of the PCB/components' thickness tolerance. Similarly, the tolerance of the spacers' width should also correspond to the tolerance of the slot width on the structure.

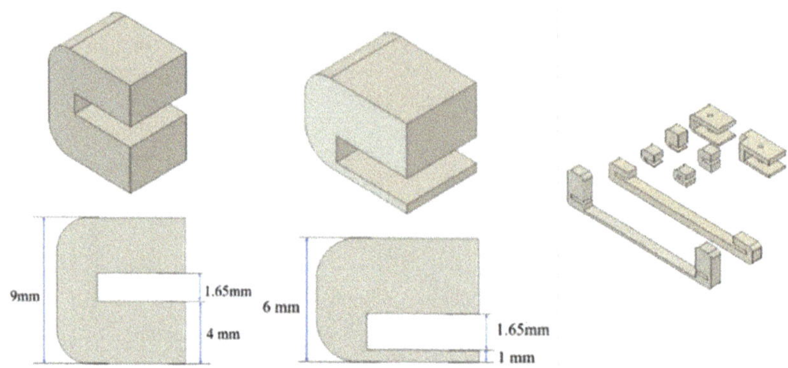

Figure 6. Standard 6 mm, 9 mm, and free-size spacers.

Regarding the spacer material, different plastic materials were assessed. However, PEEK (polyether ether ketone) was chosen due to its high strength and stiffness.

2.3. Internal Subsystem Interface Method

When the PCBs are to be inserted into the slots, the spacers are simply attached to the PCB edges at the four corners. The PCBs lock themself between the slot plates and bottom plate once the connection has been established electrically to the backplane board (BPB). The tolerance of the spacers is designed to allow the easy slide action of the PCBs during insertion and removal. The top plate fully encloses the PCB assembly, constraining the movement inside, as shown in Figure 7.

Figure 7. (**A**) PCB insertion into the slots. (**B**) The top plate encloses the satellite assembly.

2.4. Finite Element Analysis (FEA) and Environmental Test

After developing the 3D model of the STM, finite element analysis (FEA) was carried out to check the stiffness both on individual components and assembly levels using the Autodesk Fusion 360 simulation environment. The model was simplified by removing rounds and chamfers. A total weight of approximately 3.6 kg was assumed to resemble the actual total weight of the satellite at the time of the STM phase. In addition to the frequency analysis, static analysis was also conducted. All the boundary conditions were applied according to the JAXA requirement handbook [20].

The production of the STM was ordered after confirming the analysis results (Figure 8). The lesson learned when placing the production order was that the tolerance requirement of the slots as well as spacers should be carefully specified and checked, as the spacers must easily slide into the slot when assembling the PCBs. The parts were assembled using dummy internal and external subsystems, which have similar mass. Then, a vibration test was conducted to verify whether the satellite could survive the severe launch environment. Different launch loads, such as random and quasi-static vibration loads, were applied to the satellite. Since the satellite is expected to be deployed from ISS, a shock test was not necessary. Qualification-level random and sine-burst vibrations were applied to the test article. Modal surveys were taken on each occasion before and after the random and sine-burst vibration to check whether the signature of the fundamental frequency had shifted. The vibration results are discussed in Section 4.1.

Figure 8. Structural and full satellite assembly of STM 3U.

3. Design Assessment Methods

The proposed 3U STM design was critically assessed against the design goals. Since a 3U satellite has not been developed by Kyutech in the past and an actual satellite was not available to evaluate with, for assessment purposes, a demo 3U CubeSat model based on the conventional type was modeled with a backplane board electrical interface. The structure utilizes several long rods to stack up the PCBs, as shown in Figure 9. To avoid any design merits due to the design difference between the two concepts during the assessment, the quantity and shape of the basic structural parts, such as the top, bottom, and side plates, are kept similar. This helps the assessment to focus on the interface method used between the main structure and internal subassembly instead of the change in design. However, often the conventional design may only have four rails along the Z axis instead of plates and connecting rods in actual case.

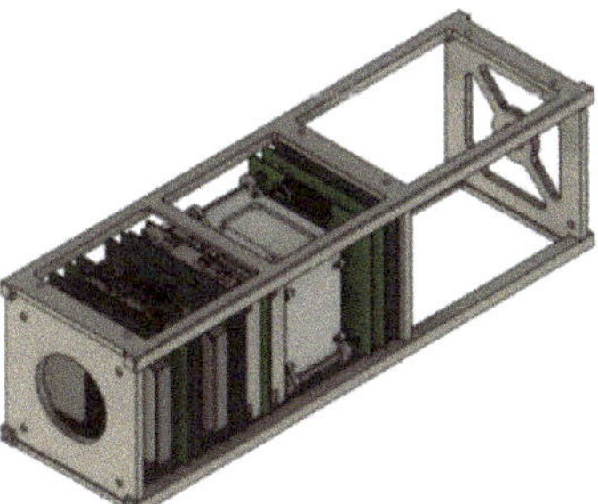

Figure 9. Conventional demo 3U model.

As previously described, important evaluation parameters, such as part and assembly count analysis and complexity analysis, were used to assess the design goals to measure the impact of the change in design on the integration efficiency of mass production missions. First, the count analysis was conducted by comparing the total quantity of only structural parts, such as plates, frames, spacers, and fasteners. The internal subsystem components were not included in this analysis because that may vary depending on the mission type. Second, complexity analysis was carried out to measure the level of difficulty encountered during the integration phase. Then, the scalability of the design concept to other CubeSat form factors was demonstrated using a 1U platform. Since this study focuses on 3U CubeSat, due to the satellite mission requirement, the new design concept was developed for that purpose. However, it is important to assess its scalability. Due to the popularity of the 1U CubeSat platform in the CubeSat community, checking the scalability of the 3U slot-based design is particularly important. A similar assessment of count and complexity analysis was conducted by comparing this platform with the previous 1U conventional-type BIRDS-3 satellite platform, which was developed by Kyutech. The BIRDS-3 satellite has already finished its two-year mission in ISS orbit. Finally, the assembly step count was used to evaluate the speed at which each assembly action was carried out. Here, the steps are for the complete satellite assembly only, without the solar panels.

3.1. Count Analysis

In this analysis, the quantity of primary structural parts of both the 3U flexible structure and conventional-type demo model is considered, as shown in Figures 9 and 10. For easy visualization, the structural parts are categorized into four groups: the main structural frames, which include rods and rails; spacers; long rods; and fasteners. This helps to visualize the major changes that contributed to the total difference. Since the quantity of screws affects the assembly process, structural screws for both designs are included in the analysis. These screws are the one that are used to secure the internal satellite components onto the structural frame. The screw count reflects the existing number of joints in the design and how parts are interconnected in the system.

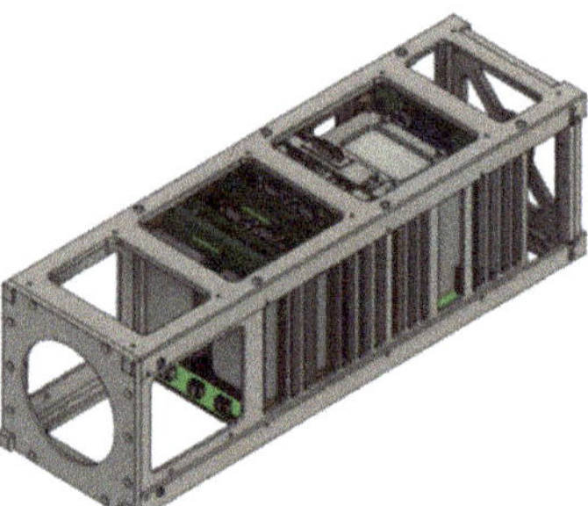

Figure 10. Slot-based 3U model.

3.2. Complexity Analysis

To evaluate design complexity, complexity metrics developed for the industrial assembly were used based on a modified version of Hückel's Molecular Orbital Theory. This metric is used to calculate the complexity of systems from different perspectives [21]. In this complexity measurement, total system complexity has three variables: The first variable, C_1^s, measures the complexities of each component on the individual level. The second complexity variable, C_2^s, measures the complexity during a pairwise interaction (liaison relationship) of two or more components. The last variable, C_3^s, is a topology complexity that measures the effect of system architecture or the arrangement of the different interfaces. Thus, the total complexity of the satellite can be calculated with a combination of these three variables using Equation (1). The detailed derivation and explanation of the metrics and each variable can be referred to in [22] (p. 48). The total complexity of the structural assembly of the satellite provides a very good piece of information about the complexity/difficulty level of the two design concepts, especially during integration. System complexity can also easily be visualized using a liaison diagram, as shown in Figure 11.

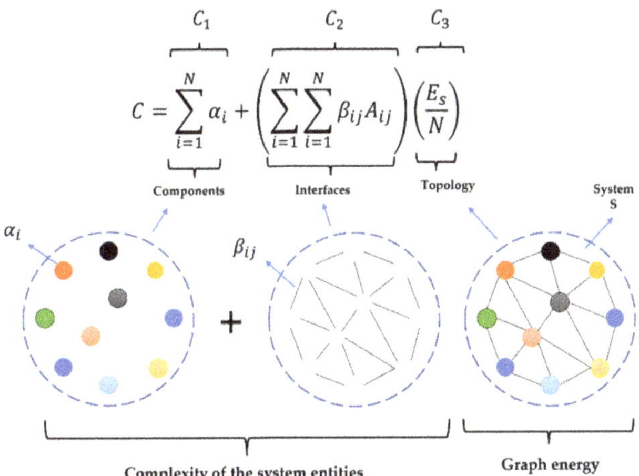

Figure 11. Elements of the overall complexity metrics shown with a liaison diagram.

The first complexity variable of the satellite is the summation of the complexity value α_i^s for all satellite components, and can be calculated using Equation (2), where α_i^s represents the complexity of the satellite components, and i and N_s denote the total number of parts forming the satellite. This indicates the technical/agronomical difficulty/efforts associated with the development and management of each of the assembly components in isolated conditions.

$$C^s = C_1^s + C_2^s C_3^s \tag{1}$$

$$C_1^s = \sum_{i=1}^{N_s} \alpha_i^s \tag{2}$$

Similarly, the liaison's complexity of assembly, C_2^s, is the sum of all the complexities in the pairwise interaction between two linked structural components and is calculated using Equation (3) below. The variables β_{ij} and ASM_{ij} denote a binary adjacency matrix derived from the components that have physical interaction in the assembled state.

$$C_2^s = \sum_{i=1}^{N_s^e} \sum_{j=1}^{N_s^e} \beta_{ij}^s ASM_{ij}^{minimal} \tag{3}$$

$$[ASM]_{ij} = \begin{cases} 1, & if\ i\ and\ j\ are\ connected \\ 0, & otherwise \end{cases} \tag{4}$$

$$\beta_{ij}^p = \frac{f_f^E + f_f^F + f_f^G + f_f^H + f_f^I + f_f^J + f_f^k}{\beta_{max}^p} \tag{5}$$

On the other hand, in the last term, C_3^s is a global measure that encapsulates the inherent arrangement of connections and is calculated by the graph energy [23]. Note that the term C_3^s requires knowledge of the complete system architecture and signifies a global effect, the influence of which could be perceived during the system integration phase [22]. The complexity of the product's topology, C_3^s, is computed by

$$C_3^s = \frac{E_{ASM^{minimal}}}{N_s^e} \tag{6}$$

$$E_{ASM^{minimal}} = \sum_{i=1}^{N_s^e} \sigma_i \tag{7}$$

The assembled system can be expressed graphically using a combination of components and liaisons. The components can be essential components that behave as a single unit, quasi-components that are used to connect the essential components, and virtual components, which are non-mechanical components such as solders or glues. However, for simplicity, the graphical expression of the satellite only shows the essential components in compact form. In the diagram, the components are denoted by circular nodes, whereas the liaisons are denoted by edges, as shown in Figures 12 and 13. The spacers and screws are considered quasi-essential components and therefore are not taken into consideration in this liaison diagram. The satellite components represented by the nodes are listed in Appendix A in Table A5.

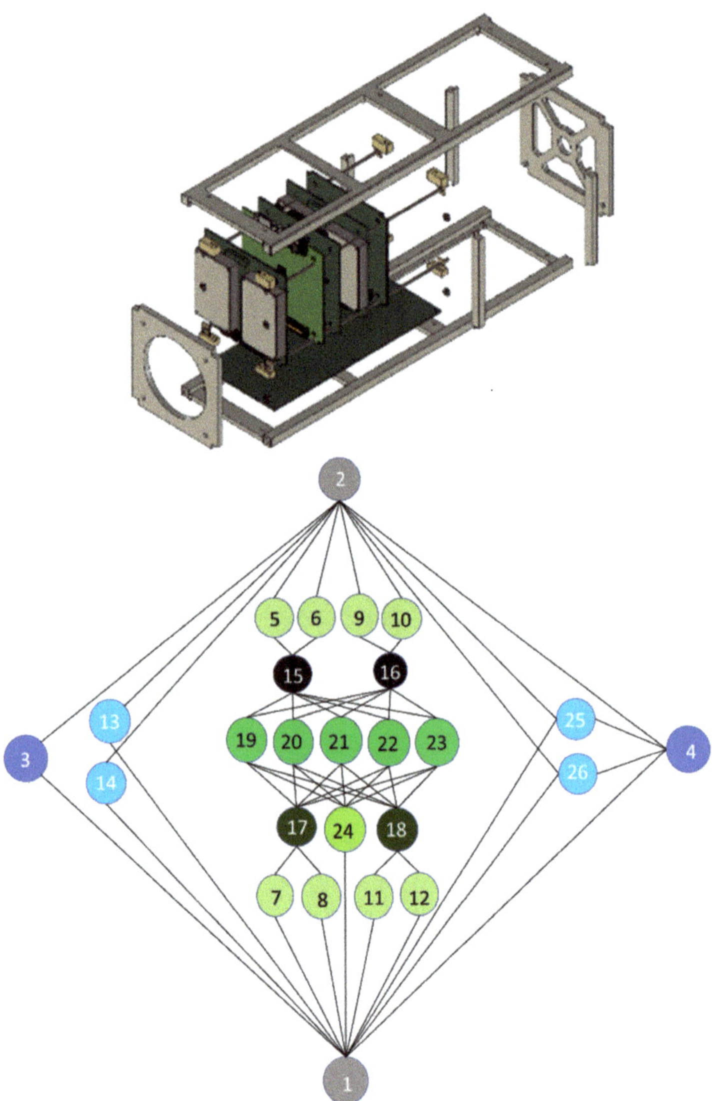

Figure 12. Liaison diagram of 3U demo conventional model.

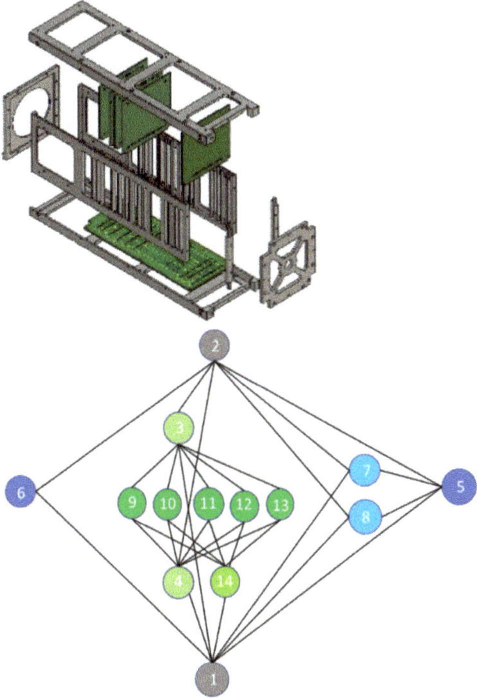

Figure 13. Liaison diagram of 3U slot-based model.

4. Results

4.1. Structural Analysis and Vibration Test Results

The results of frequency analysis show that the first mode of vibration in the fully assembled state of the satellite is 251.9 Hz, as shown in Figure 14, which is well above the minimum requirement of 60 Hz specified in the JAXA accommodation handbook document [20]. On the other hand, the maximum von Mises stress when the expected launch load is applied is 96.09 MPa, 111.8 MPa, and 99.22 MPa in the X-, Y-, and Z-axis, respectively. The safety margin is calculated using these stress values.

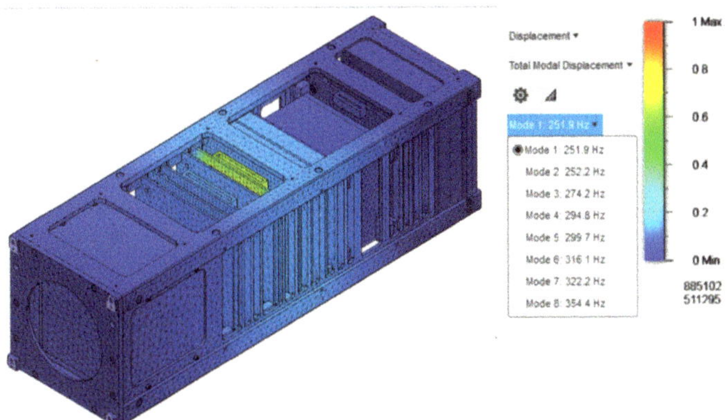

Figure 14. Frequency analysis of the 3U model.

In addition, random and sine-burst vibration tests were conducted at the Kyutech Nanosatellite development facility. All the test results show that the satellite's fundamental frequency did not change before and after the vibrations, as shown in Figure 15.

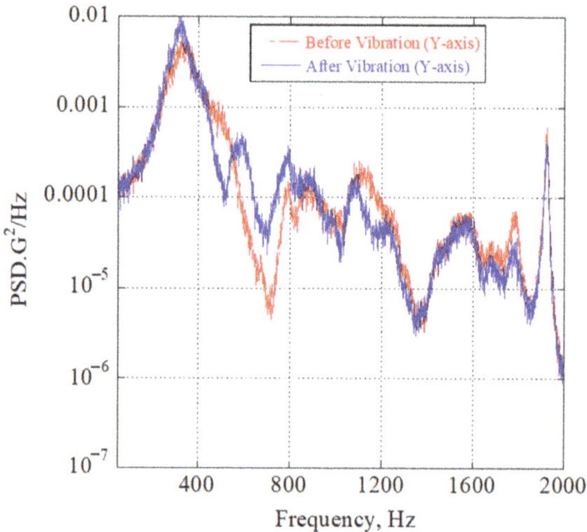

Figure 15. Modal survey before and after random vibration on Y-axis.

4.2. Evaluation Result for Count Analysis

The count analysis results for the 3U CubeSat are shown in Table 2. Both the part and screw count show a considerable amount of change in favor of the slot-based design concept. A 42% reduction is achieved by changing the interface method of internal subsystem, since the design merit is out of the equation because the designs of major structural parts in both concepts are similar in shape and quantity. Such a significant improvement, from 14 to 8, attained due to the reduction in the number of structural parts, such as the connecting frames and long rods that hold the PCB stacks and spacers. The conventional design utilizes many spacers that must be mounted between each stack, whereas in the case of the slot design, only four spacers are used for each PCB. The number of structural screws, however, does not show substantial improvement, being only 5%. This is because this STM utilizes several redundant structural screws. This point was taken as a potential area for improvement on the current EM model.

Table 2. Part count of 3U structure.

Structure Type	Structural Screws	Rails and Frames	Stacking Rods	Spacers	Total No. of Parts
Conventional	76	14	4	48	66
Slot	72	8	0	30	38
Improvement	5%	43%	100%	38%	42%

4.3. Evaluation Result for Complexity Analysis

The complexity of both satellite design concepts is computed using Equation (1). Figure 16 shows the comparative value of each complexity variable of both 3U designs. The complexity analysis reveals that the overall complexity of the slot-based design is reduced by 40.43%. All the sub-complexity variables show an improvement in comparison with the conventional-type structural design. The reduction in the component complexity by 23.98% shows a reduction in the level of difficulty when handling individual components in the case of the slot-based design. The improvement in liaison complexity is rather noticeable, being approximately 61.03%. Due to the small number of structural screws and components, as well as the way they are brought together, without the need for long screws, it is possible to reduce the liaison complexity. From a global perspective, the intricateness of the overall integration is also improved, but by a small amount.

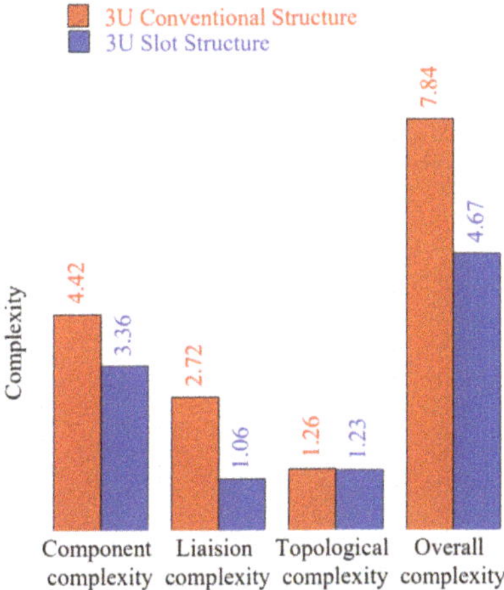

Figure 16. Complexity evaluation of 3U conventional and slot-type structures.

4.4. Evaluation Result of Assembly Steps

As described before, the steps taken to assemble all satellite components, except the structure, are counted. The process of attaching screws and spacers is not considered during the count. Assuming that the spacers are already attached to the PCBs, the total number of steps required to assemble the conventional structure is 29, whereas it is only 12 for the slot-based structure. This represents an improvement of around 58%.

4.5. Evaluation Result of Design Concept Scalability

In addition to the above evaluations performed on the 3U CubeSat, the scalability of the concept down to the lower CubeSat form factor was verified. The BIRDS-3 satellite (shown in Figure 17) is used for this assessment. BIRDS-3 is a 1U CubeSat with a structure based on the conventional design concept. Similar evaluation methods were employed to check its suitability for highly efficiency-demanding mass production applications. For the purpose of this evaluation, a slot-based 1U STM was designed (shown in Figure 18) and manufactured. Then, it was compared with an existing conventional-type BIRDS-3 platform. A similar liaison representation is used to show the relationship between components in the assembly in Figures 19 and 20.

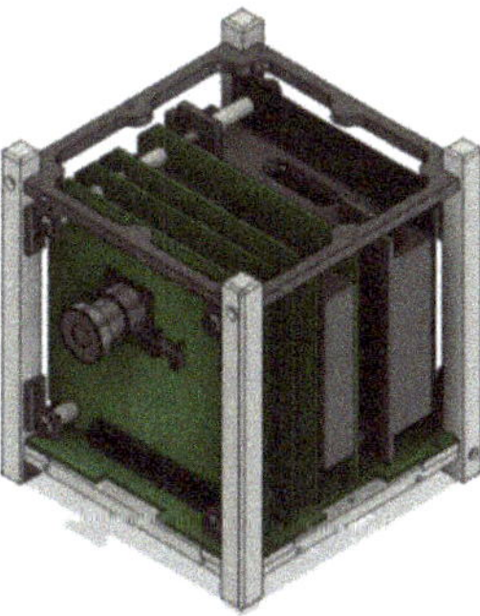

Figure 17. Conventional-type BIRDS-3 1U model.

Figure 18. Slot-based 1U model.

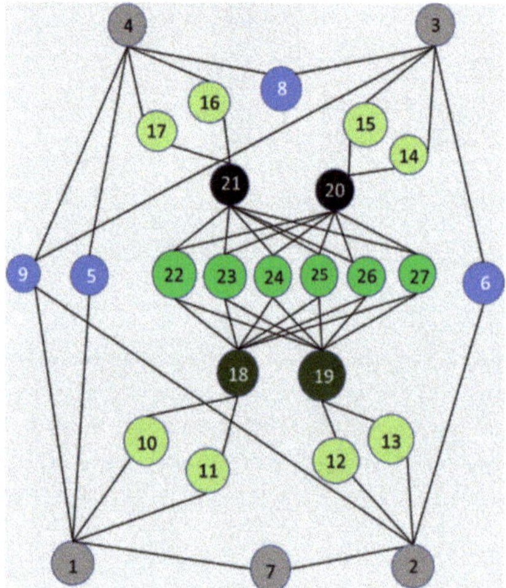

Figure 19. Liaison diagram of 1U BIRDS-3 conventional model.

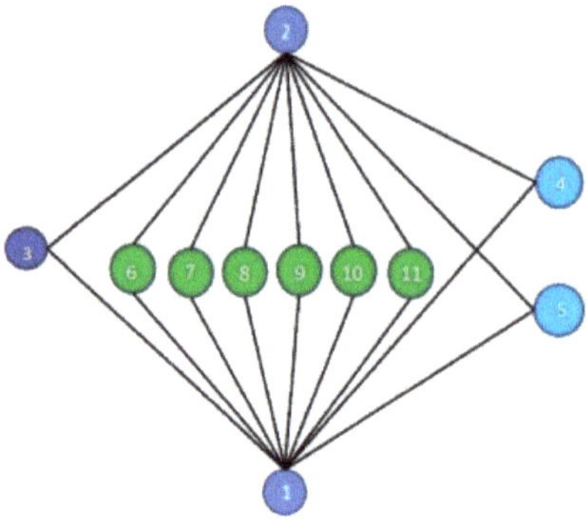

Figure 20. Liaison diagram of the 1U slot-based STM.

4.5.1. Count Analysis

In the count analysis, only the primary structural parts, such as rails, plates, rods, spacers, and fasteners, are considered. They are subcategorized accordingly in a similar way for the 3U models, as shown in Table 3.

Table 3. Count analysis of 1U structure.

Structure Type	Structural Screws	Rails and Frames	Stacking Rods	Spacers	Total No. of Parts
BIRDS-3	60	19	4	28	51
1U slot	40	7	0	18	25
Improvement	33%	63%	-	36%	51%

A similar result is obtained for the 1U-size design. The total number of screws is reduced by 33.33% for the slot structure, whereas a 50.98% reduction is achieved in the total part count, which includes structural parts and spacers. The BIRDS-3 CubeSat uses 51 structural parts, including the long mounting rods, whereas the total number of structural parts of the slot-based design is only seven. This significant reduction is due to the withdrawal of the staking rods and the associated support frames.

4.5.2. Complexity Analysis

During integration, there is also an impact on the overall reduction in complexity. In the case of the slot-type design, the individual PCBs and payload components are independently mounted in their designated slots using a special spacer, avoiding interdependency between the subsystems, which increases easiness of accessing components quickly. This effect can be easily observed with the value of the liaison parameters, which shows a 78.45% reduction. The total complexity of the conventional design of BIRDS-3 is 8.59, whereas it is only 2.73 for the slot-type concept, which is, overall, 68.22% drop. Figure 21 below illustrates the general comparison of the complexity of both 1U designs.

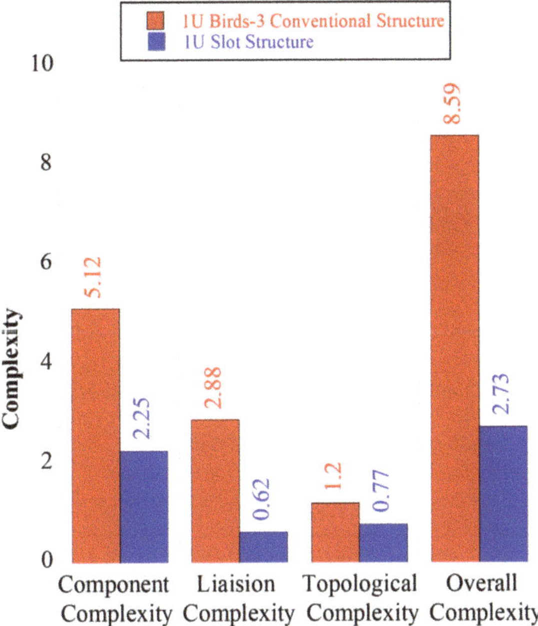

Figure 21. Complexity comparison of 1U conventional and slot-type structures.

4.5.3. Assembly Steps

In terms of the number of assembly steps, assembling the BIRDS-3 model requires a total of 16 steps. On the other hand, only five steps are required for the 1U slot-type design. This notable difference is due to the reduction in the number of structural parts.

The slot-based design utilizes 25 structural parts, which is approximately half of that for the BIRDS-3 structure.

5. Discussion

The new slot-type structural design concept was manufactured and verified with environmental tests to confirm its bottom-line requirement. Both the random and sine-burst vibration shows its compliance to the minimum mechanical strength and stiffness requirements. According to the JAXA accommodation handbook, the minimum natural frequency of satellites is defined as 60 Hz. The measured natural frequency of the STM design for 3U was around 310 Hz.

The assessment results from the count and complexity analyses strongly suggest that the new slot-based structural design showed superior advantages by providing a more efficient integration process. It was possible to reduce the structural part count of the conventional design by 42% and 51% for the 3U and 1U satellites, respectively. The improvement in the part count provides three major benefits. First, the few required assembly steps to finish integrating the satellite led to an increase in the speed of the entire process of satellite development. On the other hand, having fewer structural parts also means a reduction in manufacturing costs. The cost is also directly related to the amount of tolerance required. A higher stack-up tolerance also leads to higher tolerance on individual parts level. As the number of parts increases, the stack-up tolerance increases to achieve the tolerance requirement defined by the CubeSat design specification, which causes a dramatic increase in production costs. The last advantage acquired with the minimum part count is its contribution to the reduction in complexity during integration.

Complexity is a key factor that decides the level of simplification of the integration that impacts productivity. The complexity analysis shows a similar improvement as the part count analysis. The 3U slot design provides 40.43% reduction as compared to the conventional-type designs. On the other hand, a 68.02% change in the complexity for the case of the 1U slot design shows a significant change.

As discussed before, the complexity of the assembly increases with the number of components and joints. An increase in complexity usually leads to workmanship errors due to difficulty grasping the assembly process. Despite the overall improvement achieved, the amount varies in terms of the level of sub-complexity variables. The minimum recorded improvement is in topological complexity, which indicates the need for further refinement in design. Based on the identified issues, further refinement in the design is being carried out for the current FM model. It is expected that upon the completion of the current FM model, the count and complexity values shall improve even further. In this design, the internal subsystems are standardized within the dimensions of 90 mm (about 3.54 in) by 86 mm (about 3.39 in); however, future work will consider enhancing the flexibility to accommodate components that are not standard enough to fit into the defined standard slot widths.

Besides these major assessments, the implication of the reduction in part count and complexity, of the assembly step was evaluated. It takes 12 steps to complete the 3U slot structure assembly which is 58% reduction. The reduction in the number of assembly steps by more than half shows the method's suitability in terms of efficiency for fast integration missions. The scalability of the concept to lower the CubeSat factor demonstrates that this highly efficiency-demanding mission has potential feasibility on a 1U or higher form factor CubeSat platform.

6. Conclusions

As mass production applications using CubeSats are expanding, it can be deduced that an increase in productivity through the enhancement of flexibility and integration speed looks particularly important for the profitability and success of these missions. In this study, a highly flexible and easy-to-use 3U and 1U slot-based structure was developed, which facilitates easy and quick integration. It is believed to increase the reliability of satellite integration by reducing the complexity that case human errors during the integration and testing phases. Both structural analysis and vibration tests show that the developed model is stiffened and strong enough to withstand expected severe launch loads.

Finally, the design goals are evaluated against the conventional CubeSat structural design using the part and assembly step count, as well as complexity analysis. The results show significant improvements in the critical design parameters. The scalability of the concept is also examined by developing an equivalent 1U slot-type model showing similar design outputs. This 3U design concept is currently being implemented on one of Kyutech's mass production CubeSat projects, with additional improvements.

However, the acquired advantages are more noticeable, especially for missions where the total mass of the satellite is not critical. However, in addition to an improvement in the design, several other options are being explored to reduce the total weight of the structure. In future work, the employment of new production techniques such as additive manufacturing is believed to provide further enhancement for this design concept.

Author Contributions: Conceptualization, E.E.A. and H.M.; methodology, E.E.A., H.M. and M.C.; software, E.E.A.; validation, E.E.A., H.M. and M.C.; formal analysis, E.E.A.; investigation, E.E.A. and H.M.; resources, M.C. and H.M.; data curation, E.E.A., J.R.C.-A. and H.M.; writing—original draft preparation, E.E.A.; writing—review and editing, E.E.A., J.R.C.-A., H.M. and M.C.; visualization, E.E.A.; supervision, H.M. and M.C.; project administration, J.R.C.-A., H.M. and M.C.; funding acquisition, H.M. and M.C. All authors have read and agreed to the published version of the manuscript.

Funding: This research was funded by Ministry of Economy, Trade and Industry (METI), Japan under the grant. "International Standard Development for Energy Conservation (International Standardization of CubeSat Interface)" with grant number [FY2021-10]. The APC was funded by METI.

Institutional Review Board Statement: Not applicable.

Informed Consent Statement: Not applicable.

Data Availability Statement: Not applicable.

Acknowledgments: The authors would like to acknowledge the support of the Japanese Ministry of Economy, Trade, and Industry through the Promotion of Energy Saving and Related International Standards.

Conflicts of Interest: The authors declare no conflict of interest. The funders had no role in the design of the study; in the collection, analyses, or interpretation of data; in the writing of the manuscript; or in the decision to publish the results.

Appendix A

Table A1. The 3U conventional-type demo satellite complexity values.

	1	2	3	4	5	6	7	8	9	10	11	12	13	14	15	16	17	18	19	20	21	22	23	αip
3U Conventional Structure: CP1 of Essential Components, where α^{P}_{max}=6.9																								
fhA	1	1	1	1	1	1	1	1	1	1	1	1	1	1	1	1	1	1	1	1	1	1	1	23
Σ fhB	0	0	0	0	0	0	0	0	0	0	0	0	0	0	0	0	0	0	0.4	0.4	0.4	0.4	0.4	2
fhC	0.1	0.1	0	0	0.1	0.1	0.1	0.1	0.1	0.1	0.1	0.1	0.1	0.1	0	0	0	0	0.1	0.1	0.1	0.1	0.1	1.7
fhD	0.2	0.2	0.2	0.2	0.2	0.2	0.2	0.2	0.2	0.2	0.2	0.2	0.2	0.2	0	0	0	0	0.2	0.2	0.2	0.2	0.2	3.8
Total CP1																								4.4

Table A2. The 3U slot-based satellite complexity values.

	1	2	3	4	5	6	7	8	9	10	11	12	13	14	αip
3U Slot Structure: CP1 of Essential Components															
fhA	1	1	1	1	1	1	1	1	1	1	1	1	1	1	14
Σ fhB	0	0	0	0	0	0	0.6	0.6	0.8	0.8	0.8	0.8	0.8	0.4	5.6
fhC	0.1	0.1	0.1	0.1	0.5	0	0.5	0.5	0.1	0.1	0.1	0.1	0.1	0.1	2.5
fhD	0.2	0.2	0.2	0.2	0.4	0.2	0.4	0.4	0.2	0.2	0.2	0.2	0.2	0.2	3.4
Total CP1															3.695652

Table A3. The 1U BIRDS satellite complexity values.

1U BIRDS-3 Structure: CP1 of Essential Components																											
	1	2	3	4	5	6	7	8	9	10	11	12	13	14	15	16	17	18	19	20	21	22	23	24	25	26	27	αip
fhA	1	1	1	1	1	1	1	1	1	1	1	1	1	1	1	1	1	1	1	1	1	1	1	1	1	1	1	27
$\sum$ fhB	0	0	0	0	0	0	0	0	0	0	0	0	0	0	0	0	0	0	0	0	0	0.4	0.4	0.4	0.4	0.4	0.4	2.4
fhC	0.1	0.1	0.1	0.1	0.1	0.1	0.1	0.1	0.1	0	0	0	0	0	0	0	0	0	0	0	0	0.1	0.1	0.1	0.1	0.1	0.1	1.5
fhD	0.2	0.2	0.2	0.2	0.2	0.2	0.2	0.2	0.2	0.2	0.2	0.2	0.2	0.2	0.2	0.2	0	0	0	0	0	0.2	0.2	0.2	0.2	0.2	0.2	4.4

Total CP1: 5.115942

Table A4. The 1U slot-based satellite complexity values.

1U Slot Structure: CP1 of Essential Components											
	1	2	3	4	5	6	7	8	9	10	11	αip
fhA	1	1	1	1	1	1	1	1	1	1	1	11
$\sum$ fhB	0	0	0	0	0	0.2	0.2	0.2	0.2	0.2	0.2	1.2
fhC	0.1	0.1	0.1	0.1	0.1	0.1	0.1	0.1	0.1	0.1	0.1	1.1
fhD	0.2	0.2	0.2	0.2	0.2	0.2	0.2	0.2	0.2	0.2	0.2	2.2

Total CP1: 2.25

Table A5. Nodal definition of the 1U and 3U slots and conventional designs.

Node Number	Satellite Part Name			
	3U Conv	**3U Slot**	**1U Conv**	**1U Slot**
1	Bottom plate	Base plate	Rail-1	Slot-1
2	Top plate	Lid plate	Rail-2	Slot-2
3	Front plate	Slot-1	Rail-3	Top frame
4	Back plate	Slot-2	Rail-4	Bottom frame-1
5	Frame-1	Top plate	Bottom frame-1	Bottom frame-2
6	Frame-2	Bottom plate	Bottom frame-2	PCB-1
7	Frame-3	Frame-1	Top frame	PCB-2
8	Frame-4	Frame-2	Frame-1	PCB-3
9	Frame-5	PCB-1	Frame-2	PCB-4
10	Frame-6	PCB-2	Frame-3	PCB-5
11	Frame-7	PCB-3	Frame-4	PCB-6
12	Frame-8	PCB-4	Frame-5	
13	Frame-5	PCB-5	Frame-6	
14	Frame-6		Frame-7	
15	Rod-1		Frame-8	
16	Rod-2		Long rod 1	
17	Rod-3		Long rod 2	
18	Rod-4		Long rod 3	
19	PCB-1		Long rod 4	
20	PCB-2		PCB-1	
21	PCB-3		PCB-2	
22	PCB-4		PCB-3	
23	PCB-5		PCB-4	
24			PCB-5	
25			PCB-6	

References

1. Cappelletti, C.; Battistini, S.; Malphrus, B.K. *Cubesat Handbook from Mission Design to Operations*; Academic Press: Cambridge, MA, USA, 2021; ISBN 9780128178843.
2. Moore, G.E. Cramming More Components onto Integrated Circuits, Reprinted from Electronics, Volume 38, Number 8, April 19, 1965, Pp.114 Ff. *IEEE Solid-State Circuits Soc. Newsl.* **2006**, *11*, 33–35. [CrossRef]
3. De Carvalho, R.A.; Estela, J.; Langer, M. *Nanosatellites Space and Ground Technologies, Operations and Economics*; Wiley and Sons: Hoboke, NJ, USA, 2020.
4. Poghosyan, A.; Golkar, A. CubeSat Evolution: Analyzing CubeSat Capabilities for Conducting Science Missions. *Prog. Aerosp. Sci.* **2017**, *88*, 59–83. [CrossRef]
5. Op-Ed Satellite Bankruptcies circa 2000, vs. 2020: We've Come a Long Way! SpaceNews. 2021. Available online: https://spacenews.com/op-ed-satellite-bankruptcies-circa-2000-vs-2020-weve-come-a-long-way/ (accessed on 27 July 2022).
6. Boothroyd, G.; Dewhurst, P.; Knight, W.A. *Product Design for Manufacture and Assembly*, 3rd ed.; CRC Press: Boca Raton, FL, USA, 2010. ISBN1 1420089277. ISBN2 9781420089271.
7. Enright, J.; Jilla, C.; Miller, D. Modularity and Spacecraft Cost. *J. Reducing Sp. Mission Cost* **1998**, *1*, 133–158. [CrossRef]
8. Nanosats Database CubeSat Tables | Nanosats Database. Available online: https://www.nanosats.eu/tables# (accessed on 16 July 2022).
9. NASA Small Spacecraft Technology State of the Art. Available online: https://www.nasa.gov/sites/default/files/atoms/files/soa_2021_1.pdf (accessed on 17 July 2022).
10. Lovascio, A.; D'orazio, A.; Centonze, V. Characterization of a COTS-Based RF Receiver for Cubesat Applications. *Sensors* **2020**, *20*, 776. [CrossRef] [PubMed]
11. GOMspace GOMspace | Home. Available online: https://gomspace.com/home.aspx (accessed on 17 July 2022).
12. PUMPKINSPACE. CubeSat KitTM Structures. Available online: https://www.pumpkinspace.com/store/c4/CubeSat_KitTM_Structures.html (accessed on 17 July 2022).
13. ISIS CubeSat Structures-ISISPACE. Available online: https://www.isispace.nl/product-category/cubesat-structures/ (accessed on 17 July 2022).
14. C3S. Structures-C3S. Available online: https://c3s.hu/structures/ (accessed on 17 July 2022).
15. PC/104 Consortium-Specifications. Available online: https://pc104.org/ (accessed on 23 July 2022).
16. Tumenjargal, T.; Kim, S.; Masui, H.; Cho, M. CubeSat Bus Interface with Complex Programmable Logic Device. *Acta Astronaut.* **2019**, *160*, 331–342. [CrossRef]
17. Kramer, A.; Bangert, P.; Schilling, K. UWE-4: First Electric Propulsion on a 1U Cubesat-in-Orbit Experiments and Characterization. *Aerospace* **2020**, *7*, 98. [CrossRef]
18. Cihan, M.; Cetin, A.; Kaya, M.O.; Inalhan, G. Design and Analysis of an Innovative Modular Cubesat Structure for ITU-PSAT II. In Proceedings of the RAST 2011-Proceedings of 5th International Conference on Recent Advances in Space Technologies, Istanbul, Turkey, 9–11 June 2011; pp. 494–499.
19. CubeSat Design Specification Rev. 14.1. Available online: https://static1.squarespace.com/static/5418c831e4b0fa4ecac1bacd/t/62193b7fc9e72e0053f00910/1645820809779/CDS+REV14_1+2022-02-09.pdf (accessed on 21 July 2020).
20. JEM Payload Accommodation Handbook-Small Satellite Deployment Interface Control Document. Available online: https://iss.jaxa.jp/kibouser/library/item/jx-espc_8d_en.pdf (accessed on 5 September 2020).
21. Alkan, B.; Vera, D.; Ahmad, B.; Harrison, R. A Method to Assess Assembly Complexity of Industrial Products in Early Design Phase. *IEEE Access* **2017**, *6*, 989–999. [CrossRef]
22. Sinha, K. *Structural Complexity and Its Implications for Design of Cyber Physical Systems*; Massachusetts Institute of Technology: Cambridge, MA, USA, 2014.
23. Nikiforov, V. The Energy of Graphs and Matrices. *J. Math. Anal* **2007**, *326*, 472–1475. [CrossRef]

applied sciences

Article

An Iterative Guidance and Navigation Algorithm for Orbit Rendezvous of Cooperating CubeSats

Simone Battistini [1,*,†], Giulio De Angelis [2,†], Mauro Pontani [3,†] and Filippo Graziani [4,†]

[1] Department of Engineering and Mathematics, Sheffield Hallam University, Howard Street, Sheffield S1 1WB, UK
[2] Faculty of Civil and Industrial Engineering, Sapienza Università di Roma, Via Eudossiana 18, 00184 Rome, Italy
[3] Department of Astronautical, Electrical, and Energy Engineering, Sapienza Università di Roma, Via Salaria 851, 00138 Rome, Italy
[4] GAUSS Srl, Via Sambuca Pistoiese 70, 00138 Rome, Italy
* Correspondence: simone.battistini@mbda.it
† These authors contributed equally to this work.

Abstract: Modern space missions often require satellites to perform guidance, navigation, and control tasks autonomously. Despite their limited resources, small satellites are also involved in this trend, as in-orbit rendezvous and docking maneuvers and formation flying have become common requirements in their operational scenarios. A critical aspect of these tasks is that these algorithms are very much intertwined with each other, although they are often designed completely independently of one another. This paper describes the design and simulation of a guidance and relative navigation architecture for the rendezvous of two cooperating CubeSats. The integration of the two algorithms provides robustness to the solution, by simulating realistic levels of noise and uncertainty in the guidance law implementation. The proposed guidance law is derived based on the linearized equations of orbital motion, written in terms of spherical coordinates. The trajectory is iteratively corrected at a fixed time step, so that errors from the navigation and the initial orbital condition can be recovered. The navigation algorithm processes the bearing and range measurements from a camera and an intersatellite link through an unscented filter to provide the information required from the guidance law. A Monte Carlo campaign based on a 3-DOF simulation demonstrates the effectiveness of the proposed solution.

Keywords: small satellites; cubesat; rendezvous; docking; guidance; navigation

Citation: Battistini, S.; De Angelis, G.; Pontani, M.; Graziani, F. An Iterative Guidance and Navigation Algorithm for Orbit Rendezvous of Cooperating CubeSats. *Appl. Sci.* **2022**, *12*, 9250. https://doi.org/10.3390/app12189250

Academic Editor: Jérôme Morio

Received: 9 August 2022
Accepted: 6 September 2022
Published: 15 September 2022

Publisher's Note: MDPI stays neutral with regard to jurisdictional claims in published maps and institutional affiliations.

1. Introduction

The use of small satellites in modern space missions has become an important factor in the aerospace industry, thanks to the versatility, reliability, and low cost of these platforms. As the scenarios involving small satellites are growing in numbers and complexity, more advanced technological solutions are needed to address the new tasks of these missions. Autonomous guidance, navigation, and control (GNC) systems for small satellites, for example, are crucial for implementing challenging tasks, such as rendezvous and docking [1], formation flying [2], and space debris removal [3].

Guidance algorithms for rendezvous are usually based on the classical Hill–Clohessy–Wiltshire (HCW) equations, either under the impulsive thrust approximation or assuming finite thrust. Seminal publications on impulsive rendezvous have been written by Prussing [4,5], who investigated minimum-fuel rendezvous trajectories including two, three, or four velocity changes. By using the HCW equations, Carter [6] focused on finite thrust rendezvous with an upper bound on thrust magnitude, and proved that at most seven thrust/coast intervals can occur. Later, Carter and Humi [7] investigated fuel-optimal rendezvous relative to a point in a general Keplerian orbit, and demonstrated that no singular arc exists if the orbit is noncircular. Most recently, Pontani and Conway [8] addressed the problem of finding a variety of minimum-fuel rendezvous trajectories by using finite thrust and proved a remarkable symmetry property of optimal rendezvous paths. Although all the preceding contributions employ the HCW equations, Prussing and Chiu used the nonlinear equations of Keplerian motion, for the purpose of minimizing fuel consumption [9] under the impulsive thrust approximation. Pontani et al. [10] found both impulsive and finite thrust, fuel-optimal rendezvous trajectories by using a particle swarm algorithm. Other works use alternative deterministic and heuristic methods for rendezvous optimization and guidance, such as the glideslope multipulse technique [11], optimally timed trajectory correction maneuvers [12], $\mathcal{H}_\infty$ and μ-synthesis techniques [13]. However, in actual operational scenarios, rendezvous trajectories are affected by orbit perturbations, which make both the HCW linear equations and the Keplerian nonlinear equations relatively inaccurate for precise orbit rendezvous. Moreover, the assumption of perfect information on the relative spacecraft dynamics is not realistic, and an accurate rendezvous is likely to be infeasible without an effective navigation system, due to measurement noise or unavailable variables needed to compute the guidance command. This circumstance points out the need of an integrated, iterative guidance and navigation architecture, capable of performing correction maneuvers on the basis of the estimated state provided by the navigation system.

This paper analyzes the design of a relative guidance and navigation architecture for the rendezvous of two CubeSats, a target and a chaser, which are assumed to travel orbits in close proximity. The proposed guidance algorithm is based on the iterative application of the HCW linear equations of motion [14], written in terms of spherical coordinate displacements, under the assumption that both vehicles travel nearby orbits with small eccentricities. This simple approach aims at finding the correction maneuvers, modeled as impulsive velocity changes, that are necessary to maintain the chaser on track to the target. The orbital motion of both spacecraft is described by integrating the nonlinear equations of motion, which can include all the orbit perturbations relevant to the dynamical context.

The iterative nature of the guidance strategy allows accounting for the update of the target estimated orbit, provided by the navigation state, which is obtained from the measurements of the on-board sensors. Small space platforms have light architectures and weight constraints that require innovative navigation solutions, such as vision-based navigation. With this approach, a single [15] camera or multiple [16] cameras are employed on the chaser to measure the angular displacement to the target. A problem arising with the single-camera solution is the unavailability of relative range information [17], which affects the accuracy of the guidance law unless specific maneuvers are realized to improve the range observability [18,19]. Range becomes observable if the camera is offset with respect to the center of mass of the chaser [20], but not if the camera is placed in the line of sight between the target and the chaser, which is often the case. A thorough analysis of the performance of monocular visual navigation system is presented in [21]. Another possibility arises when the target is cooperative: in this case, direct range measurement becomes available thanks to an on-board intersatellite link (ISL), as in the case of the Milani 6U CubeSat mission [22].

The relative navigation solution is provided in real time by a recursive state estimator of the Kalman filter family of algorithms. Given the nature of the relative navigation equations, nonlinear Kalman filters are needed for this task, such as the extended Kalman filter (EKF) and the unscented filter (UF). The EKF has been a standard in relative navigation problems for a long time [23,24]. However, EKF's poor performance against highly-nonlinear systems and even divergences have led to the wide adoption of the UF, for navigation problems in either its regular [25] or square-root [26] versions. In fact, the UF provides a more accurate and robust solution to the estimation problem [27] than EKF; therefore UF is chosen for this work.

The paper is organized as follows: the mathematical model that represents the relative motion of the spacecraft is described in Section 2; the design of the guidance and navigation algorithms is given in Section 3; the setup of the numerical simulations and the results are presented in Section 4; conclusions and final remarks are given in Section 5.

2. Equations of Motion

In order to write the equations of the relative orbital motion, a reference Keplerian circular orbit is defined in an inertial reference frame with axes $\hat{N}$, $\hat{M}$, $\hat{h}$, as seen in Figure 1. Axes $\hat{N}$ and $\hat{h}$ are aligned respectively with the line of the ascending node and the angular momentum of the reference Keplerian orbit. The latter is circular and therefore has a fixed radius R_R, inclination i_R, and RAAN Ω_R. In the preceding frame, orbital motion of a space vehicle (subject to thrust and perturbing accelerations) is identified by the instantaneous radius r, right ascension ξ, declination ϕ, radial velocity v_r, transverse velocity v_t, and normal velocity v_k. Using these variables the equations of orbital motion are:

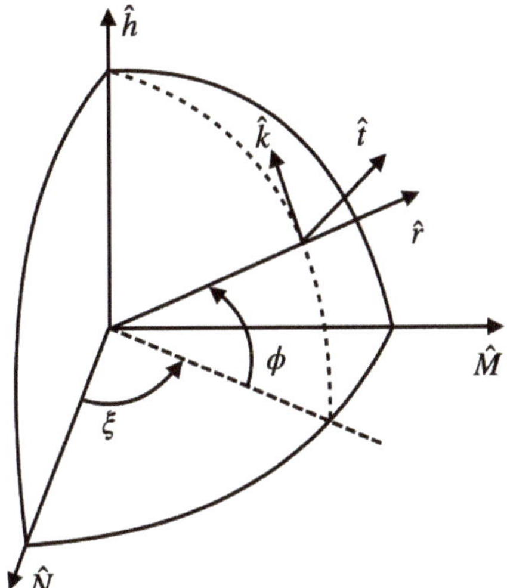

Figure 1. Representation of the inertial and moving frames.

$$\begin{cases} \dot{r} = v_r \\ \dot{\xi} = \dfrac{v_t}{r \cos \phi} \\ \dot{\phi} = \dfrac{v_k}{r} \end{cases} \qquad \begin{cases} \dot{v}_r = -\dfrac{\mu}{r^2} + \dfrac{v_t^2 + v_k^2}{r} + a_r \\ \dot{v}_t = \dfrac{v_t}{r}(v_k \tan \phi - v_r) + a_t \\ \dot{v}_k = -\dfrac{v_t^2}{r} \tan \phi - \dfrac{v_r v_k}{r} + a_k \end{cases} \qquad , \tag{1}$$

where μ is the gravitational parameter of the main attracting body and $\vec{a} = \{a_r, a_t, a_k\}$ is the vector containing the three components of the external accelerations, due to either thrust or orbital perturbations.

The guidance law is obtained by linearizing Equation (1) around the reference orbit, assuming that the variables are perturbed by small displacements, denoted with δ, from the respective values, associated with the reference circular orbit, i.e.,

$$\begin{cases} r = R_r + \delta r \\ \xi = \xi_r + \delta \xi \\ \phi = \phi_r + \delta \phi \end{cases} \qquad \begin{cases} v_r = v_{rR} + \delta v_r = \delta v_r \\ v_t = v_{tR} + \delta v_t = \sqrt{\dfrac{\mu}{R_r}} + \delta v_t \\ v_k = v_{kR} + \delta v_k = \delta v_k \end{cases} \qquad , \tag{2}$$

At this point, it is useful to define a new set of coordinates:

$$x = \delta r, \quad y = R_R \delta \xi, \quad z = R_R \delta \phi. \tag{3}$$

By using the definitions in Equations (2) and (3), expanding Equations (1) to the first order, and neglecting higher-order terms, the well-known Hill–Clohessy–Wiltshire (HCW) equations are obtained, in terms of spherical coordinates x, y, z,

$$\begin{cases} \ddot{x} - 3\omega_R^2 x - 2\omega_R \dot{y} = a_r \\ \ddot{y} + 2\omega_R \dot{x} = a_t \\ \ddot{z} + \omega_R^2 z = a_k \end{cases} . \tag{4}$$

3. Guidance and Navigation

This section describes the guidance and navigation scheme used for the rendezvous of the two satellites. The guidance law is an iterative algorithm that calculates the steering maneuvers to achieve the final position of the target, using the available information on the position and velocity of both vehicles. The navigation subsystem estimates the information needed to calculate the guidance command by using the measurements from the on-board sensors. In this paper, it is assumed that an ISL and a camera provide range and bearing measurements.

3.1. Guidance

This work assumes that both the chaser and the target travel orbits sufficiently close to the reference circular Keplerian orbit. This implies that the orbital dynamics of both vehicles can be described by using the HCW Equation (4). Let $\delta \mathbf{r} := \begin{bmatrix} x & y & z \end{bmatrix}$ and $\delta \mathbf{v} := \begin{bmatrix} \dot{x} & \dot{y} & \dot{z} \end{bmatrix}$. Due to the linear nature of the governing Equation (4), the time evolution of the displaced position and velocity vectors $\delta \mathbf{r}(t)$ and $\delta \mathbf{v}(t)$ is given by [14]

$$\begin{bmatrix} \delta \mathbf{r}(t) \\ \delta \mathbf{v}(t) \end{bmatrix} = \begin{bmatrix} M(\tau) & N(\tau) \\ S(\tau) & T(\tau) \end{bmatrix} \begin{bmatrix} \delta \mathbf{r}(t_k) \\ \delta \mathbf{v}(t_k) \end{bmatrix}, \tag{5}$$

where $\tau = t - t_k$. The closed-form expressions of the (3×3) matrices M, N, S, and T are reported in [14].

The rendezvous maneuver is assumed to have a specified duration Δt_{tot}. At a generic time t_k ($0 \leq t_k \leq \Delta t_{tot}$), the maneuver at hand can be designed by using Equation (4), by including a pair of impulsive changes of velocity, the first at time t_k and the second at time Δt_{tot}. Superscripts + and − are associated with the time instants immediately before and after each velocity changes, i.e.,

$$\begin{cases} \Delta \mathbf{v}_1 = \delta \mathbf{v}(t_k^+) - \delta \mathbf{v}(t_k^-) \\ \Delta \mathbf{v}_2 = \delta \mathbf{v}(\Delta t_{tot}^+) - \delta \mathbf{v}(\Delta t_{tot}^-) \end{cases} . \tag{6}$$

The final goal is to get the position and velocity of the target at time Δt_{tot}, i.e.,

$$\begin{cases} \delta \mathbf{r}(\Delta t_{tot}) = \Delta \mathbf{r}_T(t_{tot}) \\ \delta \mathbf{v}(\Delta t_{tot}^+) = \Delta \mathbf{v}_T(t_{tot}) \end{cases} , \tag{7}$$

where subscript T refers to the target. The magnitudes and directions of $\Delta \mathbf{v}_1$ and $\Delta \mathbf{v}_2$ are given by the following vector equations, obtained by combining the previous relations

$$\begin{cases} \Delta \mathbf{v}_1 = N_f^{-1}[\delta \mathbf{r}_T(\Delta t_{tot}) - \Delta \mathbf{r}(t_k)] - \delta \mathbf{v}(t_k^-) \\ \Delta \mathbf{v}_2 = \delta \mathbf{v}_T(\Delta t_{tot}) - \left[S_f \delta \mathbf{r}(t_k) + T_f \delta \mathbf{v}(t_k^+) \right] \end{cases} , \tag{8}$$

where subscript f denotes the value of the respective matrix, evaluated at $\tau = \Delta t_{tot} - t_k$. Let Δt_S represent the time interval between two consecutive iterations. The guidance algorithm is based on the following iterative steps, to perform at each time t_k:

- calculate $t_{k+1} = t_k + \Delta t_S$; if $t_{k+1} \geq \Delta t_{tot}$, then set $t_{k+1} = \Delta t_{tot}$;
- evaluate the displacement vectors $\delta \mathbf{r}$ and $\delta \mathbf{v}_1$ at t_k^-;
- calculate $\Delta \mathbf{v}_1$ by using Equation (8);

- using the definitions of the displaced position and velocity coordinates, obtain the spherical coordinates of position and velocity of the chaser $(r, \xi, \phi, v_r, v_t, v_k)$ at t_k;
- propagate numerically the nonlinear Equation (1) in the interval $[t_k, t_{k+1}]$;
- if $t_{k+1} = \Delta t_{tot}$, then evaluate $\delta\mathbf{v}_2$ at Δt_{tot}.

It is worth remarking that the algorithm at hand evaluates iteratively $\delta\mathbf{v}_1$, whereas $\delta\mathbf{v}_2$ is evaluated only at the final time.

3.2. Navigation

The navigation system consists of an algorithm that processes the available measurements ζ to reconstruct the target estimated state $\hat{\chi}$, defined as

$$\hat{\chi} = \{\hat{r}_T, \hat{\xi}_T, \hat{\phi}_T, \hat{v}_{r_T}, \hat{v}_{t_T}, \hat{v}_{k_T}\}. \tag{9}$$

The model assumed for the motion of the target is that of Equation (1), where the external accelerations $\vec{a}$ are considered to be null, i.e., the target is not maneuvering and no orbital perturbations are considered. The equations that describe the three measurements (one distance, and two angles) are the following:

$$\begin{cases} \zeta_1 = ||(x_2 - x_1, \quad y_2 - y_1, \quad z_2 - z_1)|| + v_{range} \\ \zeta_2 = \arctan\dfrac{y_2 - y_1}{x_2 - x_1} + v_{bear1} \\ \zeta_3 = \arctan\dfrac{z_2 - z_1}{\sqrt{(x_2 - x_1)^2 + (y_2 - y_1)^2}} + v_{bear2} \end{cases}, \tag{10}$$

where $v_{range} \in \mathcal{N}(0, \sigma_{v_{range}})$, $v_{bear1} \in \mathcal{N}(0, \sigma_{v_{bear}})$ and $v_{bear2} \in \mathcal{N}(0, \sigma_{v_{bear}})$ are three random signals representing the noise on the measurements, whereas subscripts 1 and 2 correspond to the two satellites. Because of the optical nature of the sensor, the intensity of the noise signal on the range measurement depends on the distance between the two spacecraft. Therefore, it is assumed that the variance $\sigma_{v_{range}}$ is equal to the 10% of the true range value, as proposed in [22].

Given the high non-linearity of the model in Equation (1) and of the measurements in Equation (10), the algorithm considered for the state estimation is the UF in its standard version [28]. The process noise covariance matrix Q for this problem is defined as

$$Q = \begin{pmatrix} q_0\dfrac{\Delta t_S^3}{3}I(3) & q_0\dfrac{\Delta t_S^2}{2}I(3) \\ q_0\dfrac{\Delta t_S^2}{2}I(3) & q_0\Delta t_S I(3) \end{pmatrix}, \tag{11}$$

where q_0 is a tuning parameter of the filter, Δt_S is the sampling time, and $I(n)$ is the identity matrix of dimensions $n \times n$.

4. Numerical Simulations

In this section, the effectiveness of the proposed guidance and navigation scheme is assessed by means of a numerical simulation campaign of the 3-DOF equations of motion. The first simulation consists of a nominal case in which only the guidance law is tested. The second simulation tests both the guidance and the navigation systems on a set of 100 Monte Carlo runs. The main parameters of the simulation are reported in Table 1, where Δt_F is the duration of the firing impulse of the spacecraft that generates the Δv of Equation (8). The nominal initial condition x_0 in both simulations consists in the two satellites being on the same circular, equatorial orbit at a distance of 10 km along the tangent direction.

Table 1. Simulation parameters.

Quantity	Value	Quantity	Value	Quantity	Value
σ_r	500 m	σ_ζ	1°	σ_ϕ	1°
σ_{v_r}	1 m/s	σ_{v_t}	1 m/s	σ_{v_k}	1 m/s
$\sigma_{v_{bear}}$	1°	q_0	5×10^{-7}	Δt_{tot}	600 s
Δt_F	0.25 s	Δt_S	0.25 s		

4.1. Nominal Simulation

In the nominal simulation, the guidance system is tested without the navigation solution, i.e., assuming perfect information in Equation (8). Figure 2 shows that the displacement along x and y go to zero at the end of the simulation. The obtained miss distance is indeed very small at 5 cm. It is worth noting that a distance smaller than 1 m is deemed adequate to start a docking maneuver between two small satellites [29].

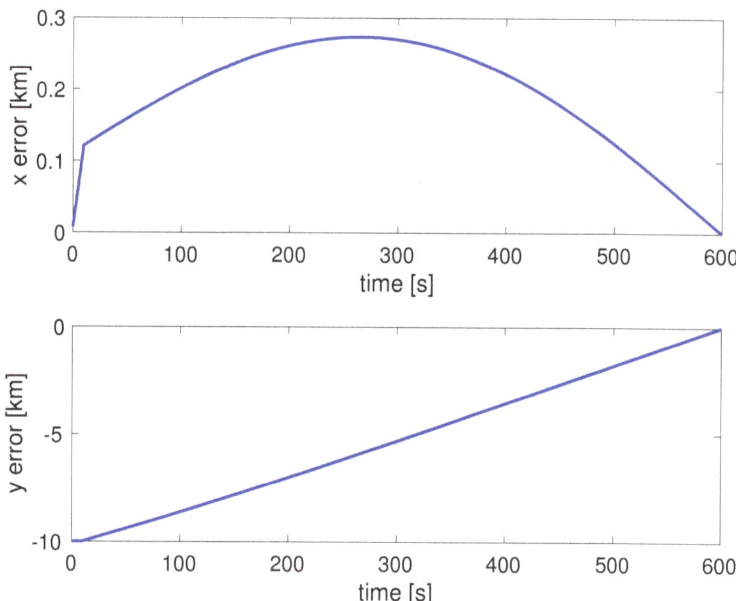

Figure 2. Position errors in the x and y directions.

The difference between the velocity of the two spacecraft is shown in Figure 3. It can be seen that the two velocities are aligned in both directions after the application of the $\delta\mathbf{v}_2$ command of Equation (8) at the end of the simulation. Having nulled the difference in position and velocity between the two spacecraft in the orbital plane, the nominal simulation can be, therefore, considered successful. The out-of-plane motion, in fact, is null in this case, as the considered orbits are Keplerian and no perturbations have been taken into account.

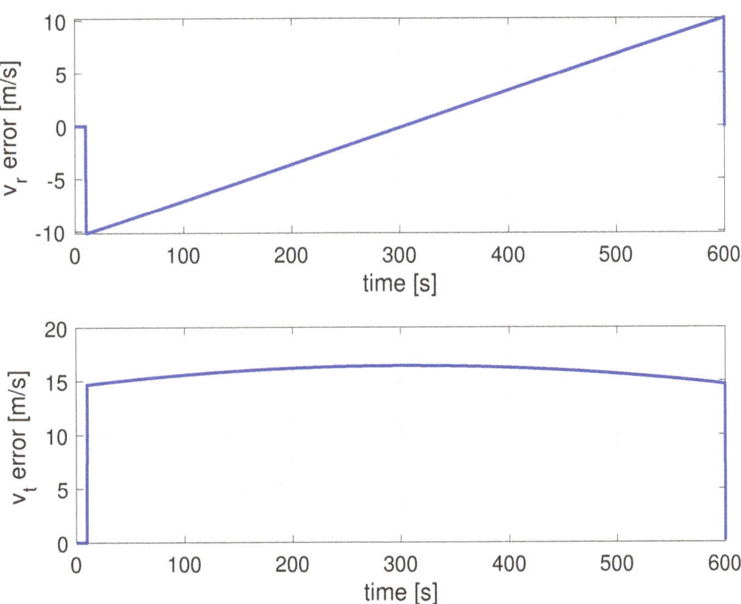

Figure 3. Velocity errors in the radial and transversal directions.

4.2. Monte Carlo Simulation

A set of 100 Monte Carlo runs has been prepared for the second simulation, varying the initial conditions $\hat{x}_{0|0} \in \mathcal{N}(x_0, \sqrt{P_{0|0}})$, with $P_{0|0}$ selected as:

$$P_{0|0} = diag[\sigma_r^2, \quad \sigma_\xi^2, \quad \sigma_\phi^2, \quad \sigma_{v_r}^2, \quad \sigma_{v_t}^2, \quad \sigma_{v_k}^2], \tag{12}$$

where the variances σ are given in Table 1.

The cumulative distribution of the miss distance obtained in the simulation campaign is shown in Figure 4. The results indicate that a miss distance shorter than 1 m is obtained in 85% of the cases, in line with the requirement on the final miss distance expressed in [29]. The accuracy obtained by the relative navigation filter in reconstructing the estimated state (9) is reported in Table 2. As can be seen by comparing the values of Table 2 with the variance of the initial guess and the level of the measurement noise, the filter is able to converge to a more accurate estimate of the variable. The only variable whose estimation accuracy has not been improved by the filter is the out-of-plane component of the velocity. This can be explained by the fact that there is no out-of-plane motion of the spacecraft in this case, as the orbit is assumed totally Keplerian. Thus, the out-of-plane motion is completely unobservable and the estimate of $\hat{v}_{k_T}$ remains bounded by the variance σ_{v_k}.

Table 2. Estimation accuracy.

Variable	Accuracy	Variable	Accuracy	Variable	Accuracy
r	20 m	ξ	0.001°	ϕ	0.001°
v_r	0.25 m/s	v_t	0.4 m/s	v_k	1 m/s

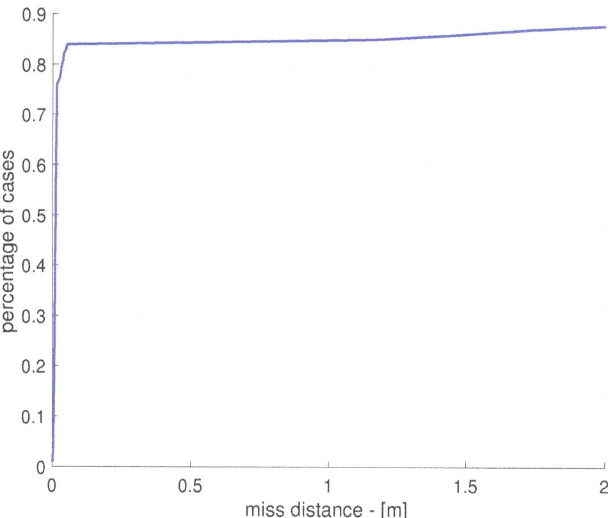

Figure 4. Miss distance cumulative distribution obtained in the simulation campaign.

5. Conclusions

This paper has presented a guidance and navigation scheme for the rendezvous of two CubeSats. Autonomous sensing and maneuvering capabilities are becoming a requirement of many space missions, even when small satellites are employed. Therefore, it is very important to investigate algorithms that can provide the necessary GNC capabilities and to test them together in simulation so that they can be validated.

The guidance law developed for this work is an iterative algorithm that calculates a discrete firing command Δv_1 at a specific sampling time, based on the linear rendezvous theory and the HCW equations. The necessary information for the guidance law is derived from the navigation solution, which is based on a UF that processes the available on-board measurements (range and bearing).

This scheme was tested in a 3-DOF simulation in both the nominal case (perfect information) and against a Monte Carlo campaign (incomplete information), assuming values compatible with CubeSat systems and applications. The Monte Carlo simulation allows us to take into account the variations in the measurement noise and the initial guess on the target's orbit. The results show that the obtained miss distance between the two spacecraft is below 1 m, a value suitable for the start of the docking operations.

Further work on this topic should include a trade-off study of the proposed solution against different sampling and firing times. Furthermore, the 3-DOF model should be improved to account for the main orbital perturbations, such as the Earth's oblateness and the aerodynamic drag.

Author Contributions: All authors contributed equally to this work. All authors have read and agreed to the published version of the manuscript.

Funding: This research received no external funding.

Institutional Review Board Statement: Not applicable.

Informed Consent Statement: Not applicable.

Data Availability Statement: Not applicable.

Conflicts of Interest: The authors declare no conflict of interest.

Abbreviations

The following abbreviations are used in this manuscript:

DOF	Degrees Of Freedom
EKF	Extended Kalman Filter
GNC	Guidance, Navigation, and Control
HCW	Hill-Clohessy-Wiltshire
ISL	Inter-Satellite Link
RAAN	Right Ascension of the Ascending Node
UF	Unscented Filter

References

1. D'Amico, S.; Bodin, P.; Delpech, M.; Noteborn, R. Prisma. In *Distributed Space Missions for Earth System Monitoring*; Springer: Berlin/Heidelberg, Germany, 2013; pp. 599–637.
2. Gasbarri, P.; Sabatini, M.; Palmerini, G. Ground tests for vision based determination and control of formation flying spacecraft trajectories. *Acta Astronaut.* **2014**, *102*, 378–391. [CrossRef]
3. Forshaw, J.L.; Aglietti, G.S.; Navarathinam, N.; Kadhem, H.; Salmon, T.; Pisseloup, A.; Joffre, E.; Chabot, T.; Retat, I.; Axthelm, R.; et al. RemoveDEBRIS: An in-orbit active debris removal demonstration mission. *Acta Astronaut.* **2016**, *127*, 448–463. [CrossRef]
4. Prussing, J. Optimal Four-Impulse Fixed-Time Rendezvous in the Vicinity of a Circular Orbit. *AIAA J.* **1969**, *7*, 928–935. [CrossRef]
5. Prussing, J. Optimal Two- and Three-Impulse Fixed-Time Rendezvous in the Vicinity of a Circular Orbit. *AIAA J.* **1970**, *8*, 1221–1228. [CrossRef]
6. Carter, T.; Humi, M. Fuel-Optimal Maneuvers of a Spacecraft Relative to a Point in Circular Orbit. *J. Guid. Control. Dyn.* **1987**, *10*, 567–573. [CrossRef]
7. Carter, T. Fuel-Optimal Rendezvous Near a Point in General Keplerian Orbit. *J. Guid. Control. Dyn.* **1984**, *7*, 710–716. [CrossRef]
8. Pontani, M.; Conway, B. Minimum-Fuel Finite-Thrust Relative Orbit Maneuvers via Indirect Heuristic Method. *J. Guid. Control. Dyn.* **2015**, *38*, 913–924. [CrossRef]
9. Prussing, J.E.; Chiu, J.H. Optimal multiple-impulse time-fixed rendezvous between circular orbits. *J. Guid. Control. Dyn.* **1986**, *9*, 17–22. [CrossRef]
10. Pontani, M.; Ghosh, P.; Conway, B.A. Particle swarm optimization of multiple-burn rendezvous trajectories. *J. Guid. Control. Dyn.* **2012**, *35*, 1192–1207. [CrossRef]
11. Hablani, H.B.; Tapper, M.L.; Dana-Bashian, D.J. Guidance and relative navigation for autonomous rendezvous in a circular orbit. *J. Guid. Control. Dyn.* **2002**, *25*, 553–562. [CrossRef]
12. Machuca, P.; Sánchez, J.P. CubeSat Autonomous Navigation and Guidance for Low-Cost Asteroid Flyby Missions. *J. Spacecr. Rocket.* **2021**, *58*, 1858–1875. [CrossRef]
13. Pirat, C.; Ankersen, F.; Walker, R.; Gass, V. $\mathcal{H}_\infty$ and μ-Synthesis for Nanosatellites Rendezvous and Docking. *IEEE Trans. Control. Syst. Technol.* **2019**, *28*, 1050–1057. [CrossRef]
14. Prussing, J.E.; Conway, B.A. *Orbital Mechanics*; Oxford University Press: Oxford, UK, 1993.
15. Pirat, C.; Ankersen, F.; Walker, R.; Gass, V. Vision based navigation for autonomous cooperative docking of CubeSats. *Acta Astronaut.* **2018**, *146*, 418–434. [CrossRef]
16. Segal, S.; Carmi, A.; Gurfil, P. Stereovision-based estimation of relative dynamics between noncooperative satellites: Theory and experiments. *IEEE Trans. Control. Syst. Technol.* **2013**, *22*, 568–584. [CrossRef]
17. Woffinden, D.C.; Geller, D.K. Observability criteria for angles-only navigation. *IEEE Trans. Aerosp. Electron. Syst.* **2009**, *45*, 1194–1208. [CrossRef]
18. Battistini, S.; Shima, T. Differential games missile guidance with bearings-only measurements. *IEEE Trans. Aerosp. Electron. Syst.* **2014**, *50*, 2906–2915. [CrossRef]
19. Mok, S.H.; Pi, J.; Bang, H. One-step rendezvous guidance for improving observability in bearings-only navigation. *Adv. Space Res.* **2020**, *66*, 2689–2702. [CrossRef]
20. Geller, D.K.; Klein, I. Angles-only navigation state observability during orbital proximity operations. *J. Guid. Control. Dyn.* **2014**, *37*, 1976–1983. [CrossRef]
21. Cassinis, L.P.; Fonod, R.; Gill, E. Review of the robustness and applicability of monocular pose estimation systems for relative navigation with an uncooperative spacecraft. *Prog. Aerosp. Sci.* **2019**, *110*, 100548. [CrossRef]
22. Ferrari, F.; Franzese, V.; Pugliatti, M.; Giordano, C.; Topputo, F. Preliminary mission profile of Hera's Milani CubeSat. *Adv. Space Res.* **2021**, *67*, 2010–2029. [CrossRef]
23. Cavenago, F.; Di Lizia, P.; Massari, M.; Wittig, A. On-board spacecraft relative pose estimation with high-order extended Kalman filter. *Acta Astronaut.* **2019**, *158*, 55–67. [CrossRef]
24. Fraser, C.T.; Ulrich, S. Adaptive extended Kalman filtering strategies for spacecraft formation relative navigation. *Acta Astronaut.* **2021**, *178*, 700–721. [CrossRef]
25. Battistini, S.; Cappelletti, C.; Graziani, F. Results of the attitude reconstruction for the UniSat-6 microsatellite using in-orbit data. *Acta Astronaut.* **2016**, *127*, 87–94. [CrossRef]

26. Tang, X.; Yan, J.; Zhong, D. Square-root sigma-point Kalman filtering for spacecraft relative navigation. *Acta Astronaut.* **2010**, *66*, 704–713. [CrossRef]
27. Menegaz, H.M.; Ishihara, J.Y.; Borges, G.A.; Vargas, A.N. A systematization of the unscented Kalman filter theory. *IEEE Trans. Autom. Control.* **2015**, *60*, 2583–2598. [CrossRef]
28. Wan, E.A.; Van Der Merwe, R. The unscented Kalman filter. In *Kalman Filtering and Neural Networks*; Wiley: Hoboken, NJ, USA, 2001; pp. 221–280.
29. Branz, F.; Olivieri, L.; Sansone, F.; Francesconi, A. Miniature docking mechanism for CubeSats. *Acta Astronaut.* **2020**, *176*, 510–519. [CrossRef]

Article

The AlfaCrux CubeSat Mission Description and Early Results

Renato Alves Borges [1,*], Andrea Cristina dos Santos [1], William Reis Silva [2], Leonardo Aguayo [1],
Geovany Araújo Borges [1], Marcelo Monte Karam [1], Rogério Baptista de Sousa [1],
Bibiano Fernández-Arruti García [3], Vitor Manuel de Sousa Botelho [3], José Manuel Fernández-Carrillo [3],
José Miguel Lago Agra [3], Fernando Aguado Agelet [4], João Vítor Quintiliano Silvério Borges [1],
Alexandre Crepory Abbott de Oliveira [1], Bruno Tunes de Mello [1], Yasmin da Costa Ferreira Avelino [1],
Vinícius Fraga Modesto [1] and Emanuel Couto Brenag [1]

[1] Faculty of Technology, Campus Universitário Darcy Ribeiro, University of Brasília,
    Brasília 70910-900, DF, Brazil
[2] Faculty of Gama, Campus Gama, University of Brasília, Brasília 72444-240, DF, Brazil
[3] Alén Space, Edificio Tecnológico Aeroespacial, Rúa das Pontes 6, Suite 2.03, 36350 Nigrán, Pontevedra, Spain
[4] atlanTTic, Escola de Enxeñaría de Telecomunicación, Universidade de Vigo, 36310 Vigo, Pontevedra, Spain
*   Correspondence: raborges@ene.unb.br; Tel.: +55-31-3107-5556

Abstract: On 1 April 2022, the AlfaCrux CubeSat was launched by the Falcon 9 Transporter-4 mission, the fourth SpaceX dedicated smallsat rideshare program mission, from Space Launch Complex 40 at Cape Canaveral Space Force Station in Florida into a Sun-synchronous orbit at 500 km. AlfaCrux is an amateur radio and educational mission to provide learning and scientific benefits in the context of small satellite missions. It is an opportunity for theoretical and practical learning about the technical management, systems design, communication, orbital mechanics, development, integration, and operation of small satellites. The AlfaCrux payload, a software-defined radio hardware, is responsible for two main services, which are a digital packet repeater and a store-and-forward system. In the ground segment, a cloud-computing-based command and control station has been developed, together with an open access online platform to access and visualize the main information of the AlfaCrux telemetry and user data and experiments. It also becomes an in-orbit database reference to be used for different studies concerned with, for instance, radio propagation, attitude reconstruction, data-driven calibration algorithms for satellite sensors, among others. In this context, this paper describes the AlfaCrux mission, its main subsystems, and the achievements obtained in the early orbit phase. Scientific and engineering assessments conducted with the spacecraft operations to tackle unexpected behaviors in the ground station and also to better understand the space environment are also presented and discussed.

Keywords: AlfaCrux LEOP; CubeSat; educational mission; amateur radio

Citation: Borges, R.A.; dos Santos, A.C.; Silva, W.R.; Aguayo, L.; Borges, G.A.; Karam, M.M.; de Sousa, R.B.; García, B.F.A.; Botelho, V.M.D.S.; Fernández-Carrillo, J.M.; et al. The AlfaCrux CubeSat Mission Description and Early Results. *Appl. Sci.* **2022**, *12*, 9764. https://doi.org/10.3390/app12199764

Academic Editor: Cristian De Santis

Received: 28 July 2022
Accepted: 21 September 2022
Published: 28 September 2022

Publisher's Note: MDPI stays neutral with regard to jurisdictional claims in published maps and institutional affiliations.

1. Introduction

The last two decades have seen an enormous rise in the number of launched small satellite missions [1–3]. The growth of launch opportunities, the technological advances, and the creation of standards such as the CubeSats for subsystems and devices has allowed passing from exclusively academic-based projects to missions that involve research centers, private companies, and public institutions [4]. Small satellites have been shown to share many characteristics of disruptive innovations and can be employed in a range of different missions, such as scientific and educational missions, remote sensing, and communications [5–7]. In this context, the need for capacity building and allocation of human resources toward better management and technological innovation for the space field is critical in countries where space activities are not sufficiently established yet, and the CubeSat represents a great platform for education and research [8,9]. Moreover, CubeSats have reduced entry-level costs for space missions, and due to the fact that innovative miniaturized hardware is being developed for deep-space missions, opportunities for lunar orbit and interplanetary space are already becoming available [10–12]. Consequently, educational training is a very important and strategic step in order to further increase the expertise of the scientific community, for instance in Brazil, with the goal of mastering the operation and development of critical and restricted-access technologies in satellite missions.

Elaborated missions are able to address different needs of our society, which imposes strict requirements on the execution of the mission phases. For instance, the need for improving the terrestrial communication coverage providing massive connectivity to under-served areas is driving communication satellites development to a new era of data-based services and applications [13,14]. If one considers the development and operation of reliable data and voice communication systems in remote areas, or those with difficult access, it still represents a challenge in the modern world that affects not only civil society but also a country's defense. In the civil area, it can be said that a few decades ago there were no online applications to improve agriculture and its accuracy, for direct communication between devices, or even interaction and data exchange in order to elevate data connectivity to a more comprehensive level. In the defense field, remote regions with poor infrastructure lack vital communication capabilities beyond line-of-sight to serve tactical land users. When operating over long distances, over rough terrain, or in a jungle environment, tactical users cannot maintain radio line-of-sight. This creates gaps in situational awareness, increasing the possibility that criminal threats or activities go unnoticed, not monitored, not reported, and reactions by local authorities are not triggered.

The AlfaCrux mission is the first space mission financed by the Federal District Government, via the Federal District Research Support Foundation, with institutional support from the Brazilian Space Agency, and under the coordination of the University of Brasília, the owner and operator of AlfaCrux. It is an amateur radio and educational mission with in-orbit technological demonstration and high-level tasks such as management, planning, and risk analysis along the typical life cycle of a space mission. It aims to provide its participants with an excellent technical and intercultural experience and to become a network of students, professors, researchers, radio amateurs, and other partners for the technological advancement of space. The AlfaCrux system is a satellite communication solution with practical and research implications for general society, allowing the responsible team to obtain knowledge about aerospace technology, with innovative experiences and skills, when working on the development of a satellite from the initial phase to its operation. Research and development involving typical satellite and payload architectures will provide experience with respect to the use and applications of nanosatellite technology.

Within this context, this paper provides an overview of the AlfaCrux mission and early achievements obtained so far by the satellite operation. It is intended to describe the methods and materials used to achieve the mission goals, that is, an educational platform for science, technology, engineering, and mathematics (STEM) training and learning within the aerospace field, exploring active methodologies with early in-orbit experiences. It also describes the framework built to explore its scientific capabilities gathered around the topic of digital twin (DT) modeling, which includes, for instance, on-board data handling, data processing and visualization, attitude determination and reconstruction, environment modeling, orbit dynamics, and space weather impact in the communication system. The scientific and engineering assessments conducted during the early orbit phase with the AlfaCrux operations were important not only to check and validate the health of the satellite, but also its functional characteristics for the proposed digital twin architecture. Digital twin for CubeSat applications is still an incipient topic that is driving the development and application of different methods to improve our knowledge based on digital models. A simple example of a first result is provided, but due to the initial stage of operation, detailed comparisons and performance analyses are still to come. The first step was to build, integrate, and run a pilot digital twin architecture for the AlfaCrux kinematics reconstruction, described in the paper, and as the knowledge database increases, successive refinements and performance improvements are implemented, increasing the quality of the proposed model.

Educational Mission

Many engineering programs have used CubeSat development projects as mechanisms for hands-on experience [9]. The benefits of small spacecraft projects in engineering education using the Project-Based Learning (PBL) methodology to engage students in the learning process have been shown by several authors [9,15–17]. PBL is a model that organizes learning around complex tasks based on challenging questions or problems that involve students in design, problem-solving, decision-making, or investigative activities [18]. Compared to traditional forms of instruction, PBL enhances students' ability to transfer knowledge to new problems and to achieve more coherent understanding [9,15–17]. Moreover, the use of Systems Engineering (SE) approach may help to minimize the risks, especially in highly complex projects such as a CubeSat mission development. The importance of SE involves a paradigm shift from an emphasis on designing functional technical systems (electrical systems, control systems, and others) to an emphasis on meeting technical, economic, learning, and management requirements.

For the AlfaCrux mission, the documentation was prepared based on the standards provided by the ECSS [19] and NASA [20], and the spiral model was chosen in the development due to its recursive nature. This model defines a sequence of steps to achieve a viable product taking into account its whole life. Even though the project is currently in Phase E, i.e., in operation, the project team had to go through each phase in the early development process. This is important for a comprehensive definition of the scope of the mission and its requirements. As the project evolves, more knowledge is gathered, and the product quality increases. Figure 1 illustrates the proposed approach, in which the phases are defined as bellow:

- Phase 0—Mission analysis and needs identification: definition of the user needs and requirements, identification of systems concepts;
- Phase A—Feasibility: validation of the stakeholders requirements, identification of solutions that meet the requirements, definition of the mission architecture;
- Phase B—Preliminary design: definition of preliminary system design, validation solutions based on mission requirements;
- Phase C—Final design and Fabrication: definition of final system design (ground and space segments), system breakdown into subsystems;
- Phase D—Integration, Tests, and Launch: integration, tests and launch of the system;
- Phase E—Operations: operation and maintenance of the system to conduct the mission;
- Phase F—Disposal: planning of decommissioning and disposal.

The AlfaCrux mission involved multiple agents and actors. The team members had the opportunity to learn from activities and experiences during different stages of the mission design life cycle. For instance, graduate students were challenged beyond technical development, they were part of the project management team. Undergraduate students were involved in specific problems under the supervision of one of the guiding professors. During the project development, the involvement of graduate students took place outside a formal context of the discipline, through engagement and participation. On the other hand, the involvement of undergraduate students took place within the context of final graduation projects. At the current stage, a formal learning analysis and comparison with other methods was not yet conducted. However, on 6 June 2022, the AlfaCrux operation, command, and control activities entered the syllabus of the course named Engineering Topics, offered by the Electrical Engineering Department of the University of Brasilia. In this course, engineering students had the opportunity to learn in a hands-on approach about the daily activities of operating a CubeSat. Results are yet to come, but so far, it is possible to observe the motivation of the students when facing new challenges concerned with, for instance:

- The mission life cycle;

- The functional requirements and physical architecture of the system;
- The assembly and integration of the systems components;
- The verification and validation process throughout the system development;
- Orbital dynamics, attitude determination and control;
- Elementary concepts of systems engineering.

Finally, concerning the AlfaCrux operational life cycle, it is expected at least 3 years of nominal activities, being possibly extended up to 10 years (depending on the Solar cycle activities impact, the regulation of spectrum use, among others). During the 3 years of operation, the first one is dedicated to the commissioning of the space and ground segments, hardware and software improvements, training, data management system validation, and estimation of the AlfaCrux DT maturity and quality during the early acquisition phase. Consistent and large in-orbit open access database to support a better analysis of the technical maturity of the proposed DT architecture is expected during the second year. Moreover, improvements in the telemetry viewer and mission planner considering the AlfaCrux main services are also a desired outcome at this stage. Finally, the third year should be dedicated to high-quality operation and data analysis in full compliance with the user segment, based on lessons learned from the first two years. It is important to emphasize that during the whole life cycle, the daily activities and operation of the AlfaCrux CubeSat may be covered, or at least motivated, in the syllabus of different courses offered at the University of Brasilia.

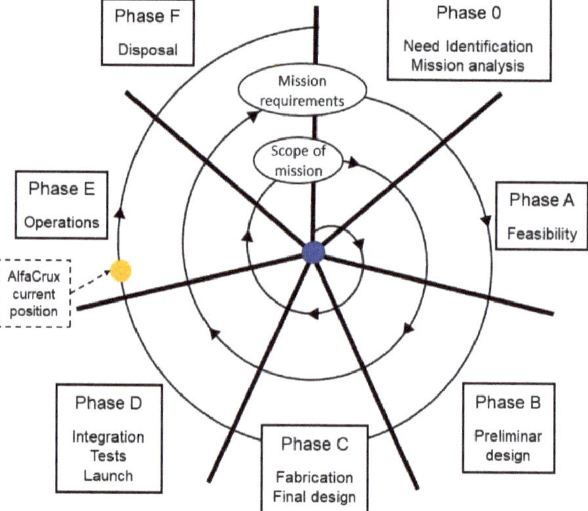

Figure 1. AlfaCrux mission life cycle in the development process.

2. Methods and Materials

2.1. Mission Description

2.1.1. User Needs

The main motivation for carrying out the AlfaCrux mission is to provide an opportunity for space educational training and scientific benefits in the context of small communication satellite missions. The AlfaCrux project fosters the educational collaboration and technical investigations carried out by researchers, professors, students, and radio amateurs, with the sole purpose of self-training in topics such as radio communications, technical management, system-level design, and the development, integration, and operation of small satellites. The need for training and allocation of human resources for better management and technological innovation in the space sector is a critical need in our country. Space activities need to be established in a sustainable model in which constant training plays a fundamental role.

The specific context of the AlfaCrux mission also arises from another critical need in our society, which is to provide reliable narrow-band communication coverage. For instance, narrow-band communication via satellite can meet the needs of data collection systems in places without terrestrial infrastructure, such as underdeveloped areas, uninhabited areas, and areas that have been devastated by a natural disaster as well as provide tactical communication links. Several applications can be constructed based on a system that provides low data rate communication services.

To achieve this goal, the following set of objectives are proposed:

- Launch, operate, command, and control an amateur radio satellite.
- Provide a communication application to interact with the amateur radio community (demonstration of digital data repeater system).
- Provide communication between users and a set of generic data platforms (demonstration of data store-and-forward system).
- Implement and run an online open access database of in-orbit data for general investigations.
- Analyze narrow-band signal attenuation due to the effect of ionospheric scintillation (critical in low latitude regions, as in the case of Brazil).
- Analyze the satellite dynamic based on computer simulations and in-orbit data for studies and validation of Attitude Determination and Control System (ADCS) solutions and applications.
- Analyze and investigate digital twin modeling for aerospace applications.

2.1.2. Mission Concept

AlfaCrux is an amateur radio and educational mission to provide a hands-on experience with technological demonstration in orbit. The main purpose is the operation, command, and control of a satellite-based system for providing communication applications. Digital twin modeling, scientific experiments related to the impact of space weather in the communication channel, as well as the dynamic analysis for ADCS applications, are also part of the mission objectives.

The AlfaCrux payload implements a low-rate, bidirectional, short-message-based store-and-forward communication system between terrestrial and satellite terminals. This system allows, among many other applications, communication with weather stations in remote areas, environmental sensors, communications in areas affected by disasters, etc. For retrieving data from the satellites, the ground station will be the core component of the data distribution system. The system end users can access, through an internet connection, the data collected by these sensors and send the desired information. Cooperation in the ground segment will be established among the AlfaCrux collaborators.

In addition to the communication service itself, the system demonstrates:

- Digipeater for amateurs radio data transfer;
- User-developed scientific applications based on GNU Radio libraries for communication channel characterization;
- Data relay for real-time communication between terminals in the same area.

The operational scenario schematic can be seen in Figure 2, and it is described as follows:

- The satellite transmits beacons while orbiting the Earth, sending telemetry information.
- When the satellite passes through the ground station, the operation team verifies and set the proper operational mode, as well as schedules the next services to be available for the users. This information is available online in the AlfaCrux mission planner.
- When the store-and-forward service is available, a user terminal receives the beacon and sends the available information to the satellite. The satellite, in turn, will transmit the stored information to it.

- The satellite can send the received information in real time to another user terminal or ground station that is in the same area.
- When the satellite passes through a ground station, it exchanges messages from the collected user terminals and those that the system has for the terminals.
- The users will be able to interact with the system through the online official web page, receiving messages from their terminals or delivering the data they wish.
- When the digipeater service is available, the users can exchange digital data when they are in the footprint of the satellite. The current service available is the Automatic Packet Reporting System (http://www.aprs.org, accessed on 20 September 2022), an amateur radio-based system for real time digital communications of information of immediate value in the local area.

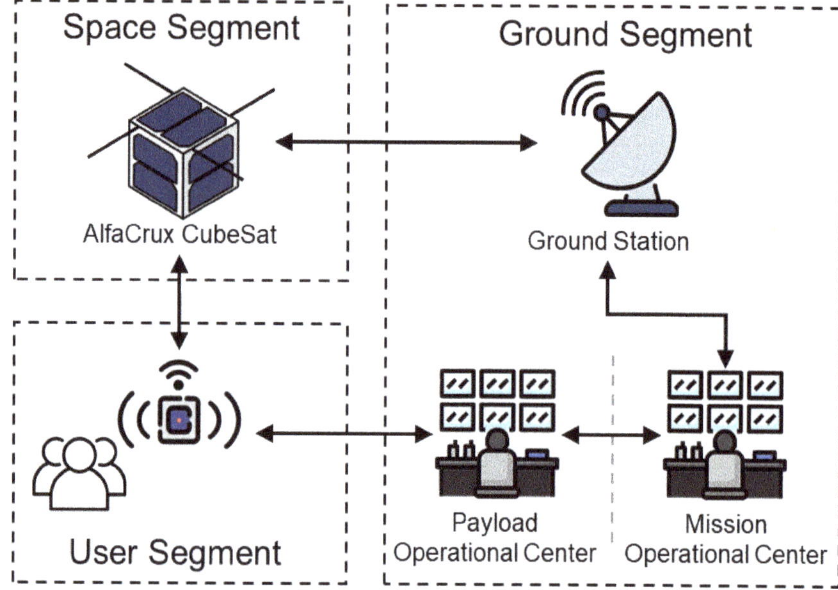

Figure 2. AlfaCrux operational concept.

2.2. Alfacrux Spacecraft

The AlfaCrux is a standard 1U CubeSat with dimensions of 10 by 10 by 10 cm. The final configuration can be seen in the CAD rendering shown in Figure 3 and the mass properties in Table 1. The summary of AlfaCrux orbit and designator is as follows:

- Satellite name: AlfaCrux;
- NORAD ID: 52160;
- International Designator (INTLDES): 2022-033D;
- Country: BRAZ;
- Launch: 1 April 2022;
- Site: Air Force Eastern Test Range;
- Radar cross-section (RCS): Small;
- Period: 94.64 min;
- Inclination: 97.39°;
- Apogee: 508 km;
- Perigee: 494 km.

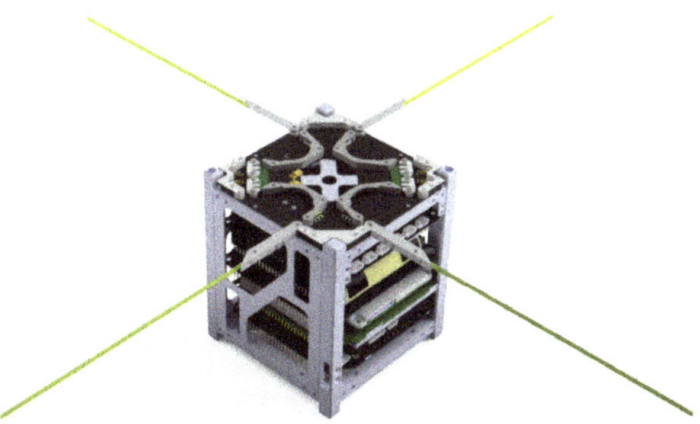

Figure 3. AlfaCrux CAD isometric view.

Table 1. AlfaCrux mass properties.

Parameter	Value	Tolerance	Unit
Mass	1.0650	+/−5 g	kg
Center of gravity	$x_{CG} = 52.27,$ $y_{CG} = 50.81,$ $z_{CG} = 61.81$	+/−10%	mm
Moment of Inertia	$I_{xx} = 0.00183508557,$ $I_{yy} = 0.00185284505,$ $I_{zz} = 0.00184586679$	+/−10%	kg·m^2
Product of Inertia	$I_{xy} = -0.00000456992,$ $I_{yz} = 0.00000153435,$ $I_{zx} = 0.00000877987$	+/−10%	kg·m^2

2.2.1. Alfacrux Subsystems

A short description of the components can be seen in Table 2. The AlfaCrux subsystem configuration can be seen in Figure 4.

Table 2. AlfaCrux subsystems.

Subsystem	Description
Electric Power System (EPS)	3V3, 5V, and VBAT rails. Six switchable power outputs. Latch-up and over-current protection. Watchdog for monitoring GS contact.
Battery	4 × Li-Ion cells (40 Wh). Integrated heaters.
Payload	Alén Space TOTEM-SDR with UHF 437 MHz front-end.
Onboard Computer (OBC)	AVR32 MCU; 128 MB Flash; RTC; I2C, CAN and UART. Three-axis gyroscope. Three-axis magnetometer.
Telemetry, Tracking, and Control (TTC)	UHF 435–438 MHz; Forward error correction, 30 dBm output, 4.8/9.6 kbps.

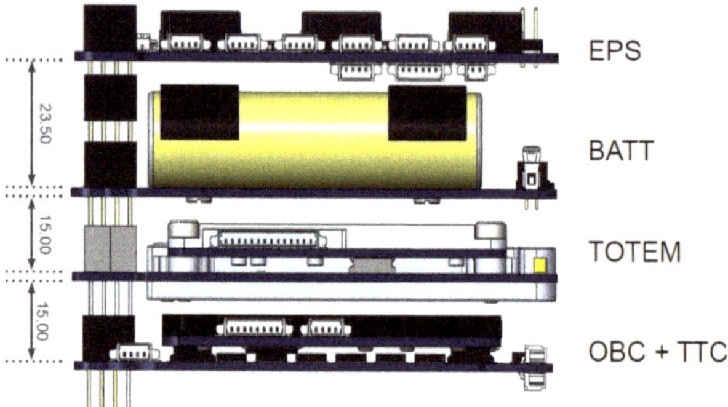

Figure 4. AlfaCrux configuration.

2.2.2. Operational Modes

The AlfaCrux operational modes schematic can be seen in Figure 5. A short description of each mode is as follows.

- Boot-up: During the first boot up of the system, the software shall execute a sequence of first tests and actions, including health check and memories and file system initialization. These actions shall be executed only once. After the completion of the boot up actions, the satellite shall enter Init mode and continue.
- Init: In this mode, the software checks if the antenna was deployed. If not, it shall immediately enter Startup mode; further health checks will be executed once the antenna is deployed.
- Startup: The main objective of this mode is to deploy the antenna and to move to Survival mode once the antenna deployment sequence finalizes. This mode will also collect telemetry from the first minutes of the mission. The antennas shall be autonomously deployed by the onboard software 30 min after ejection. At the end of the antenna deployment sequence, the TTC is configured to allow transmissions. The file system will not be enabled in Startup to avoid errors that could avoid the completion of the Startup sequence. It will be enabled after the completion of Startup, just before the change to Survival mode.
- Survival: This mode is intended to be the fallback mode in case of critical problems. This mode shall have a positive power budget and shall guarantee the possibility to access the satellite from ground. Payloads and non-critical subsystems are turned off in this mode. A beacon with basic health information is transmitted periodically to ease the location of the satellite and to provide a first assessment of its status. TTC is configured with its default configuration. The only way to exit Survival mode is to set a mode change by telecommand (TM) and reboot the software.
- Nominal: The main tasks for the mission execution are performed in Nominal mode. Payloads are allowed to operate in this mode, providing a digital data repeater at 437.225 MHz based on AX.25 protocol, and also a store-and-forward service at 437.125 MHz using Gaussian Minimum Shift Keying modulation. The main scheduler will be in charge of the execution of the different tasks. Health checks are executed periodically. If a critical problem is found, the software will change to Survival mode autonomously.

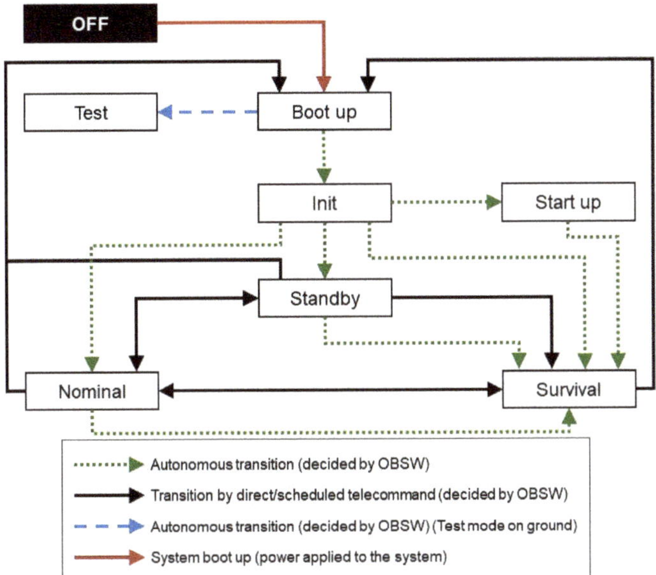

Figure 5. AlfaCrux operational modes schematic.

2.2.3. User Interface with OSI Layers

A general overview of the user interface with the AlfaCrux satellite open systems interconnection (OSI) layers is presented, specifically for the telemetry downlink. It is worth mentioning that the AlfaCrux downlink signal is not encrypted, in accordance with the amateur satellite service regulation, specifically article 25.2A 1A of the International Telecommunication Union (ITU) radio regulation RR25-1.

The TTC physical layer is as follows. TTC frequency of 437.100 MHz, effective isotropic radiated power of 30 dBm, Turnstile antenna, and mixed circular polarization (right and left hand).

The data link layer is summarized in Table 3. The use of Gaussian Minimum Shift Keying (GMSK) modulation provides a good compromise between speed and bandwidth. This type of modulation has no phase discontinuities thus the side lobes of the signal spectrum are reduced. Apart from minimizing channel interference, it provides efficient use of spectrum and enables high efficiency.

Table 3. AlfaCrux data link layer.

Parameter	Description
Modulation	Gaussian Minimum Shift Keying.
Bitrate	4800/9600 bps.
Sync word	0x930B51DE.
Frame format	ASM+Golay (AX100 mode 5).
Bit encoding	NRZ, most significant bit first.
Data randomization	Consultative Committee for Space Data Systems (CCSDS) randomization.
Error-correcting code	Reed-Solomon (255, 223).

All the packets transmitted by the AlfaCrux use the Cubesat Space Protocol (CSP) (https://github.com/libcsp/libcsp, accessed on 20 September 2022). Inside the CSP data field, telecommand transfer frames are used to encapsulate upper layer data. The TM transfer frames are an adaptation of CCSDS TM transfer standards. When the satellite is in communication with the ground station, a reliable channel is established between the station and the satellite using CSP Reliable Data Protocol.

The TM transfer frames transport standard CCSDS Space Packets implementing ECSS Packet Utilization Standard (PUS) services. A detailed description of PUS packets can be found in the ECSS, Telemetry and telecommand packet utilization, ECSS-E-ST-70-41C of 15 April 2016.

The satellite transmits a set of 5 packets every 30 s. Each packet contains a TM frame with a single space packet inside. Standard PUS service 3 (Housekeeping) is used to format these space packets. Five different beacons are transmitted, each one with a different ID, as follows:

- B1-OBC: telemetry from the main on-board computer;
- B2-EPS: telemetry from the electric power subsystem;
- B3-TTC: telemetry from the telemetry and telecommand subsystem;
- B4-UHF: telemetry from the antenna deployment subsystem;
- B5-Temps: temperature telemetry of different subsystems of the satellite.

Apart from the beacons, the satellite can generate other telemetry packets, the majority of them only under specific telecommand from the ground station, that is related with command and control of the satellite, with some packets out of the PUS standard. More information can be found at the official website of the mission at https://lodestar.aerospace. unb.br/projects/alfacrux (accessed on 20 September 2022).

2.3. Alfacrux Ground Segment

2.3.1. Ground Station Architecture

The ground station information technology (IT) architecture is composed of physical and logical components, shown in Figure 6. The physical components aim to provide the computational power and make the internal and external interconnection with the Internet, namely:

- Two desktop computers, with I5 CPU, 8 GB of RAM, and 500 GB SATA Hard Disk;
- One Dell server with Intel(R) Xeon(R) E-2224 CPU @ 3.40GHz, 8 GB of RAM, and 1 TB SATA Hard Disk;
- One load balancer, which allows balancing the links to the Internet;
- CAT 6 A cabling, used to interconnect computers, servers, and load balancer;
- Scripts for automation;
- One uninterruptible power source (UPS), which allows system autonomy for about 35 min without power from the electrical grid.

The logical components are formed by operating systems, firmware, software, applications, and virtual machines, as follows:

- Computers or operating stations: its purpose is to operate and control the software-defined-radio and antenna rotors using an orbit propagator and Doppler shift calculator along with the mission control software (MCS). The VirtualBox was installed for extra functions such as virtual private network (VPN) and monitoring with Zabbix. Some examples of monitored parameters are processing load, RAM memory consumption, and SWAP memory consumption, among others.
- Server dedicated to the subsystems necessary for antenna control and Doppler correction. It aims to provide the necessary hardware resources for Rotcld, a Hamlib rotator control daemon, and MCS applications.
- General-purpose scripts to automate internal operation tasks.

The AlfaCrux team implemented an architecture for the ground station that allowed remote and secure operation. The computers were configured not to be a single point of failure, having replicated the virtual machines and applications, allowing them to be switched quickly, in a matter of minutes. To monitor the IT architecture and its components in order to identify possible points of attention, such as overloads, unavailability, and loss of performance, the Zabbix monitoring solution was implemented.

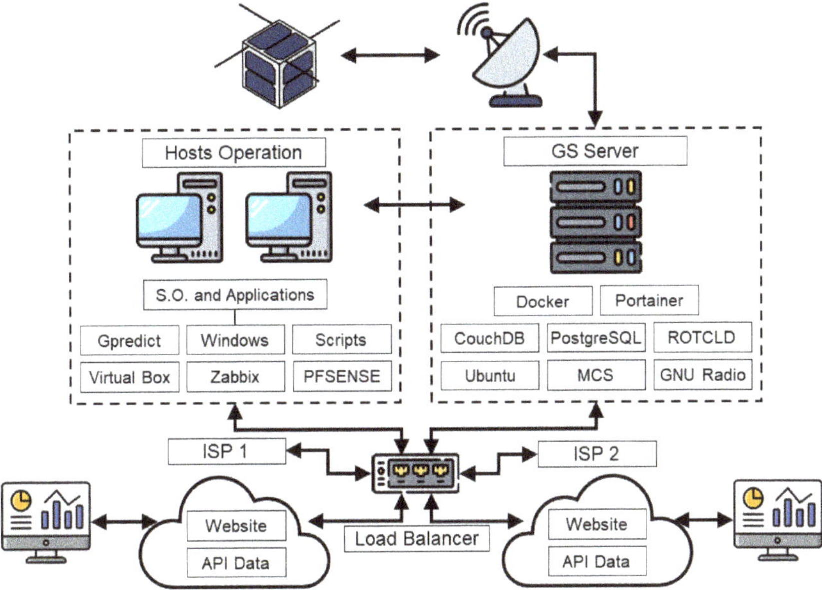

Figure 6. Ground station architecture.

The Zabbix is an open-source and free solution with several applications in different contexts including, for instance, the aerospace field. The basic resources of the server and computers are monitored, such as CPU, RAM, and SWAP memories, disk writing and reading latency, bandwidth consumption, status of processes and containers. For the administration and remote operation of the ground station, an open-source VPN solution, the OpenVPN, was implemented, embedded in the pfSense Linux distribution. The encryption used is based on public and private keys, with each user having their unique access key. pfSense integrated solutions with the OpenVPN and the key export package were chosen for their ease in generating and managing keys, being considered mature and a reliable security solution used by several companies in the world.

In order to mitigate the impact of power outage in the operation, a UPS capable of supporting the entire environment was implemented, being able to maintain the operation with the AlfaCrux and data management system for 30 min when the main power fails. Another important point to be highlighted is the need to configure the power schemes of the computers and the server. Specifically, the following characteristics were activated in the basic input–output system: power on after power outage, scheduled power on, and wake on local area network (LAN).

To increase efficiency and reliability in the operation tasks, some scripts were created, and additional configurations were set up on the server as well as the desktop computers. On the client side, a batch script was developed with an interactive menu for connectivity testing, activation, deactivation, restart, and collection of information from the rotors and the software-defined radio (SDR), as well as a bash script to control and manage containers. Time update enforcement twice a day has been added to the task schedulers. This procedure was implemented after observing the impact caused by the delay or advance of a few minutes during the operation with AlfaCrux. Moreover, it guarantees a reliable and satisfactory satellite clock synchronization.

The satellite clock synchronization is performed by the operator through specific telecommands, and it is regulated from the main server at the ground station. In this framework, the precision of the satellite clock is given by the precision of the ground station's main server and computers. The clock synchronization between the server and computer systems is achieved using the Network Time Protocol (NTP), specifically a server provided by the https://ntp.br/ (accessed on 20 September 2022). The maximum error of a clock set through an NTP client varies from fractions of milliseconds to a few tens of milliseconds, depending on the quality of the network. Asymmetric routes and network congestion can cause errors of 100 ms or more. Quantitatively speaking, for the proposed solution, values less than 100 ms were noticed, which should provide an idea of the precision magnitude.

The impact of these delays is twofold: orbit propagation for the tracking system and data analysis and use for different applications. In the first case, considering the pointing tolerance of 5 degrees and the precision of the tracking system, it was noticed that a mismatch of less than 30 s is affordable. In the case of data use and analysis, a regularization procedure is executed at least once per day. The regularization of the satellite's current time is progressively performed in order to avoid jumps in time. Consequently, the time is not changed immediately. The duration of the regularization must be longer than the time span to be advanced or delayed. For example, if the on-board time is 20:30 and we want to set it to 20:15, the duration of the regularization should be set to greater than 900,000 ms. Moreover, in case higher precision is needed for data use, time corrections can also be applied in the mission control software.

Another innovation in the architecture of the ground station was the use of containers, allowing for greater compartmentalization and isolation of solutions. To facilitate the container management, the free graphical manager Portainer was installed. In all, 5 containers are configured, namely, CouchDB for the MCS database, PostgreSQL for the MCS database, MCS server, Rotctld rotator control daemon, and an SDR container.

Data registration is performed in an encrypted form in a CouchDB database and in a PostgreSQL database exclusive to the application. To comply with the need for data study and analysis, the MCS software has an API that allows access to data through web services. For this reason, a Python 3.6 program was developed with the objective of performing three operations during its execution: (i) update all existing parameter tables; (ii) update all executed remote controls; (iii) update all telemetry received. The result of this operation, in a way, is a backup of the data while creating a security layer for data access since, after updating, the data are provided through another API, developed by the AlfaCrux team to be accessed by students, researchers, and the amateur radio community.

In addition to this API, the AlfaCrux team also developed an interface that allows the collaborative collection of telemetry from anyone that have received packages from the AlfaCrux satellite, recorded the telemetry and generated a KISS frame file. This interface, an online form, allows recording latitude, longitude, email, name, call sign, and KISS frame file. Finally, data duplication analysis mechanisms will be necessary and even, without prejudice to data quality, the creation of a composite primary key that allows differentiating the data source, which may lead to the possibility of telemetry duplication.

2.3.2. User Segment

Currently, the AlfaCrux team is developing and providing the online telemetry viewer (https://lodestar.aerospace.unb.br/projects/alfacrux/radio#telemetry-viewer, accessed on 20 September 2022), a web platform that selects the main parameters of the satellite beacons to the user. The data provided by the collaborators will also be made available, and a main time series of the parameters will be implemented after checking and validating the received package. All frames available online are properly associated with the responsible call sign in such a way that the amateur radio can search in our database for its specific contribution in the AlfaCrux mission.

2.4. Alfacrux Digital Twin Architecture

One of the main innovative contributions of the AlfaCrux mission is concerned with the development of a digital twin model, see Figure 7. The DT can be defined as a high-fidelity and up-to-date digital representation of an physical counterpart. This representation is achieved by combining different models (physics-based and data-driven models), simulations, sensor updates and historical data [21]. Although this topic has been studied in the aerospace field [22], the use and applications of digital twin in the context of small satellites are still incipient, with very few examples and systems implemented so far [23,24].

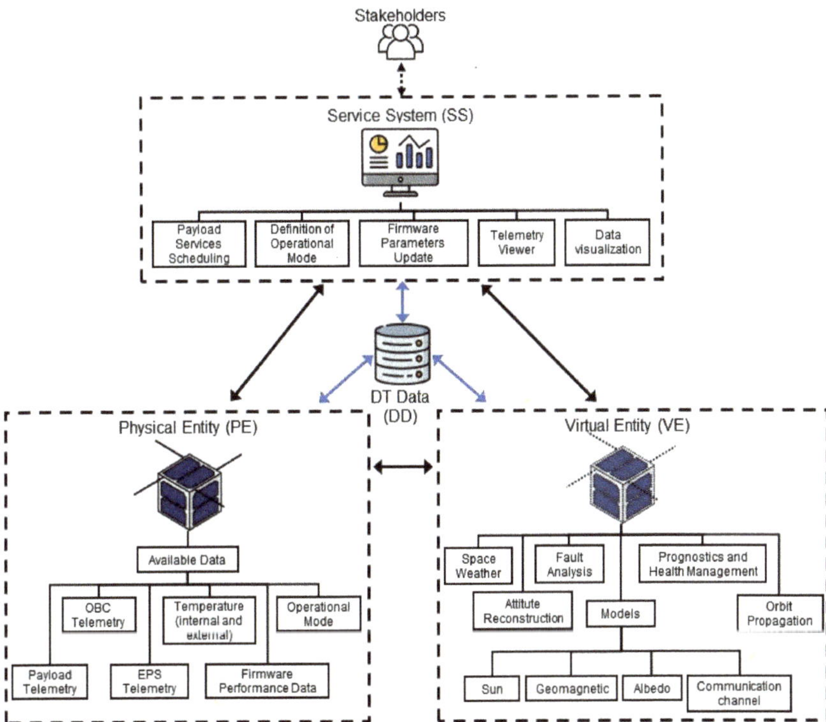

Figure 7. Five dimensional architecture for the CubeSat digital twin framework.

In this first approach, the goal is to set the start point and the guidelines to establish the concept and a methodology to build a reliable framework for the operation and main services of nanosatellites, in a first moment CubeSat standard, fulfilling the necessary space standards (such as health checks, dynamic characteristics, and safety mechanisms, among others) and conforming to the necessary payload requirements.

By innovative framework, we mean a new architecture in which the development of a digital model of the nanosatellite is considered. In this innovative approach, the digital model is constantly under improvements based on the data collection system (laboratory tests, computer simulations, and in-orbit data), and more than that, the digital model can improve and update the physical entity, i.e., the satellite in orbit). In other words, we are proposing a digital twin framework for nanosatellite mission applications.

The digital twin can be modeled in a five-dimensional architecture composed of the physical entity (PE), the virtual entity (VE), the DT data (DD), the system services provided (SS) and the connections between the parts (CN) [25]. Figure 7 provides a conceptual model of the DT specifically for the AlfaCrux. The sensors in the PE gather data that will be used as input for the analysis, models, and simulations of the VE. In this way, the VE can mirror the physical geometry, physical properties, behaviors, and rules of the PE. The SS consists of the services for the physical and virtual entities, and for the project stakeholders. These services include the scheduling of the payload services, the change in operational mode, the telemetry viewer, and firmware updates. The DD is composed of data collected from the PE, VE, and SS through the connections. The connections are also used to link the PE, VE, and SS among themselves [25].

In the center of the five-dimensional architecture is the digital twin data, due to their critical role in the DT as the knowledge representation [26]. The process of creating the knowledge involves, among other things, the data collection and pre-processing of different sources. These sources can be divided into three main categories: synthetic data (computer simulation), experimental data (laboratory tests and validation), and in-operation data (satellite telemetry) [22]. Specifically, the synthetic data are generated by simulations in the virtual world, the experimental data are gathered through experiments conducted in the physical entity, and the last category, the in-operation data, are collected during regular operation of the physical entity, in our case the AlfaCrux CubeSat.

The first part of the DT that has been developed is dedicated to kinematic reconstruction, as seen in Figure 8. The models were initially validated with simulated and real telemetry data. This pilot DT is explained below, considering the five dimensions shown in Figure 7.

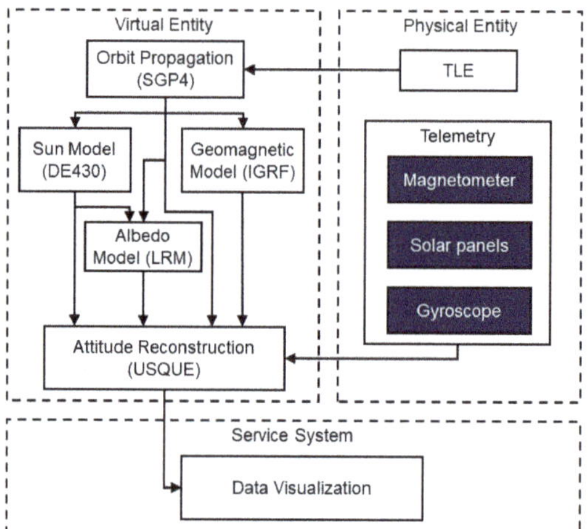

Figure 8. AlfaCrux kinematic reconstruction DT.

1. Physical entity: acquires measurements of the magnetometer, solar panels, and gyroscope, and the two-line elements (TLE);
2. Virtual entity: uses the Sun, Geomagnetic, and Albedo models; the orbit propagation; and the telemetry data to reconstruct the satellite attitude;
3. Connection: responsible for the data collection and management system;
4. Service system: responsible for the attitude visualization.

3. Results and Discussions

The preliminary orbital data (injection) given by the launch provider team were used to propagate the orbit (Simplified General Perturbations-4 (SGP4)) and predict the first passages over the command and control station. The first access report provided the estimates presented in Table 4.

Table 4. Ground station to AlfaCrux first access report.

Access	Start Time (UTC)	Stop Time (UTC)	Duration (s)
1	2 April 2022 01:14:35.727	2 April 2022 01:19:56.946	321.218
2	2 April 2022 02:46:49.656	2 April 2022 02:55:04.360	494.704
3	2 April 2022 13:12:37.268	2 April 2022 13:19:58.469	441.201
4	2 April 2022 14:46:26.445	2 April 2022 14:53:21.925	415.480
5	3 April 2022 02:28:02.961	3 April 2022 02:37:04.707	541.746
6	3 April 2022 12:55:30.532	3 April 2022 13:00:24.969	294.438
7	3 April 2022 14:27:14.360	3 April 2022 14:35:37.980	503.620
8	4 April 2022 02:09:32.490	4 April 2022 02:18:45.596	553.106
9	4 April 2022 14:08:27.400	4 April 2022 14:17:31.124	543.724

As a pre-launch setup, the ground station equipment, configuration, and procedures were verified. The mission control software was updated, and the control sequences loaded. The pass procedure for the first days based on the access report was revised and ready to be executed.

In the first passage over the command and control ground station, we were able to receive beacons from the AlfaCrux, in a first moment not able to demodulate, but as soon as the orbit position became more precise, the quality of the tracking, and consequently of the operation, were improving. During the first days, the operation team was focused on the correct identification of the AlfaCrux spacecraft in cooperation with the 18th Space Control Squadron of the United States Space Force. As an operator organization, the AlfaCrux team started using the TLEs from the Space-Track (https://www.space-track.org/, accessed on 20 September 2022) platform, exchanging information about the AlfaCrux operation, checking, and analyzing safety information provided by the conjunction data message, and after some days, the AlfaCrux was properly identified with the international designator 2022-033D.

3.1. Satellite and Ground Station Commissioning

During the AlfaCrux satellite and ground station commissioning, the SDR start/stop procedures were constantly tested and verified before the passage, leading to new scripts to improve the efficiency of the operation. The elevation and azimuth rotors' start/stop procedures and the remote control interface were also tested and checked constantly, leading to adjustments in the antenna mounting, pointing calibration, soft start/stop ramp, and creation of preventive maintenance routines.

For the first contact, the main actions were to receive beacons; verify communications (Ping command to the satellite); check preliminary TLE by taking notes of time/frequency of Doppler shifts, especially at the moment in which the Doppler shift equals 0 Hz; and check the overall health of the satellite. The critical housekeeping information for those days were battery level, solar panels' voltages and currents, battery charge current, EPS output currents, and internal and external temperatures.

For the next contacts, the procedure includes receiving beacons, verifying communications (Ping/Pong), checking if operational mode was set to Survival, establishing a reliable communications channel, obtaining telemetry errors (if any), and obtaining telemetry since the first hours, storing it, and analyzing the critical parameters.

Recently, the clock was synchronized with the ground station, and a new procedure was included in the daily passage, that is, a time reference update. As of 27 July 2022, 01:56 UTC, the AlfaCrux changes to Nominal mode.

3.2. Alfacrux Digital Twin Pilot

The first elements of the proposed digital twin, shown in Figure 7, were developed and tested in the current phase with synthetic and preliminary telemetry data, specifically the attitude reconstruction, environment, and communication channel models; the telemetry viewer; and risk analysis.

3.2.1. Attitude Reconstruction and Dynamic Analysis

The attitude reconstruction system is responsible for the spacecraft pose determination using the in-orbit data stored in the DT database, generating an offline time-series of the attitude parameters. This feature allows the assessment of onboard sensors performance and their characterization; the analysis of filtering methods, such as Kalman filter and its derivations; the study of environment models; and the estimation of perturbations. For the reconstruction, not only the attitude sensors, for instance gyroscopes, magnetometers, and Sun sensors, can be considered but also other onboard data that may be affected by the satellite orientation, such as solar panels, optical payloads, or even the antenna signal power received by the ground station.

For the AlfaCrux attitude estimation, the onboard sensors available are one gyroscope and one magnetometer, both with tree-axis measurements. In order to improve the filter performance, the solar-panel-related data are also considered so that the Sun line-of-sight can be estimated as a Sun sensor. Specifically, the AlfaCrux CubeSat has a total of six solar panels, each one mounted on one spacecraft face. Solar panels from opposite sides are connected in parallel and the onboard system collects one current and voltage measurements from each pair along with individual temperature measurements. As shown in [27], the Sun vector can be estimated through computations with individual current measurements. However, in the AlfaCrux case, the power is considered instead of the current, and the highest temperature of one solar panel pair indicates which one is generating such power.

Once the measurements are available, an initial attitude determination method is executed in order to compute the full attitude profile. The simplest one is the Triaxial Attitude Determination (TRIAD), which is based on finding an attitude matrix that rotates an orthonormal base in the inertial frame to the base expressed in the spacecraft body frame from two not aligned vector measurements, such as, for instance, the Sun position and the Geomagnetic field [27]. After the definition of the initial attitude, stochastic filtering techniques provide the solution for the attitude propagation over time through the dynamical model and sensor data. Such a filtering problem consists of a sub-optimal estimation of the attitude state along with related parameters, such as sensor biases, misalignment, and orthogonality. In this context, the state vector and sensor observations are defined as random variables, so that their statistics represent the confidence degree in both estimated and measured quantities in the form of error covariance.

The filter structure for AlfaCrux attitude estimation is formulated in [28], called Unscented Quaternion Estimator (USQUE), which is based on the Unscented Kalman Filter, a nonlinear filter that uses selected samples from the state probability density function instead of computing a first-order linear approximation using the Jacobian matrix, as in the Extended Kalman Filter. The USQUE algorithm considers the multiplicative quaternion approach, where a global attitude state is the spacecraft estimated pose, and a local attitude state represents the error to be computed in the filtering process. After an iteration, the global attitude is updated with the estimated error from the filter.

The system is modeled by the following equations:

$$\mathbf{x}_k = \mathbf{f}(\mathbf{x}_{k-1}) + G_k \mathbf{w}_k,$$
$$\mathbf{y}_k = \mathbf{h}(\mathbf{x}_k) + \mathbf{v}_k,$$

(1)

where $\mathbf{x}$ is the 6×1 state vector; $\mathbf{y}$ is the 6×1 output vector; $\mathbf{f}(\cdot)$ is the attitude kinematic equation; $\mathbf{h}(\cdot)$ is the measurement model equation; $\mathbf{w}$ and $\mathbf{v}$ are the process and observation errors, both modeled as uncorrelated Gaussian random variables with zero mean; and Q and R are covariance matrices, respectively.

The state vector is composed of attitude-error angles parameterized using the Generalized Rodrigues Parameters (GRP) and the gyroscope bias, as follows: $\mathbf{x} \equiv [\delta\mathbf{p}^T \ \beta^T]^T$. The GRP represents the local attitude-error that will be incorporated in the global representation, once $\| \varrho\mathbf{p} \|$ is equal to the rotation angle, ϑ, for small errors. The sigma-points, $\chi(i)$ for $i = 0, 1, \cdots, 12$, are generated according to the following equations:

$$\sigma_i \longleftarrow \text{i-th column of } \pm \sqrt{(n+\lambda)(P^+_{k-1} + \bar{Q}_{k-1})},$$

(2)

$$\chi(0) = \hat{\mathbf{x}}^+,$$
$$\chi(i) = \sigma_i + \hat{\mathbf{x}}^+,$$

(3)

where P^+_{k-1} is the state error covariance, $\bar{Q}_{k-1}$ is the process error covariance, and λ is a scalar chosen arbitrarily.

For the prediction step where the state vector is propagated from time instant $k-1$ to k, the attitude kinematic equation is considered. The GRP vector of each sigma-point is converted to an error-quaternion, $\delta\mathbf{q}(i)$, and incorporated to the full attitude representation, $\hat{\mathbf{q}}$, through a quaternion multiplication operation. Then, the attitude kinematic for the sigma-points quaternions, $\mathbf{q}(i)$, is executed using the gyroscope angular velocities measurements for propagation in time.

After the prediction, the sigma-points in GRP representation are retrieved from the propagated quaternions. The mean predicted state vector is computed along with the predicted state error covariance matrix:

$$\hat{\mathbf{x}}_k^- = \frac{1}{n+\lambda}\left\{ \lambda\chi_k(0) + \frac{1}{2}\sum_{i=1}^{2n}\chi_k(i) \right\},$$

(4)

$$P_k^- - \frac{1}{n+\lambda}\left\{ \lambda[\chi_k(0) - \hat{\mathbf{x}}_k^-][\chi_k(0) - \hat{\mathbf{x}}_k^-]^T \ + \frac{1}{2}\sum_{i=1}^{2n}[\chi_k(i) - \hat{\mathbf{x}}_k^-][\chi_k(i) - \hat{\mathbf{x}}_k^-]^T \right\} + \bar{Q}_k.$$

(5)

The correction step is performed once the mean state vector and covariance are available. This procedure considers the observation model, where reference vectors (such as Sun line-of-sight and geomagnetic field) in the inertial frame are transformed to the spacecraft body frame and compared with the magnetometer and solar panels data observations in order to compute the innovation vector, v. Such measurement model is given by

$$\tilde{\mathbf{y}}_k = \begin{bmatrix} A(\mathbf{q})\mathbf{r}_1 & A(\mathbf{q})\mathbf{r}_2 & \cdots & A(\mathbf{q})\mathbf{r}_n \end{bmatrix}^T + \begin{bmatrix} v_1 & v_2 & \cdots & v_n \end{bmatrix}^T,$$

(6)

where $A(\mathbf{q})$ is the attitude matrix built with the quaternion obtained from the GRP, $\mathbf{r}_j$ is a reference vector observation provided by a environment model, and v_j is the respective measurement error.

With the obtained quaternions, the mean observation vector can be computed as well as the measurement vector, innovation covariance matrices and the cross-correlation matrix:

$$\hat{\mathbf{y}}_k^- = \frac{1}{n+\lambda}\left\{ \lambda\gamma_k(0) + \frac{1}{2}\sum_{i=1}^{2n}\gamma_k(i) \right\},$$

(7)

where $\gamma_k(i) = \mathbf{h}[\chi_k(i), k]$

$$P_k^{yy} = \frac{1}{n+\lambda}\left\{ \lambda[\gamma_k(0) - \hat{\mathbf{y}}_k][\gamma_k(0) - \hat{\mathbf{y}}_k]^T + \frac{1}{2}\sum_{i=1}^{2n}[\gamma_k(i) - \hat{\mathbf{y}}_k][\gamma_k(i) - \hat{\mathbf{y}}_k]^T \right\}, \tag{8}$$

$$P_k^{vv} = P_k^{yy} + R, \tag{9}$$

$$P_k^{xy} = \frac{1}{n+\lambda}\left\{ \lambda[\chi_k(0) - \hat{\mathbf{x}}_k^-][\gamma_k(0) - \hat{\mathbf{y}}_k]^T + \frac{1}{2}\sum_{i=1}^{2n}[\chi_k(i) - \hat{\mathbf{x}}_k^-][\gamma_k(i) - \hat{\mathbf{y}}_k]^T \right\}. \tag{10}$$

The innovation is computed as the difference between the sensor observations and the predicted output:

$$v_k \equiv \tilde{\mathbf{y}}_k - \hat{\mathbf{y}}_k^- = \tilde{\mathbf{y}}_k - \mathbf{h}(\hat{\mathbf{x}}^-, k). \tag{11}$$

The state vector and error covariance matrix are updated following the equations:

$$\hat{\mathbf{x}}_k^+ = \hat{\mathbf{x}}_k^- + K_k v_k, \tag{12}$$

$$P_k^+ = P_k^- - K_k P_k^{vv} K_k^T. \tag{13}$$

The attitude-error from $\hat{\mathbf{x}}_k^+$ is transformed into a quaternion, and it is used to update the global attitude representation, which is the new attitude estimate. Later, the attitude error in the state vector is set to zero before the beginning of the next iteration.

Finally, the in-orbit data will be used as input information for the small satellite simulator available at the Laboratory of Simulation and Control of Aerospace Systems (LODESTAR) [29–31]. Moreover, a new air-bearing based simulator dedicated to the CubeSat standard was built, and it is already operational to be used for the reconstruction of dynamics characteristics of the AlfaCrux. The CAD view is shown in Figure 9. The results of the experiments in the simulator are stored in the DT database and can be used in future simulations.

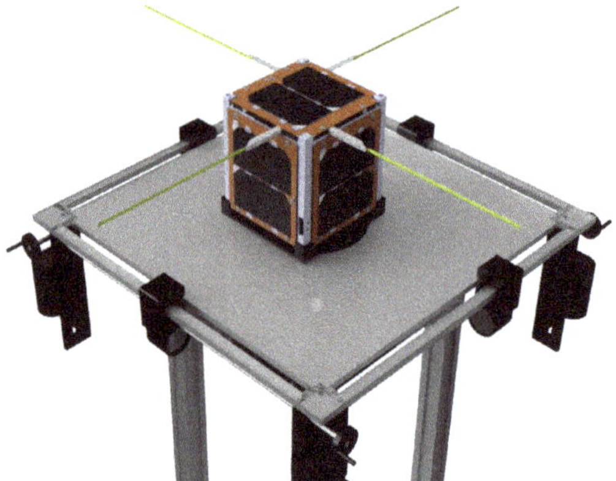

Figure 9. CubeSat air-bearing simulator CAD view with the AlfaCrux.

3.2.2. Environment Models

For the proposed DT model, the spacecraft motion is a major concern. The formulation considered in this work uses the SGP4 model for orbit propagation, uses the TLE file as the only orbital data source, and provides a prediction for the satellite position and velocity over time. The SGP4 considers perturbations such as the Earth's ellipsoidal shape effect over the gravitational field, third-body influence (Sun and Moon) and the atmospheric drag.

As described in [32], the Earth Magnetic Field is modeled as the gradient of the magnetic potential, which is approximated by a spherical harmonic series expansion. One of the standard models is the International Geomagnetic Reference Field (IGRF), and it consists of the mathematical description along with the coefficient values for the series expansion. The model allows the computation of the geomagnetic field at any point in space given its spherical coordinates. Furthermore, it allows not only computer simulations but also serves as input data for laboratory experiments in Helmholtz cages.

The Sun position is another important model, once it is the primary power source for most of the spacecrafts. According to [32], a more precise approach to determine the position is through the data set DE430, a set of celestial bodies' ephemerides provided by the Jet Propulsion Laboratory and generated by laser observations and the numerical integration of a dynamic system. However, this model requires access to the data and numerical methods. A simpler and less precise model is based on analytical equations that may be considered in simulations and is also useful for implementing in the onboard software. This model considers the Sun elliptic motion, containing polynomials and trigonometric terms only.

With respect to the problem of attitude determination, the Earth albedo model may be considered if the spacecraft has Sun sensors onboard, once the solar radiation reflected by the Earth surface can introduce errors in the Sun position vector computing. Albedo is measured as the reflectivity index, which varies according to the cloud coverage, geographic location, season of the year, and type of surface. In [33], the reflectivity index is provided by the NASA Total Ozone Mapping Spectrometer (TOMS) satellite, which records the measurements observations constituting a grid of data points. The total energy reflected to the satellite is formulated as the sum of each data point contribution, that is, illuminated by the Sun and visible by the satellite. The model presented depends on the reflectivity index, the angle of incidence, satellite altitude, and the angle of reflection to the spacecraft.

For the results obtained so far in the early stage, the Sun position and Geomagnetic field described in the inertial frame are provided by the software System Tool Kit (STK) from Ansys Inc, along with the AlfaCrux orbit propagation with the TLE. The Earth albedo model was implemented based on the Lambertian reflectance model (LRM), which finally guarantees the attitude reconstruction process.

3.2.3. A Preliminary Outcome Example

As a simple example to illustrate the AlfaCrux DT model outcome, a first analysis of the attitude reconstruction by means of stored telemetry data is shown in Figure 10, where the Euler angles are parameterized in the 3-1-3 rotation sequence. Currently, the AlfaCrux ground station is undergoing technical adjustments and hardware upgrade in order to improve the signal downlink. Moreover, for the early orbit phase and commissioning performed in these first months of operation, the in-orbit data are gathered into parameter report structures using a low data sampling rate. Both scenarios lead to missing information in the sensors;' time series, impacting the quality of the attitude reconstruction. In this case, to evaluate the DT kinematic pilot model, preliminary studies were carried out considering telemetry data sampled within short time intervals. In order to illustrate the first results, consider the AlfaCrux data from the 4 June 2022 within a time interval of 180 s, where the telemetry is uniformly sampled at each 30 s during a passage over the command and control station. A quadratic interpolation was performed in order to generate new points at each 5 s time interval providing a smooth and more appropriate time series for attitude reconstruction. For future analysis, the AlfaCrux sampling rate will be properly adjusted, increasing the knowledge database, and consequently improving the quality of the estimation.

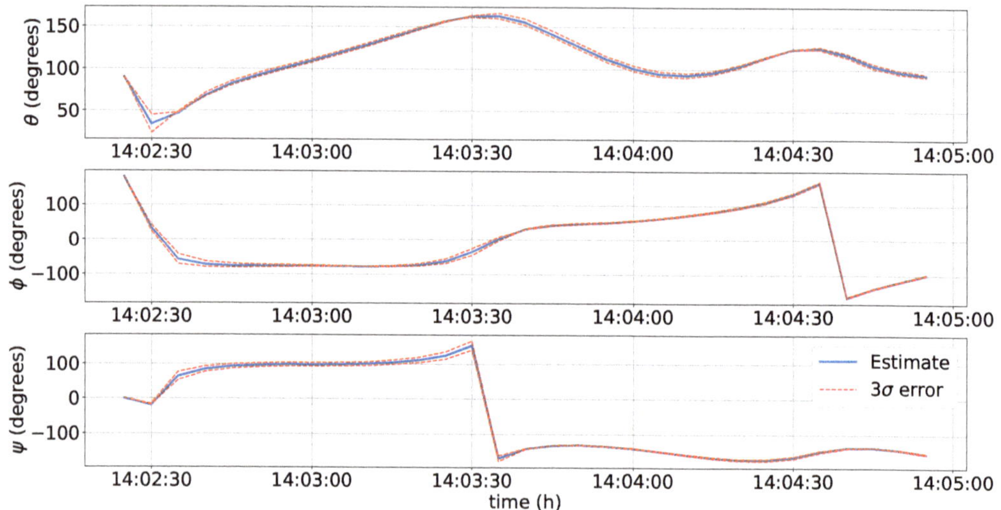

Figure 10. Euler angles from attitude reconstruction.

Since the AlfaCrux satellite does not have an attitude control subsystem, the only torques acting on the CubeSat are the disturbance ones, such as those due to the gravity-gradient and residual magnetic dipole. For these first analyses, the influence of a possible residual magnetic dipole was not considered. This will be estimated later along with the magnetometer data analysis and calibration. By using the least squares method, it was possible to estimate an initial magnetometer bias with an order of magnitude of 10^4 nT. With respect to the solar panels, since they do not have available information about sensor characterization, the uncertainties must be estimated. As regular coarse sensors are inaccurate in the order of degrees, the same order of magnitude is expected for the estimation of the Sun vector direction based on the power and thermal analysis. Finally, the evolution of the state error standard deviations can be calculated, providing the results shown in Figure 11.

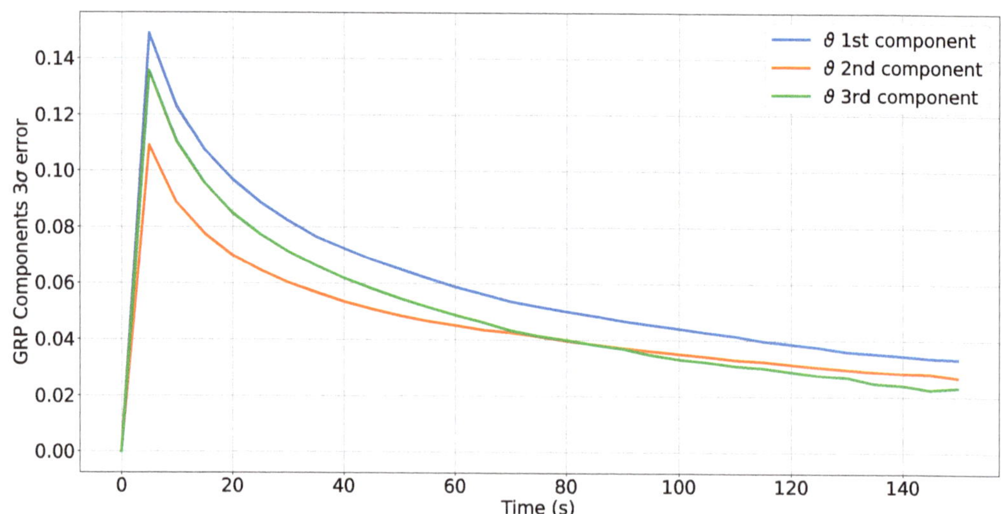

Figure 11. Estimated state vector error for the angles parameters.

3.2.4. Data Management and Telemetry Viewer

The ground station information technology team is responsible for keeping the data management system and online open access database. Currently, the data are updated after a set of operations, and they are included as a procedure to be executed after a passage. The current architecture can be seen in Figure 12.

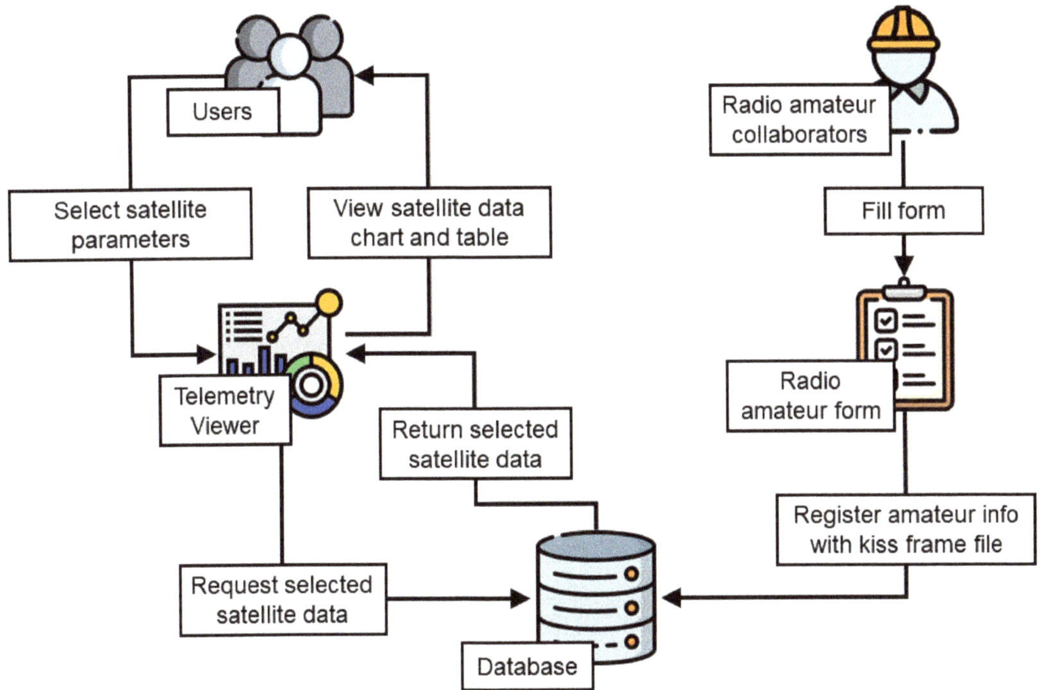

Figure 12. User segment architecture: data management and viewer.

3.2.5. Communication Channel Modeling and Risk Analysis

The elaboration of a detailed model for the communication link in both directions (uplink and downlink) is relevant to simulating adverse situations, which helps the planning and assessment of mission risks. Within the AlfaCrux scope, the joint use of a communication simulator and the digital twin framework allows the elaboration of worst-case scenarios and the testing of new features that can be added in future missions. The measurement and modeling of scintillation are particularly relevant for sub-GHz frequencies when ground stations are positioned at low latitudes. The effects of space weather in the planning of the AlfaCrux ultra-high frequency (UHF) communication system was previously analyzed, see [34] for more details, and should be further investigated considering now the in-orbit data and radio frequency signals received at the ground station.

Currently, the simulation framework is the GNU Radio platform, where two diagram blocks implement distinct aspects of the link modeling. The first block focuses on the dynamic channel model, considering the satellite passage above the ground station. Within the visibility window, the simulator provides fundamental channel aspects for different orbit altitudes and frequencies, such as free space attenuation and Doppler frequency deviation. Following the ITU-Regulations (ITU-R) recommendations, the insertion of channel impairments from the troposphere (such as attenuation from gases, rain, and clouds) and ionosphere (such as scintillation, Faraday rotation, and propagation delay) is currently under development. At the transmitting and receiving nodes, we also consider local parameters such as antenna gain-to-noise-temperature (G/T), transmitted power, Effective

Isotropic Radiated Power (EIRP), and equivalent noise temperature of the receiver. In this block, the simulator updates the satellite position (and its effects on the channel) once a second.

The second block diagram simulates the GMSK digital transmission in baseband under controlled conditions of signal-to-noise ratio. The simulation rate is milliseconds, and dynamical channel effects are considered here under the assumption of slow changes in the channel. With this module, it is possible to apply different techniques in the receiver chain, such as carrier/symbol synchronization, robust error-correcting codes, and advanced digital signal processing methods. This module also allows future prototyping by embedding the algorithms in a parallel software-defined radio platform.

4. Conclusions

This paper presents the AlfaCrux space mission and the main results obtained so far. AlfaCrux was launched on April 1, 2022, and it entered its nominal mode on 27 July 2022, 01:56 UTC. Fundamental aspects of the early orbit phase operation and the development of a digital twin model are discussed, and the lessons learned are presented. A project-based learning methodology has been adopted for developing students professional knowledge and transferable skills. In these first 5 months of operation, and from the engineering point of view, the AlfaCrux satellite is perfectly healthy with a high performance level, and the mission is considered a complete success.

Author Contributions: Conceptualization, R.A.B., A.C.d.S., B.F.-A.G., V.M.d.S.B., J.M.L.A., F.A.A., J.V.Q.S.B., Y.d.C.F.A., A.C.A.d.O., B.T.d.M. and V.F.M.; data curation, M.M.K. and R.B.d.S.; formal analysis, R.A.B., W.R.S., L.A., G.A.B., B.F.-A.G., J.M.F.-C., J.M.L.A., F.A.A., A.C.A.d.O., B.T.d.M., Y.d.C.F.A. and E.C.B.; funding acquisition, R.A.B.; investigation, R.A.B., A.C.d.S., W.R.S., L.A., G.A.B., M.M.K., R.B.d.S., J.V.Q.S.B., A.C.A.d.O., B.T.d.M., Y.d.C.F.A. and E.C.B.; methodology, R.A.B., A.C.d.S., V.M.d.S.B., J.M.F.-C. and A.C.A.d.O.; project administration, R.A.B., A.C.d.S., B.F.-A.G., V.M.d.S.B. and F.A.A.; resources, R.A.B., A.C.d.S., M.M.K., R.B.d.S. and V.F.M.; software, R.A.B., L.A., M.M.K., R.B.d.S., J.M.F.-C., J.M.L.A., B.T.d.M., E.C.B. and V.F.M.; supervision, R.A.B., A.C.d.S. and V.M.d.S.B.; validation, R.A.B., W.R.S., G.A.B., B.F.-A.G., V.M.d.S.B., J.V.Q.S.B. and B.T.d.M.; writing—original draft, R.A.B. and A.C.A.d.O.; writing—review and editing, R.A.B., A.C.d.S., W.R.S., L.A., G.A.B., M.M.K., R.B.d.S., J.V.Q.S.B., A.C.A.d.O., B.T.d.M. and V.F.M. All authors have read and agreed to the published version of the manuscript.

Funding: The AlfaCrux space mission was funded by the Brazilian agency Fundação de Apoio à Pesquisa do Distrito Federal (FAPDF).

Data Availability Statement: The AlfaCrux Telemetry Viewer can be accessed at https://lodestar. aerospace.unb.br/projects/alfacrux/radio (accessed on 20 September 2022).

Acknowledgments: The authors would like to thank the Brazilian agencies Higher Education Personnel Improvement Coordination (CAPES), National Council for Scientific and Technological Development (CNPq), and the Federal District Research Support Foundation (FAPDF) that supported this work, and also the Brazilian Space Agency (AEB) and the Brazilian National Telecommunications Agency (ANATEL) for technical support. They would also like to thank the International Amateur Radio Union (IARU) for the frequency coordination of the AlfaCrux satellite.

Conflicts of Interest: The authors declare no conflict of interest.

References

1. Suhadis, N.M. Statistical overview of CubeSat mission. In Proceedings of the International Conference of Aerospace and Mechanical Engineering 2019, Penang, Malaysia, 20–21 November 2019; Springer: Berlin/Heidelberg, Germany, 2020; pp. 563–573.
2. Kopacz, J.R.; Herschitz, R.; Roney, J. Small satellites an overview and assessment. *Acta Astronaut.* **2020**, *170*, 93–105. [CrossRef]
3. Villela, T.; Costa, C.A.; Brandão, A.M.; Bueno, F.T.; Leonardi, R. Towards the Thousandth CubeSat: A Statistical Overview. *Int. J. Aerosp. Eng.* **2019**, *2019*, 5063145. [CrossRef]
4. National Academies of Sciences, Engineering, and Medicine. In *Achieving Science with CubeSats: Thinking Inside the Box*; The National Academies Press: Washington, DC, USA, 2016.

5. Cappelletti, C.; Battistini, S.; Malphrus, B.K. *CubeSat Handbook—From Mission Design to Operations*; Elsevier: Amsterdam, The Netherlands, 2020.
6. Pelton, J.N.; Madry, S. *Handbook of Small Satellites—Technology, Design, Manufacture, Applications, Economics and Regulation*; Springer: Berlin/Heidelberg, Germany, 2020.
7. Pelton, J.N.; Madry, S.; Lara, S.C. *Handbook of Satellite Applications*, 2nd ed.; Springer: Berlin/Heidelberg, Germany, 2017.
8. Chin, A.; Coelho, R.; Nugent, R.; Munakata, R.; Suari, J.P. CubeSat: The Pico-Satellite Standard for Research and Education. In Proceedings of the AIAA Space 2008 Conference and Exhibition, San Diego, CA, USA, 9–11 September; American Institute of Aeronautics and Astronautics: Reston, VA, 2008.
9. Kang, J.; Gregory, J.; Temkin, S.; Sanders, M.; King, J. Creating future space technology workforce utilizing CubeSat platforms: Challenges, good practices, and lessons learned. In Proceedings of the AIAA Scitech 2021 Forum, Online, 11–15 January 2021 ; American Institute of Aeronautics and Astronautics: Reston, VA, 2021.
10. Cervone, A.; Topputo, F.; Speretta, S.; Menicucci, A.; Turan, E.; Di Lizia, P.; Massari, M.; Franzese, V.; Giordano, C.; Merisio, G.; et al. LUMIO: A CubeSat for observing and characterizing micro-meteoroid impacts on the Lunar far side. *Acta Astronaut.* **2022**, *195*, 309–317. [CrossRef]
11. Baker, J.; Colley, C.N.; Essmiller, J.C.; Klesh, A.T.; Krajewski, J.A.; Sternberg, D.C. MarCO: The First Interplanetary CubeSats. In Proceedings of the EPSC-DPS Joint Meeting 2019, Geneva, Switzerland, 15–20 September 2019; Volume 13; Abstract Number #EPSC–DPS2019–2009 .
12. Walker, R.; Binns, D.; Bramanti, C.; Casasco, M.; Concari, P.; Izzo, D.; Feili, D.; Fernandez, P.; Fernandez, J.G.; Hager, P.; et al. Deep-space CubeSats: Thinking inside the box. *Astron. Geophys.* **2018**, *59*, 5.24–5.30. [CrossRef]
13. Kodheli, O.; Lagunas, E.; Maturo, N.; Sharma, S.K.; Shankar, B.; Montoya, J.F.M.; Duncan, J.C.M.; Spano, D.; Chatzinotas, S.; Kisseleff, S.; et al. Satellite Communications in the New Space Era: A Survey and Future Challenges. *IEEE Commun. Surv. Tutorials* **2021**, *23*, 70–109. [CrossRef]
14. Saeed, N.; Elzanaty, A.; Almorad, H.; Dahrouj, H.; Al-Naffouri, T.Y.; Alouini, M.S. CubeSat communications: Recent advances and future challenges. *IEEE Commun. Surv. Tutorials* **2020**, *22*, 1839–1862. [CrossRef]
15. Straub, J.; Marsh, R.A.; Whalen, D.J. *Small Spacecraft Development Project-Based Learning*, 1st ed.; Springer International Publishing: Basel, Switzerland, 2017.
16. Castaldi, P.; Mimmo, N. An experience of project based learning in aerospace engineering. *IFAC-PapersOnLine* **2019**, *52*, 484–489. [CrossRef]
17. Honore-Livermore, E.; Birkeland, R. Managing product development and integration of a university CubeSat in a locked down world. In Proceedings of the 2021 IEEE Aerospace Conference (50100), Big Sky, MT, USA, 6–13 March 2021.
18. Thomas, J.W. *A Review of Research on Project-Based Learning 2000*; The Autodesk Foundation: San Rafael, CA, USA, 2010.
19. *ECSS-EST-10C Rev. 1*; Space Engineering-System Engineering General Requirements. ESA Requirements and Standards Division: Noordwijk, The Netherlands, 2017.
20. Hirshorn, S.R.; Voss, L.D.; Bromley, L.K. *Nasa Systems Engineering Handbook*; Technical Report; NASA: Washington, DC, USA, 2017.
21. Glaessgen, E.H.; Stargel, D.S. The digital twin paradigm for future NASA and U.S. Air force vehicles. In Proceedings of the Collection of Technical Papers—AIAA/ASME/ASCE/AHS/ASC Structures, Structural Dynamics and Materials Conference, Honolulu, HI, USA, 23–26 April 2012 ; pp. 1–14. [CrossRef]
22. Li, L.; Aslam, S.; Wileman, A.; Perinpanayagam, S. Digital Twin in Aerospace Industry: A Gentle Introduction. *IEEE Access* **2022**, *10*, 9543–9562. [CrossRef]
23. Kontaxoglou, A.; Tsutsumi, S.; Khan, S.; Nakasuka, S. Towards a Digital Twin Enabled Multifidelity Framework for Small Satellites. In Proceedings of the PHM Society European Conference, Jeju, Korea, 8–10 September 2021; Volume 6, p. 10.
24. Capon, C.; Lorrain, P.; Smith, B.; Brown, M.; Kurtz, J.; Boyce, R. Numerical Predictions for On-Orbit Ionospheric Aerodynamics Torque Experiment. In Proceedings of the 2020 IEEE Aerospace Conference, Big Sky, MT, USA, 7–14 March 2020; pp. 1–12.
25. Tao, F.; Zhang, M.; Liu, Y.; Nee, A. Digital twin driven prognostics and health management for complex equipment. *CIRP Ann.* **2018**, *67*, 169–172. [CrossRef]
26. Tao, F.; Zhang, H.; Liu, A.; Nee, A.Y.C. Digital Twin in Industry: State-of-the-Art. *IEEE Trans. Ind. Inform.* **2019**, *15*, 2405–2415. [CrossRef]
27. Markley, F.; Crassidis, J. *Fundamentals of Spacecraft Attitude Determination and Control*; Space Technology Library; Springer: New York, NY, USA, 2014.
28. Crassidis, J.L.; Markley, F.L. Unscented Filtering for Spacecraft Attitude Estimation. *J. Guid. Control. Dyn.* **2003**, *26*, 536–542. [CrossRef]
29. da Silva, R.C.; Ishioka, I.S.; Cappelletti, C.; Battistini, S.; Borges, R.A. Helmholtz cage design and validation for nanosatellites HWIL testing. *IEEE Trans. Aerosp. Electron. Syst.* **2019**, *55*, 3050–3061. [CrossRef]
30. da Silva, R.C.; Guimarães, F.C.; Loiola, J.V.L.; Borges, R.A.; Battistini, S.; Cappelletti, C. Tabletop testbed for attitude determination and control of nanosatellites. *J. Aerosp. Eng.* **2019**, *32*, 04018122. [CrossRef]
31. da Silva, R.C.; Borges, R.A.; Battistini, S.; Cappelletti, C. A review of balancing methods for satellite simulators. *Acta Astronautica* **2021**, *187*, 537–545. [CrossRef]

32. Vallado, D.A. *Fundamentals of Astrodynamics and Applications*, 4th ed.; Space technology Library, Microcosm Press: Hawthorne, CA, USA, 2013.
33. Bhanderi, D.; Bak, T. Modeling Earth Albedo for Satellites in Earth Orbit. In Proceedings of the AIAA Guidance, Navigation, and Control Conference and Exhibit, San Francisco, CA, USA, 15–18 August 2005. [CrossRef]
34. Ferreira, A.A.; Borges, R.A.; Reis, L.R.; Borries, C.; Vasylyev, D. Investigation of Ionospheric Effects in the Planning of the AlfaCrux UHF Satellite Communication System. *IEEE Access.* **2022**, *10*, 65744–65759. [CrossRef]

Article

Modelling the Stages of Pre-Project Preparation and Design Development in the Life-Cycle of an Investment and Construction Project

Azariy Lapidus, Dmitriy Topchiy, Tatyana Kuzmina * and Polina Bolshakova

Department of Technology and Organization of Construction Production, Moscow State University of Civil Engineering (National Research University) (MGSU), 26 Yaroslavskoe Shosse, 129337 Moscow, Russia
* Correspondence: kuzminatk@mgsu.ru

Abstract: The stages of pre-project preparation and design development are the fundamentals to the further implementation of an investment and construction project. The success of construction and commissioning of facilities depends on the smooth completion of these stages. The duration of stages depends on the competence and coordination of interaction between participants of an investment and construction project, rational decision making by the project manager and a number of other factors. Identifying these factors and finding rational planning options in terms of the stages of pre-project preparation and design development is a highly relevant task. With this in mind, the authors have developed an organizational and management model of a rational procedure to be implemented by the project manager at the stages of pre-project preparation and design development for a facility to be constructed. The model takes into account the influence of negative factors. The authors have also developed a method for selecting rational solutions at the stages in question. The analysis of the research literature and regulatory documents was performed for this purpose. The method of expert evaluations, elements of numerical analysis, mathematical processing of practical results and methods of mathematical modeling were applied. As a result, the authors have developed an organizational and management model for a rational combination of procedures at the stages of pre-project preparation and design development for a facility to be constructed and derived a formula determining the duration of implementation of each stage and a technique designated for selecting rational solutions at the stages in question. The findings of this study can simplify project planning and process management at the stages of pre-project preparation and design development in terms of information modeling.

Keywords: pre-project preparation; design development; organizational and management model; life cycle of building

Citation: Lapidus, A.; Topchiy, D.; Kuzmina, T.; Bolshakova, P. Modelling the Stages of Pre-Project Preparation and Design Development in the Life-Cycle of an Investment and Construction Project. *Appl. Sci.* **2022**, *12*, 12401. https://doi.org/10.3390/app122312401

Academic Editor: Paulo Santos

Received: 11 November 2022
Accepted: 30 November 2022
Published: 4 December 2022

Publisher's Note: MDPI stays neutral with regard to jurisdictional claims in published maps and institutional affiliations.

1. Introduction

A commissioned building is a finished product of the construction industry, characterized by the successful completion of the investment and construction project, which life cycle begins long before its construction.

Completion and commissioning of facilities within defined timeframes depends not only on the production of construction and installation works but, above all, on the duration of the stages of pre-project preparation and design development.

Currently, the analysis of statistical and reporting data and the effectiveness of the stages of pre-project preparation and design development are assessed by the authorities according to the number and timing of administrative procedures.

Reputable international rankings, such as the World Bank's Doing Business ranking, play a key role for many businessmen and choosing a project and country for investment [1].

The World Bank's Doing Business ranking was calculated according to the results of a survey of developers in 189 countries from 2002 to 2020. The countries were ranked by the number, duration and cost of procedures and required to obtain a construction permit, with account taken of the construction control quality index. At the moment, the evaluation is suspended, and the methodology is being updated.

Since 2012, the Russian Federation has participated in the World Bank's Doing Business score, namely, in its "Dealing with Construction Permits" section. It enables to assess the results of measures taken to optimize procedures annually (Figure 1). Over 8 years, Russia has risen in the ranking from the 178 to the 26th position, which shows the positive dynamics of procedures and their optimization. It is noteworthy that, during this period, both the number of procedures and the time to execute them have decreased (more than twice) [1–8].

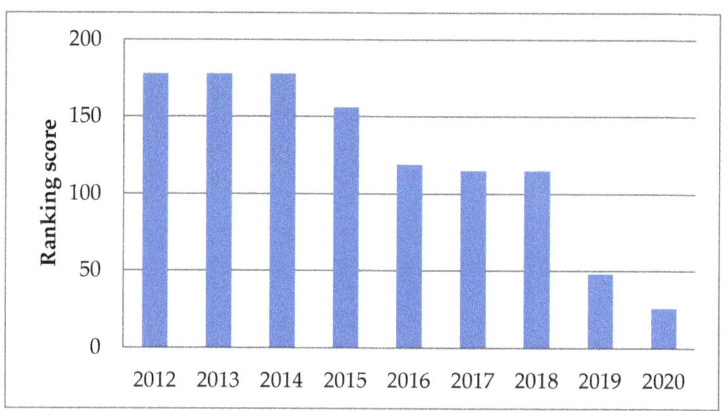

Figure 1. Russian Federation's score in the World Bank's Doing Business ranking titled "Dealing with Construction Permits" (2012–2020).

The task of reducing excessive administrative barriers and improving the entrepreneurial climate in construction is set not only by the Russian Federation but also by many countries. Figure 2 shows the ranking score of a number of countries, such as Bulgaria, the Czech Republic, China, Great Britain, India, Finland, France, Germany, Italy, Poland, Spain and the United States of America, in obtaining construction permits for 2020.

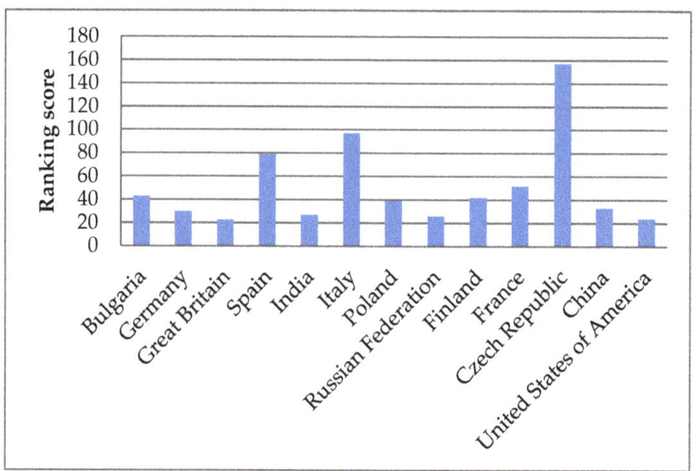

Figure 2. The ranking scores of a number of countries in the World Bank's Doing Business ranking of construction permits for 2020.

In addition to excessive administrative barriers, the duration of the stages of pre-project preparation and design development may be affected by many other factors, such as the diversity of interests of construction participants, incompetence of employees recruited by the project manager, poorly drafted design assignments and project designs, etc. [1,3,9–12].

Different authors have analyzed the factors that influence the implemented stages of investment and construction projects. For example, Soliman [13] conducted a study to identify the factors, including those dealing with communication issues, that affected the duration and quality of a state-funded investment and construction project in Kuwait and gave recommendations to eliminate a number of factors by introducing information technologies in construction and improving communication channels.

In their research paper, Fernando et al. [14] identified a number of factors that contribute to the introduction of innovations in the operations of project managers in Australian construction companies. Banihashemi et al. developed a multipurpose planning model to minimize the time, cost and environmental impacts and improve the implementation quality of an investment and construction project in Iran [15].

Shubham Sharma et al. [16] identified a number of factors that influence the time limit overrun and other costs of implementation of investment and construction projects in India. These authors substantiated the most important factors and proposed their classification. Bennett, Clough et al. and Selen Mubarak address the planning and control of investment and construction projects, taking into account information modeling in the United States and Canada [17–19].

Various scientists considered such an external factor as the COVID-19 pandemic [20]. The pandemic has had a major impact on the construction market. The activity of developers has decreased several times. Most companies have had to face the forced digitalization of many processes.

Additionally, our analysis of other works has shown that the study of the duration required for the procedures flow, the identification of factors affecting their flow and the possibility of combining the procedures flow in time during the stages of pre-project preparation and design development is relevant [13–16,21–30].

Therefore, the stages of pre-project preparation and design development were selected as the objects of the study.

At the stage of pre-project preparation, the main tasks include the building site approval, obtaining the title documents to the building site, conducting engineering surveys and collecting the initial permitting documentation.

The stage of design development includes the development and approval of design documentation and detailed engineering documents for capital construction projects.

The study of the stages of pre-project preparation and design development is necessary to generate a competent and clear algorithm for the project management system within the framework of information modeling in construction, contributing to effective scheduling and making organizational and managerial decisions [17–19,31]. It is at these stages that project decisions are made, and the implementation of an investment and construction project is scheduled. The quality of project development and the duration of the stages determine the overall duration and success of construction and commissioning.

At these stages, the project manager carries out the coordination of actions between design and survey organizations, external contractors (including executive authorities), support of design and the procedures flow for the implementation of investment and construction projects.

Purpose of the study—development of a method for choosing rational solutions at the stages of pre-project preparation and design development, ensuring a reduction in the time and quality of procedures for the project manager.

In this regard, the authors have considered the peculiarities of the flow of administrative procedures, tasks and functions performed by the project manager at the stages under study.

The list of tasks and functions performed by the project manager varies depending on the function of the facility. Therefore, within the framework of the study, it was decided to choose facilities with the same function. The authors have chosen residential houses, being part of residential complexes in the Russian Federation, as, for today, one of the main objectives of the country within the national development objectives is to increase the volume of housing construction at least up to 120 million square meters a year. According to the Federal State Statistics Service (Rosstat), 285,821 residential buildings with a total area of 111.6 million square meters were commissioned in 2019 [32].

In the course of the study, a number of methods were used, such as the method of expert assessment, elements of numerical analysis and mathematical processing of the experience results and the method of mathematical modeling. As a result, the main functions of the project manager are specified and grouped into a system of consolidated procedures.

Additionally, the authors have substantiated the boundary intervals of duration of procedures set by the project manager and identified the factors that influence the duration of stages of pre-project preparation and design development. This allowed developing an organizational and management model to combine procedures at the stages of pre-project preparation and design development of the facility for construction and to develop a method for selecting rational solutions at the stages under consideration.

The method simplifies the process of planning the stages of pre-project preparation and design development to identify and prevent, if possible, the influence of factors on the duration of stages and to simulate procedures within these stages and choose the most rational version of the model.

2. Methods of Research

Primarily, the authors implemented an analysis of the scientific and technical literature; technical standard documents and regulatory, legal, organizational and methodological documentation [9–11,33]. As a result of the analysis, the main functions performed by the project manager at the stages of pre-project preparation and design development were specified [34,35]. The authors have also made an initial list of factors influencing an increase in the duration of the stages.

Within the framework of the research, the term "Procedure" is defined. Procedure means any interaction between the project manager and external and internal contractors in order to obtain a document, permission, conclusion, seal, signature or other result necessary to implement the stages of pre-project preparation and design development. In addition, the authors have identified consolidated procedures within which the functions of the project manager are grouped.

Due to the lack of statistical data for determining the boundary intervals of procedures and factors affecting the duration of the stages of pre-project preparation and design development, the method of expert examination was selected.

A questionnaire was developed for the survey. The research was planned to have two stages. At the first stage, the authors identified boundary intervals of the duration of the project manager's functions, performed during the flow of certain previously consolidated procedures at the stages under study, for which regulation terms are not developed or time consumption exceeds the standard ones due to a number of factors. Additionally, the authors determined the factors affecting the duration of the procedures.

At the second stage, the ranking of factors, obtained as a result of the first stage of the study, the weigh characteristics of the factors were determined.

The authors examined the implementation of the stages of pre-project preparation and design development for the housing facilities to be constructed as part of residential complexes.

To confirm the validity of the results of the expert examination, the requirements for the experts' qualification were formulated [36,37]: representatives of the project manager (not less than 50% of the total number of experts), investor, designer and surveyor; general construction experience of more than 5 years and higher education. The sphere of project implementation—representatives of the state-owned enterprises (40–60% of the total number of experts) and commercial ones (40–60% of the total number of experts).

The first phase of the survey involved 121 experts, whereas the second phase involved 24 experts, correspondingly. The experts met the preset requirements, which was confirmed by the results of the processing of "Section 1. General Issues" of the questionnaire (Figures 3–6).

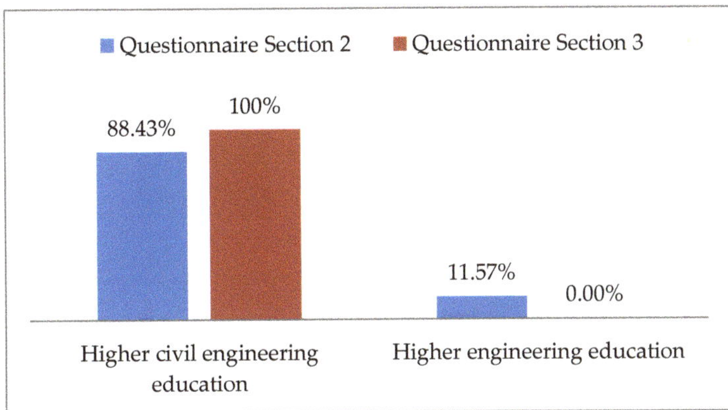

Figure 3. Processed results of Section 1 of the questionnaire. The question "Education of the experts".

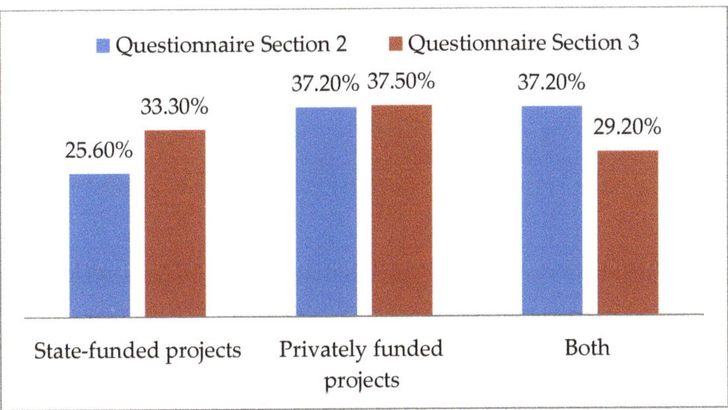

Figure 4. Processed results of Section 1 of the questionnaire. The question "Types of projects implemented by the experts".

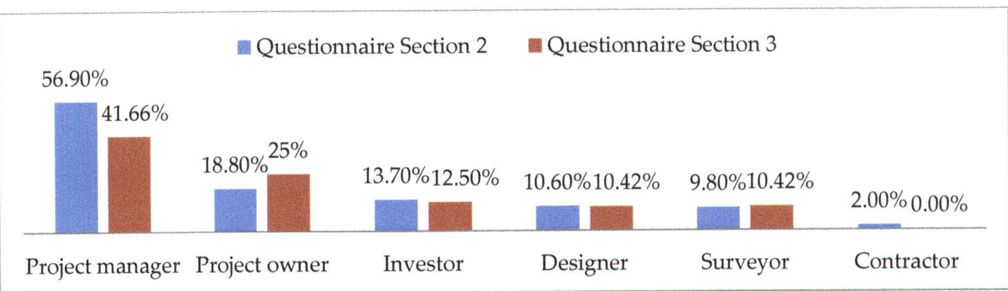

Figure 5. Processed results of Section 1 of the questionnaire. The question "Identification of involvement of experts in the construction process".

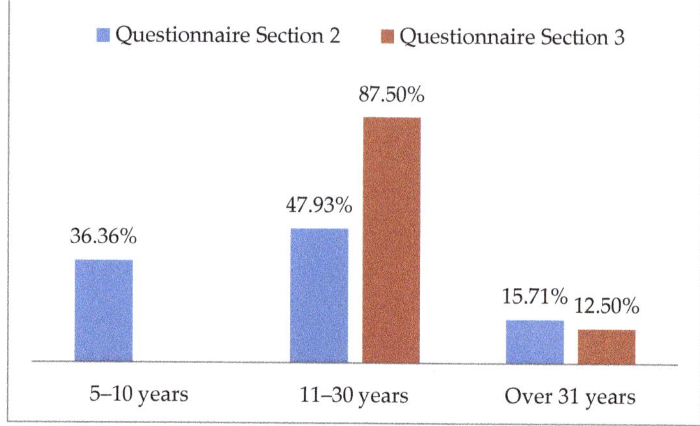

Figure 6. Processed results of Section 1 of the questionnaire. The question "The general construction experience of experts".

The processing of the survey results was carried out in the following sequence:

1. Determination of the weighted mean values of the boundary intervals of the procedure duration and factors affecting the duration of the procedures at the considered stages:
   - determination of linear regression coefficients by the least squares method and the construction of linear regression equations (Equation (1));

$$y = a + bx \tag{1}$$

   where:

   y—dependent variable (effective feature);
   x—independent variable (feature-factor);
   a and b—linear regression coefficients.

   - determination of Pearson's linear correlation coefficients (Equation (2)) and mean approximation errors (Equation (3));

$$r_{xy} = b\frac{\sigma_x}{\sigma_y} \tag{2}$$

   where:

   r_{xy}—Pearson's correlation coefficient;
   b—linear regression coefficient;
   σ_x—mean square deviation on x;
   σ_y—mean square deviation on y.

$$\overline{A} = \frac{1}{n}\sum\left|\frac{y - \hat{y}}{y}\right| \times 100\% \tag{3}$$

   where:

   $\overline{A}$—mean approximation error; the permissible limit of the value does not exceed 8–10%.

   - evaluation of regression equation quality by means of Fisher's criterion.

2. Ranking of the factors affecting the duration of the procedures:
   - forming a consolidated table containing the number of experts and the results of factor ranking;
   - determination of the sum of the ranks;

- finding the arithmetic mean of the ranks;
- detecting the deviation of the sum of ranks from the arithmetic mean sum of the ranks;
- Kendall's correlation coefficient determination (Equation (4));

$$W = \frac{12 \times S}{m^2(n^3 - n)} \tag{4}$$

where:

m—the number of experts;
n—ranks;
S—the sum of the squares of the deviations of the sums of ranks (n) from the average sum of ranks.

- composing a transformed rank matrix;
- compiling a matrix of normalized weights and determining individual and group ranks according to the weights of all factors;
- establishing distances between individual preferences using the Spearman correlation coefficient (Equation (5)).

$$p = 1 - 6\frac{\sum d^2}{n^3 - n} \tag{5}$$

where:

d^2—the square of the differences between the ranks;
n—the number of ranked objects.

3. Conducting a multiple regression analysis to establish a mathematical relationship between the experts' evaluation of the factors and the actual deviation of the planned duration of the stages of pre-project preparation and design development for the construction of facilities from the actual one.

The factors are independent variables of a multiple regression, the totality of which affects the total assessed dependent variable Y (the increment of the duration ΔT of pre-project preparation and design development).

The unknown regression coefficients and the shift (intercept term) were estimated by the least squares method using matrix calculus and determining the residual function from the original ranking matrix obtained from the expert survey.

The analysis was carried out in the following sequence:

- compilation of the initial matrix X of independent regression variables (with the additionally entered column "δ1") and the vector column of dependent variable Y (total increment of the duration of the stages of pre-project preparation and design development for the facilities to be constructed by respective construction organizations);
- determining the product of the transposed and source matrices;
- calculation of the inverse matrix;
- determining the product of the transposed matrix X and matrix Y;
- determination of regression equation coefficients.

Mathematical processing of the questioning results enabled to determine marginal intervals of procedure duration and deviation from the standard duration of design works to identify factors influencing the duration of procedures and to estimate the duration increment (ΔT) of pre-project preparation and design development under the effect of these factors. The results of the study are presented in Section 3.

3. Results and Discussion

Based on the results of the analysis of the scientific and technical literature and current legislative regulations, the main functions of the project manager were specified and grouped into a system of consolidated procedures (Figures 7–9) [38–44].

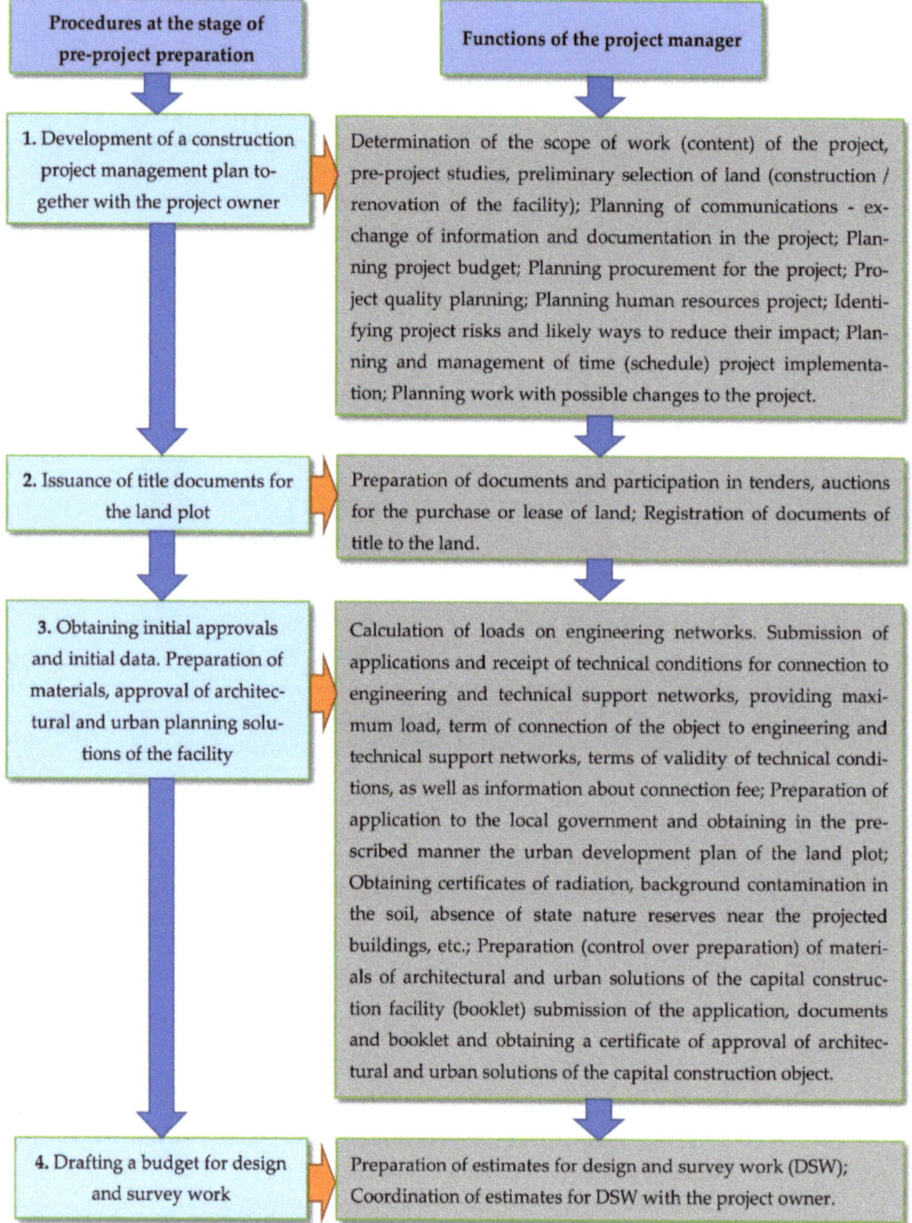

Figure 7. Consolidated procedures and functions of the project manager at the stage of pre-project preparation of a facility for construction. Part one.

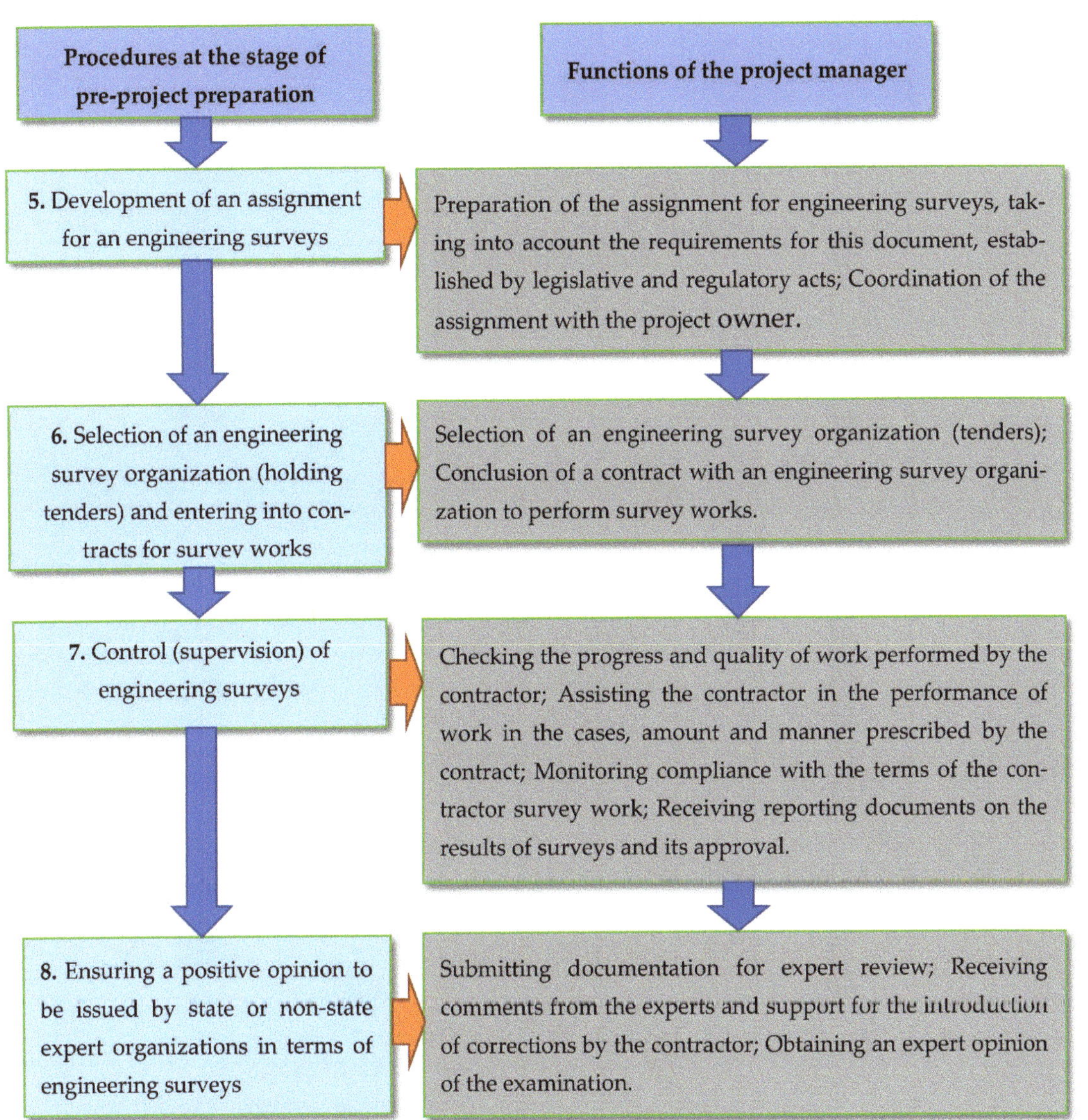

Figure 8. Consolidated procedures and functions of the project manager at the stage of pre-project preparation of a facility for construction. Part two.

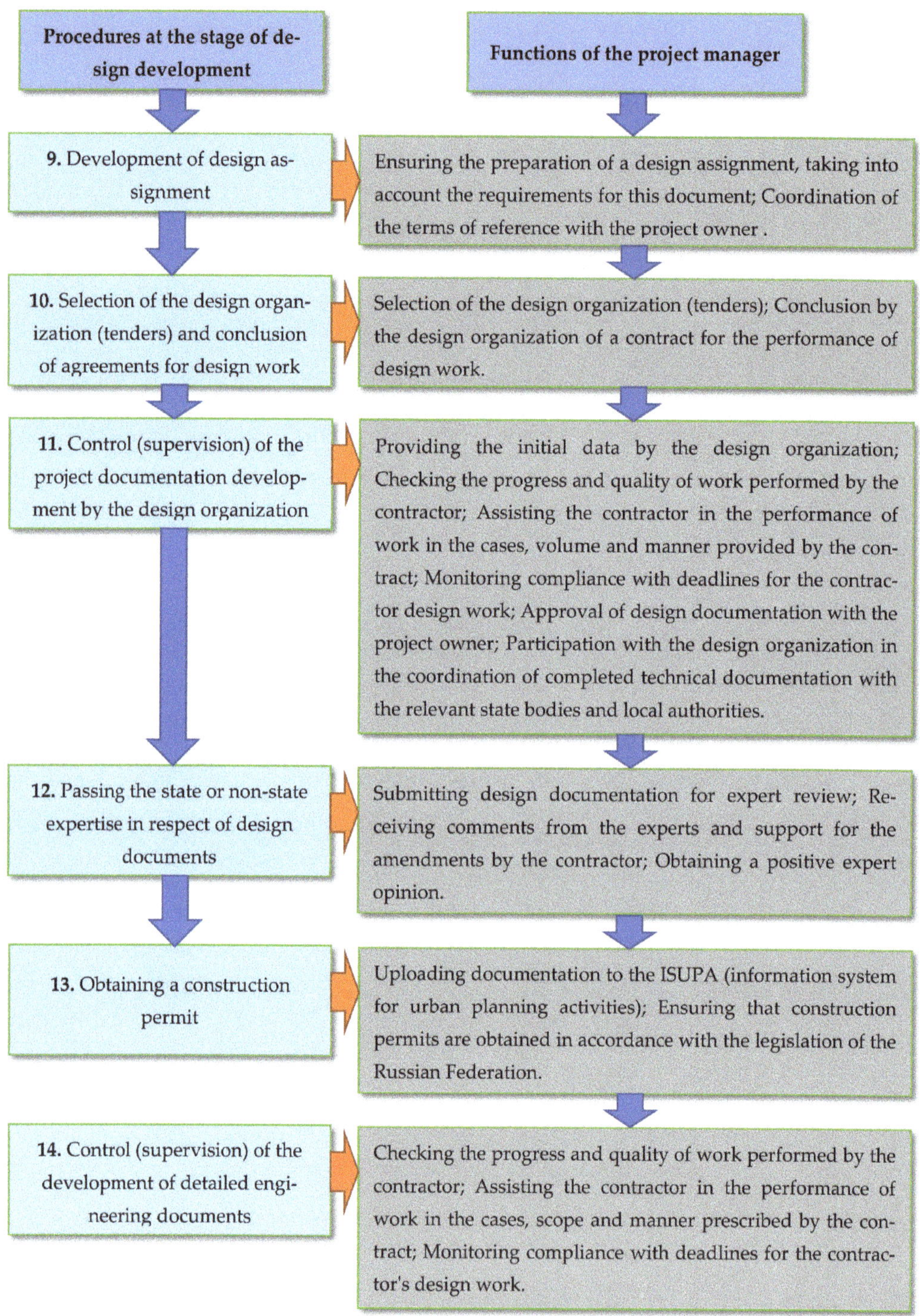

Figure 9. Consolidated procedures and functions of the project manager at the stage of design preparation for the facility to be constructed.

Mathematical processing of the results of the survey enabled to determine the weighted mean limit intervals of the procedures and deviations from the standard duration of design work (Tables 1 and 2).

Table 1. Boundary intervals for the duration of procedures at the stages of pre-project preparation and design development for the facilities to be constructed.

Procedures Cipher	Procedues		Boundary Intervals of the Duration (Months)
p_1	Development of a construction project management plan together with the project owner		$1 \leq t_{p1} \leq 6$
p_2	Issuance of title documents for the land plot		$1 \leq t_{p2} \leq 4$
p_3	Obtaining initial approvals and initial data. Preparation of materials, approval of architectural and urban planning solutions of the facility	Calculation of loads on engineering networks; submitting applications and obtaining technical conditions for connecting the facility to the networks of cold water supply, external firefighting, heat supply, sewage, electricity, gas supply, telephony, radio, etc.	$1.5 \leq t_{p3} \leq 3$
		Preparation (control over preparation) of materials of architectural and urban planning solution (AGR/AGO) of the capital construction facility (booklet); submission of application, documents and booklet to the Committee for Architecture and Urban Planning of Moscow (Moscow Region); receipt of a certificate of approval of AGR (AGO) of the capital construction facility.	$2 \leq t_{p3} \leq 4$
		Obtaining certificates on radiation, background contamination in the soil, the absence of state nature reserves near the projected buildings, etc.	$1 \leq t_{p3} \leq 3$
p_4	Drafting a budget for for design and survey work		$0.5 \leq t_{p4} \leq 1.5$
p_5	Development of an assignment for engineering surveys		$0.25 \leq t_{p5} \leq 2$
p_6	Selection of an engineering survey organization (holding tenders) and entering into contracts for survey works		$0.5 \leq t_{p6} \leq 3$
p_7	Control (supervision) of engineering surveys		$1.5 \leq t_{p7} \leq 4$
p_8	Ensuring a positive opinion to be issued by state or non-state expert organizations in terms of engineering surveys		$1 \leq t_{p8} \leq 3$
p_9	Development of design assignment		$0.25 \leq t_{p9} \leq 2$
p_{10}	Selection of the design organization (tenders) and conclusion of agreements for design work		$1 \leq t_{p10} \leq 3$
p_{12}	Passing the state or non-state expertise in respect of design documents		$1 \leq t_{p12} \leq 3$
p_{13}	Obtaining a construction permit		$0.66 \leq t_{p13} \leq 2$

Table 2. Boundary intervals of deviations from the standard duration of design work.

Procedure Cipher	Procedures	Boundary Intervals of Duration Deviations (Months)
P_{11}	Control (supervision) of the project documentation development by the design organization	1–3
P_{14}	Control (supervision) of the development of detailed engineering documents	0.25–2

Sample calculations of the weighted average value of the boundary interval of the duration of the procedure "Development of the project management plan by the project manager together with the project owner" are provided in Appendix A. The algorithm can be applied to other procedures performed at other stages of an investment and construction project or when analyzing construction facilities that have other functions.

As a result of the first phase of the study, the authors substantiated 29 factors, affecting the duration of procedures performed by the project manager at the stages of pre-project preparation and design development. The experts have picked these factors from the original list made on the basis of an analysis of the research literature and current legislation. In the course of the questionnaire survey, the experts could formulate their own factors, influencing the duration of the stages. However, the experts picked the factors from the list. Alternatively, the experts formulated the factors which meanings are similar to those available in the initial list, although they were worded in a different way. The responses were processed using the method of least squares. The sequence is provided in Section 2 of this article.

The substantiated factors are systematized into four groups:

- Organizational and managerial factors (δ^{oy});
- Procedural and executive factors ($\delta^{\Pi u}$);
- Economic factors ($\delta^{\scriptstyle\ni}$);
- Design factors (δ^{Π}).

At the second stage of the study, as a result of ranking, the values of the factors' affect the duration of the procedures at the stages of pre-project preparation and design development for the facilities to be constructed (Table 3).

The duration increment Y (ΔT) under the impact of the group of factors was determined with due account to the established dependencies (Equations (6) and (7)):

$$Y = A0 + A1 \cdot \delta^{oy} + A2 \cdot \delta^{\Pi u} + A3 \cdot \delta^{\ni} + A4 \cdot \delta^{\Pi} \tag{6}$$

$$\Delta T = 16 + 0.043\delta^{oy} + 0.013\delta^{\Pi u} + 0.042\delta^{\ni} + 0.027\delta^{\Pi} \tag{7}$$

Provided $\Delta T \geq 0$,
where:

δ^{yo}, $\delta^{\Pi u}$, $\delta^{\ni}$ and δ^{Π}—point assessment of the impact of the group factors for the object in a given range of values from the minimum to maximum (dimensionless value). Conducted by a specialist who performs the functions of the developer (technical customer) for a private facility:

$A0$—shift value;

$A1$ and $A2$—multiple regression coefficients;

ΔT—increment of the duration under the influence of factors.

Table 3. Weighting characteristics of the factors affecting the duration of procedures at the stages of pre-project preparation and design development for facilities to be constructed.

Factor Cipher	Factor	Group of Factors with a Denomination in the Mathematical Model	Denomination of the Factor in the Mathematica Model	Factor's Weight %
X_1	Disaccord of interaction between the project manager and other participants in the creation of the object	Organizational management $\delta^{оу}$	$\delta^{оу}_1$	6.22
X_2	Failure to comply with the procedure for developing, coordinating and approving documents		$\delta^{оу}_2$	6.27
X_3	Incompetence of employees of the project manager		$\delta^{оу}_3$	5.22
X_4	Obtaining the right of limited use of neighboring land plots (easements) for the period of construction		$\delta^{Пи}_4$	4.65
X_5	Complaints from other participants of the auction, tender about potentially illegal actions of the seller considered by the Office of the Federal Antimonopoly Service (FAS) in the course of the auction (tender) on the selection of a land plot		$\delta^{Пи}_5$	4.65
X_6	Obtaining additional input data		$\delta^{Пи}_6$	5.53
X_7	Increasing the time for the development of materials required for the approval of architectural and urban planning solutions/appearance (AGR/AGO)		$\delta^{Пи}_7$	5.04
X_8	Correction of architectural and urban planning solutions		$\delta^{Пи}_8$	4.80
X_9	Increasing the time frame for obtaining technical conditions for connecting to the existing utility networks, factoring in the excessive requirements of energy supplying organizations		$\delta^{Пи}_9$	5.10
X_{10}	Provision of technical specifications (TS) issued on the basis of outdated/unspecified data on networks		$\delta^{Пи}_{10}$	4.63
X_{11}	Extending the duration of the development of an engineering survey task by the project manager		$\delta^{Пи}_{11}$	4.19
X_{12}	Complaints from other bidders about potentially unlawful actions of the project manager considered by the FAS during tenders (competitions) for selection of a surveying organization	Procedural-executive $\delta^{Пи}$	$\delta^{Пи}_{12}$	3.46
X_{13}	Extending period of the preparation of reporting documents on engineering surveys by the contractor		$\delta^{Пи}_{13}$	3.71
X_{14}	Correction of reporting documentation of engineering survey results in case of deficiencies during the state or non-state expertise		$\delta^{Пи}_{14}$	3.52
X_{15}	Making corrections to the set of documentation of the results of engineering surveys, identified when submitting an application for expert review		$\delta^{Пи}_{15}$	3.54
X_{16}	Increasing the duration of the development of the design assignment by the project manager		$\delta^{Пи}_{16}$	3.01
X_{17}	Complaints from other bidders about potentially unlawful actions of the project manager considered by the FAS during tenders (competitions) for selection of a design organization		$\delta^{Пи}_{17}$	3.60
X_{18}	Revision of project documentation (the collection of additional initial data) in the case of deficiencies (insufficient initial data) in the course of passing the state or non-state expertise		$\delta^{Пи}_{18}$	3.60
X_{19}	Making corrections to the set of project documentation identified in the application for expert review		$\delta^{Пи}_{19}$	2.66

Table 3. *Cont.*

Factor Cipher	Factor	Group of Factors with a Denomination in the Mathematical Model	Denomination of the Factor in the Mathematica Model	Factor's Weight
				%
X_{20}	Incorrect execution of the design assignment by the project manager		$\delta^{\Pi u}{}_{20}$	2.17
X_{21}	Late submission of project documentation by the project organization within the time frame set in the contract		$\delta^{\Pi u}{}_{21}$	2.68
X_{22}	Untimely correction of documents		$\delta^{\Pi u}{}_{22}$	2.23
X_{23}	Transfer of documentation is not complete		$\delta^{\Pi u}{}_{23}$	2.01
X_{24}	Refinement (elimination of technical errors) when loading documentation into the ISUPP (information system for urban planning provision)		$\delta^{\Pi u}{}_{24}$	1.54
X_{25}	Incorrect estimate of contract prices		$\delta^{\text{э}}{}_{25}$	1.05
X_{26}	Failure to fulfill contractual obligations by contractors	Economic $\delta^{\text{э}}$	$\delta^{\text{э}}{}_{26}$	1.46
X_{27}	Failure to conclude contracts in a timely manner		$\delta^{\text{э}}{}_{27}$	1.06
X_{28}	Instability of funding		$\delta^{\text{э}}{}_{28}$	1.57
X_{29}	Poor quality of design solutions (volume-planning, structural, organizational and technological)	Design-engineering δ^{Π}	$\delta^{\Pi}{}_{29}$	1.54

Various options for combining procedures in time at the stages of pre-project preparation and design development were analyzed, and the most rational sequence of stages was proposed.

The organizational and management model of a rational combination of procedures by the project manager in the network (Figure 10) reflecting the sequence of stages of pre-project preparation and design development was developed, where: p_1, p_2, p_3, $\ldots$, p_{14}—procedures' cipher and tp_1, tp_2, tp_3, $\ldots$, tp_{14}—the duration of the procedures.

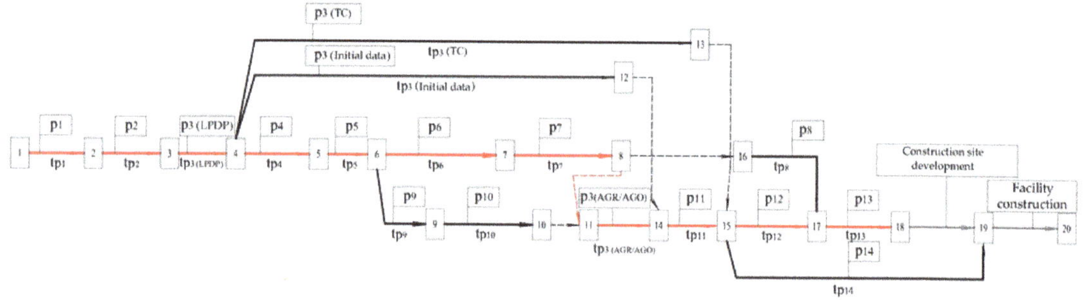

Figure 10. Organizational and management model of the rational combination of procedures at the stages of pre-project preparation and design development in the network form.

Figure 11 shows a sample organizational and management model in the linear form, where the duration of procedures tp is taken as a share of the total duration of the stages of pre-project preparation and design development (T_{p}). Durations are the minimum durations of the procedures.

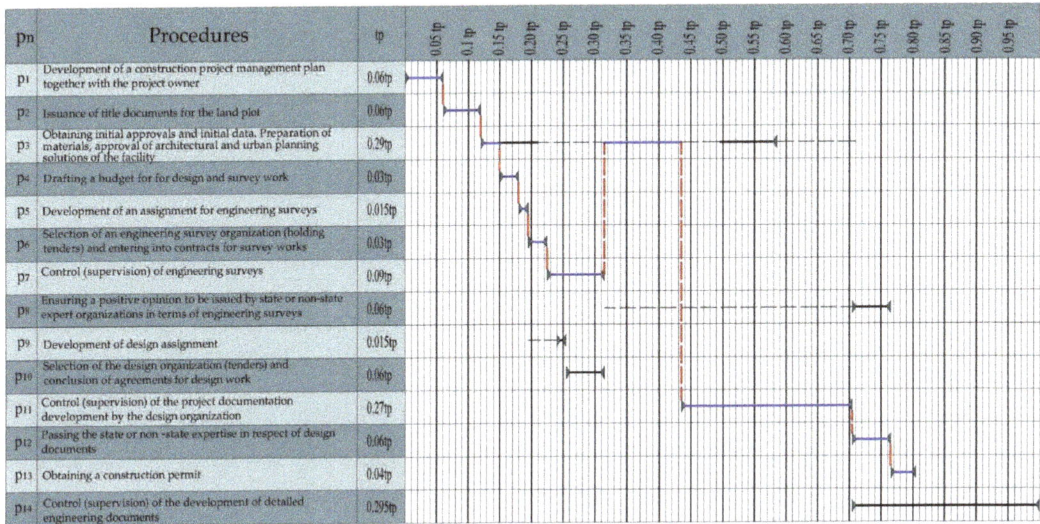

Figure 11. Organizational and management model of the rational combination of procedures at the stages of pre-project preparation and design development in linear form.

The combination of procedures at the stages of pre-project preparation and design development in time and between each other is performed as follows: The first procedure launched is developing a plan for project management in construction that is done in the collaboration of a project manager and a project owner.

After approval of the project management plan, the project owner, assisted by the project manager, selects a land plot for the construction site and draws it up in accordance with the established procedure. Having title documents for the land, the project manager proceeds to the collection of initial data for the design and obtaining primary permitting documentation.

Initially, the project manager obtains a land plot development plan. Then, he begins to collect background data: certificates of radiation, background pollution of the soil, the absence of state nature reserves near the projected buildings, etc. and permits to connect to facilities and existing utilities (technical conditions for connecting the object to cold water supply networks, for external firefighting, to networks of heat, water disposal, power supply, gas supply, telephone, radio, etc.). The processes of collecting various certificates go along with obtaining technical specifications [45].

Along with the collection of the initial data, immediately after receipt of the land plot development plan (LPDP), the development of cost estimate documentation for design and survey work (DSW) begins (initially consolidated costs for engineering survey are already laid in the plan of project management) by specialists of the service of the project manager. The next process is the development of a task on engineering survey, then there is the choice of organization, which conducts them, and then there is a control over their performance and presentation of the report within the specified period and its approval.

Along with selecting the organization and control of engineering surveys, the development of the assignment for the design of the facility and the selection of design organization is carried out [46]. By the completion of engineering surveys, the design organization is selected and the process of developing an album with materials for coordination of architectural and urban planning solutions (AUPS/AUPA) by the design organization (control and coordination of architectural solutions with the project owner is the responsibility of the project manager). Next, the process of coordination AUPS (AUPA) commences. Upon completion of coordination of architectural and urban solutions, the design organization begins to develop project documentation, and the control of development is carried out by the project manager.

Next, there are procedures for passing the examination of engineering surveys and design documentation. Reporting the documentation of engineering surveys can pass expertise earlier, immediately after its development.

The procedure for developing working documentation begins after the development of design documentation.

Planning permission is obtained upon the receipt of the positive expert opinion. Before applying for a planning permission, it is necessary to upload the project documentation and other necessary documents into the information system for urban planning.

The duration of the stages of pre-project preparation and design development, taking into account the influencing factors is determined (Equation (8)):

$$T = Tp + \Delta T \tag{8}$$

where:

T—the duration of the stages of pre-project preparation and design development, taking into account the influencing factors;

Tp—estimated duration of the stages of pre-project preparation and design development, taken as a result of combining procedures in time and building a graphical model;

ΔT—increment of duration under the impact of factors.

The organizational and management model of the rational combination of procedures at the stages of pre-project preparation and design development can facilitate the transition to information modeling of construction at these stages.

The results, obtained in the course of the research, contributed to the development of a method for selecting rational solutions at the stages of pre-project preparation and design development. The proposed method contains general provisions and the procedure for selecting rational solutions at the stages of pre-project preparation and design development. Additionally, the new method establishes the procedure for selecting rational solutions at the stages of pre-project preparation and design development.

The new method focuses on residential buildings as part of residential complexes analyzed as capital construction facilities.

This method is applicable to construction facilities that have other functions or construction conditions. However, in this case, the operator in charge of scheduling should specify a list of functions and procedures depending on the conditions of construction in accordance with the characteristic.

If it is necessary to take into account additional destabilizing factors, the operator in charge of scheduling, acting independently or in cooperation with groups of experts, collects and processes initial data on destabilizing the factors and boundary intervals in the duration of procedures and performs a generalized evaluation of objects, identification of consistency of expert opinions, ranking of objects and assessment of the reliability of the data processing results.

The selection of rational solutions at the stages of pre-project preparation and design development has stages and steps (Figure 12) completed by the conventional operator of the organization performing the functions of the project manager.

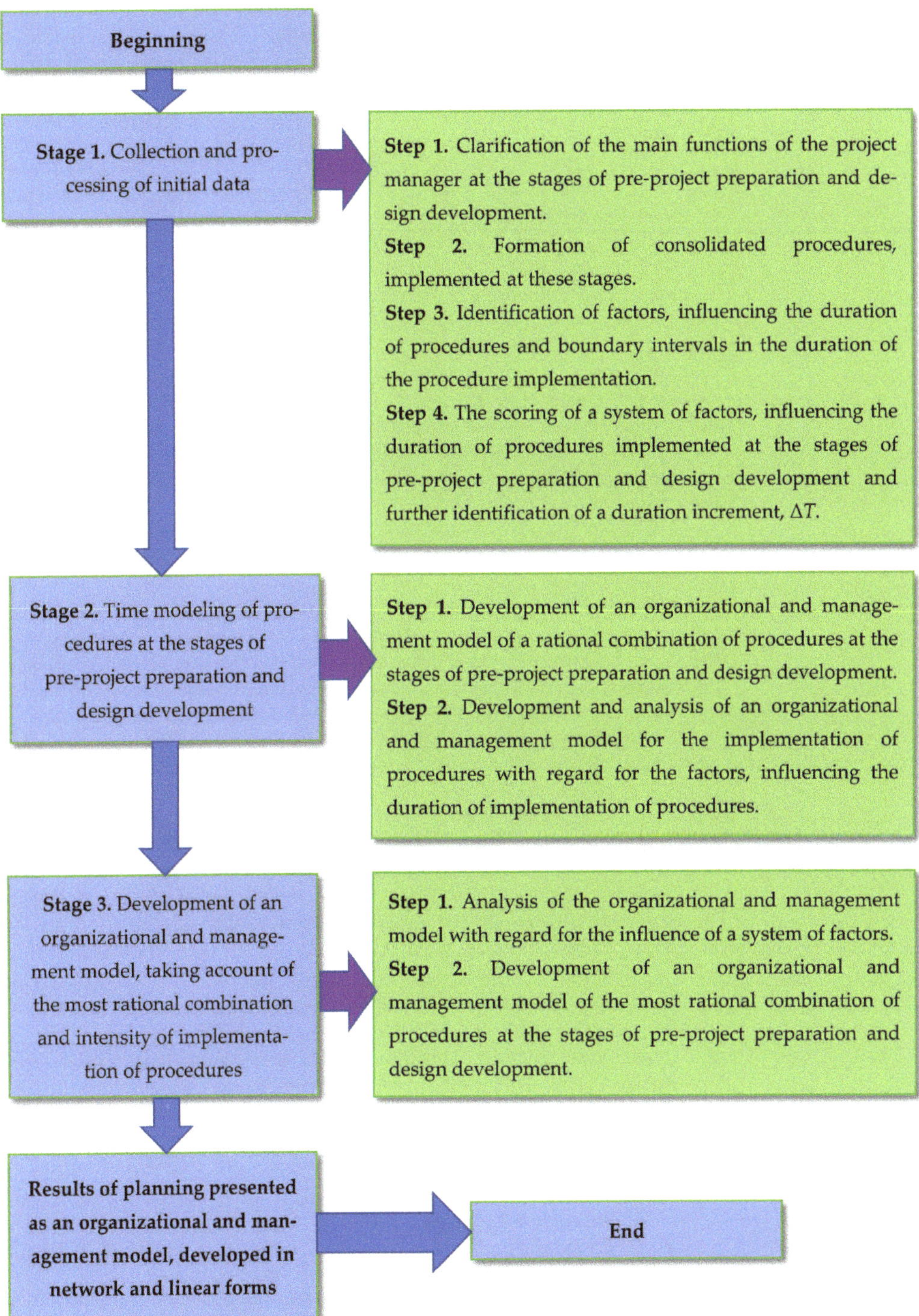

Figure 12. Selection of rational solutions at the stages of pre-project preparation and design development for the facilities to be constructed (stages and steps of the method).

Stage 1. Collection and processing of the initial data signifies identifies the main functions of the project manager and the formation of extensive procedures at the stages of pre-project preparation and design development based on the initial characteristics of the future facility, followed by the identification of factors affecting the duration of the procedures, the definition of limits on the duration of procedures and increments of duration ΔT.

Stage 2. Time modeling of the procedures at the stages of pre-project preparation and design development of the facility for construction includes the development of organizational and management model of rational combination of procedures at the stages of pre-project preparation and design developmentof the object for construction, which take into account the factors affecting the duration of the procedures.

Stage 3. Building up of an organizational and management model, factoring in the most rational combination and intensity of procedures, is carried out by assessing the increase in the durations of the stages' development under the influence of factors and by selecting the most rational option of combining procedures in time.

Time modeling of the procedures at the stages of pre-project preparation and design development is based on the search of options for combined procedures, accounting for the impact on the duration of their implementation combinations of factors. As a result of the analysis of the obtained variants, the most rational combination of procedures in time is accepted.

The case of the use of stages 2 and 3 of the method is discussed in Appendix B.

4. Conclusions

For successful implementation of the considered stages, the project manager needs to ensure passing through a set of procedures. In this case, the functions and tasks of the project manager depend on the set of procedures required to implement these stages, depending on the established parameters, and the effectiveness of their passing determines the duration of the stages of pre-project preparation and design development.

In this article, there was proposed a method for selecting the most rational solutions when planning the stages of pre-project preparation and design development for the facilities to be constructed, accounting for the impact of factors on the duration of the stages' development. The scientific and technical literature, standard technical, normative–legal and organizational–methodical documents in the sphere of development of the stages of pre-project preparation and design development were analyzed for the facilities to be constructed. The research was done by the method of expert estimates followed by mathematical processing of the questionnaire results and the use of mathematical modeling methods.

As the result, the following conclusion can be drawn:

1. Fourteen consolidated procedures, simplifying the planning process, were established by the project manager to obtain a construction permit.
2. The boundary intervals of the duration of the procedures to be passed by the project manager were defined, but no regulatory time limits were developed for these intervals or the time input exceeds the standard one due to a number of factors.
3. There were 29 identified factors affecting the duration of the stages of pre-project preparation and design development. The factors are classified into groups:
   - Organizational and management factors (δ^{oy})—the share of the impact on the duration of the stages is 17.71%;
   - Procedural-executive factors ($\delta^{\Pi_{\text{H}}}$)—the share of the impact on the duration of the stages is 75.61%;
   - Economic factors ($\delta^{\text{э}}$)—the share of the impact on the duration of the stages is 5.14%;
   - Design factors (δ^{Π})—the share of the impact on the duration of the stages is 1.54%.

4. The dependence of the influence of groups of factors on the duration of the stages of pre-project preparation and design development (T) and an increment in duration (ΔT) were identified.

5. The authors developed linear and network versions of the organizational and management model for the rational combination of procedures at the stages of pre-project preparation and design development. The model takes into account the boundary intervals in the duration of procedures and, thereby, simplifies the task of planning.

6. The method for choosing rational solutions at the stages of pre-project preparation and design development was devised. These methods have general provisions, a procedure for planning the stages of pre-project preparation and design development, taking into account the influence of negative factors. Residential buildings, constructed as part of residential complexes, are considered as capital construction facilities.

This method encompasses a mathematical apparatus that allows, finding the boundary intervals for the duration of procedures, substantiating the factors affecting the duration of procedures and finding an increment in the duration of stages.

The method can be applied to construction facilities, with other functions or constructed under different conditions. However, in this case, the operator in charge of project planning should clarify a list of functions and procedures depending on the conditions of construction in accordance with the characteristic. An organizational and management model can be used for planning works at the stages of pre-project preparation and design development using information technologies.

Any further research might include amplifying the selection of rational solutions at the stages of pre-project preparation and design development for facilities to be constructed, depending on the natural and climatic conditions, function, space planning and structural solutions of capital construction facilities.

Author Contributions: Conceptualization, A.L.; methodology, A.L. and D.T.; software, P.B. data analysis, D.T.; investigation, P.B. and T.K.; data curation, D.T. and T.K.; writing—original draft preparation, P.B. and T.K.; writing—review and editing, P.B. and T.K. and final conclusions, P.B. and T.K. All authors have read and agreed to the published version of the manuscript.

Funding: The Moscow State University of Civil Engineering (National Research University) (MGSU) funded this research.

Institutional Review Board Statement: Not applicable.

Informed Consent Statement: Not applicable.

Data Availability Statement: Not applicable.

Conflicts of Interest: The authors declare no conflict of interest.

Appendix A

Let's consider one of the questions in the second section of the research survey as an example. The purpose of the question is to identify the most appropriate boundary interval for the duration of the procedure "Development of a construction project management plan together with the project owner". Table A1 shows the findings of the survey, conducted among the experts.

Table A1. Survey findings: identification of the interval for the duration of the procedure "Development of a construction project management plan together with the project owner".

№	Answer Option	Number of Experts	Share
1	1–2 months	12	9.9 %
2	1–3 months	23	19 %
3	2–3 months	23	19 %
4	3–4 months	7	5.8 %
5	4–5 months	7	5.8 %
6	5–6 months	25	20.7 %
7	1–6 months	22	18.2 %
8	Your answer	2	1.6 %

Based on the survey findings, data ranking was performed and the results were obtained (Table A2). The results were used to calculate the initial statistical data (Table A3).

Table A2. Calculated ranked data of the survey results.

№	The Time Interval Defining the Duration of the Procedure	Number of Experts Who Selected the Answer
1	Your answer	2
2	3–4 months	7
3	4–5 months	7
4	1–2 months	12
5	1–6 months	22
6	1–3 months	23
7	2–3 months	23
8	5–6 months	25

Table A3. Source statistical data.

	x	y	yx	x^2	y^2	yTeop	y-yTeop	$\overline{A}$
	1	2	2	1	4	2.666667	−0.66667	0.333333
	2	7	14	4	49	6.22619	0.77381	0.110544
	3	7	21	9	49	9.785714	−2.78571	0.397959
	4	12	48	16	144	13.34524	−1.34524	0.112103
	5	22	110	25	484	16.90476	5.095238	0.231602
	6	23	138	36	529	20.46429	2.535714	0.110248
	7	23	161	49	529	24.02381	−1.02381	0.044513
	8	25	200	64	625	27.58333	−2.58333	0.103333
Total	36	121	694	204	2413			1.443637
Average	4.5	15.125	86.75	25.5	301.625			
σ^2	5.25	72.85938						

Let's make a linear regression equation. We will use the method of least squares to find coefficients a and b.

Coefficient a = −0.89; b = 3.56.

The equation of linear regression looks like: $y = -0.89 + 3.56x$.

The closeness of the relationship of the phenomenon under study is assessed by Pearson's linear correlation coefficient r_{xy}.

The average error of approximation (average deviation of calculated values from the actual ones) $\overline{A}$ gives an estimate of the quality of the model built.

Evaluation of the quality of the regression equation is performed using Fisher's F-criterion. We compare actual F_{actual} and critical (tabulated) F_{table} values of F-criterion to test the hypothesis of statistical insignificance of the regression equation and the closeness of the relationship.

Using the formulas, presented in the second section of the article, we obtain the following results: $r_{xy} = 0.96$, $\overline{A} = 9.03$, $F_{actual} = 62.93$, $F_{table} = 5.99$.

According to the average error of approximation, the calculated values deviate from the actual values by 9.03%.

The correlation coefficient $r_{xy} = 0.96$ means that the relationship is good and direct.

Since $F_{actual} > F_{table}$, the hypothesis H_0 is rejected. Therefore, the parameters of the equation are statistically significant.

As a result, let's plot the linear regression (Figure A1). The graph shows experimental and theoretical values.

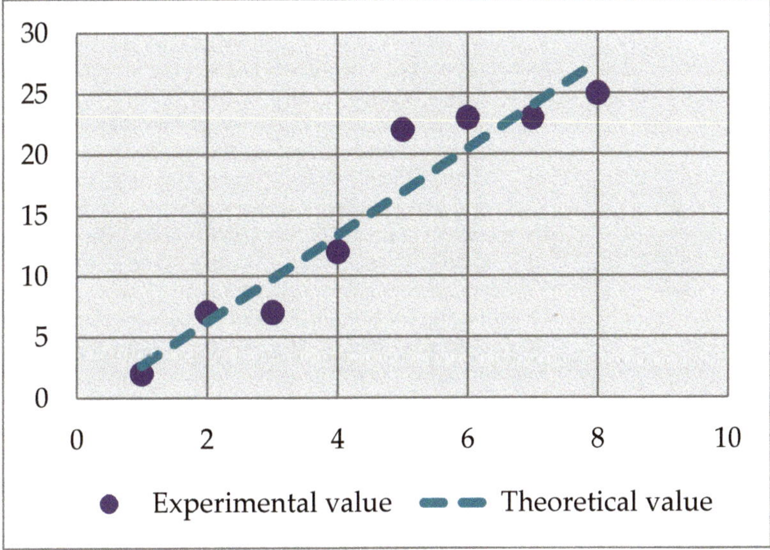

Figure A1. Linear regression.

The weighted average value of the indicator $\overline{y} = 19.28$. It follows from this that at the level of significance $\alpha = 0.05$ the following answer is significant: "1–6 months".

Appendix B

Let's consider stages 2 and 3 of the method, using a combination of factors (Table A4) influencing the duration.

Table A4. The effect of a combination of factors on the duration of procedures at the stages of pre-project preparation and design development.

№	Procedure Number	Procedure Name	Factor Number	Factor
1	p3	Obtaining initial approvals and initial data. Preparation of materials, approval of architectural and urban planning solutions of the facility	$\delta^{\text{Пи}}_6$	Obtaining additional input data
			$\delta^{\text{Пи}}_7$	Increasing the time for the development of materials required for the approval of architectural and urban planning solutions/appearance (AGR/AGO)
			$\delta^{\text{Пи}}_8$	Correction of architectural and urban planning solutions

Let's make a model reflecting the influence of factors on the duration of procedures, based on the organizational and management model with the minimum values of the duration of procedures (Figure A2).

Let's add a column with values of durations of procedures, taking into account the influence of factors (t), to the tabular part of the model. In the graphical part of the model, let's first display the motion of procedures in time, taking into account the influence of a combination of factors, linked to the sequence provided in the integrated model, and then we offer an option of the time optimization by changing sequence of links between procedures.

The following symbols have been introduced in the graphical part of the model:

blue bars: minimum values of duration of procedures on the critical path;

red bars: motion of procedures in time to the maximum values of procedures durations;

green bars: motion of procedures in time, taking into account the time optimization of the model by changing the sequence of links between procedures;

red dotted line: linking procedures on the critical path.

In the course of finding the duration of the stages of pre-project preparation and design development, the authors consider procedures, starting from the development of a plan for the management of the project and ending with obtaining a construction permit. Hence, the duration of the Control (supervision) procedure of development of detail documentation (p_{14}) is partially taken into account.

When influenced by a combination of factors, the total duration of the stages of pre-project preparation and design development before obtaining a construction permit is 15.91 months.

The duration increment ΔT is 2 months.

Let's consider a combination of factors affecting the duration of stages.

The factor "Obtaining additional initial data ($\delta^{\text{Пи}}_6$)" influences the duration of initial data collection. According to the network form of the organizational and management model (Figure 10 in the article), this function, which is a component of procedure p_3, is not on the critical path and has time reserves. The execution of the function should be completed before the start of the procedure "Control (supervision) of the development of design documentation by the design organization (p_{11})". Consequently, this factor does not affect the duration of the stages of pre-project preparation and design development.

The factors "Increasing the time for the development of materials required for the approval of architectural and urban planning solutions/appearance (AGR/AGO) ($\delta^{\text{Пи}}_7$)" and "Correction of architectural and urban planning solutions ($\delta^{\text{Пи}}_8$)" affect the duration of development of materials and approval of AGR/AGO, which is a function of the project manager in procedure p_3. This function is on the critical path. Consequently, factors $\delta^{\text{Пи}}_7$ and $\delta^{\text{Пи}}_8$ influence the duration of the stages.

The model is corrected through the parallel performance of the functions of procedure p_3 and the launch of the development of project documentation before obtaining a

Certificate of Approval AGR/AGO, or from the time when the planned deadline for the procedure expires.

After correcting the model, the total duration of the stages of pre-project preparation and design development before obtaining a construction permit is 13.91 months, which corresponds to Tp. As a result, the duration of the stages after optimization was reduced by 13%.

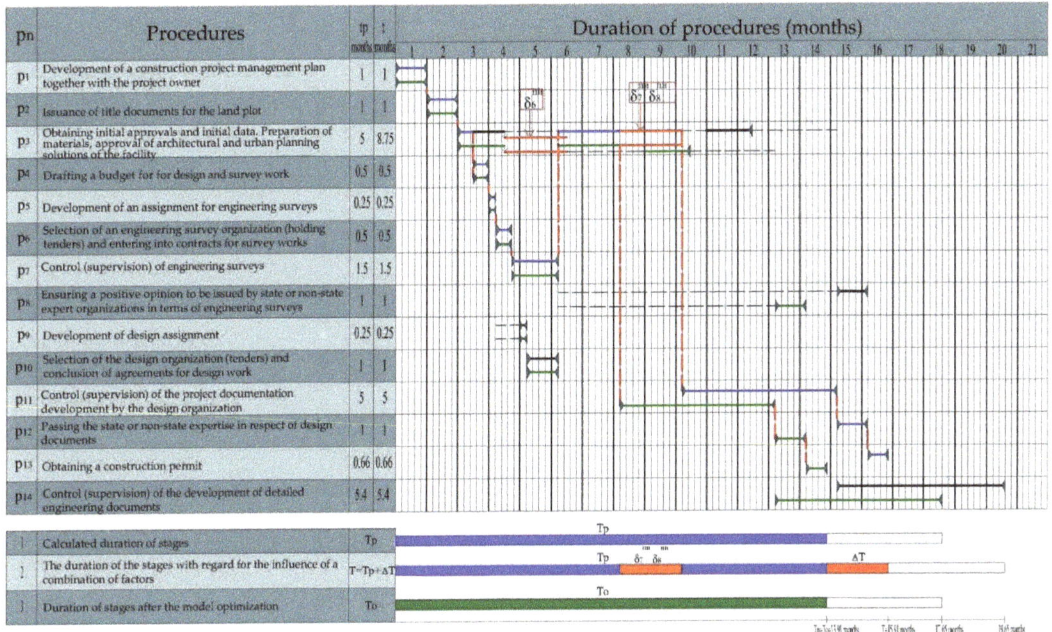

Figure A2. Organizational and management model of procedures based on the minimum values of the duration of stages of pre-project preparation and design development (a combination of factors).

References

1. Kuzmina, T.; Bolshakova, P.; Zueva, D. Completion of administrative procedures by the developer (technical customer). *E3S Web Conf.* **2021**, *258*, 09004. [CrossRef]
2. Kostrikin, P.N. Changes in the organization of residential construction as a result of the introduction of a comprehensive list of administrative procedures. *Real Estate Econ. Manag.* **2017**, *1*, 12–18.
3. Kuzmina, T.K. On some problems of the investment climate in the field of construction. *Sci. Rev.* **2016**, *21*, 192–195.
4. Peshkov, A.V. Administrative procedures in the implementation of investment and construction projects: A management mechanism. *Izv. Vuzov. Investig. Constr. Real Estate* **2014**, *6*, 45–53.
5. Galyamova, A.V.; Kuzmina, T.K. Analysis of procedures for obtaining permits for construction. *Technol. Organ. Constr. Prod.* **2018**, *2*, 6–9.
6. Ginzburg, A.V. Information model of the life cycle of a construction project. *Ind. Civ. Eng.* **2016**, *9*, 61–65.
7. Ginzburg, A.V. Technologies of information modeling of the life cycle of the object of capital construction. In Proceedings of the Actual Problems of the Construction Industry and Education, First National Conference, Moscow, Russia, 30 September 2020; pp. 936–939.
8. Ginzburg, A.V. BIM-technologies during the life cycle of the construction facility. *Inf. Resour. Russ.* **2016**, *5*, 28–31.
9. Zelentsov, L.B.; Kravchenko, A.I.; Pirko, D.V. The problems of organization and regulation of the relationship between participants of the investment and construction project. In Proceedings of the Construction and Architecture, International Scientific and Practical Conference, Rostov-on-Don, Russia, 19–31 October 2020; pp. 290–291.
10. Zelentsov, L.B.; Mailyan, L.D.; Shogenov, M.S. Management of temporal parameters in complex dynamic building systems. *Eng. Her. Don.* **2018**, *2*.
11. Zelentsov, L.B.; Shogenov, M.S.; Pirko, D.V.; Assaira, M.M. Dynamic methods of forecasting time and cost parameters of construction projects. *Constr. Prod.* **2020**, *3*, 61–64. [CrossRef]

12. Chugunov, V.I. Attracting investment to the regions through public-private partnerships: Problems, status, prospects. *Compet. Glob. World Econ. Sci. Technol.* **2017**, *3*, 156–159.
13. Soliman, E. Communication problems causing governmental projects delay–Kuwait case study. *Int. J. Constr. Proj. Manag.* **2017**, *9*, 1–18.
14. Fernando, S.; Panuwatwanich, K.; Thorpe, D. Introducing an innovation promotion model for construction projects. *Eng. Constr. Archit. Manag.* **2021**, *28*, 728–746. [CrossRef]
15. Banihashemi, S.A.; Mohammad, K.; Shahraki, A.; Rostami Malkhalifeh, M.; Ahmadizadeh, S.S.R. Optimization of environmental impacts of construction projects: A time–cost–quality trade-off approach. *Int. J. Environ. Sci. Technol.* **2021**, *18*, 631–646. [CrossRef]
16. Sharma, S.; Gupta, A.K. Analysis of Factors Affecting Cost and Time Overruns in Construction Projects. *Adv. Geotech. Struct. Eng. Lect. Notes Civ. Eng.* **2021**, *143*, 55–63. [CrossRef]
17. Kuzmina, T.K.; Bolshakova, P.V.; Ledovskikh, L.I.; Zueva, D.D. Features of the work of the technical customer with the use of BIM-technologies. In Proceedings of the Actual Problems of the Construction Industry and Education, First National Conference, Moscow, Russia, 30 September 2020; pp. 960–964.
18. Bennett, F.L. *The Management of Construction: A Project Life Cycle Approach*; Butterworth-Heinemann: Boston, MA, USA, 2003; 316p.
19. Clough, R.H.; Sears, G.A.; Sears, S.K.; Rounds, J.L.; Segner, R.O. *Construction Project Management: A Practical Guide to Field Construction Management*, 6th ed.; Wiley: Hoboken, NJ, USA, 2015; 419p.
20. Gajić, T.; Petrović, M.D.; Blešić, I.; Vukolić, D.; Milovanović, I.; Radovanović, M.; Vuković, D.B.; Kostić, M.; Vuksanović, N.; Malinović Milićević, S. COVID-19 certificate as a cutting-edge issue in changing the perception of restaurants' visitors-Illustrations from Serbian urban centers. *Front. Psychol.* **2022**, *13*, 914484. [CrossRef] [PubMed]
21. Golotina, Y.I.; Ryzhkova, Y.I.; Arutyunyan, M.S. Factors affecting the timing of construction. *Sci. Proc. Kuban State Tech. Univ.* **2018**, *9*, 65–73.
22. Kuzmina, T.K.; Bolshakova, P.V. Identification and systematization of factors in the preparation of facilities for construction by the technical customer (developer). *Constr. Prod.* **2020**, *4*, 38–43.
23. Opekunov, V.A.; Shcherbinin, I.V. Research of influence of preparation of initial permissive documentation on terms of realization of investment-construction projects. *Univ. Her.* **2016**, *12*, 59–64.
24. Opekunov, V.A.; Shcherbinin, I.V. The study of the existing system of development and implementation of investment and construction projects. *Univ. Bull.* **2018**, *9*, 122–126. [CrossRef]
25. Abramov, I.L.; Lapidus, A.A. Implementing large-scale construction projects through application of the systematic and integrated method. *Constr. Form. Living Environ. Conf. Proc. IOP Conf. Ser. Mater. Sci. Eng.* **2018**, *365*, 062002. [CrossRef]
26. Abramov, I.L.; Lapidus, A.A. Systemic integrated method for assessing factors affecting construction timelines. *MATEC Web Conf. Int. Sci. Conf. Environ. Sci. Constr. Ind. ESCI* **2018**, *193*, 05033. [CrossRef]
27. Yu, S.; Zhang, Q.; Hao, J.L.; Ma, W.; Sun, Y.; Wang, X.; Song, Y. Development of an extended STIRPAT model to assess the driving factors of household carbon dioxide emissions in China. *J. Environ. Manag.* **2023**, *325*, 116502. [CrossRef] [PubMed]
28. Popescu, C.R.G. (Ed.) Measuring Progress towards Sustainable Development Goals: Creativity, Intellectual Capital and Innovation. In *Handbook of Research on Novel Practices and Current Successes in Achieving the Sustainable Development Goals*; University of Bucharest: Bucharest, Romania; IGI Global: Hershey, PA, USA, 2021; Chapter 6; pp. 125–136.
29. Measuring the Impacts of Business on Well-Being and Sustainability. Available online: https://www.oecd.org/fr/investissement/measuring-business-impacts-on-peoples-well-being.htm (accessed on 26 November 2022).
30. Toward Sustainable Economic Development through Promoting and Enabling Responsible Business Conduct. Available online: https://www.oecd-ilibrary.org/sites/f7813858-en/index.html?itemId=/content/component/f7813858-en (accessed on 26 November 2022).
31. Mubarak, S. *Construction Project Scheduling and Control*; John Wiley & Sons: Toronto, ON, Canada, 2010; 480p.
32. Federal Service State Statistics. Available online: https://rosstat.gov.ru/ (accessed on 29 June 2022).
33. Kuzmina, T.K. Strengthening the functions of financing, accounting and reporting in the activities of the services of the developer (customer) with the transition to market relations. *Technol. Organ. Build. Prod.* **2012**, *1*, 50–53.
34. Kuzmina, T.; Cherednichenko, N. Systematization of the major stages of the client in certain branches of construction production. *MATEC Web Conf.* **2016**, *86*, 05012. [CrossRef]
35. Oleinik, P.; Kuzmina, T. Modelling the Reduction of Project Making Duration. *MATEC Web Conf.* **2017**, *117*, 00129. [CrossRef]
36. Danelyan, T.Y. Methods of Expert Evaluations. *Econ. Stat. Inform.* **2015**, *1*, 183–187. [CrossRef]
37. Zagorskya, A.V.; Lapidus, A.A. Application of methods of expert evaluation in scientific research. The required number of experts. *Constr. Prod.* **2020**, *3*, 21–34.
38. Kuzmina, T.K.; Sinenko, S.A.; Slavin, A.M. Combining the functions of the main participants of investment and construction activities at the present stage. *Ind. Civ. Eng.* **2016**, *6*, 71–75.
39. Morozenko, A.A. Peculiarities of the life cycle and stages of development of an investment and construction project. *Vestn. MGSU* **2013**, *6*, 223–228. [CrossRef]
40. Oleynik, P.P. *Organization of Construction Production*; Scientific Publication, ASV: Moscow, Russia, 2010; 576p.
41. Oleynik, P.P.; Brodsky, V.I. Organization of planning of building production. *Technol. Organ. Build. Prod.* **2013**, *2*, 40–43.
42. Telichenko, V.I.; Korol, E.A.; Shirshikov, B.F.; Yarovenko, S.M.; Bachurina, S.S. Application of network planning in the design, preparation and construction of urban housing programs. *Build. Mater. Equip. Technol. XXI Century* **2004**, *12*, 40–41.

43. Topchiy, D.V.; Yurgaytis, Y.S.; Popova, A.D. Optimization of planning processes of design works and approval of design estimates for capital construction, reconstruction and re-profiling facilities. *Bull. Civ. Eng.* **2019**, *2*, 93–98.

44. Potkany, M.; Vetrakova, M.; Babiakova, M. Facility Management and Its Importance in the Analysis of Building Life Cycle. *Procedia Econ. Financ.* **2015**, *26*, 202–208. [CrossRef]

45. Kuzmina, T.K.; Bolshakova, P.V.; Popova, A.D. Peculiarities of the initial data collection for non-productive objects in the technical customer's activity at the stage of pre-project preparation. *Sci. Bus. Ways Dev.* **2019**, *1*, 31–35.

46. Sinenko, S.A.; Ivanov, V.A.; Efimov, V.V. Features of the organization and conduct of competitive bidding for the implementation of investment and construction projects. *Sci. Rev.* **2017**, *13*, 104–107.

Article

Design and Validation of a U-Net-Based Algorithm for Star Sensor Image Segmentation

Marco Mastrofini, Ivan Agostinelli and Fabio Curti *

School of Aerospace Engineering, Sapienza University of Rome, 00138 Roma, Italy
* Correspondence: fabio.curti@uniroma1.it

Abstract: The present work focuses on the investigation of an artificial intelligence (AI) algorithm for brightest objects segmentation in night sky images' field of view (FOV). This task is mandatory for many applications that want to focus on the brightest objects in an optical sensor image with a particular shape: point-like or streak. The algorithm is developed as a dedicated application for star sensors both for attitude determination (AD) and onboard space surveillance and tracking (SST) tasks. Indeed, in the former, the brightest objects of most concern are stars, while in the latter they are resident space objects (RSOs). Focusing attention on these shapes, an AI-based segmentation approach can be investigated. This will be carried out by designing, developing and testing a convolutional neural network (CNN)-based algorithm. In particular, a U-Net will be used to tackle this problem. A dataset for the design process of the algorithm, network training and tests is created using both real and simulated images. In the end, comparison with traditional segmentation algorithms will be performed, and results will be presented and discussed together with the proposal of an electro-optical payload for a small satellite for an in-orbit validation (IOV) mission.

Keywords: artificial intelligence; star trackers; resident space objects; convolutional neural network; segmentation; space situational awareness; space surveillance and tracking; optical images; in orbit validation

Citation: Mastrofini, M.; Agostinelli, I.; Curti, F. Design and Validation of a U-Net-Based Algorithm for Star Sensor Image Segmentation. *Appl. Sci.* **2023**, *13*, 1947. https://doi.org/10.3390/app13031947

Academic Editor: Theodore E. Matikas

Received: 29 November 2022
Revised: 15 January 2023
Accepted: 27 January 2023
Published: 2 February 2023

1. Introduction

In the framework of on-board autonomous star sensors algorithms, star segmentation and accurate centroid estimation are critical problems to face. The number of actual detected stars and the relative centroids' estimation accuracy have a huge impact on the success of star-identification-based attitude determination routines [1,2]. Taking into account that star sensor images generally contain several noise sources (stray light noise, single point noise, single events upsets (SEUs) and so on), good initial processing of image data is mandatory. In this way, it is possible to improve the percentage of success in the star identification process [3,4] and the possibility of retrieving accurate attitude information from images. A good image segmentation process is also needed for RSO detection, where often the image noise and the weak nature of an RSO streak can lead to an incorrect detection of the object and loss of precious information for the orbit determination (OD) modules.

The process of image segmentation is needed to highlight the useful image information which is geometrically encoded in the image pixels. Often, not all the pixels in an image are needed, and a process of selection and exclusion is mandatory to focus attention on just that image part. Image segmentation faces and solves this problem bringing as a result a binarized image which is often called Mask. Through the convolution of this Mask with the original image, it is possible to extract photometric and geometrical information about the target pixels. In this case, we will not be interested in the photometric information, but this will be discussed in the next sections.

Regarding image segmentation, in the last decades several traditional approaches have been proposed, classified according to the element they take into account to discern foreground and background pixels: image histograms, detection of image gradients, detection of object edges and complexity analysis techniques [1].

Among them, commonly used methods are the following:

- Iterative threshold method [5]: It performs an iterative optimization. An initial threshold is set, and then the algorithm improves the estimation at every step with a suitable improvement strategy. The strategy should be fast enough in convergence and should improve the quality of the segmented image at each step.
- Local threshold methods: They are based on the selection of an initial threshold value plus a margin. For a generic examined pixel, its value is compared with a reference one. This reference value is updated continuously and takes into account the local energy value of the surrounding pixels in such a fashion that characterizes the method. A local threshold method is described in [6], and it is based on a moving streak average, while in [7], the approach uses rectangular areas and the local contrast level.
- Otsu's algorithm [8]: It performs a more refined approach where the intraclass variance of the foreground and background pixels is considered as a cost index and where the selected threshold comes from its maximization.
- Niblack's method [9]: It is a localized thresholding algorithm in which the threshold is varied across the image as a function of the local intensity average and standard deviation.

Each reported traditional method presents limitations. Iterative threshold methods rely on all the energy levels in the global image, and these constitute their limit: a simple strong stray-light scenario can badly affect the predictions of these methods. Moreover, when the histogram of a star image is unimodal, the traditional iterative threshold method is time-consuming and cannot reach a suitable segmentation. A local threshold method would fail in the same way if the difference between the star signal and the local noise intensity is lower than the margin. The Otsu method, instead, can only provide satisfactory results for thresholding if the histogram is not unimodal or close to unimodal [10]. Artificial intelligence could be an appealing and useful tool to face this challenging problems due to the recent development and application of AI in image processing.

Star segmentation is a task that has many aspects in common with the segmentation of dim small targets, whose primary goal is to enhance the contrast between the target itself and the background, regardless of its nature; segmentation must be performed in every possible situation (noise, blurring, angular motion of the camera and/or motion of the target).

All these issues also occur in star tracker images. Dim small target segmentation has been studied for both synthetic aperture radar (SAR) and optical images. Jin et al. [11] proposed the application of a lightweight patch-to-pixel (P2P) CNN for ship detection in PolSAR (polarimetric SAR) images. Their approach prefers the use of dilated convolutional layers rather than conventional convolutional layers in order to expand the receptive field without adding any model parameter. Zhao and Jia [12] employed a more general-purpose target segmentation using a CNN on infrared images, testing it on both real and synthetic images; however, their study suffers from the necessity for a large amount of training data to be fed to the network. Fan et al. [13] overcame the issue of large training data by using the MNIST [14] database in such a way that it simulates images having similar properties to the long-range infrared images. Nasrabadi [15] proposed an autonomous target recognition in forward looking infrared (FLIR) images based on different deep CNN architectures and thresholding. However, this approach suffers from a lack of reliability in a real-world scenario. Shi and Wang [16] based their studies on the use of a CNN and a denoising autoencoder network by treating the small targets as background noise and therefore transforming the segmentation task into a denoising task.

The use of the famous biomedical-based U-Net for small target segmentation tasks has not been broadly investigated as of this writing, but Tong et al. [17] proposed an enhanced symmetric attention (EAA) U-Net, which employs information extracted in the same layer to focus on the target and cross-layer information to learn higher level features.

Xue et al. [18,19] investigated a more specialized version of the small target segmentation problem by going deeper into the problem of segmenting star images by proposing StarNet. This network is particularly complex due to the fact that its training is carried out in a multistage approach, an aspect that affects the training time in a negative way. In addition to this issue, StarNet is pre-trained on weights taken from the first three stages of VGG16 [20], whose initial input size must be a three-channel image. This forced choice translates into a computationally heavy process that affects the prediction time and could require top choice GPUs. All these aspects seem to make real-time image processing unfeasible. The U-Net proposed in this work tries to overcome this problem by working with mono-channel images by exploiting a simpler network architecture and training strategy.

The present work is focused on an AI-based image segmentation module of a proposed star-tracker-based payload for attitude determination and extended RSO detection functions. This payload is thought to investigate and validate the AI behaviour for onboard optical sensor applications for small satellite missions. This choice can guarantee a high segmentation quality of star tracker images against a variety of noise scenarios without the need for intensive calibration activity. Indeed, it represents a robust processing solution for commercial and cheap star sensors which can be used in small platform missions.

The Discussion section will be devoted to a description of the integrated hardware and software payload (design, optical head specifications, component choice and validation mission). The Materials and Methods section, together with the Results, will be about the design of an AI-based star and RSO segmentation algorithm capable of filtering the faintest objects from background in several signal-to-noise (SN) level scenarios (stray light noise, ghost noise, SEUs). In particular this algorithm aims to provide information about the brightest objects within the FOV in order to facilitate star identification and object detection and reduce the memory storage burden of a star tracker for successive operations. By brightest objects, we mean the ones that stand out most inside an image. Here, the formulation of the segmentation problem is similar to the creation of a saliency map. In the predicted mask, just the most salient objects will appear. This philosophy seems to be suitable to detect stars and objects in low SN environments as strong stray light noise that could affect the star sensor image. The algorithm design aims to have few modules to reduce the algorithm size, complexity and hardware implementation difficulties in order to provide a reliable star detection product.

The machine learning (ML) world was investigated to take advantage of the neural networks (NNs) capability of learning specific tasks, performing well against unpredictable situations and reducing algorithm complexity for segmentation purposes; this could avoid the problem of algorithm re-calibration to successfully process images in different noise conditions. Moreover, the CNNs are considered for this kind of task. They have the advantage of processing the image directly without iterative steps and without a local approach that would be power- and time-consuming. All of these features are designed along with the capability of resolving objects in strong stray-light environments if a suitable dataset for training is available.

Contributions of this work are as follows:

- Creation and sharing of a balanced dataset of real and simulated star images for training, validation and testing of the CNN block.
- Design of an AI-based brightest object detection algorithm capable of providing the most significant information for star identification and object tracking algorithms.
- Comparison of our proposed AI-based algorithms with traditional segmentation algorithms.
- Validation of the AI algorithm both with star tracker simulated images and real images.
- Proposal of a an electro-optical payload for brightest objects segmentation in the onboard night sky images and a validation mission on a small platform.

The work is organized as follows: a brief overview of the CNN and U-Net model is described in the Materials and Methods section, together with the proposed dataset and our proposed segmentation-clustering algorithm scheme. Then, a description of the traditional algorithms used for comparison will follow, along with a description of the considered performance indices. Results of U-Net's training, validation and testing are then reported and compared with traditional algorithms. They are followed by a Discussion section about the obtained results with a payload concept proposal and a possible validation mission. In the end, a Conclusion section summarizes the achievements and future goals.

2. Materials and Methods

To face the segmentation of star tracker images, an ML approach has been considered. This AI area is being extensively explored and applied for image processing due to the great improvements achieved in several tasks. In particular, CNNs have been developed for computer vision algorithms because of their capability of image data processing and information extraction. Actually, lots of them are available to be used as Vgg-16 and Vgg-19 [20], Inception-v3 [21] from Google, the U-Net [22], AlexNet [23] and many others. CNNs have shown great performances in achieving many tasks such as object detection and classification, semantic segmentation and pose problems involving both cooperative and non-cooperative targets.

The problem faced in this work is semantic segmentation. Indeed, the scope is dividing the original image pixels in two groups: foreground pixels and background pixels belonging, respectively, to the white area and the black area of the segmented output. This result will be achieved through the application of the U-Net to obtain a final output which contains as white areas just the most salient objects in the sky, avoiding stray light noise, faintest stars and objects on the background and reducing the overall noise signal in the image. The reason why this is performed through an ML-based algorithm is due to the CNN capability to handle objectives and targets which are easily understandable for a human being but sometimes very difficult to formulate in a rigorous fashion; the problem of segmenting just stars and RSOs which stand out the most in a night sky image is one of them. This is the goal pursued in this work.

2.1. Convolutional Neural Networks

CNNs are heavily applied whenever complex information extraction from images is required. In a rough approximation, they are stacks of convolution and maxpooling layers. They are capable of learning high- and low-level features of an image regardless of their position and orientation. A great advantage of a CNN is that the weights to be learned are shared between the layers, minimizing the needed memory storage.

The convolutional layer works in the following way: a convolution operation is performed by sliding a small window (typically 3×3 pixels) over height and width of the image and over its color channels. The window creates a patch feature map, which is multiplied with a learned weights matrix called convolution kernel. This operation could cause image size reduction, which is handled by applying padding, which means to add border pixels in order to make the output size the same as the input. The neurons contained in the network layers are activated through an activation function; ReLU (rectified linear unit), softmax and sigmoid are typically used.

The pooling layer has the aim to downsample the image in order to optimize the learning of features. In particular, the max pooling layer takes a portion of the image (typically a 2×2 window) and substitutes it with the maximum pixel value.

The performance of a CNN is represented by values of loss and accuracy, which are computed during training, validation and test processes. It is desirable to obtain similar accuracies in all these phases to avoid overfitting and generalize network performances against never seen data. There are different ways to overcome this issue. These include L2 weight regularization [24], batch normalization [25], data augmentation [26] and dropout [27].

2.2. U-Net

The U-Net is a fully convolutional neural network (FCN) which was first used for segmentation of biomedical images and was later adapted to space-based applications such as crater detection [28]. The network architecture in this paper is chosen to be a slightly modified version of the original one as shown in Figure 1, and it has a symmetric encoder–decoder structure: the encoding, downsampling path is a stacked sequence of two ReLU 3×3 convolutional layers followed by a 2×2 max pooling layer. At each level, the number of filters doubles, reaching its maximum value at the bottom.

The decoding upsampling path contains 2×2 up-convolutional layers concatenated with layers coming from the corresponding downsampling level in order to preserve already learned features. The concatenated layers are followed by dropout and a sequence of two 3×3 convolutional layers up to the output layer in which a final sigmoid-activated 1×1 convolution produces data ready to be binarized. The present U-Net will be fed with 512×512 images and their corresponding masks.

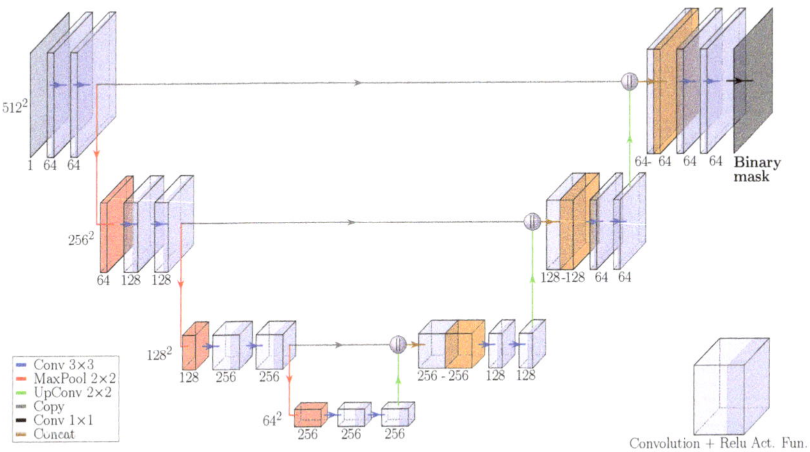

Figure 1. U-Net model with features dimensions and number of features [29].

Configuration parameters used for this network to prevent overfitting are as follows:

- Learning Rate: It is an optimizer's parameter that fixes the step size at each iteration during minimization of the loss function.
- Regularization Factor: It is a parameter needed for the l2 regularization based on penalization of the cost function.
- Dropout Rate: It regulates the percentage of inactive network elements in the dropout layers during training and the validation process.
- Kernel initializer: It sets the weight initialization method; in this case, it is set on he_normal [30].

In this work, Adam was chosen as the optimizer and Binary Crossentropy was chosen as the loss function.

2.3. Dataset Creation and Input Data Preprocessing

The dataset created and used for the U-Net training phase is provided at [31]. It is composed of 5600 squared monochromatic jpeg images and 5600 associated jpeg masks. All of them have a size of 512×512 px and a bit depth of 8. Masks' pixels can have one of two possible values: 0 for background and 255 for foreground pixels.

The samples for the training and validation phases are the first 5300 ones. They are generated from a batch of 600 images to which 180°clockwise rotation, 90° clockwise rotation, added noise, increased stray light, blur effects, increased luminosity, contrast and further modifications have been applied. This original batch is made of 300 samples coming from several night sky acquisition campaigns and 300 samples randomly acquired using Stellarium [32] software in night sky conditions considering different FOV and attitudes (Figure 2). Acquisition campaigns were performed using several kinds of cameras and objectives: ZWO ASI 120MM-S camera with default optics, reflex Nikon D3100 equipped with a Nikkor 18-105 mm and a ProLine PL16803 camera coupled with an Officina Stellare RiFast 400 telescope.

Figure 2. Dataset samples: real night sky image (on the **left**) and Stellarium-generated night sky (on the **right**).

Last. 300 dataset images were obtained with a high fidelity star tracker image simulator (HFSTIS). It is used to simulate realistic night sky images, simulating all the physical, functional and geometrical characteristics of the star sensor, along with all the instrumental and environmental noises [33–37]. Simulator images were used only in the test phase to provide independent samples the network has never been trained on to monitor the generalization capability of the trained network.

The dataset contains images of night sky where stars are the main actors but where planets, light pollution caused stray light noise, SEUs, clouds, airplanes' streaks, satellites, comets, nebulae and galaxies appear as well. Masks were obtained using specific thresholds for different groups of images from different campaigns, sensors and software. In particular, mask creation was carried out using Adobe Photoshop CC 2019 with the scope of obtaining just the most salient objects in the FOV and filtering all the noise and undesired objects cited above. An example of a dataset image and mask is provided in Figure 3, where it is clearly visible that just the brightest points appear inside the mask. Input images and mask preprocessing is performed in order to organize these data in two float 4D tensors for NN training and testing. Every image is converted into floating point arrays and normalized using its maximum value to carry out an image adaptive normalization process; every most significant image pixel will have an associated value close to 1. This will help the network detect the most significant pixels and make the learning process easier for the net by constraining the interval of signal values between 0 and 1. Moreover, this normalization process improves the algorithm capability of working with images that have different range of energy levels. The same process applies for mask data vectorization and normalization, with the only difference being that the normalizing factor is constant and equal to 255 for every mask.

Figure 3. Dataset samples: real night sky image (on the **left**) and relative mask (on the **right**).

It must be pointed out that the dataset does not contain only point-like stars but also even images where they appear as streaks due to the angular speed of the sensor. Considered angular speeds are within the $[0°/s, 1°/s]$. This allows the proposed AI-based algorithm to work properly even in situations where the star sensor is rotating with respect to the fixed stars or where a sidereal pointing of the camera occurs together with RSO passages.

Concerning the noise, 5300 images for training and validation are generated from a batch of 300 images. This batch is composed of several acquisitions performed with different sensors and so, different noise conditions. In less-noise-corrupted images, an added source of uniform noise was added using Adobe Photoshop 2019 (Filter->Noise->Add Noise->Uniform 6.3%). For different stray light conditions, +0.5, +1 and +1.5 stop in image exposure were added randomly. It is quite impossible to retrieve the sensors' noise level characterization due to the different kind of sensor used and the unavailability of their noise features. For the last 300 test images, it is possible to characterize the noise:

- Shot Noise: Poisson probability model and proportional to square root of the detected signal;
- ReadOut Noise (RON): Normal distribution with mean $0\ e-$ and standard deviation of $82\ e-$;
- Dark Current (DC): Constant value of $550\ \frac{e-}{px \times sec}$;
- Dark Signal Non-Uniformity (DSNU): Normal distribution with 1 mean value and a standard deviation of 0.065;
- Photo Response Non-Uniformity (PRNU): Normal distribution with 1 mean value and a standard deviation of 0.01;
- Stray Light: $30000\ \frac{e-}{sec}$.

2.4. Algorithm Design and Configuration

In this section, our proposed AI-based algorithm's scheme is described and shown (Figure 4). It will be referred to it as the brightest objects sky segmentation (BOSS) algorithm.

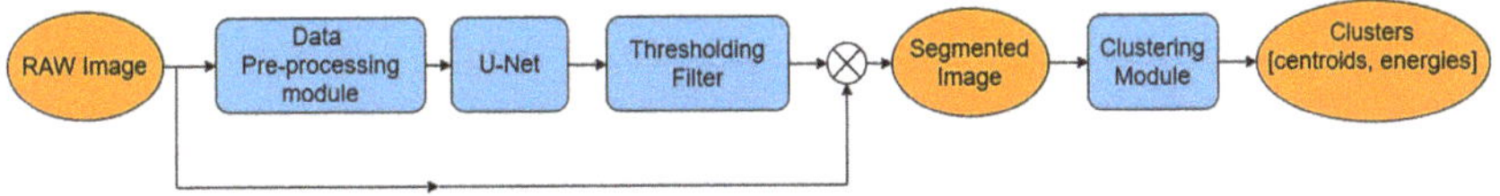

Figure 4. BOSS algorithm scheme: data are in orange, and modules are in light blue.

The monochromatic raw image from the camera arrives at the preprocessing module where it will be resized, normalized and vectorized into a floating point array. This module will be tailored according to specific sensors' bit depth, while the size must be the one the NN is trained on. Then, the array is processed by the trained U-Net which gives its output prediction. This is a 2D array where every pixel has an associated value of probability p of being foreground. If $p \geq p_{min}$, then this value is rounded up to 1. After the thresholding filter, a binary image is obtained, and its element product with the original image will provide an array with just active pixels' energy values. This output will be used by the clustering algorithm to detect clusters and compute their centroids and total energies. The clustering module developed within this work takes all the active pixels in the segmented image and organizes them in a list. Then. it associates the first pixel in the list to the first cluster and starts to verify if the next pixels are part of the same cluster or not through a distance-based criterion. Whenever the next active pixel does not belong to the already identified clusters, then a new cluster is identified.

The distance criterion uses the Euclidean norm and the condition to assess the membership of two active pixels ($\mathbf{p}_i$ and $\mathbf{p}_j$) to the same cluster as expressed by Equation (1).

$$\|\mathbf{p}_i - \mathbf{p}_j\| \leq \sqrt{2} \tag{1}$$

The clustering algorithm then first performs a filtering action with a minimum and maximum dimension and then a sorting of all the clusters. A minimum dimension filter is needed to avoid a possible noise signal in the clustering outputs (ex. hot pixels), while the maximum dimension filter is needed to avoid great detected clusters due to non-stars objects.

The sorting operation is then performed in descending order and considers the first N clusters in output from the previous filtering actions. It is based on a combination (Equation (2)) of the clusters' dimension ($dim_{cluster}$), energy ($E_{cluster}$) and maximum energy value ($max(E(\mathbf{p}_i))$) over the cluster because this sorting index proved to be suitable to select the best clusters for star pattern recognition purposes with the conducted tests.

$$Sorting\ Index = \frac{E_{cluster}}{dim_{cluster}} \times max(E(\mathbf{p}_i))\ \ with\ i\ over\ cluster \tag{2}$$

N is computed by applying a 1.25 factor to the average number of stars which depends on the FOV and cut-off magnitude of the selected sensor [2]. The maximum number of clusters in output is fixed a priori to process just the most significant ones, and the 25% margin factor is selected to take into account possible uncertainties of the average number of stars formula.

The U-Net is capable of resolving background and foreground pixels after training, validation and test phases, but it still gives continuous values that have to be discretized. This is the reason why the thresholding filter and the selection of a reasonable value for p_{min} will be described in the next sections in order to obtain the best correspondence between algorithm prediction and targeted masks. Once the U-Net is trained and p_{min} selected, the BOSS algorithm will be compared with traditional image segmentation ones, both described in the next section in terms of segmentation performance.

2.5. Traditional Image Processing Algorithms Description

In this framework, several traditional algorithms for image segmentation were selected for comparison with the AI-based proposed algorithm. In the following, five algorithms are considered. They are based on Niblack's method, Otsu's method, the weighted iterative threshold approach (WIT) [1] and two local threshold (LT) approaches.

2.5.1. Niblack's Algorithm

Niblack's method is based on the computation of a threshold value T which is a function of the energy mean value and standard deviation computed over a rectangular window. Through its sliding over the whole image, it is possible to then identify which are the foreground and background pixels by comparing the window's pixels energy values with the local threshold.

$$T(x,y) = m(x,y) + k \times \sigma(x,y) \tag{3}$$

where k is a parameter which varies between -0.2 and -0.1, while the other parameter is the dimension of the window (supposed squared in our paper with side d_{Nb}).

2.5.2. Otsu's Method

It is based on the gray level image's histogram. It aims to find the threshold value which minimizes the intraclass variance [38] of the thresholded black and white pixels. The main advantage of this method is its unparametric nature which avoids the need for configuring it according to the image noise level.

2.5.3. Weighted Iterative Threshold Approach

This method is an iterative way to compute the optimal threshold for a given image. At each step, the threshold is computed using the following formula:

$$T = \frac{(1+\delta) \times \mu_1 + (1-\delta) \times \mu_0}{2} \tag{4}$$

where μ_1 and μ_0 are, respectively, the average gray level values of foreground and background pixels. With this new threshold, new sets of foreground and background pixels can be computed with the following mean values, and the process repeats with the computation of a new threshold. When the difference between two successive thresholds is lower than a certain tolerance Δ, the process stops. This algorithm is dependent on a scalar parameter δ which can vary in the range $[-1.0, +1.0]$.

2.5.4. Local Threshold Approach Based on Rectangular Areas (LTA)

The algorithm used is from MATLAB. It scans the whole image and computes a local threshold. The used size of the window is given by the following formula:

$$Size = 2 \times floor\left(\frac{Image\,size}{16}\right) + 1 \tag{5}$$

Over this window, a Bradley's mean [39] is computed. By comparison between a pixel's energy value and this local mean, the classification of the pixel as background or foreground occurs. This process is carried out using a scalar parameter called sensitivity (S). It varies in the range $[0, 1]$ and indicates sensitivity towards thresholding more pixels as foreground.

2.5.5. Local Threshold Approach Based on a Moving Streak Average (LTS)

The last one, the segmentation algorithm [40], can be configured using a static or a dynamic approach [41] for the background noise estimation [38]. In this work, a dynamic approach based on a zigzag local thresholding with moving average is used [42]. In particular, a pixel is saved if its energy value is greater than the background noise, evaluated through a moving average line-by-line, plus a threshold τ_{pre}.

The information relative to the segmented image is used for the clustering process, i.e., the combination of segments belonging to the same star streak [34–37]. The most important step required by the clustering algorithm is the evaluation of the primitive clusters containing pixels which share at least one corner. A suitable technique has been developed in order to relate streaks in the image to the same star if the corresponding streak is broken into multiple streaks due, for example, to the high rate of the spacecraft. Three conditions will be satisfied related to the minimum distance, the direction and the density of the considered primitive clusters. Centroids' coordinates of each cluster are evaluated, taking into account the energies of segments merged to build the cluster itself according to Equation (6):

$$\mathbf{P}_{cluster,i} = \sum_{j} \left(\frac{E_{seg,j} \cdot \mathbf{P}_{seg,j}}{E_{cluster,i}} \right) \tag{6}$$

where $\mathbf{P}_{seg,j}$ are the baricenter coordinates of the j-th segment in the image plane, while $E_{cluster,i}$ is the sum of the energies $E_{seg,j}$ which forms the cluster i-th:

$$E_{cluster,i} = \sum_{j} E_{seg,j} \tag{7}$$

2.6. Comparison Indices

The performance comparison between the BOSS algorithm and other algorithms will be detailed in the Results section. The compared algorithms will be tested with a set of gray scale images and relative masks; in order to assess the accuracy of the algorithm's predictions against test masks, the use of suitable indices is necessary. These indices are heavily used in the segmentation field to compare two image masks as the U-Net output and ground truth from the dataset actually are. These indices can also be used to compare predictions of any couple of segmentation algorithms. Before introducing them, the notion of true positive (TP), false positive (FP) and false negative (FN) must be given:

- TP: A counter that increases by one unit every time both the predicted pixel and the reference one belong to the foreground set;
- FP: A counter that increases by one unit every time the predicted pixel belongs to the foreground set while the corresponding mask one is a background pixel;
- FN: A counter that increases by one unit every time the predicted pixel belongs to the background set while the corresponding mask one is a foreground pixel.

A good prediction has background pixels and foreground ones almost in the same position inside an array as the associated ground truth mask. The goodness of the prediction can be assessed through the evaluation of Precision, Recall and F1 index.

$$Precision = \frac{TP}{TP + FP} \times 100 \tag{8}$$

It represents the percentage of the TP with respect to the sum of TP and FP. The lower the FP is, the higher the Precision will be (with maximum value equal to 100% under ideal conditions).

$$Recall = \frac{TP}{TP + FN} \times 100 \tag{9}$$

It represents the percentage of the TP with respect to the sum of TP and FN. The lower FN is, the higher the Recall is (with maximum value equal to 100% under ideal conditions). Recall and Precision are similar but with a slight difference. This is due to the distinction of FP and FN. Every time an FP or FN occurs, there is a discrepancy between the prediction and the reference mask, but the difference between FP and FN helps us better understand the behaviour of the trained network.

Another index to be defined is F1:

$$F1 = \frac{2 \times Precision \times Recall}{Precision + Recall} \tag{10}$$

This index combines the Precision and Recall and lets us understand when the best compromise between FP and FN occurs. It reaches its maximum value when the discrepancy between the prediction and the mask is minimized. Ideally, the maximum value of F1 is 100%, but the best prediction will be the one with the highest value of the F1 index.

3. Results

In this section, results of U-Net's training, validation and testing will be shown, together with a reasonable choice for p_{min} and number of initial filters. Then, a comparison test of the BOSS algorithm with traditional ones is conducted against the same set of images.

3.1. U-Net Training and Test

The U-Net tested here is configured with the values shown in Table 1.

Table 1. U-Net configuration parameters.

Parameter	Value
Image size	512
Learning Rate	10^{-4}
Regularization Factor	10^{-5}
Dropout Rate	0.25
Kernel Size	3
Kernel initializer	*'he_normal'*

Training and validation were performed in Tensorflow Keras [43] considering three epochs and a dimension of the training batch equal to three. This has been conducted for four different values of initial filters: 16, 32, 64 and 128. This choice was made to see how performance in terms of accuracy and loss varies with the increasing network complexity in order to select the minimum required number of filters while achieving satisfactory performance. Every trained network reached accuracies higher than 99.80% after the first epoch and remained constant for training, validation and testing. The same behavior applies for the final losses which are not higher than 0.016. Moreover, no overfitting phenomenon was indicated.

The training, validation and tests were performed on the following workstation:

- CPU: AMD Ryzen Threadripper PRO 3975WX 32 Cores 3.50 GHz
- RAM: 128 GB
- GPU: Nvidia Quadro RTX 5000.

3.2. Tuning of Thresholding Filter

From the previous phase, it seems that every model has learned its task, but their output is not yet a binary mask. To achieve this, a thresholding operation has to be performed to select a suitable value of p_{min}. Its tuning is conducted in a range of values from 0.01 to 0.9 and different models over the 300 images test set.

For every p_{min} and model, the values (over the whole test set) of Precision, Recall and F1 are computed and reported in Figure 5. It can be seen (Figure 6) that the network with 128 filters performs better in terms of output mask. The value around 69% for the F1 index means a minimized number of discrepancies between the BOSS algorithm's prediction and the mask. This value is the highest one if the best achieved F1 values are considered among all the networks. Because of this, a model with 128 filters and $p_{min} = 0.04$ was chosen for the BOSS algorithm design.

Precision behaviour for low p_{min} values is due to an increasing number of FPs because as the p_{min} decreases, the number of FPs becomes greater if compared to TPs. For higher p_{min}, the number of FPs tends to 0, while TPs tend to a finite value, and the 100% Precision is achieved. This maximum value of Precision still does not mean that the prediction is good.

A similar discussion can be conducted for the Recall: as the p_{min} decreases, the risk of predicting FNs decreases with respect to TPs. This is why Recall grows. For higher p_{min}, the increasing number of FNs causes the Recall value to fall towards 0%.

A test with a 256 initial filters model was conducted, but performances do not show relevant improvements with respect to the 128 initial filters model, while the required storage memory and computational time increment is not negligible (and also undesired). This is the reason why they are not reported in Figure 5.

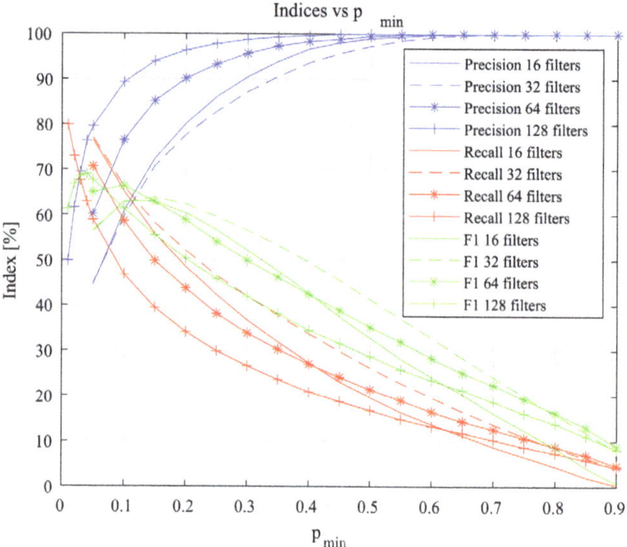

Figure 5. Performance curves vs. p_{min} and models.

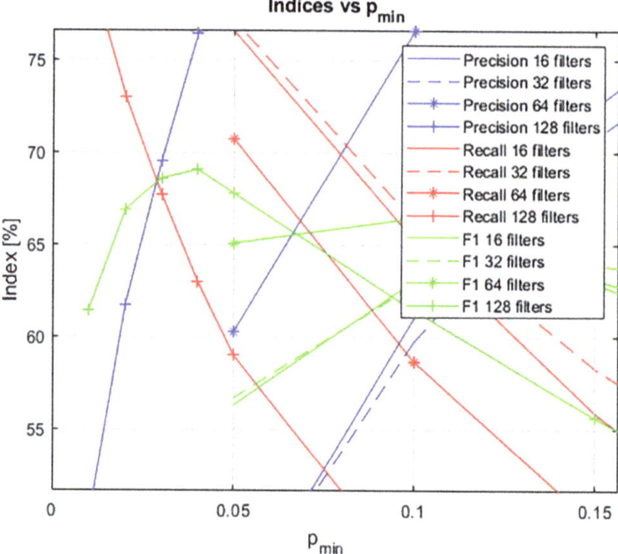

Figure 6. Details of performance curves vs. p_{min} and models.

3.3. BOSS Algorithm Tests

In this section, examples of BOSS algorithm applications for stray light removal and ghost noise removal in night sky images are shown. The aim of a good segmentation in this case is to retrieve the point-like stars and streaks (RSOs) while removing background pixels that are often badly affected by several noise sources. By noise, we mean that part of the image signal that does not represent the target scene and must be filtered in a certain way. It corrupts the image actors, and if the ratio between the useful signal and noise is higher than but close enough to one, the useful signal could be filtered out as noise and lost for further processing. The image noise is due to several causes briefly divided into the following:

- Inner noise: It is due to the sensor components, electronics and realization technique.
- External noise: It is due to planets, the Sun, the Moon, the Earth, hosting platform structures and camera baffle together with lenses non-uniform reflection.

3.3.1. Sidereal Pointing Camera and Different Exposure Times

This test aims to investigate the BOSS algorithm behaviour against an increasing stray light noise source. Stray light noise is due to external light sources which cause a flare all over the image. It can be experienced when an optical-space-based sensor has a boresight direction close to the Sun, the Moon, or another high luminosity source. This stray light can have a uniform intensity or uniform gradient. If it is too strong, it can hide the useful signal associated to stars and RSOs and make them difficult or even impossible to be segmented and detected. A change in the stray light intensity during the mission may cause the need for continuous sensor calibration. This could be avoided with a calibration-less algorithm such as the U-Net-based proposal in this work. This noise can increase with the increase of the sensor exposure time as the amount of collected photons proportionally increases with this time interval. In this test, images acquired with a night sky simulator facility (NSSF) (Figure 7) were used. This facility is made of a darkroom where a ZWO ASI 120MM-S camera points towards a screen. Here, Stellarium software is used to simulate the sky as seen by a specific camera with a selected FOV and optics (2.8 mm of focal length, f/1.4). All the simulated stars have a magnitude lower than 6.5 in Figures 8 and 9. With this software and the real camera, it is possible to simulate the working of a camera pointing at the night sky both for stars and RSO segmentation and detection for preliminary results. Images' noise sources are the real camera noises plus a further stray light source due to the screen technology (backlit screen).

Figure 7. NSSF at ARCALab, School of Aerospace Engineering, Sapienza University of Rome.

A sidereal pointing was considered in this test just to see the effect of the image stray light noise increasing with the increase of the camera exposure time. Here 100, 200, 500 and 1000 ms of exposure were considered, and the original images and segmented ones are reported below (Figure 8). From this figure, it is possible to assess that the segmentation algorithm removes the stray light and correctly segments the stars. As the noise increases, the foreground pixels increase because of the increasing magnitude of the targets.

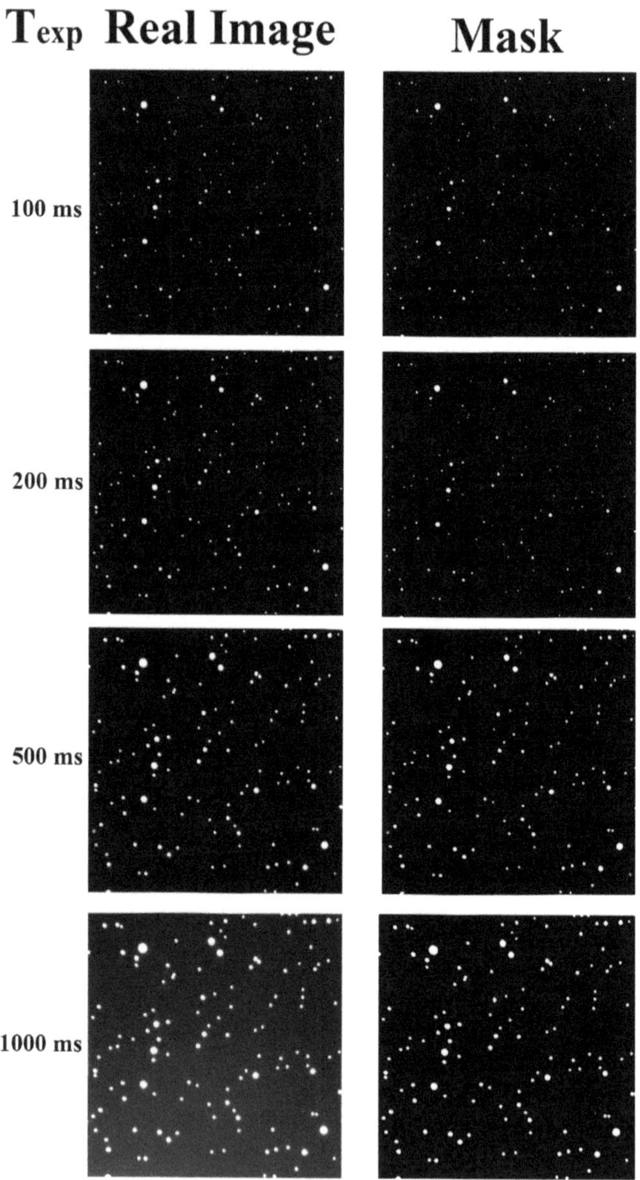

Figure 8. Left column: images from NSSF; right column: segmented images by the BOSS algorithm; from top to bottom: 100, 200, 500 and 1000 ms of exposure time.

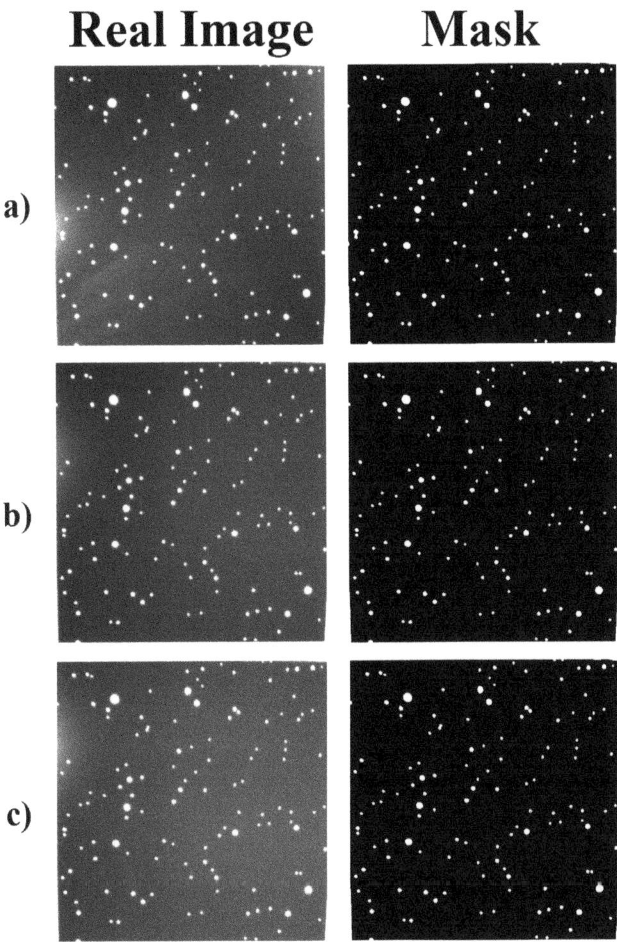

Figure 9. Left column: images from NSSF with ghost noise; right column: segmented images by the BOSS algorithm; from top to bottom three different ghost noise scenarios (**a–c**).

3.3.2. Ghost Noise Removal

This test aims to investigate the BOSS algorithm behaviour against a nonuniform gradient noise: ghost noise. This undesired noise source is due to camera lens reflections, spacecraft structure reflection and dust all over the camera lenses. The ghost noise is named in such a way due to its appearance: it is a vague and partially transparent shape superposed on the background and image objects and a nonuniform gradient signal which may cover the whole image or part of it. If it is too strong, it can hide the useful signal associated with stars and RSOs and make them difficult or even impossible to be segmented and detected like the stray light. A change in the ghost noise intensity during the mission may cause the loss of information if no calibration is performed. This could be avoided with a calibration-less algorithm such as the U-Net-based proposal in this work, and it is shown with simulated images in Figure 9 and with real images in the next sections. This noise does not increase with the increase of the sensor exposure time, while the gradient level tends to be more uniform. Here, the same facility has been used but with an added external flashlight source. The stray light and ghost noise were produced with three different positions of an external flashlight source which was moved at different heights on the left side of the camera and target screen inside the darkroom. Even here, it is possible to see a good segmentation quality provided by BOSS (Figure 9). A particular thing that can be noted is the white edges at the images' corners. They are due to a displayed undesired box in the night sky images made by Stellarium.

3.3.3. Real Image Test on BOSS Algorithm Output

In this section, a real night sky image is considered to test the BOSS algorithm against it. The aim of the test was the demonstration of the BOSS capability of providing a suitable segmentation and stars localization with a real image and real noise sources. The real image is shown in Figure 10 together with the segmented image. The camera used is a Nikon D3100 whose image has been cropped and resized to the U-Net input size of 512 pixels. The image contains, besides real sensor noises due to the electronics, a strong stray light source coming from the bottom and due to nearby city light pollution (Acquisition site: Tusculum/Frascati, Rome, Italy). Parameters of the camera are listed in Table 2. Clustering algorithms extracted the centroids of the eight brightest stars in the FOV and correctly localized them. They are shown in Figure 11.

Table 2. Nikon D 3100 camera features. The image has been resized to 512 px and 1:1 aspect ratio.

Feature	Value
Size	512 px
Aspect Ratio	1:1
focal length	30 mm
FOV	28.79°
pixel size	30.07 μm
Cut Off Magnitude	4.0
f-number	f/2.0

Figure 10. Ursa Major constellation segmentation provided by the BOSS algorithm. **Left:** real Ursa Major image; **right:** BOSS output of the Ursa Major real image.

Figure 11. Ursa Major constellation stars localization performed by the BOSS algorithm. Names and visual magnitudes of the segmented objects have been reported in Table 3.

Table 3. IDs, names and visual magnitudes of the segmented objects in Figure 11.

ID	Name	Magnitude
1	Dubhe	1.95
2	Phecda	2.42
3	Merak	2.34
4	Mizar	2.25
5	Alioth	1.75
6	Psi Ursae Majoris	3.16
7	Megrez	3.33
8	Alkaid	1.80

3.4. BOSS Algorithm Comparison with Traditional Image Segmentation Algorithms

Here, a comparison between BOSS and traditional algorithms is conducted. The F1 index (Equation (10)) is chosen to compare the performance of the algorithms. Every algorithm was tested against the same test set of 300 simulated images [31]. Otsu's algorithm did not need to be configured, while LTA, WIT and LTS algorithms did. For these configurable algorithms, a tuning of their parameters was performed, and the F1 index was computed for every combination of their internal parameter.

3.4.1. Algorithms Comparison Procedure

The rationale behind the algorithm comparison is described in this section. The comparison dataset was composed of 300 images and 300 reference masks. Each reference mask represented the desired result of the segmentation process. Every reference mask was manually obtained using Adobe Photoshop 2019 (Image->Adjustments->Threshold), selecting a suitable threshold level because of the varying noise conditions over the 300-image test set.

Now, the scope compared the output of the generic segmentation algorithm with the corresponding reference mask. The procedure was as follows:

- The output mask was compared with the reference mask pixel by pixel;
- The number of FP, FN and TP were updated during the mask comparison;
- FP, FN and TP values were used to compute the Precision, Recall and F1 values (these indices were described in the previous Section 2.6 *Comparison Indices*).

This process was repeated for all the 300 images, and a final averaged value for the F1 index was obtained for the considered algorithm.

This procedure was directly applied for the BOSS and Otsu algorithms because they do not need to be configured: Otsu does not need any configuration parameter, and the BOSS algorithm has its fixed value of p_{min} which was frozen during its design in Section 3.2.

Niblack, LTA, WIT and LTS require an additional step before computing the final F1 value: the configuration parameters' optimization. Indeed all of them have at least one configuration parameter to be selected with a suitable criterion:

- Niblack's configuration parameters are k and d_{Nb};
- LTA's configuration parameter is the *Sensitivity*;
- WIT's configuration parameter is δ;
- LTS's configuration parameters are τ_{pre} and BKG_0.

By considering the generic configurable algorithm, the averaged F1 index was computed for every combination of the configuration parameters varying in their specific ranges. Indeed, every selected value for the configuration parameter brings the algorithm to be more or less severe in terms of segmentation performance and changes the final F1 index value. Results of this process, using commonly used values for the parameters, have been collected for each configurable algorithm in Tables 4–7. In the end, the six averaged F1 values for Otsu, BOSS and the optimized algorithms can be obtained and were reported in Table 8.

Table 4. Niblack's algorithm tuning. F1 index vs. k and d_{Nb}. The first parameter varies along the rows from -0.2 to -0.1, while the second varies one along the columns from 1 to 10.

F1 (%)	$d_{Nb} = 1$	$d_{Nb} = 3$	$d_{Nb} = 5$	$d_{Nb} = 7$	$d_{Nb} = 10$
$k = -0.2$	0.041	0.061	0.064	0.065	0.065
$k = -0.15$	0.041	0.062	0.065	0.066	0.067
$k = -0.1$	0.041	0.061	0.066	0.067	0.068

Table 5. LTA algorithm tuning. F1 index vs. S. The Sensitivity varies along the columns from 0.1 to 1.0.

S	0.1	0.3	0.5	0.7	1.0
F1 (%)	79.63	68.97	3.62	0.04	0.04

Table 6. WIT algorithm tuning. F1 index vs. δ. Here, δ varies along the columns from -1.0 to $+1.0$.

δ	-1.0	-0.7	-0.5	-0.3	-0.1	0	$+0.1$	$+0.3$	$+0.5$	$+0.7$	$+1.0$
F1 (%)	0.07	1.19	3.35	19.35	37.43	50.86	27.17	16.96	12.08	7.58	3.92

In these tables, the configuration parameters which maximize the F1 index were considered, and the maximum value of F1 is then reported in Table 8 for the final comparison. In this way, every configurable algorithm was optimized against the 300-image test set in order to make the comparison more challenging for the BOSS algorithm.

Table 7. LTS algorithm tuning. F1 index vs. τ_{pre} and BKG_0. The first threshold varies along the rows from 5 to 55, while the second one varies along the columns from 1000 to 2000.

F1 (%)	$BKG_0 = 1000$	$BKG_0 = 1500$	$BKG_0 = 2000$
$\tau_{pre} = 5$	13.39	13.39	13.39
$\tau_{pre} = 15$	72.75	72.75	72.75
$\tau_{pre} = 25$	60.85	60.85	60.85
$\tau_{pre} = 35$	46.55	46.55	46.55
$\tau_{pre} = 45$	35.15	35.15	35.15
$\tau_{pre} = 55$	26.92	26.92	26.92

Table 8. Summary of algorithms' best F1 scores.

Algorithm	Niblack	LTA	WIT	LTS	Otsu	BOSS
F1 (%)	0.068	79.63	50.86	72.75	0.07	69.05

3.4.2. Comparison Results

Otsu's algorithms is the only traditional one which does not need any configuration of parameters. Its F1 score is 0.07%.

The best achieved F1 values for every algorithm are summarized in Table 8 for a fast comparison.

From previous tables, it can be seen that the best F1 score was achieved by the LTA algorithm followed by the LTS one and the BOSS algorithm. Niblack and Otsu behaved worse than the others, while the WIT achieved high but not satisfying performances.

Both LTA and LTS behave better on the test set if compared to the BOSS algorithm. As a first impression, it would seem that the use of BOSS algorithm does not bring any advantage. However, the performances achieved by LTA and LTS were obtained via a tuning of their parameters, while the BOSS algorithm was not tuned after its design. LTA and LTS have to be calibrated every time the noise level changes inside the selected scenario to achieve the best segmentation quality output, while the BOSS algorithm does not need this calibration because it has been trained to segment well with several SN levels. This consideration means that the 69% of F1 index is a generalized performance value, while the 79% and 72% values for the traditional algorithms are optimized and not generalized. LTA and LTS were tuned against the test set, while the BOSS did not need any tuning.

This would mean that a star tracker based on the BOSS algorithm would not require any calibration or segmentation performance degradation during its lifetime in orbit. Even if the sensor's noise increases with the increasing of the lifetime, the U-Net would be able to adapt itself to several levels of SN ratio because it has been trained to do so. A stray light or a higher radiative region would not affect the quality of the segmentation product very much. With this consideration, the strength and meaning of the BOSS performance can be more appreciated and understood.

In Figures 12–14, it is possible to visually compare the segmentation algorithms output quality of these LT approaches and the BOSS algorithm to understand the limit of a traditional parameterized segmentation algorithm. Three real images of the night sky both from Earth and space were considered with different kinds and levels of noises. A discussion follows for each of them.

Figure 12. Italian Space Agency Matera Telescope's image segmentation: (**a**) real image; (**b**) LTA's prediction; (**c**) LTS' prediction; (**d**) the BOSS algorithm's prediction.

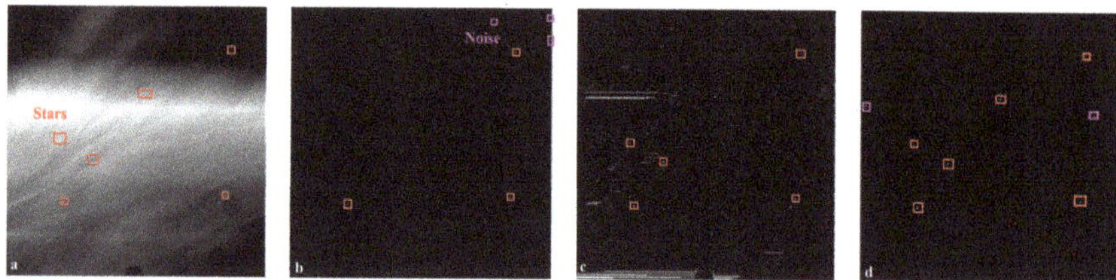

Figure 13. E and B EXperiment mission [44] star camera's image detail segmentation: (**a**) real image; (**b**) LTA's prediction; (**c**) LTS' prediction; (**d**) the BOSS algorithm's prediction.

Figure 14. Juno Star Reference Unit's [45] image segmentation (Credits to NASA): (**a**) real image, (**b**) LTA's prediction; (**c**) LTS' prediction; (**d**) the BOSS algorithm's prediction.

Considering the processing of these three figures:

- Figure 12: In the real image, two geostationary satellites (green boxes) on the bottom part of the image are clearly visible with the surrounding star field. This image was obtained via the telescope of the Italian Space Agency Matera Observatory. The noise that mostly appears in this image is a nonuniform flare all over the image stronger in the bottom part and weaker in the top. It is due both to telescope lenses and nearby city light pollution. It is an example of ghost noise. The noise sources associated with the sensor electronics are negligible due to the sensor cooling system which was keeping the camera at $-20\,°C$. There are foreground horizontal and short streaks on the top left and top right edges of the image produced by the LTA algorithm. Moreover, there is some noise that has not been filtered in the whole image, together with the brightest and weakest objects. The same problem is present in LTS' prediction with a greater percentage of noise. Here, the best output quality is provided by the BOSS algorithm which correctly returns the most salient star-like objects, together with precious RSOs information. All the algorithms return the two small streaks of the geostationary. The stars in the background have a magnitude lower than 12. The used optics is an Officina Stellare RiFast 400 Telescope with a focal length of 1520 mm and an f-number of f/3.8.
- Figure 13: The real image is a detail of a frame from the EBEX mission star camera [44]. The camera is mounted onto a stratospheric balloon carrying a telescope. It is pointed towards the above sky and contains a lot of noise. Indeed, here the dust over the sensor's lenses, mesosphere wind turbulence and stray light from the Antarctic continent's albedo are clearly visible and cause a strong ghost noise which covers the few

stars in the background. Recognizing stars in this condition is important because star sensors are not used just in space where the absence of atmosphere avoids many noise sources, but also on ships or cruise missiles for navigation purposes, where being able to remove ghost noise due to clouds and other atmospheric effects is mandatory. The LTA (b) algorithm's prediction shows that few of the brightest stars have been correctly segmented with a not complete filtering action of the noise in the top right part of the image noise. The LTS behaves worst because the strong noise scenario does not allow it to correctly segment the image. There is a lot of noise, with horizontal foreground streaks in addition to the correct segmented stars. This bad behavior is due to the algorithm's inability to adapt itself to different SN conditions. Here, a re-calibration of the algorithm threshold would provide a better segmentation quality for the mask. In the last image (d), the BOSS algorithm prediction shows the segmentation of all the brightest stars (red boxes) with just a little portion of noise in the most corrupted region. Both the LTA and the BOSS algorithm show some problems with the strong ghost noise conditions but in different regions of the image and in different ways: the BOSS algorithm seems to detect as stars the strong nonuniform brightest corrupted region due to the mesosphere winds in the image, while the LTA seems to have an opposite problem with the darker region in the top right part. The explanation for the BOSS algorithm is that the whitest spots in the ghost noise are recognized as possible embedded foreground objects, while in the LTA case, the gray level gradients in the darker localized region mislead the algorithm in properly segmenting the image. Among the two false positives cases, this LTA behaviour could greatly affect the output mask quality because the gradient in the darker region of the image (generally the 99 % of these samples) increases the percentage of noise in the output mask (with possible negative effects both for attitude determination routines and RSO detection).

- Figure 14: This is a frame from a video [45] from Juno's Star Reference Unit (SRU) camera. The image contains stars and a huge number of SEU's crossing the sensor with different impact angles as the Juno spacecraft crosses Jupiter's high radiation polar regions. SEUs over images are caused by ionizing radiation which hits the sensor pixels. The more perpendicular to the sensor plane the ionizing radiation is, the more point-like the footprint of the high energy particle on the sensor will be. This noise source can be reduced by shielding the electronics properly, but it cannot be removed. In the analysed image, many SEUs cross the sensor due to the high energetic region where the Juno spacecraft is. It is quite difficult to distinguish the white spots nature, but the purpose of this image processing is to show how the weakest elements on the detector (certainly SEUs under spacecraft inertial pointing condition) are removed. By a rapid inspection of the figures, it is possible to assess that LTA and LTS show similar segmentation outputs, and the SEUs which cross the detector almost tangentially are segmented as streaks. They are segmented by the LTA and LTS algorithms, while they are removed by BOSS. This is due to the dataset used to train the U-Net algorithm. The NN learned to detect just the the brightest star-like objects, filtering all the less salient other ones. Here, the weak streaks of the SEU are removed, but this does not mean that all the star-like objects which appear in the BOSS prediction of Figure 14 are stars; they can be that part of SEUs with a high impact angle.

These examples show that the BOSS algorithm has performances that are slightly lower but acceptable and generalized if compared to the calibrated traditional algorithms. Moreover, it has an intrinsic robustness to SEUs which a traditional algorithm does not normally have.

4. Discussion

Previous results have proved the high quality of night sky image segmentation provided by the algorithm. The unparametric nature of it makes the calibration operation unnecessary and maintains the high robustness to strong stray light variation scenarios. These features are welcome both for attitude determination and RSO detection through electro-optical sensors such as star sensors. Moreover, the absence of a calibration activity would make an optical camera based on this technology ready to work when the platform is released into its orbit. This algorithm could be used as an image processing main module for lost in space [36] or star tracking routines in common star sensors or RSO detection modules for onboard space surveillance camera or star sensors [40,46]. Both in the former and latter applications, good noise filtering capability is crucial for the success of the star and objects information extraction. This suggests a star sensor based on the U-Net image processing algorithm can be used to validate this technology in a real space mission and increase its technology readiness level (TRL) up to eight. This module will serve both the AD routines and the RSO detection modules. As the image is acquired by the optical head (OH), it is fed into the U-Net-based image segmentation module and then the mask will be used both by lost in space (LIS)/star tracking and RSO detection routines (Figure 15).

The selected components of the payload are the following:

- ARCSEC SPACE Sagitta Star Tracker as OH;
- PHOEBE board as Payload Processing Unit.

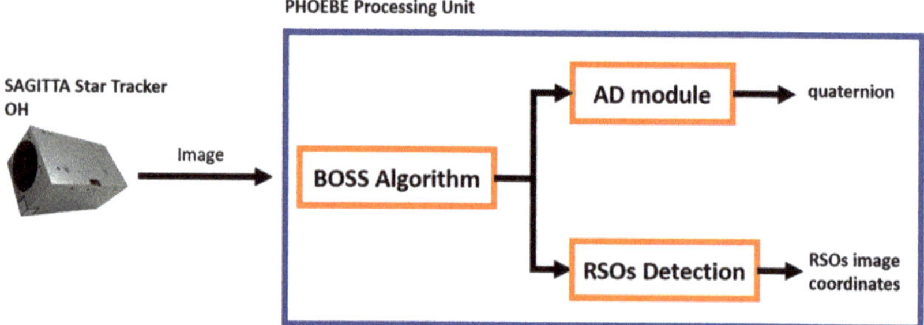

Figure 15. Payload architecture and the role of the BOSS algorithms (for the OH SAGITTA, Credits to ARCSEC SPACE).

4.1. ARCSEC SPACE SAGITTA Star Tracker

The SAGITTA Star Tracker from ARCSEC SPACE (Figure 16) was chosen due to its good features both for AD and RSO detection purposes. Its most important features for these purposes are the following:

- FOV: 24.8 deg (squared);
- Sensor Size: 2048 × 2048 px
- Cut Off Magnitude: 7;
- Working Frequency: Up to 10 Hz;
- Accuracy: 2 arc seconds (1 sigma) cross-boresight, 10 arc seconds (1 sigma) around boresight.

It is compact and suitable for microsatellites and cubesats missions with dimensions of 45 × 50 × 95 mm^3, a mass of 275 g and a low power consumption of 1.4 W. This device will be customized for the intended missions and used just as OH.

Figure 16. Sagitta Star Tracker, (Credits to ARCSEC SPACE).

4.2. PHOEBE Board

PHOEBE Board [47] (Figure 17) was used by School of Aerospace Engineering, (Sapienza University of Rome) for the STECCO (Space Travelling Egg-Controlled Catadioptric Object) mission. PHOEBE Board Features include the following:

- Microcontroller to handle data and tasks;
- FPGA for fast image data processing;
- Low power consumption: up to 0.3 W;
- Small Satellites Protocol: CAN, I^2C and RS485;
- Low Mass: 54 g;
- Dimensions: $99 \times 19 \times 29$ mm^3.

Figure 17. PHOEBE board, (Credits to School of Aerospace Engineering).

The algorithm will be implemented on the PHOEBE system on Chip/Field Programmable Gate Array (SoC/FPGA) hardware for taking advantage of the parallel computation capabilities of the FPGA which would boost the image processing time and make the proposed algorithm suitable for real-time applications. Indeed, as already demonstrated in another work [48], the SoC/FPGA hardware is able to reduce the processing time by a factor of 70 when compared with a personal computer based solution of the algorithm itself. At the moment, a personal computer processing time of this algorithm requires no more than 5 seconds per image. A materialization on the cited hardware would mean a realistic processing time, not higher than 100 ms, and thus a possibility of working with real 10 Hz star trackers and onboard cameras.

4.3. Small Satellite Mission Proposal

The payload will be able to process camera images up to 10 Hz. This would be good both for AD and RSO detection functions. It will have a total weight not higher than 330 g, a total power consumption not higher than 2 watts with a total volume lower than 0.3 U. It can be installed both on Cubesats and micro satellite platforms, and once in orbit, it can work without any limitations of pointing except for the luminosity conditions: Sun, Moon and Earth to be avoided within the FOV.

- The orbit: The RSOs spatial and mass distribution is strictly related to the former space activities and the collisions and fragmentation of orbiting objects. Several studies have described and modeled the debris evolution during the years, showing that LEO is the most densely populated region around the Earth, especially in 600–800 km altitudes with high inclinations [49]. For this reason, the onboard SST mission will be focused on LEO debris detection at a 600–800 km altitude range, and the orbit selection will be constrained by the sun illumination. To guarantee the correct detection functioning, the optical sensor (on-board camera or star sensor) cannot be oriented against the Sun. Figure 18 shows possible sun-illumination conditions. In particular, $\hat{d}_{s_i t_i}$ represent the target directions, with s_i denoting the observer satellite and t_i one of the debris elements. Satellite s_2 cannot detect target t_2 because it is backlit, while the sun illumination allows for a correct detection of target t_1 by satellite s_1. Another limiting factor to be considered is the presence of the Earth in the FOV, because of the consequent reduced FOV of the optical sensor. To guarantee the visibility of the target by the observer, the vector $\hat{d}_{s_i t_i}$ cannot cross the Earth. Finally, to ensure the same sun-illumination condition during the mission, the best candidate orbits for object detection are the Sun-Synchronous Orbits (SSOs). In particular, the dawn–dusk orbits can ensure a continuous and constant utilization of the optical sensors, while the noon/midnight orbits are not recommended for this kind of application due to the possibility of having eclipses. Due to all these considerations, the selected orbit for the mission will be a circular dawn–dusk SSO with a height of 700 km.
- The purpose: Detection of RSO populations in the LEO region will be with an onboard optical sensor and an AI-based image segmentation routine.
- The Validation: The satellite will be equipped with an attitude control system (ACS) capable of orienting the SAGITTA boresight direction in the along-track direction and anti-Sun direction to maximize the time the RSOs spend inside the FOV. Ground commands will be sent to the platform to schedule the pointing operation to switch between these two targeted directions. An input image will be stored together with the corresponding mask for validation purposes during a test phase. Once on the ground, the raw image will be processed and compared with the onboard produced mask. A match score based on Precision, Recall and F1 indices will be used to assess the correctness of the onboard payload routine. During normal operation, the payload will be responsible for the provision of the attitude for the onboard platform. The platform will be equipped with a reliable and independent attitude determination module to compare the payload outputs in term of quaternion.

The mission intended to be proposed has as a goal the detection of space debris in LEO. The payload will be mounted onto a 3U Cubesat, where an entire unit will be dedicated for our payload.

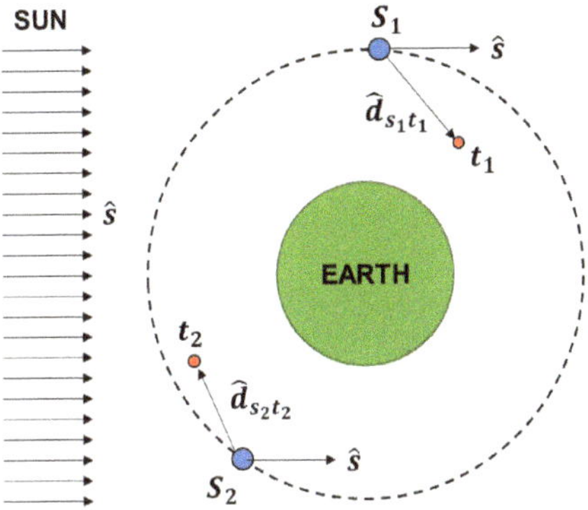

Figure 18. Visibility conditions.

5. Conclusions

In conclusion, this work has shown several aspects:

- The increment of initial filters in the U–Net increases the accuracy of the model predictions in terms of image segmentation quality and brightest objects detection.
- The BOSS algorithm is capable of achieving satisfying segmentation performance against different signal-to-noise scenarios. It does not need any re-calibration activity after its design in contrast to traditional LT segmentation algorithms.
- The comparison shows that the BOSS algorithm has segmentation performances which are comparable with respect to the optimized LT algorithms.
- The BOSS algorithm is able to properly remove the uniform stray light noise, most of the SEUs, weakest objects in an image and ghost noise and is able to provide a good product for star identification routines.
- A balanced CNN dataset for night sky images segmentation has been realized and provided.
- The algorithm works fine, both with a simulated star tracker and with real images of the night sky (both from ground and in orbit platforms).
- Once trained with images from specific star sensors, the algorithm does not need to be calibrated again.
- The simple structure in the algorithm makes it simple to analyze, to implement and to validate.
- Early versions of this algorithm proved to be suitable for RSO detection when working with real images [46].
- The optical head: cut off magnitude, FOV and working frequency were selected.
- The SoC/FPGA processing hardware was selected.
- An IOV mission for validating the BOSS algorithm against the AD and RSO detection purposes was proposed.

As future goals, the development of a real payload for a future IOV mission will be pursued:

- The algorithm will be implemented in the SoC/FPGA environment [50].
- We will design a buffer to store both RAW images and algorithm masks for their download and for on-ground checks of the algorithm behaviour in space.
- We will implement classical attitude determination routines and RSO detection modules [40,48] to assess the feasibility of the BOSS algorithm for these purposes.

Author Contributions: Conceptualization, M.M. and F.C.; methodology, M.M. and I.A.; validation, M.M. and I.A.; data curation, M.M. and I.A.; writing—original draft preparation, M.M. and I.A.; writing—review and editing, M.M., I.A. and F.C.; supervision, I.A. and F.C. All authors have read and agreed to the published version of the manuscript.

Funding: This research received no external funding and is conducted within the main author's Ph.D. research activity in astronomy astrophysics and space sciences XXXVI cycle.

Institutional Review Board Statement: Not applicable.

Informed Consent Statement: Not applicable.

Data Availability Statement: U-Net Dataset link [31].

Acknowledgments: The author and co-authors want to thank Cosimo Marzo of *Italian Space Agency "Centro di Geodesia Spaziale Giuseppe Colombo"* Observatory for having provided night sky image data both for building the dataset and testing the algorithms.

Conflicts of Interest: The authors declare no conflict of interest.

References

1. Xu, W.; Li, Q.; Feng, H.j.; Xu, Z.h.; Chen, Y.T. A novel star image thresholding method for effective segmentation and centroid statistics. *Optik* **2013**, *124*, 4673–4677. [CrossRef]
2. Liebe, C.C. Accuracy performance of star trackers-a tutorial. *IEEE Trans. Aerosp. Electron. Syst.* **2002**, *38*, 587–599. [CrossRef]
3. Rijlaarsdam, D.; Yous, H.; Byrne, J.; Oddenino, D.; Furano, G.; Moloney, D. Efficient Star Identification Using a Neural Network. *Sensors* **2020**, *20*, 3684. [CrossRef] [PubMed]
4. Xu, L.; Jiang, J.; Liu, L. RPNet: A Representation Learning-Based Star Identification Algorithm. *IEEE Access* **2019**, *7*, 92193–92202. [CrossRef]
5. Wang, Z.; Zhang, Y.L. Algorithm for CCD star image rapid locating. *Chin. J. Space Sci.* **2006**, *26*, 209–214.
6. Spiller, D.; Curti, F. A geometrical approach for the angular velocity determination using a star sensor. *Acta Astronautica* **2022**, *196*, 414–431. [CrossRef]
7. Bernsen, J. Dynamic thresholding of gray-level images. In Proceedings of the International Conference on Pattern Recognition, Berlin, Germany, January 1986.
8. Liu, D.; Yu, J. Otsu method and K-means. In Proceedings of the 2009 Ninth International Conference on Hybrid Intelligent Systems, Shenyang, China, 12–14 August 2009; Volume 1, pp. 344–349.
9. Niblack, W. *An Introduction to Image Processing*; Strandberg Publishing Company: Copenhagen, Denmark, 1986; pp. 115–116.
10. Fan, J.L.; Lei, B. A modified valley-emphasis method for automatic thresholding. *Pattern Recognit. Lett.* **2012**, *33*, 703–708. [CrossRef]
11. Jin, K.; Chen, Y.; Xu, B.; Yin, J.; Wang, X.; Yang, J. A Patch-to-Pixel Convolutional Neural Network for Small Ship Detection With PolSAR Images. *IEEE Trans. Geosci. Remote Sens.* **2020**, *58*, 6623–6638. [CrossRef]
12. Zhao, D.; Zhou, H.; Rang, S.; Jia, X. An Adaptation of Cnn for Small Target Detection in the Infrared. In Proceedings of the IGARSS 2018—2018 IEEE International Geoscience and Remote Sensing Symposium, Valencia, Spain, 22–27 July 2018; pp. 669–672. [CrossRef]
13. Fan, Z.; Bi, D.; Xiong, L.; Ma, S.; He, L.; Ding, W. Dim infrared image enhancement based on convolutional neural network. *Neurocomputing* **2018**, *272*, 396–404. [CrossRef]
14. Deng, L. The mnist database of handwritten digit images for machine learning research. *IEEE Signal Process. Mag.* **2012**, *29*, 141–142. [CrossRef]
15. Nasrabadi, N.M. DeepTarget: An Automatic Target Recognition Using Deep Convolutional Neural Networks. *IEEE Trans. Aerosp. Electron. Syst.* **2019**, *55*, 2687–2697. [CrossRef]
16. Shi, M.; Wang, H. Infrared dim and small target detection based on denoising autoencoder network. *Mob. Netw. Appl.* **2020**, *25*, 1469–1483. [CrossRef]
17. Tong, X.; Sun, B.; Wei, J.; Zuo, Z.; Su, S. EAAU-Net: Enhanced asymmetric attention U-Net for infrared small target detection. *Remote Sens.* **2021**, *13*, 3200. [CrossRef]
18. Xue, D.; Sun, J.; Hu, Y.; Zheng, Y.; Zhu, Y.; Zhang, Y. StarNet: Convolutional neural network for dim small target extraction in star image. In Proceedings of the 2018 IEEE Fourth International Conference on Multimedia Big Data (BigMM), Xi'an, China, 13–16 September 2018; pp. 1–7.
19. Xue, D.; Sun, J.; Hu, Y.; Zheng, Y.; Zhu, Y.; Zhang, Y. Dim small target detection based on convolutinal neural network in star image. *Multimed. Tools Appl.* **2020**, *79*, 4681–4698. [CrossRef]
20. Simonyan, K.; Zisserman, A. Very deep convolutional networks for large-scale image recognition. *arXiv* **2014**, arXiv:1409.1556.
21. Szegedy, C.; Vanhoucke, V.; Ioffe, S.; Shlens, J.; Wojna, Z. Rethinking the inception architecture for computer vision. In Proceedings of the IEEE Conference on Computer Vision and Pattern Recognition, Las Vegas, NV, USA, 27–30 June 2016; pp. 2818–2826.
22. Ronneberger, O.; Fischer, P.; Brox, T. U-net: Convolutional networks for biomedical image segmentation. In Proceedings of the International Conference on Medical Image Computing and Computer-Assisted Intervention, Munich, Germany, 5–9 October 2015; Springer: Berlin/Heidelberg, Germany, 2015; pp. 234–241.
23. Krizhevsky, A.; Sutskever, I.; Hinton, G.E. Imagenet classification with deep convolutional neural networks. *Adv. Neural Inf. Process. Syst.* **2012**, *25*, 1097–1105. [CrossRef]

24. Cortes, C.; Mohri, M.; Rostamizadeh, A. L2 Regularization for Learning Kernels. *arXiv* **2012**, arXiv:1205.2653.
25. Santurkar, S.; Tsipras, D.; Ilyas, A.; Madry, A. How does batch normalization help optimization? *arXiv* **2018** arXiv:1805.11604.
26. Mikołajczyk, A.; Grochowski, M. Data augmentation for improving deep learning in image classification problem. In Proceedings of the 2018 International Interdisciplinary PhD Workshop (IIPhDW), Świnoujście, Poland, 9–12 May 2018; pp. 117–122.
27. Srivastava, N.; Hinton, G.; Krizhevsky, A.; Sutskever, I.; Salakhutdinov, R. Dropout: A simple way to prevent neural networks from overfitting. *J. Mach. Learn. Res.* **2014**, *15*, 1929–1958.
28. Silburt, A.; Ali-Dib, M.; Zhu, C.; Jackson, A.; Valencia, D.; Kissin, Y.; Tamayo, D.; Menou, K. Lunar crater identification via deep learning. *Icarus* **2019**, *317*, 27–38. [CrossRef]
29. Iqbal, H. HarisIqbal88/PlotNeuralNet v1.0.0. 2018. Available online: https://zenodo.org/record/2526396#.Y9tKOepBxPY (accessed on 15 September 2021).
30. He, K.; Zhang, X.; Ren, S.; Sun, J. Delving deep into rectifiers: Surpassing human-level performance on imagenet classification. In Proceedings of the IEEE International Conference on Computer Vision, Santiago, Chile, 7–13 December 2015; pp. 1026–1034.
31. Mastrofini, M. nightskyUnet Repository. 2022. Available online: https://github.com/marco92m/nightskyUnet (accessed on 10 November 2022).
32. Zotti, G.; Wolf, A. Stellarium 0.19.0 User Guide. Technical Report. Available online: https://github.com/Stellarium/stellarium (accessed on 15 November 2022).
33. Curti, F.; Spiller, D.; Ansalone, L.; Becucci, S.; Procopio, D.; Boldrini, F.; Fidanzati, P. Determining high rate angular velocity from star tracker measurements. In Proceedings of the International Astronautical Conference, Jerusalem, Israel, 12–16 October 2015; pp. 1–13.
34. Curti, F.; Spiller, D.; Ansalone, L.; Becucci, S.; Procopio, D.; Boldrini, F.; Fidanzati, P.; Sechi, G. High angular rate determination algorithm based on star sensing. *Adv. Astronaut. Sci. Guid. Navig. Control* **2015**, *154*, 12.
35. Schiattarella, V.; Spiller, D.; Curti, F. Star identification robust to angular rates and false objects with rolling shutter compensation. *Acta Astronaut.* **2020**, *166*, 243–259. [CrossRef]
36. Schiattarella, V.; Spiller, D.; Curti, F. Efficient star identification algorithm for nanosatellites in harsh environment. *Adv. Astronaut. Sci.* **2018**, *163*, 287–306.
37. Schiattarella, V.; Spiller, D.; Curti, F. A novel star identification technique robust to high presence of false objects: The Multi-Poles Algorithm. *Adv. Space Res.* **2017**, *59*, 2133–2147. [CrossRef]
38. Otsu, N. A threshold selection method from gray-level histograms. *IEEE Trans. Syst. Man Cybern.* **1979**, *9*, 62–66. [CrossRef]
39. Bradley, D.; Roth, G. Adaptive Thresholding using the Integral Image. *J. Graph. Tools* **2007**, *12*, 13–21. [CrossRef]
40. Spiller, D.; Magionami, E.; Schiattarella, V.; Curti, F.; Facchinetti, C.; Ansalone, L.; Tuozzi, A. On-orbit recognition of resident space objects by using star trackers. *Acta Astronaut.* **2020**, *177*, 478–496. [CrossRef]
41. Kazemi, L.; Enright, J.; Dzamba, T. Improving star tracker centroiding performance in dynamic imaging conditions. In Proceedings of the 2015 IEEE Aerospace Conference, Big Sky, MT, USA, 7–14 March 2015; pp. 1–8.
42. Gonzalez, R.C.; Woods, R.E. *Digital Image Processing*; Prentice Hall: Hoboken, NJ, USA, 2002.
43. Chollet, F. *Deep Learning with Python*; Manning: Edmonton, AL, Canada, 2017.
44. Chapman, D.; Aboobaker, A.M.; Araujo, D.; Didier, J.; Grainger, W.; Hanany, S.; Hillbrand, S.; Limon, M.; Miller, A.; Reichborn-Kjennerud, B.; et al. Star camera system and new software for autonomous and robust operation in long duration flights. In Proceedings of the 2015 IEEE Aerospace Conference, Big Sky, MT, USA, 7–14 March 2015; pp. 1–11.
45. NASA-Juno Star Reference Unit Camera. Available online: https://www.jpl.nasa.gov/images/pia24436-high-energy-and-junos-stellar-reference-unit (accessed on 11 January 2022).
46. Mastrofini, M.; Goracci, G.; Agostinelli, I.; Salim, M. Resident Space Objects Detection And Tracking Based On Artificial Intelligence. In Proceedings of the Astrodynamics Specialist Conference AAS/AIAA, Charlotte, NC, USA, 7–11 August 2022.
47. PHOEBE, School of Aerospace Engineering. Available online: https://sites.google.com/uniroma1.it/stecco-sia/home (accessed on 20 November 2022).
48. Farissi, M.S.; Mastrofini, M.; Agostinelli, I.; Goracci, G.; Curti, F.; Facchinetti, C.; Ansalone, L. Real-Time Image Processing Implementation For On-Board Object Detection And Tracking. In Proceedings of the Astrodynamics Specialist Conference AAS/AIAA, Charlotte, NC, USA, 7–11 August 2022.
49. Schaub, H.; Jasper, L.E.; Anderson, P.V.; McKnight, D.S. Cost and risk assessment for spacecraft operation decisions caused by the space debris environment. *Acta Astronaut.* **2015**, *113*, 66–79. [CrossRef]
50. labis7/UNET-FPGA. Available online: https://github.com/labis7/UNET-FPGA (accessed on 20 October 2022).

Article

Elevation Angle Characterization for LEO Satellites: First and Second Order Statistics

Juan Misael Gongora-Torres *,†, **Cesar Vargas-Rosales** †, **Alejandro Aragón-Zavala** †
and **Rafaela Villalpando-Hernandez** †

Tecnologico de Monterrey, School of Engineering and Science, Monterrey 64849, Mexico
* Correspondence: misael.gongora@tec.mx
† These authors contributed equally to this work.

Abstract: The elevation angle θ is relevant for the Low Earth orbit (LEO) satellite communications since it is always changing its relative position with respect to fixed Earth stations (ES's), and this affects the link length and received power, P_R. This article provides a new methodology to compute the probability density function (PDF) and cumulative distribution function (CDF) of the elevation angle, θ, for diverse ES locations. This methodology requires as input parameters an ES latitude, ϕ, an orbit inclination value, i, and an orbit altitude, h. The elevation angle is characterized through a well known random variable, which facilitates the computation of the first and second-order statistics, and helps to determine the expected value and measures of dispersion of the angle θ for a particular ES location. The proposed methodology allows an easy and quick calculation of the elevation angle's CDF, facilitating comparisons against CDF's of more ES's located at different latitudes, and longitudes, λ; as well as the comparisons of CDF's of the elevation angle produced by different orbits. Extensive simulation results are summarized in a small table, which allows computation of the elevation angle's CDF and PDF for multiple ES locations without requiring of simulations and statistical fitting. Finally, the proposed methodology is validated through an extensive error analysis that show the suitability of the obtained results to characterize the elevation angle.

Keywords: Low Earth orbit (LEO); elevation angle; satellite communications

Citation: Gongora-Torres, J.M.;
Vargas-Rosales, C.; Aragón-Zavala,
A.; Villalpando-Hernandez, R.
Elevation Angle Characterization for
LEO Satellites: First and Second
Order Statistics. *Appl. Sci.* **2023**, *13*,
4405. https://doi.org/10.3390/
app13074405

Academic Editors: Simone Battistini,
Filippo Graziani and Mauro Pontani

Received: 26 February 2023
Revised: 18 March 2023
Accepted: 21 March 2023
Published: 30 March 2023

1. Introduction

The characterization of the elevation angle, θ is relevant for low Earth orbit (LEO) satellite communications since this parameter is directly related to the varying distance between the satellite and Earth station (ES), and affects the link total attenuation. The elevation angle description through an analytical expression is a difficult problem addressed in [1,2] which has received less attention in the literature.

Nonetheless, this parameter is directly related to the link performance and channel characterization of LEO satellites. The LEO channel characterization has also received less attention than the geostationary (GEO) satellite channel and just few models such as [3] have been specifically developed and published to consider the elevation angle variations introduced at LEO. The lack of LEO channel models accounting for the always-changing elevation angle have resulted in just a few channel models available in the literature for LEO satellites and specially for small satellites as mentioned in [4].

LEO satellites are increasing their numbers and role as an enabling technology for Internet of Things (IoT), 5G, 6G, and next generation wireless networks aimed to provide global coverage with very low latency [5]. Then it is relevant to develop methodologies to analyze and compare the link performance and channel characteristics considering the variations introduced by the always-changing elevation angle.

The always-changing elevation angle condition has been a limitation to analyze the link budget and channel of LEO satellites, and common approaches to characterize it have followed segmentation in best and worst-case of the elevation angle [6], instead of analyzing the short and long term behavior of the elevation angle as an analytical function. However, some emerging problems, such as efficient power management [7], related to LEO satellites have made evident the necessity of having a way to characterize the elevation angle as an analytical function.

The observed elevation angle, θ, is always-varying for LEO satellites and can be described as a function of time, $\theta(t)$, for intervals in which a satellite is visible. The calculation of θ for a given position can be deterministically obtained with mathematical procedures developed in orbital mechanics books [8] and widely implemented in software for space dynamics simulations. However, it is important to note that even with accurate predictions of θ (accounting for contacts in several days or months), the relations between one contact and another, as well as the long term behavior of the elevation angle can be hardly described without randomness. Thus, appropriate use of probability theory becomes a great tool to analyze the behavior of the elevation angle, θ, in the long term.

The elevation angle, θ, is usually defined for an ES as the angle (above the local horizon) at which the satellite is visible, and within the interval of $0°$ to $90°$, $\theta \in (0°, 90°]$, regardless of its azimuth. The minimum value of θ at which communication is possible is often called θ_{min} (subject to $\theta_{min} \geq 0°$), similarly, the maximum value of θ that can be observed from the ES is called θ_{max} (subject to $\theta_{min} \leq \theta_{max} \leq 90°$). We can define a random variable, r.v., to take possible values of θ, such that $\theta \in (\theta_{min}, \theta_{max}]$ as Θ.

The probability density function, PDF, and cumulative distribution function, CDF, of the elevation angle, $f_\Theta(\theta)$ and $F_\Theta(\theta)$, respectively, are useful functions to characterize the LEO channel since the elevation angle affects the received power level, P_R. The reasoning behind that is that θ depends on the distance from the satellite to the ES, $r_{S,E}$. Then, at greater link distances, it can be expected to have a lower received power, P_R, (of the transmitted signal from the satellite) than at shorter distances.

The distance between a LEO satellite and an ES, $r_{S,E}$, can vary several thousands of kilometers from a low value of θ to a high value of θ. Figure 1a shows the extreme cases for the elevation angle in a LEO satellite link. Those extreme cases are not necessarily met at every contact, but represent the best and worst length-scenarios in a long period. Path lengths between the ES and satellite, $r_{S,E}$, are described in this figure using the variable r_i, $i \in \{1, 2, 3\}$. The figure also shows that at low values of θ the path length is larger and the received power, P_R is at its minimum value. This last implication between the elevation angle, θ, the received power, P_R, and the link length, $r_{S,E}$, assumes some constraints explained in detail in [9].

Figure 1b shows the typical-case for the elevation angle, where the value of θ is not necessarily at its minimum nor at its maximum, but it can be at any value within the range $\theta \in [\theta_{min}, \theta_{max}]$. Since the elevation angle of a LEO satellite appears as always-varying from an ES's, it is convenient to determine statistical indicators of its behavior, such as its expected value, $E[\cdot]$, median, $MED[\cdot]$, standard deviation, $SD[\cdot]$; and how often does the elevation angle will be above or below a threshold, or within a region of interest, for example, using its quantiles nQ. In addition, since the elevation angle variations are directly related to the link length, the elevation angle for LEO satellites is also directly related to the variations of the received power P_R at an ES, as described in [9].

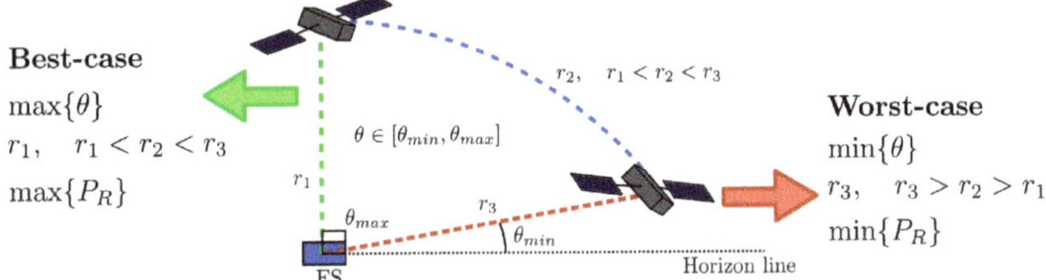

(**a**) Extreme cases for the elevation angle, θ, in a satellite link

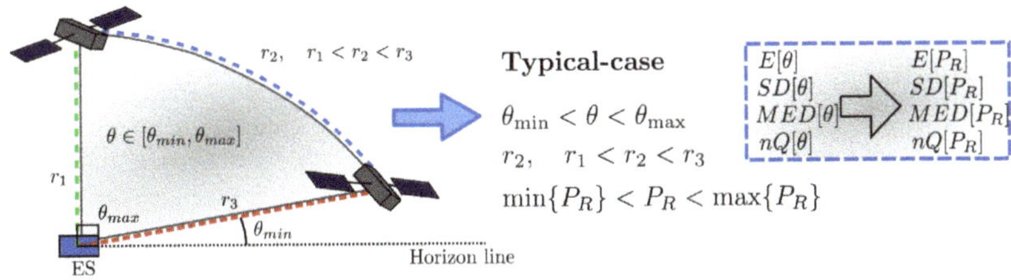

(**b**) Typical-case for the elevation angle

Figure 1. Elevation angle cases for LEO satellite systems: (**a**) The values of θ_{min} and θ_{max} are usually related to the best and worst case of the received power, P_R, at an ES; nonetheless, those cases are rare and represent the extremes of θ. (**b**) The elevation angle of a LEO satellite system as observed from an ES is always-varying; then, it is more convenient to analyze its behavior through statistical tools.

1.1. Contributions

The elevation angle characterization is relevant for LEO satellites because it is an always-changing variable for those communication systems. The effects of the elevation angle variations are observable in the received power, link quality, and channel behavior; then, an accurate characterization of this variable is a topic of interest for planning and implementation of LEO satellite systems.

In this article, we have addressed the characterization of the elevation angle for LEO satellites, by obtaining its PDF and CDF, as well as the derivation of its first and second order statistics. First and second order statistics are relevant to evaluate the suitability of different LEO orbits and to determine which orbits are more convenient to provide coverage for a particular application or Earth station location.

In addition, we have developed an extensive analysis to validate our results, and we include supplementary materials containing elevation angle times series. The supplementary materials will facilitate reproducibility of our work, and will allow future research based on our proposed methodology and results.

1.1.1. Contribution 1

This document shows the feasibility of using a random variable to characterize the elevation angle behavior for LEO satellites with different orbit configurations. The suitability of using the proposed random variable is verified through an extensive error analysis with diverse orbit configurations.

1.1.2. Contribution 2

The PDF and CDF of the elevation angle can be obtained as proposed in [1,2]. Nonetheless, this document describes a methodology to obtain the PDF and CDF parameters of the elevation angle distribution for different orbit configurations and ES locations using a well-known random variable. The proposed random variable facilitates analytical manipulation as well as computation of the probabilities of occurrence of the elevation angle at specific values. A small table containing the resultant parameters to characterize multiple LEO orbits configurations as observed from multiple ES's is included.

1.1.3. Contribution 3

The expected value of the elevation angle and its measures of dispersion are relevant for planning and implementation of LEO satellite systems, since the received power varies according to the elevation angle. This document describes an analytical methodology to obtain the first and second order statistics of the elevation angle for different orbit configurations and ES locations, which, to the best of our knowledge are not available in the literature and were just utilized for a specific case in [9].

1.2. Outline

The remaining of the article is organized as follows: Section 2 introduces the theoretical fundamentals required to understand the subsequent sections; Section 3 describes the developed methodology to characterize the elevation angle behavior through $f_\Theta(\theta)$ and $F_\Theta(\theta)$; Section 4 contains the main results; and Section 5 concludes analyzing the obtained results and opportunities for future work. Figure 2 shows the main structure of the document.

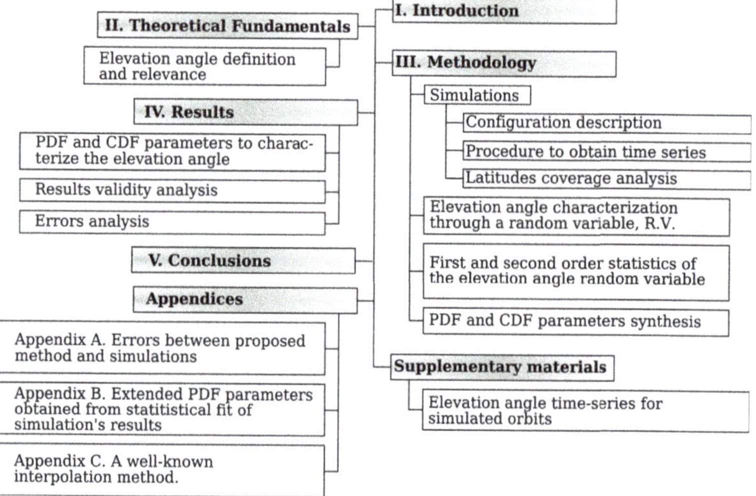

Figure 2. Basic structure and contents of this article.

2. Materials and Methods

2.1. Elevation Angle Definition

The elevation angle, θ, for an ES can be defined as the angle between the local horizon and the satellite. Several sources such as [10–12] contain expressions to calculate θ from geometrical relations between the satellite and the ES instantaneous positions. One of those definitions is as follows

$$\theta = \arctan\left(\frac{\cos\Delta\cos\phi_{ES} - (r_{E,O}/r_{S,O})}{\sqrt{1 - \cos^2\Delta\cos^2\phi_{ES}}}\right),\tag{1}$$

where $r_{E,O}$ and $r_{S,O}$ are the distances from the center of the Earth, O, to the ES and the satellite, respectively; ϕ_{ES} is the latitude at which the ES is located (in degrees); M is the subsatellite point, which corresponds to the latitude and longitude of the satellite instantaneous position; and Δ is the difference in longitude between the ES and M (in degrees). Figure 3a shows a derivation of θ based on (1).

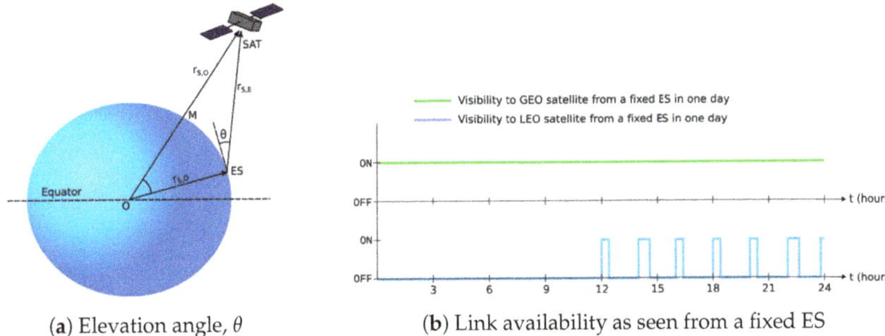

(**a**) Elevation angle, θ (**b**) Link availability as seen from a fixed ES

Figure 3. LEO characteristics: (**a**) Graphical representation of the always-varying elevation angle, θ, and (**b**) an On-Off model showing the link availability for a LEO satellite vs a GEO satellite during a random day.

2.2. Relevance of the Elevation Angle for the Satellite Channel

LEO satellites differ to those at GEO in its relative position as seen from a fixed ES. Whereas GEO satellites appear as fixed, those at LEO appear as always-moving points in the sky or are absent causing link unavailability. Link availability can be seen as an ON-OFF process, where the link is ON if the satellite is visible in the sky, and it is OFF if the satellite is absent. Figure 3b shows the ON-OFF process for both a LEO and GEO satellite.

In addition to the short visibility that LEO satellites have, conditions in this lapse are always varying as a consequence of the changing position of the satellite. When the satellite is first visible from an ES, it is at the largest distance supported by the link, then, the satellite starts approaching the ES until reaching the shortest distance in that particular contact. Finally, the satellite moves away after being at its minimum contact distance. It is important to note that the minimum distance from an ES to a LEO satellite, $\min(r_{E,S})$, varies from one contact to another, as well as $\max(r_{E,S})$ does.

Differences between the Earth's rotation rate and LEO orbiting velocities cause an always-varying link length (as seen from a fixed ES), and then, an always-varying elevation angle.

An important relation that needs to be accounted for the elevation angle, θ, and the link length, $r_{S,E}$ is given by the implication that when the angle is minimum, the $r_{S,E}$ distance is maximum, i.e.,

$$\theta \to \min(\theta) \iff r_{S,E} \to \max(r_{S,E}), \tag{2}$$

similarly

$$\theta \to \max(\theta) \iff r_{S,E} \to \min(r_{S,E}); \tag{3}$$

where the maximum link length, $\max(r_{S,E})$, occurs at some value of θ such that $\min(\theta) < \theta \leq 90°$; and $\min(r_{S,E})$ occurs at some value of $\theta \geq 0°$ lower than θ_{max}. The implications in (2) indicate that when values of θ diminish, $r_{S,E}$ increases; similarly, in (3), when values of θ increase then $r_{S,E}$ decreases.

As a satellite moves away from an ES, $r_{S,E}$ and the free-space path loss, L_{FS}, increases. Longer paths occur for lower values of θ, then, the atmospheric attenuation, A_{Atm}, increases for larger paths (assuming similar atmospheric conditions for different link paths). On the other hand, when θ takes values close to $90°$, $r_{S,E}$ will be at its minimum, and so does A_{Atm}. Then, low elevation angles cause a greater L_{FS}, and more atmospheric attenuation, A_{atm} (assuming that the atmospheric conditions are approximately the same for two distinct paths).

For LEO it is important to consider link interference, I, usually addressed as noise in link-budget calculations, and also dependent on the elevation angle as mentioned in [13]. Furthermore, low values of θ are also associated in practice with non-line-of-sight (NLOS) conditions, increasing ground interference [14], and increasing multipath fading in the land mobile satellite (LMS) channel [15–17].

There are several models available in the literature to characterize the received signal from satellite systems. Most of these works have been elaborated for the LMS channel, which describes the received signal at a land-moving ES; nonetheless, most of those works focus on GEO systems. Extensive reviews for the LMS channel can be found at [16,18,19].

Among the available channel models for LMS systems, a few of them focus on characterizations for LEO and non-GEO (NGEO) satellites; but, as noted by [4,20], those are few and there are still many challenges. Most of channels for NGEO systems have been developed based on previous models for GEO satellites, specially on well-known models, such as those by Loo [21] and Lutz [22].

Even though the Loo and Lutz models were not originally intended for LEO satellites, those were later used as a base for much of the non-geosynchronous (NGSO) channel models. The Loo channel assumes that the complex envelope of the received signal, r_T, contains a LOS, r_D, and a multipath component, r_M, which in its phasor notation can be represented as

$$r_T \exp(j\phi_T) = r_D \exp(j\phi_D) + r_M \exp(j\phi_M) \tag{4}$$

where ϕ_T, ϕ_D, and ϕ_M indicate the phase of the total received signal, of the direct or LOS component, and of the multipath component, respectively.

The channel developed by Lutz [22] models a varying received signal affected by different levels of shadowing. This channel is characterized by a Markov chain with a state corresponding to light shadowing, G, and another corresponding to deep shadowing conditions B. Figure 4 illustrates the Lutz channel.

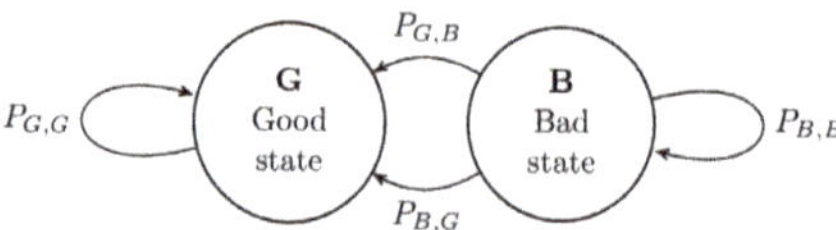

Figure 4. First order Markov chain illustrating the channel model developed by Lutz [22].

The Loo channel [21] was adapted by Corazza and Vatalaro [3] to formulate one of the best-known channel models for LEO satellites; similarly, it was modified by Abdi et al. [23], to achieve straightforward analytical expressions for the first and second order statistics of the received signal based on the Nakagami distribution. However, current models for GEO and NGEO systems have recognized the limitations of a single distribution and a single state to describe the received signal, and mutistate models with different distributions at each state have became popular.

Channel models for LEO with multiple states as in [22] have been developed containing the same kind of distribution at each state. For example, a combination of the Lutz and Loo models was implemented for NGSO by Perez-Fontan et al. [24], focusing on time series generation. This model used a three-state Markov chain, as shown in Figure 5, to indicate different levels of shadowing; the first state, S_1, indicates deep shadow; the second, S_2, indicates moderate shadowing; and the third state, S_3, indicates line-of-sight (LOS) conditions. Each state describes the received signal according to a Loo distribution with a different Loo triplet, α, ψ, MP; and a different Markov chain for distinct elevation angle values. The so-called Loo triplet, which includes the mean of the direct signal amplitude, α, the standard deviation of the direct signal amplitude, ψ, and the multipath power, MP, is the set of parameters required by the Loo distribution.

Figure 6 shows two Loo distributions to characterize the satellite channel at two elevation angles; the CDF's were computed assuming presence of a LOS component, and a multipath power component very close to the LOS level. The Loo distributions in Figure 6 can be understood as two of the states of Figure 5 at different elevation angles; e.g., a state transition from S_{3,θ_1} to S_{3,θ_2}.

In addition to the shadowing level at a receiver environment, which is mainly determined by natural and human made objects at the surroundings of the ES, the received signal level will be changing according to the elevation angle value. For LEO satellites, the elevation angle will change rapidly, and with a different rate at every contact.

As observed in Figure 6, the Loo CDF curves of the received signal rely on the elevation angle value since those depend on the LOS component of the received signal, α. Then, in order to predict different curves we need to know the expected value and measures of dispersion of the elevation angle.

Even though we illustrate the received signal through the well-known Loo model in Figure 6, there are more LEO channel models depending on a line-of-sight component, and thus, directly depending on the elevation angle. The, elevation angle characterization proposed in this article can be a valuable tool to determine CDF curves of the received signal, based on the first and second order order statistics of the elevation angle. For example, to determine the expected CDF of the Loo model based on the elevation angle expected value, and determine how far apart other curves will be based on the measures of dispersion of the elevation angle.

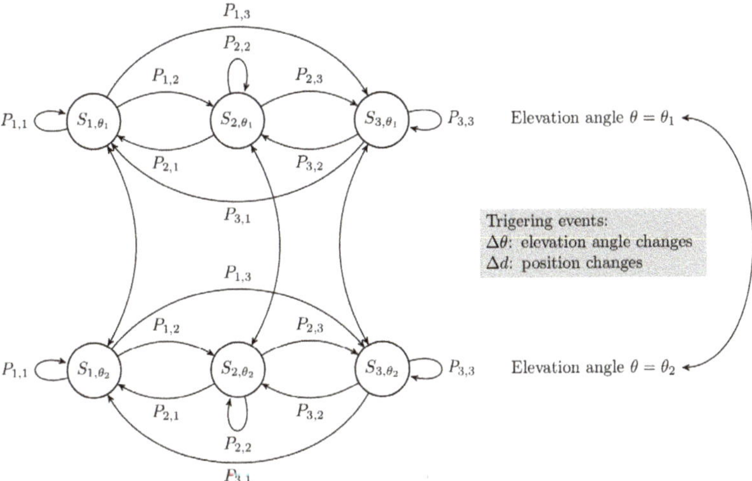

Figure 5. Generalization of a three-state channel model as proposed by [24] considering state transitions based on elevation angle changes.

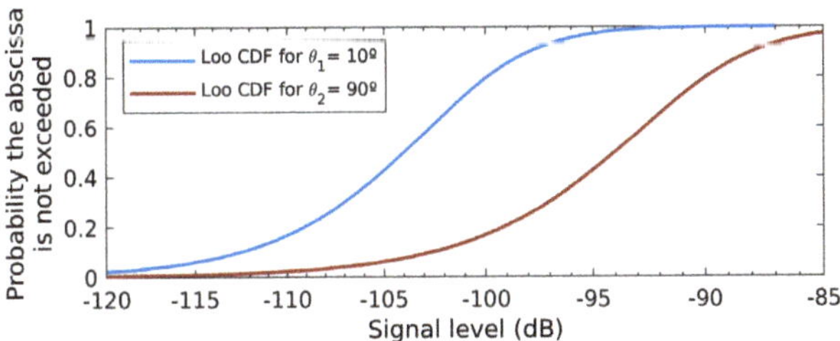

Figure 6. Loo CDF of the received signal for two elevation angle values. The Loo triplet for $\theta_1 = 10°$ is $\alpha = -105$ dB, $\psi = 5$ dB, and $MP = -108$ dB; and $\alpha = -95$ dB, $\psi = 5$ dB, and $MP = -98$ dB for $\theta_2 = 90°$.

2.3. Analytical Characterization of the Elevation Angle

The elevation angle was analytically characterized in [2] through its probability density function (PDF), f_Θ, which is defined as a marginal distribution from the joint PDF of Θ and the maximum value of the elevation angle, Θ_{max}, as follows

$$f_\Theta(\theta) = \int_\theta^{\theta_M} f_{\Theta,\Theta_{max}}(\theta, \theta_{max}) d\theta_{max} \tag{5}$$

where $f_{\Theta,\Theta_{max}}$ is given by

$$f_{\Theta,\Theta_{max}}(\theta,\theta_{max}) = f_{\Theta|\Theta_{max}}(\theta|\theta_{max})f_{\Theta_{max}}(\theta_{max}) \tag{6}$$

or equivalently

$$f_{\Theta,\Theta_{max}}(\theta,\theta_{max}) = \frac{G(\theta)\sin\gamma(\theta)}{\sqrt{\cos^2\gamma(\theta_{max}) - \cos^2\gamma(\theta)}} \cdot \frac{f_{\Theta_{max}}(\theta_{max})}{\int_{\theta_{min}}^{\theta_M} f_{\Theta_{max}}(x)\cos^{-1}\left(\frac{\cos\gamma(\theta_{min})}{\cos\gamma(x)}\right)dx} \tag{7}$$

The integral dividing the right hand term is a constant, say C_1, then, (7) can be rewritten as

$$f_{\Theta,\Theta_{max}}(\theta,\theta_{max}) = \frac{f_{\Theta_{max}}(\theta_{max})G(\theta)\sin\gamma(\theta)}{C_1\sqrt{\cos^2\gamma(\theta_{max}) - \cos^2\gamma(\theta)}} \tag{8}$$

and the auxiliary functions $\gamma(\cdot)$ and $G(\cdot)$ require an extra parameter a defined as $\alpha = r_E/r_S$, where $r_E = 6378$ km is the radius of the Earth; and $r_S = r_E + h$ depends on the altitude h (from the Earth's surface) of the circular LEO orbit in kilometers. The functions $\gamma(\cdot)$ and $G(\cdot)$ are defined as follows

$$\gamma(\theta) = \cos^{-1}(\alpha\cos\theta) - \theta \tag{9}$$

$$G(\theta) = \frac{1 + \alpha^2 - 2\alpha\cos\gamma(\theta)}{1 - \alpha\cos\gamma(\theta)} \tag{10}$$

The term $f_{\Theta_{max}}(\theta_{max})$ is defined as follows

$$f_{\Theta_{max}}(\theta_{max}) = \frac{G(\theta_{max})}{K_2} \cdot f_\Phi(\phi_0 - \gamma(\theta_{max})), \quad \text{for } \theta_{min} \le \theta_{max} < \theta_c \tag{11}$$

$$f_{\Theta_{max}}(\theta_{max}) = \frac{G(\theta_{max})}{K_2} \cdot [f_\Phi(\phi_0 - \gamma(\theta_{max}) +$$
$$f_\Phi(\phi_0 + \gamma(\theta_{max}))], \quad \text{for } \theta_c < \theta_{max} < \theta_M \tag{12}$$

where $f_\Phi(\phi)$ is defined from the orbital inclination i; and ES latitude, ϕ, as follows

$$f_\Phi(\phi) = \frac{\cos\phi}{\pi\sqrt{\sin^2 i - \sin^2\phi}}, \quad \text{for } |\phi| < i \tag{13}$$

and $\theta_M = \pi/2$, and K_2 is given by

$$K_2 = \frac{1}{2} - \frac{1}{\pi}\sin^{-1}\left(\frac{\sin(\phi + \gamma(\theta_{min}))}{\sin(i)}\right) \tag{14}$$

where θ_{min} is the minimum elevation angle that will be considered and is in the range $0° < \theta_{min} < 90°$.

Although results shown in [2] are highly accurate with respect to the actual elevation angle PDF, the analytical expression f_Θ is cumbersome to obtain and evaluate. Furthermore, analytical expressions for second or higher order statistics, are a missing result in the literature to the best of our knowledge.

3. Methodology for the Elevation Angle Characterization

In this section, we describe the procedure to characterize the elevation angle behavior through a random variable in a manageable analytical expression. Also, we include a procedure to obtain useful expressions of the elevation angle, such as measures of central tendency and dispersion.

The elevation angle characterization for different LEO orbit's configuration will help to predict which orbits are more convenient for a particular ES location. In addition, the measures of central tendency and dispersion will help to choose suitable orbits based on the expected elevation angle, and to calculate the link attenuation based on the distribution of the elevation angle, and on its expected value and variance.

3.1. Simulations Configuration

Simulations were performed for a LEO satellite with characteristics mentioned in Table 1, and for circular orbit configurations listed in Table 2. The LEO region goes from a few hundred kilometers above the Earth's surface up to 2000 km; but the performed simulations and methodology cover just the upper part of this region since orbit perturbations effects are much lower at those altitudes than at orbital heights closer to the surface, and facilitates operation for longer periods of time without a complex propulsion system.

Table 1. Satellite characteristics for the ephemeris generation.

Dry Mass	Drag Area	Solar Radiation Pressure Area
5 kg	1 m^2	1 m^2

Three initial values of orbital parameters not included in Table 2 are right ascension of the ascending node, Ω, argument of perigee, ω, and true anomaly, ν, which were all set to zero. From those simulations, ephemeris files with the satellite position and velocity in five-second steps were generated. Then, ES's were placed (simulated) at 18 different and arbitrarily chosen locations listed in Table 3. Finally, time series of the observed elevation angle from the ES to the satellite were calculated for each ground location using the previous generated ephemeris files and geometrical relations as in (1).

The orbital mechanical equations required to calculate the position and velocity of the satellite as a function of time, were solved using an open source program developed and maintained by the NASA, GMAT [25]. The time series for the elevation angle discussed in this paper can be generated in GMAT using the data of Tables 1–3; also, these can be generated using other software, for example STK as in [2]. The time series mentioned in this paper were generated by the authors specially for this work, and are available at [26].

Table 2. Orbit characteristics for the ephemeris generation.

	Semi-Major Axis, *a*	Orbit Inclinations, *i*
1	7378 km	20°, 25°, ..., 85°
2	7578 km	20°, 25°, ..., 85°
3	7778 km	20°, 25°, ..., 85°
4	7978 km	20°, 25°, ..., 85°
5	8178 km	20°, 25°, ..., 85°
6	8378 km	20°, 25°, ..., 85°

Table 3. ES locations for the performed simulations specified through their latitude, ϕ, and longitude, λ.

	Latitude	Longitude		Latitude	Longitude
ES_1	0°	276.7121°	ES_{10}	45°	259.7121°
ES_2	5°	276.7121°	ES_{11}	50°	259.7121°
ES_3	10°	276.7121°	ES_{12}	55°	259.7121°
ES_4	15°	267.7121°	ES_{13}	60°	259.7121°
ES_5	20°	276.7121°	ES_{14}	65°	259.7121°
ES_6	25°	261.7121°	ES_{15}	70°	265.7121°
ES_7	30°	259.7121°	ES_{16}	75°	265.7121°
ES_8	35°	259.7121°	ES_{17}	80°	265.7121°
ES_9	40°	259.7121°	ES_{18}	-85°	259.7121°

3.2. Elevation Angle Time Series Analysis

Each elevation angle data set (obtained from the simulations) was individually analyzed to observe its statistical values of interest (mean, median, standard deviation, quantiles). Those time series were fitted to several distributions using the maximum likelihood estimation method. The goodness of fit of the proposed distributions was evaluated for each elevation angle time series using the Kolmogorov-Smirnov test, and it was observed a better performance of the gamma distribution, $Gamma(a, b)$, for most of the data sets. Other distributions that showed good fit based on the Kolmogorov-Smirnov test were the beta, $B(a, b)$, and Weibull, $Weib(a, b)$. Figure 7 shows the values of θ as well as the PDF and CDF for one elevation angle time series, $\theta(t)$ using the proposed distributions. Figure 8 shows the Kolmogorov-Smirnov statistic, often indicated as D, for the time series obtained for different altitudes. The Kolmogorov-Smirnov statistic is an indicator of the maximum distance between the actual time series distribution and the proposed gamma distribution. A low value of the Kolmogorov-Smirnov is often desirable to indicate a better goodness-of-fit.

The gamma PDF, $f_\Theta(\theta)$, and CDF, $F_\Theta(\theta)$, are defined as follows

$$f_\Theta(\theta) = \frac{1}{b^a \Gamma(a)} \theta^{a-1} \exp(-\theta/b),\tag{15}$$

and

$$F_\Theta(\theta) = \frac{1}{\Gamma(a)} \gamma\left(a, \frac{\theta}{b}\right),\tag{16}$$

respectively, where $\Gamma(\cdot)$ is the gamma function, $\gamma(\cdot)$ is the incomplete gamma function, a is the shape parameter, and b is the scale parameter. By definition both a and b are greater than zero.

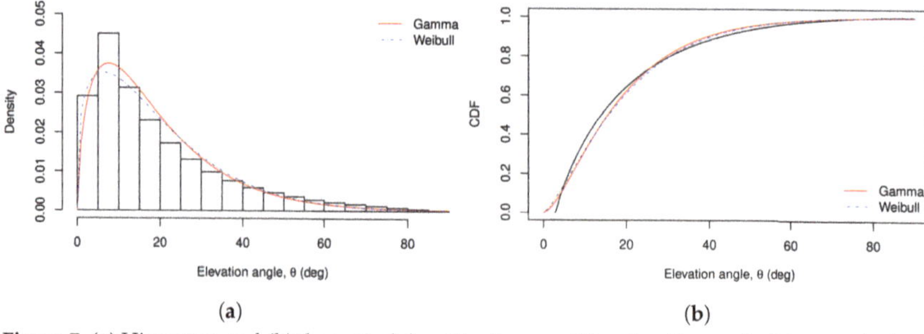

Figure 7. (**a**) Histogram and (**b**) theoretical densities for one of the elevation angle data sets obtained from the performed simulations.

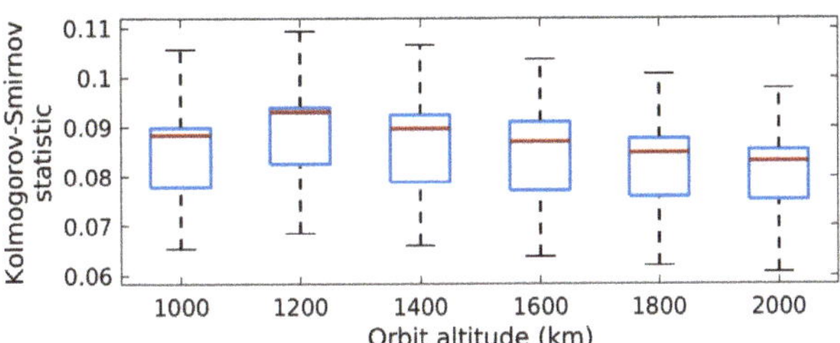

Figure 8. Kolmogorov-Smirnov statistic for the performed simulations at different altitudes.

The shape, a, and scale, b, parameters for the gamma distribution were obtained for each of the elevation angle time series of each ES. Those parameters were organized into several matrix arrays as shown in Table 4, which contains the shape parameter for the simulated orbits as seen from the ES_1, and as in Table 5, which contains the scale parameter also for ES_1.

Table 4. Rate parameter, a, of the gamma distribution to characterize the elevation angle distribution at ES_1, for different altitudes, h, and orbit inclinations, i.

ES_1	20°	25°	30°	35°	40°	45°	50°	55°	60°	65°	70°	75°	80°	85°
7378 km	1.39	1.50	1.48	1.31	1.30	1.34	1.35	1.35	1.35	1.37	1.37	1.36	1.36	1.37
7578 km	1.62	1.79	1.80	1.65	1.56	1.61	1.62	1.63	1.63	1.64	1.64	1.64	1.64	1.64
7778 km	1.57	1.77	1.82	1.73	1.55	1.61	1.63	1.64	1.64	1.65	1.65	1.66	1.66	1.66
7978 km	1.63	1.74	1.83	1.79	1.62	1.60	1.63	1.65	1.65	1.66	1.66	1.66	1.66	1.67
8178 km	1.70	1.66	1.83	1.82	1.70	1.55	1.63	1.65	1.66	1.67	1.67	1.67	1.70	1.67
8378 km	1.75	1.67	1.82	1.84	1.75	1.58	1.63	1.66	1.67	1.68	1.68	1.68	1.68	1.68

Table 5. Scale parameter of the gamma distribution, $b \times 10^{-2}$, to characterize the elevation angle distribution at ES_1, for different altitudes, h, and orbit inclinations, i.

ES_1	20°	25°	30°	35°	40°	45°	50°	55°	60°	65°	70°	75°	80°	85°
7378 km	7.51	8.71	9.34	8.88	8.40	8.42	8.40	8.36	8.34	8.39	8.38	8.35	8.35	8.36
7578 km	7.86	9.13	9.91	9.74	9.05	9.02	8.99	8.97	8.93	8.95	8.94	8.93	8.91	8.91
7778 km	7.45	8.55	9.41	9.56	8.83	8.73	8.70	8.67	8.64	8.66	8.64	8.64	8.63	8.64
7978 km	7.41	8.08	8.98	9.33	8.88	8.48	8.46	8.44	8.40	8.41	8.39	8.39	8.38	8.39
8178 km	7.38	7.59	8.60	9.07	8.88	8.18	8.26	8.24	8.22	8.22	8.20	8.18	8.26	8.18
8378 km	7.34	7.47	8.27	8.82	8.83	8.21	8.09	8.08	8.06	8.05	8.03	8.02	8.00	8.02

3.3. Orbit Coverage

Figure 9a illustrates (without scale) passes of a satellite in each one of the six altitudes and for all the inclinations mentioned in Table 2. This figure also shows dashed lines indicating the ES's latitudes mentioned in Table 3.

In Figure 9a as in Table 2 the orbit inclination parameter, i, for each altitude goes from 20° to 85°, then, this figure can be redrawn as the grid shown in Figure 9b, where the curved lines of Figure 9a are replaced by vertical lines indicating the orbit inclination. Additionally, the ES latitude lines were inverted to start with the lowest value at the top.

From Figure 9a it can be observed that satellites with certain orbit inclination i, reach (in its orbit trajectory) at most the latitude that coincide with the inclination value. For example, a satellite with $i = 20°$ will not be able to pass over and ES located at $\phi = 80°$, since the maximum reached latitude in that orbit will be $\phi \approx i \approx 20°$, then, that satellite will be best suited to provide coverage at ES's located at latitudes $\phi \leq i$ or at most $\phi \approx i$. This coincides with basic knowledge of satellite coverage; whereas polar orbits can cover almost all the globe, lower inclination orbits cover smaller portions of the Earth. Then, we can redraw Figure 9b to have a coverage grid by discarding points corresponding to cases of $\phi_{ES} > i_{SAT}$, this resultant grid is shown in Figure 9c.

From the simulation results obtained for the grid points of Figure 9c, matrices A_h and B_h containing the shape and scale parameters of $f_\Theta(\theta)$ and $F_\Theta(\theta)$ can be defined as follows

$$
A_h = \begin{bmatrix}
a_{1,1} & a_{1,2} & a_{1,3} & \cdots & a_{1,14} \\
a_{2,1} & a_{2,2} & a_{2,3} & \ddots & a_{2,14} \\
a_{3,1} & a_{3,2} & a_{3,3} & \ddots & \vdots \\
a_{4,1} & a_{4,2} & a_{4,3} & \ddots & \vdots \\
a_{5,1} & a_{5,2} & a_{5,3} & \ddots & \vdots \\
NaN & a_{6,2} & a_{6,3} & \ddots & \vdots \\
NaN & NaN & a_{7,3} & \ddots & \vdots \\
\vdots & \vdots & \vdots & \ddots & \vdots \\
NaN & NaN & NaN & \cdots & a_{18,14}
\end{bmatrix}_{18\times14}, \quad
B_h = \begin{bmatrix}
b_{1,1} & b_{1,2} & b_{1,3} & \cdots & b_{1,14} \\
b_{2,1} & b_{2,2} & b_{2,3} & \ddots & b_{2,14} \\
b_{3,1} & b_{3,2} & b_{3,3} & \ddots & \vdots \\
b_{4,1} & b_{4,2} & b_{4,3} & \ddots & \vdots \\
b_{5,1} & b_{5,2} & b_{5,3} & \ddots & \vdots \\
NaN & b_{6,2} & b_{6,3} & \ddots & \vdots \\
NaN & NaN & b_{7,3} & \ddots & \vdots \\
\vdots & \vdots & \vdots & \ddots & \vdots \\
NaN & NaN & NaN & \cdots & b_{18,14}
\end{bmatrix}_{18\times14}, \quad (17)
$$

where NaN's are placed for the case $\phi_{ES} > i_{SAT}$, and $h \in h_s$ corresponds to the simulation altitudes set defined in Table 2 as

$$
h_s = \{1000 \text{ km}, 1200 \text{ km}, \ldots, 2000 \text{ km}\}. \tag{18}
$$

Tables A1–A6 in Appendix B, show the shape and scale parameters obtained from simulations with configurations shown in Tables 1–3. From those values, matrices A_h and B_h can be constructed, and linear interpolation can be applied to those in order to obtain shape and scale parameters for different orbit inclinations, ES's latitudes, and altitudes that were not simulated, but are in the range of h_s, i_s, ϕ_s.

3.4. Reducing the Shape and Rate Matrices

From the grid of Figure 9c, an array as shown in Figure 9d can be created for each simulated orbit altitude. Figure 9d shows the diagonal numbers of a rectangular matrix D_h defined as

$$
D_h = \begin{bmatrix}
d_0 & d_1 & d_2 & \cdots & d_{13} \\
d_{-1} & d_0 & d_1 & \ddots & d_{12} \\
d_{-2} & d_{-1} & d_0 & \ddots & \vdots \\
d_{-3} & d_{-2} & d_{-1} & \ddots & \vdots \\
d_{-4} & d_{-3} & d_{-2} & \ddots & \vdots \\
NaN & d_{-4} & d_{-3} & \ddots & \vdots \\
NaN & NaN & d_{-4} & \ddots & \vdots \\
\vdots & \vdots & \vdots & \ddots & \vdots \\
NaN & NaN & NaN & \cdots & d_{-4}
\end{bmatrix}_{18\times14}, \quad (19)
$$

for each altitude $h \in h_s$, with NaN's are placed for cases where $\phi_{ES} > i_{SAT}$.

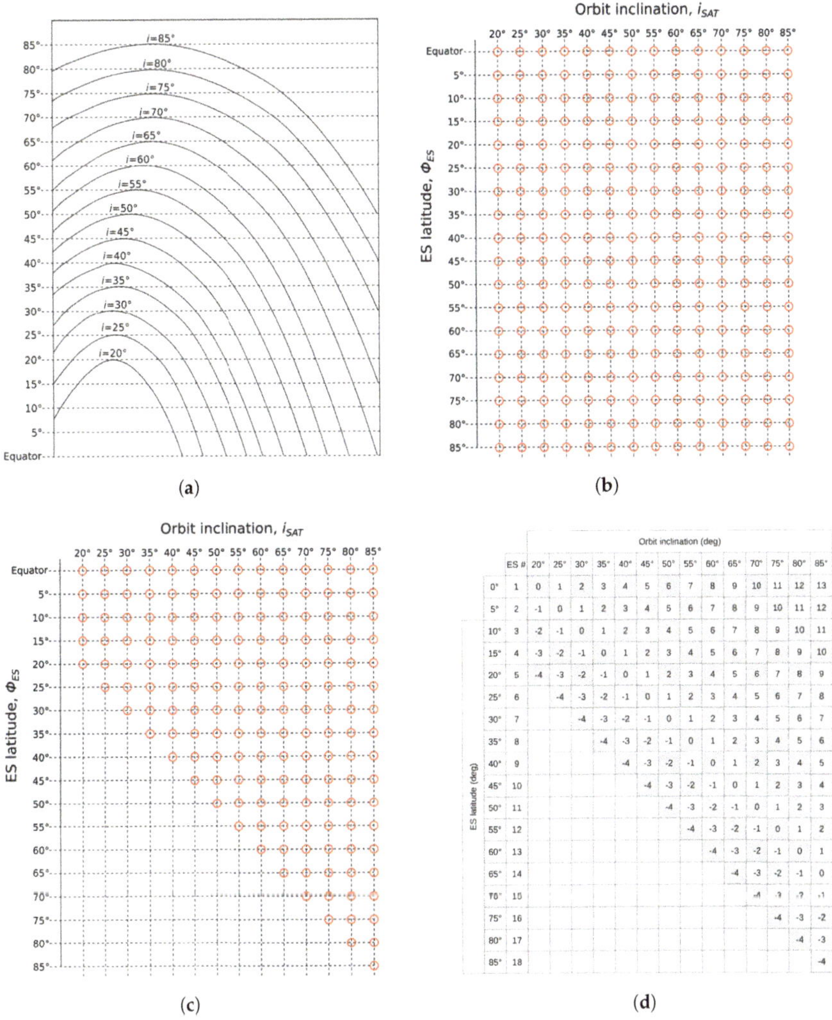

Figure 9. Simulated orbits: (**a**) Model without scale illustrating satellite passes at different orbit inclinations, *i*, over different ES's latitudes. (**b**) Simulated orbit inclinations and ES's latitudes as points in a grid. (**c**) Reduction of the grid showing in each column the ES's below the simulated orbit inclinations. (**d**) Simulated points as diagonal elements within a matrix.

3.5. PDF and CDF of the Elevation Angle

In addition to h_s, we can define two additional sets containing the orbit inclinations (as in Table 2) and ES's latitudes (as in Table 3) at which simulations were performed, as

$$i_s = \{20°, 25°, \ldots, 85°\} \tag{20}$$

and

$$\phi_s = \{0°, 5°, \ldots, 85°\}, \tag{21}$$

respectively.

Depending on the input values of ϕ_k, h_k, and i_k, if those are part of the previously defined sets (i_s, h_s, ϕ_s) eight cases can arise as follows:

- Case 1. $i_k \in i_s$ and $\phi_k \in \phi_s$ and $h_k \in h_s$

- Case 2. $i_k \in i_s$ and $\phi_k \in \phi_s$ and $h_k \notin h_s$
- Case 3. $i_k \in i_s$ and $\phi_k \notin \phi_s$ and $h_k \in h_s$
- Case 4. $i_k \in i_s$ and $\phi_k \notin \phi_s$ and $h_k \notin h_s$
- Case 5. $i_k \notin i_s$ and $\phi_k \in \phi_s$ and $h_k \in h_s$
- Case 6. $i_k \notin i_s$ and $\phi_k \in \phi_s$ and $h_k \notin h_s$
- Case 7. $i_k \notin i_s$ and $\phi_k \notin \phi_s$ and $h_k \in h_s$
- Case 8. $i_k \notin i_s$ and $\phi_k \notin \phi_s$ and $h_k \notin h_s$

Case 1 occurs when all the input values ϕ_k, i_k and h_k are contained in ϕ_s, i_s, and h_s, respectively; and Case 8 occurs when none of ϕ_k, i_k and h_k parameters coincide with previously simulated values. Cases 2 to 7 occur when at least one of the input values (any of ϕ_k, i_k, and h_k) coincides with a value within its corresponding sets ϕ_s, i_s and h_s. Those cases illustrate all possible scenarios, ranging from not-required interpolation to interpolation inside a four-point mesh.

We can represent the ES latitude, ϕ_k, orbit inclination, i_k, or both (for which we want to determine their PDF) as a query point between two-known points, inside a three-point mesh, or inside a four-point mesh. Then we can use an interpolation method, such as those proposed in Appendix C, to find an appropriate value for the gamma distribution parameters at that query coordinates. In addition to those proposed in Appendix C, more interpolation methods are widely available in the literature.

3.6. First and Second Order Statistics of $f_\Theta(\theta)$

The elevation angle expected value can be calculated from (15) as

$$E[\Theta] = \int_\theta \theta f_\Theta(\theta) d\theta, \tag{22}$$

Nonetheless, sometimes a satellite link requires to operate above a minimum value of θ, θ_{min}, then, elevation angle values below θ_{min} are not of interest. Recent satellite systems operating above the Ku band consider θ_{min} values above some tens of degrees, e.g., the Starlink constellation considered a $\theta_{min} = 40°$ as mentioned in [27]. We can obtain the conditional expected value for values above θ_{min} using conditional probability as follows

$$E[\Theta|\Theta \geq \theta_{min}] = \int_\theta \theta f_\Theta(\theta|\theta \geq \theta_{min}) d\theta, \tag{23}$$

which can be rewritten as in [28] as

$$E[\Theta|\Theta \geq \theta_{min}] = \frac{\int_{\theta_{min}}^{\theta_{max}} \theta f_\Theta(\theta) d\theta}{F_\Theta(\theta_{max}) - F_\Theta(\theta_{min})}, \tag{24}$$

The variance of Θ, is obtained as follows

$$\mathrm{Var}[\Theta|\Theta \geq \theta_{min}] = E[\Theta^2|\Theta \geq \theta_{min}] - E[\Theta|\Theta \geq \theta_{min}]^2, \tag{25}$$

And the standard deviation of Θ is then defined as

$$\mathrm{SD}[\Theta|\Theta \geq \theta_{min}] = \mathrm{Var}[\Theta|\Theta \geq \theta_{min}]^{1/2} \tag{26}$$

3.7. Choosing Orbits to Maximize Mean Value of θ

It is possible to choose an orbit configuration to maximize the elevation angle expected value, $E[\Theta|\Theta \geq \theta_{min}]$, for given ranges of ϕ, h, and i. The problem can be stated as

$$\max_{i,\phi,h} \quad E[\Theta|\Theta \geq \theta_{min}]$$
$$\text{s.t.} \quad i_1 \leq i \leq i_2 \tag{27}$$
$$\phi_1 \leq \phi \leq \phi_2$$
$$h_1 \leq h \leq h_2$$

where $i_1 \leq i_2$, $\phi_1 \leq \phi_2$, and $h_1 \leq h_2$. This problem can be solved by optimization methods or using computational tools to perform an iterative evaluation within a `for` or `while` cycle.

Reduction of the Elements in the Diagonals

It was found that functions $f_\Theta(\theta)$ and $F_\Theta(\theta)$ were very similar for some diagonals, d shown in Figure 9d. Then, from the diagonals as shown in Figure 9d, we propose a reduction first, by obtaining the expected value $E[\cdot]$ of each diagonal, d_l, (note the use of subindex $l \in \{-4, -3, \ldots, 13\}$ as an indicator of the diagonal number) as $E[d_l]$. Then, (19) can be rewritten as the following row vector

$$E[\boldsymbol{D_h}] = [E[d_{-4}], E[d_{-3}], \ldots, E[d_{13}]]_{1 \times 18} \tag{28}$$

where the diagonals with just $NaN's$ elements are omitted.

From observation, we find that some adjacent elements in $E[\boldsymbol{D_h}]$ are approximately equal, then, we reduced $E[\boldsymbol{D_h}]$ by taking the mean value of adjacent similar values according to

$$\frac{1}{|L-l|} \sum_{l}^{L} E[d_l], \quad L > l \tag{29}$$

where l, and L are, respectively, the lower and upper diagonal numbers for the interval in which the values $E[d_l], E[d_{l+1}], \ldots, E[d_L]$ were observed approximately the same (with a maximum-distance criteria between the fitted curves of 5%). After this reduction $E[\boldsymbol{D_h}]$ ends as a row vector with only four elements, and we call this vector $\boldsymbol{p_h}$ for $h \in h_s$. This procedure can be repeated for each altitude in h_s, and after grouping all the row vectors $\boldsymbol{p_h}$ for all simulated altitudes h we arrive to the matrix $\boldsymbol{P}$ given by

$$\boldsymbol{P} = \begin{bmatrix} \boldsymbol{p}_{1000\text{km}} \\ \boldsymbol{p}_{1200\text{km}} \\ \boldsymbol{p}_{1400\text{km}} \\ \boldsymbol{p}_{1600\text{km}} \\ \boldsymbol{p}_{1800\text{km}} \\ \boldsymbol{p}_{2000\text{km}} \end{bmatrix} = \begin{bmatrix} p_{1,1} & p_{1,2} & p_{1,3} & p_{1,4} \\ p_{2,1} & p_{2,2} & p_{2,3} & p_{2,4} \\ p_{3,1} & p_{3,2} & p_{3,3} & p_{3,4} \\ p_{4,1} & p_{4,2} & p_{4,3} & p_{4,4} \\ p_{5,1} & p_{5,2} & p_{5,3} & p_{5,4} \\ p_{6,1} & p_{6,2} & p_{6,3} & p_{6,4} \end{bmatrix} \tag{30}$$

Matrix $\boldsymbol{P}$ will be called $\boldsymbol{P_a}$ if it contains the shape parameters, a, for $f_\Theta(\theta)$, and $\boldsymbol{P_b}$ if it contains the scale parameter, b, for $f_\Theta(\theta)$.

From matrices $\boldsymbol{P_a}$ and $\boldsymbol{P_b}$, it is possible to recover any of the points of A_h and B_h and then estimate the CDF for a query orbit altitude h_k, orbit inclination, i_k, and ES's latitude, ϕ_k. However, we observed that some $F_\Theta(\theta)$ curves for high latitudes were not fitting very well after the reduction from matrices A_h and B_h to $\boldsymbol{P_a}$ and $\boldsymbol{P_b}$, then, two weight matrices were obtained through manual tuning to improve the correspondence between $\boldsymbol{P_a}$ and

A_h, and between P_b and B_h. The weight matrix for the shape parameter, w_a and w_b, is defined as

$$
w_a = \begin{bmatrix}
w_{a_{1,1}} & w_{a_{1,2}} & w_{a_{1,3}} & \cdots & w_{a_{1,14}} \\
w_{a_{2,1}} & w_{a_{2,2}} & w_{a_{2,3}} & \ddots & w_{a_{2,14}} \\
w_{a_{3,1}} & w_{a_{3,2}} & w_{a_{3,3}} & \ddots & \vdots \\
w_{a_{4,1}} & w_{a_{4,2}} & w_{a_{4,3}} & \ddots & \vdots \\
w_{a_{5,1}} & w_{a_{5,2}} & w_{a_{5,3}} & \ddots & \vdots \\
NaN & w_{a_{6,2}} & w_{a_{6,3}} & \ddots & \vdots \\
NaN & NaN & w_{a_{7,3}} & \ddots & \vdots \\
\vdots & \vdots & \vdots & \ddots & \vdots \\
NaN & NaN & NaN & \cdots & w_{a_{18,14}}
\end{bmatrix}_{18 \times 14}, \quad
w_b = \begin{bmatrix}
w_{b_{1,1}} & w_{b_{1,2}} & w_{b_{1,3}} & \cdots & w_{b_{1,14}} \\
w_{b_{2,1}} & w_{b_{2,2}} & w_{b_{2,3}} & \ddots & w_{b_{2,14}} \\
w_{b_{3,1}} & w_{b_{3,2}} & w_{b_{3,3}} & \ddots & \vdots \\
w_{b_{4,1}} & w_{b_{4,2}} & w_{b_{4,3}} & \ddots & \vdots \\
w_{b_{5,1}} & w_{b_{5,2}} & w_{b_{5,3}} & \ddots & \vdots \\
NaN & w_{b_{6,2}} & w_{b_{6,3}} & \ddots & \vdots \\
NaN & NaN & w_{b_{7,3}} & \ddots & \vdots \\
\vdots & \vdots & \vdots & \ddots & \vdots \\
NaN & NaN & NaN & \cdots & w_{b_{18,14}}
\end{bmatrix}_{18 \times 14}, \quad (31)
$$

Both w_a and w_b are the same for all altitudes, and their numerical values are shown in Appendix B.

4. Results

The shape, a, and rate, b, parameters for matrices P_a and P_b, respectively, are shown in Tables 6 and 7 for the orbit altitudes $h \in h_s$ and for the diagonals as in D_h, (19).

Table 6. Shape parameter, a, for matrix P_a, using the diagonal notation as in (19), and as shown in Figure 9d.

Altitude	Diagonals Range			
	(d_{-4}, d_{-2})	(d_{-1}, d_0)	(d_1, d_2)	(d_3, d_{13})
1000 km	1.470954	1.638425	1.450794	1.488802
1200 km	1.707323	1.935321	1.721572	1.752709
1400 km	1.712923	1.951520	1.746205	1.754760
1600 km	1.723436	1.955254	1.782190	1.757023
1800 km	1.731110	1.952362	1.796549	1.774827
2000 km	1.741450	1.954794	1.836833	1.760709

Table 7. Rate parameter, $b \times 10^{-2}$, for matrix P_b, using the diagonal notation as in (19), and as shown in Figure 9d.

Altitude	Diagonals Range			
	(d_{-4}, d_{-2})	(d_{-1}, d_0)	(d_1, d_2)	(d_3, d_{13})
1000 km	7.037052	9.459459	9.075805	8.805617
1200 km	7.391963	9.840010	9.640436	9.284945
1400 km	7.130587	9.319436	9.412412	8.952740
1600 km	6.943998	8.879456	9.242146	8.693598
1800 km	6.791048	8.510588	9.041604	8.632867
2000 km	6.675182	8.224503	8.941424	8.305444

Figure 10a–c show the PDF obtained from the empirical data and the ones obtained from the parameters listed in Tables 6 and 7. Table 8 shows the shape, a, and scale, b, parameters as well as the first and second order statistics for the $f_\Theta(\theta)$ functions of Figure 10a–d.

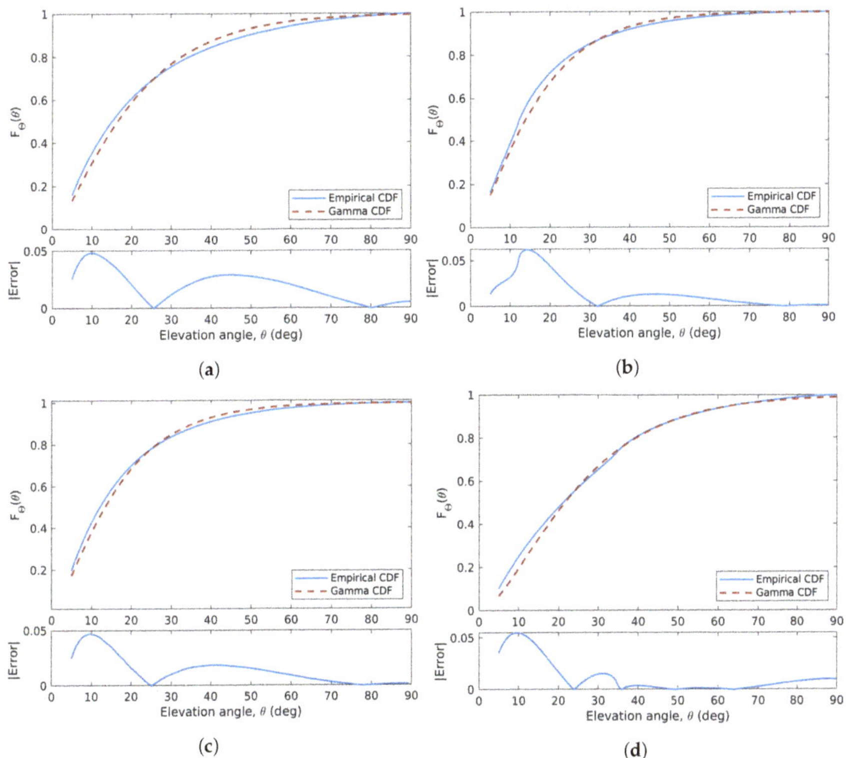

Figure 10. Comparison and error of the empirical and Gamma elevation angle CDF's for different ES latitudes, ϕ_{ES}, orbit inclinations of the satellites, i_{SAT}, and, orbit altitudes, h_k. (**a**) CDF's for $\phi_{ES} = 30°$ and $i_{SAT} = 30°$ at $h_k = 1000$ km; (**b**) CDF's for $\phi_{ES} = 45°$ and $i_{SAT} = 65°$ at $h_k = 1000$ km; (**c**) CDF's for $\phi_{ES} = 20°$ and $i_{SAT} = 75°$ at $h_k = 1000$ km; (**d**) CDF's for $\phi_{ES} = 85°$ and $i_{SAT} = 85°$ at $h_k = 1000$ km.

Table 8. First and second order statistics of the elevation angle for the same orbit configurations and ES's locations that in Figure 10a–d.

	Empirical Result	Proposed Method's Result	Empirical Result	Proposed Method's Results	Empirical Result	Proposed Method's Results	Empirical Result	Proposed Method's Results
ES latitude, ϕ_{ES}	30°		45°		20°		85°	
Orbit inclination, i_{SAT}	30°		65°		75°		85°	
Orbit altitude, h_{SAT}				1000 km				
Shape parameter, a	–	1.4710	–	1.6384	–	1.4888	–	1.8387
Rate parameter, $b \times 10^{-2}$	–	7.0371	–	9.4595	–	8.8056	–	7.0371
$E[\Theta > 5°]$	24.55°	23.09°	19.43°	19.74°	20.24°	19.70°	27.70°	26.66°
$SD[\Theta > 5°]$	18.40°	16.86°	14.19°	13.13°	14.94°	13.48°	18.12°	18.90°
$E[\Theta > 15°]$	34.77°	31.13°	29.77°	27.96°	30.55°	28.24°	35.48°	33.23°
$SD[\Theta > 15°]$	17.46°	16.26°	14.24°	12.45°	14.49°	12.93°	16.32°	17.96°
$E[\Theta > 25°]$	44.33°	39.66°	40.12°	37.03°	40.39°	37.38°	42.73°	40.91°
$SD[\Theta > 25°]$	15.93°	15.88°	13.40°	12.05°	13.50 °	12.61°	14.79°	17.29°

A case study was developed for a satellite with an altitude of 1500 km, an orbit inclination of 43°, and ES latitude of 22°. Using the numerical results of Tables 6 and 7, a mesh as shown in Figure 11a can be created for altitudes of 1400 km and 1600 km (shape and scale values were obtained for those altitudes and are shown in Tables 6 and 7). Then, using an interpolation method as that described in Appendix C, parameters a and b can be obtained to characterize the elevation angle curve.

The shape and scale parameters for the case study are shown in Table 9. First, parameters for the four-points (as in Figure 11b) are listed for each altitude (1400 km and 1600 km). Then, the resultant parameters obtained from interpolation were computed. Finally, the shape and scale parameters for the query altitude can be obtained using a weighted arithmetic mean were the weights can be obtained from the distance of the query altitude to the closest characterized orbit altitudes. In this case, the distances from the query altitude to the closest characterized altitudes are the same, then we obtained the shape and scale for the query altitude using the arithmetic mean of the shape and scale values at 1400 km and 1600 km.

Both the elevation angle CDF obtained with the proposed methodology and the empirical CDF are shown in Figure 11b. The distance magnitude between those two curves is also shown in Figure 11b and labeled as absolute error ($|\text{Error}|$). The distance (or error) between the gamma CDF and the empirical CDF is very small, and its maximum value is around 0.025 (2.5%).

Table 9. Shape and scale parameters for the elevation angle PDF and CDF for $\phi_{ES} = 22°$ and $i_{SAT} = 43°$ at $h_k = 1500$ km, corresponding to the four-point mesh shown in Figure 11a.

	Four-Point Mesh Setup				
	1400	1600		1400	1600
$a(\phi_1, i_1)$	1.9515	1.9552	$b(\phi_1, i_1)$	9.3194×10^{-2}	8.8794×10^{-2}
$a(\phi_1, i_2)$	1.9515	1.9552	$b(\phi_1, i_2)$	9.3194×10^{-2}	8.8794×10^{-2}
$a(\phi_2, i_1)$	1.9515	1.9552	$b(\phi_2, i_1)$	9.3194×10^{-2}	8.8794×10^{-2}
$a(\phi_2, i_2)$	1.7462	1.7822	$b(\phi_2, i_2)$	9.4112×10^{-2}	9.2215×10^{-2}
	Obtained parameters from interpolation				
$a(\phi, i)$	1.8776	1.8929	$b(\phi, i)$	9.3529×10^{-2}	9.0025×10^{-2}
$a(\phi, i)$		1.8853	$b(\phi, i)$		9.1777×10^{-2}
	Empirical parameters				
$a(\phi, i)$		1.9804	$b(\phi, i)$		9.5987×10^{-2}

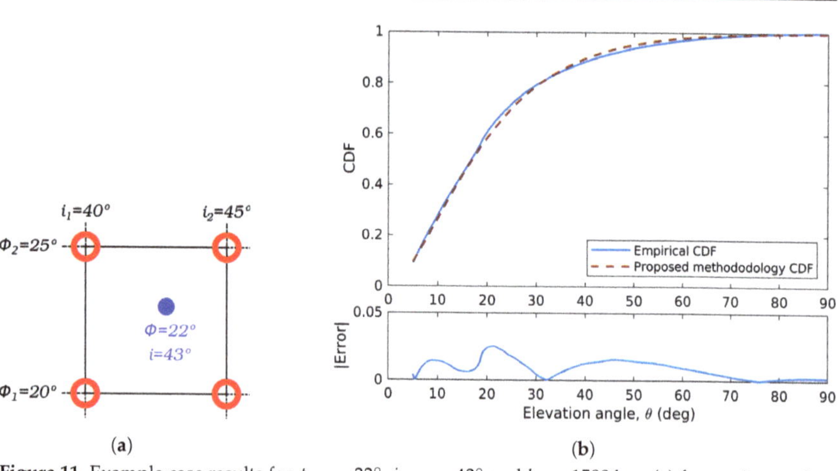

Figure 11. Example case results for $\phi_{ES} = 22°$, $i_{SAT} = 43°$ and $h_k = 1500$ km: (**a**) four-point mesh to calculate the shape, a, and scale, b parameters of $f_\theta(\theta)$ and (**b**) empirical CDF vs. CDF obtained with the proposed methodology.

4.1. Results Validity

4.1.1. Effects of the Orbit Configuration in the Elevation Angle Distribution

The Kepler orbital elements were chosen to describe the orbits, those parameters include the semi-major axis, a, eccentricity, e, orbit inclination, i, right ascension of the ascending node, Ω, argument of perigee, ω, and true anomaly, ν. The values of the right ascension of the ascending node, Ω, argument of perigee, ω, and true anomaly, ν, do not affect the statistical properties of the elevation angle, then, those values are not further discussed. Figure 12a,b show the PDF and CDF elevation angle curves observed from the same location for two simulated satellites with different orbit configurations which share the same initial semi-major axis, a, eccentricity, e, and orbit inclination, i; but, differ in the initial values of the right ascension of the ascending node, Ω, argument of perigee, ω, and of the true anomaly, ν.

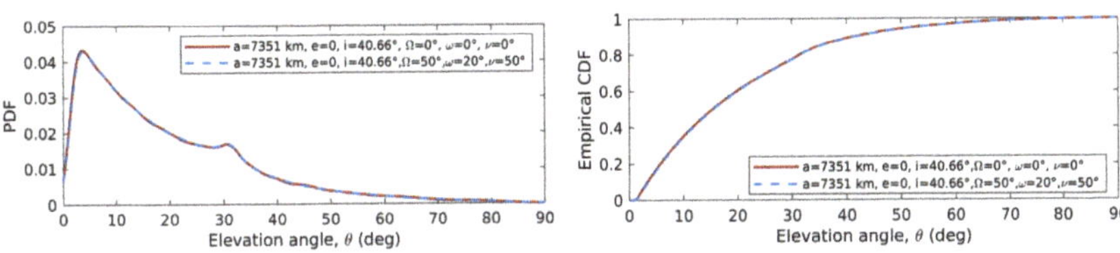

(**a**) Probability density function, $f_\theta(\Theta)$ (**b**) Cumulative distribution function, $F_\theta(\Theta)$

Figure 12. PDF and CDF for two orbit configurations sharing the same altitude, eccentricity, and orbit inclination, but with the remaining orbital parameters being different.

4.1.2. Elevation Angle CDF for Different Longitudes

ES's located at different longitudes share approximately the same elevation angle distribution for the same LEO satellite configuration. The reasoning behind this fact is discussed in [2], and it is illustrated in Figure 13a, which shows the elevation angle PDF for three ES's located at the same latitude, ϕ, but at different longitudes, λ's. Figure 13a shows that the elevation angle PDF is the same for ES located at different longitudes, λ's, when they share the same latitude, ϕ, and those are served by satellites with the same orbit configuration (semi-major axis, a, eccentricity, e, and, inclination, i). The PDF's shown in Figure 13a were generated for a satellite with a circular orbit and an inclination of $i = 60^\circ$, and an altitude of $h = 1800$ km; in addition, all ES's were located at the same latitude $\phi = 25^\circ$ N.

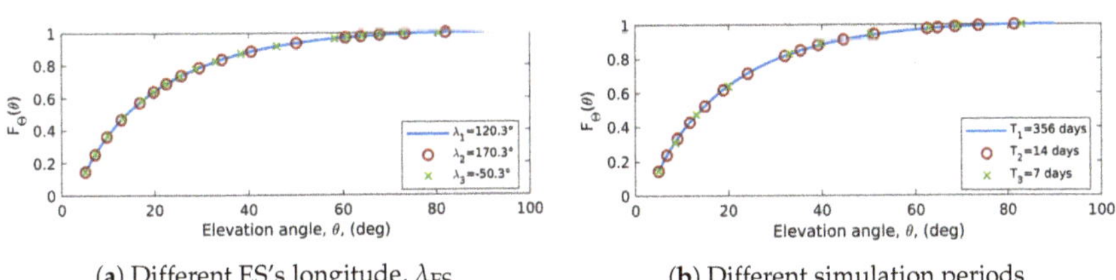

(**a**) Different ES's longitude, λ_{ES} (**b**) Different simulation periods

Figure 13. CDF $F_\Theta(\theta)$ for the same orbit configuration and ES latitude but for (**a**) different ES longitude and (**b**) different simulation period.

4.1.3. Long Term Validity of $f_\Theta(\theta)$ and $F_\Theta(\theta)$

From a few days $f_\Theta(\theta)$ can be observed very similar compared with the same PDF is a much greater period, e.g., one year. Figure 13b shows the functions $F_\Theta(\theta)$ for three different simulation periods ranging from one week to one year. The CDF curve is almost identical for all the simulation periods. The orbit configuration and ES's location were choose to be the same that for Figure 13a.

4.1.4. Elevation Angle CDF for Different Satellite Characteristics

Simulations for this paper where performed for a satellite with the mass and drag areas listed in Table 1, nonetheless, satellites with different values in those characteristics will share very similar functions $F_\Theta(\theta)$. Figure 14 shows the empirical CDF's for two satellites with the same orbital parameters but with different mass and drag areas. For both satellites, the orbit altitude, h, was set to 1800 km and the orbit inclination, i, to 60°, and both ES's were located at 25° N. As shown in Figure 14, variations in the satellite mass and drag area characteristics will have a very little impact in the elevation angle CDF's. For lower altitudes the effect is greater, but not significant enough for the altitudes covered in this paper. Table 10 show the mass and areas for satellites shown in Figure 14.

Table 10. Satellites with different mass and drag area characteristics for additional performed simulations.

	Satellite 1	Satellite 2
Dry mass	5 kg	200 kg
Drag area	$1\ \mathrm{m}^2$	$10\ \mathrm{m}^2$
Solar radiation pressure area	$1\ \mathrm{m}^2$	$10\ \mathrm{m}^2$

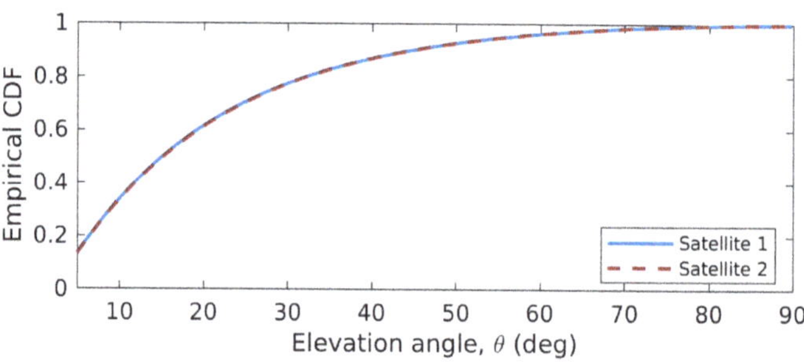

Figure 14. CDF $F_\Theta(\theta)$ for different satellite mass and drag area characteristics.

4.2. Error Analysis

The error between the empirical CDF curves of the elevation angle and the ones obtained with the parameters in Tables 6 and 7, was quantified by means of the absolute error, ϵ, and its CDF. Figure A1a–f in Appendix A show the mean absolute error, $\bar{\epsilon}$, between the empirical CDF's for the simulated points of Figure 9c, and for the orbit altitudes $h_k \in h_s$.

The maximum observed mean error for the individual characterizations shown in Appendix A was about 5% ($\bar{\epsilon}_{max} \approx 0.05$) for the worst case. However, the greatest error values were observed to occur at lower elevation angles (below 20° were the link is often unreliable since line-of-sight is harder to find) that are often not considered by satellite communication systems due to greater attenuation.

The absolute errors ϵ were also analyzed through their individual CDF's, showing that most of the errors where below 5%. Figure 15a–d show the CDF's of the absolute error, ϵ, corresponding to Figure 10a–d. In addition, ϵ was analyzed for all the simulated satellite orbits and according to their elevation angle range.

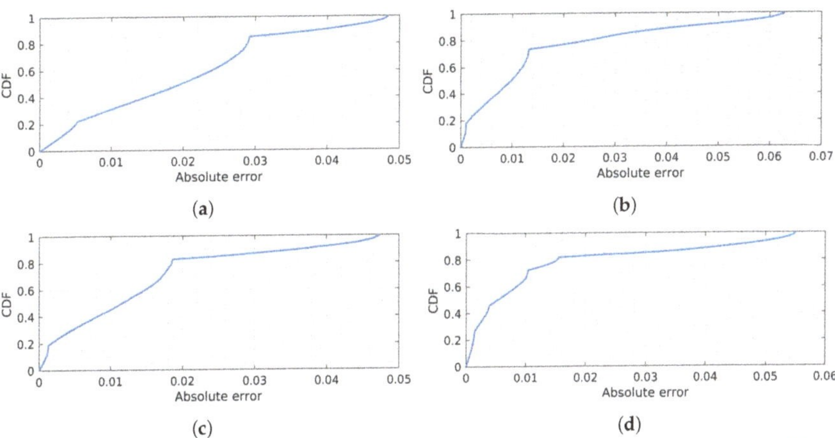

Figure 15. CDF's for the absolute errors at different orbit configurations and ES's as in Figure 10a–d. Absolute error CDF's for: (**a**) $\phi_{ES} = 30°$ and $i_{SAT} = 30°$ at $h_k = 1000$ km; (**b**) $\phi_{ES} = 45°$ and $i_{SAT} = 65°$ at $h_k = 1000$ km; (**c**) $\phi_{ES} = 20°$ and $i_{SAT} = 75°$ at $h_k = 1000$ km; (**d**) $\phi_{ES} = 85°$ and $i_{SAT} = 85°$ at $h_k = 1000$ km.

The PDF, CDF, and the complementary CDF (CCDF) showing the resultant absolute errors, ϵ, for all the simulations are shown in Figure 16, where it can be observed that around 95% of the values of ϵ are below 0.05. Figure 17 shows the same absolute errors as in Figure 16b, but now expanded and classified by elevation angle range; in this figure, it can be observed that most of the absolute errors are below 5% for all the elevation angle ranges, and that errors' CDF distributions are approximately the same for all the elevation angle ranges. Figure 18 expands the 90th-percentile of the CDF's in Figure 17, showing that in fact, most of the errors are below 2%; the 90th-percentile mean value is shown as a red cross for each boxplot.

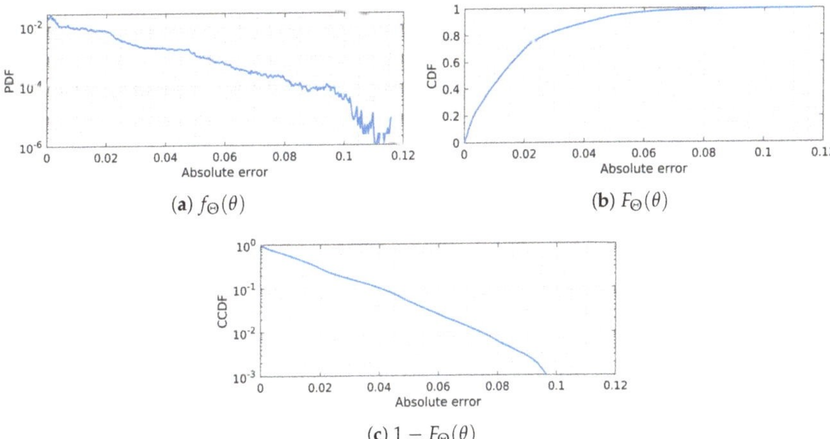

Figure 16. Distribution of the absolute errors behavior of all the performed simulations: (**a**) PDF, (**b**) CDF, and (**c**) CCDF.

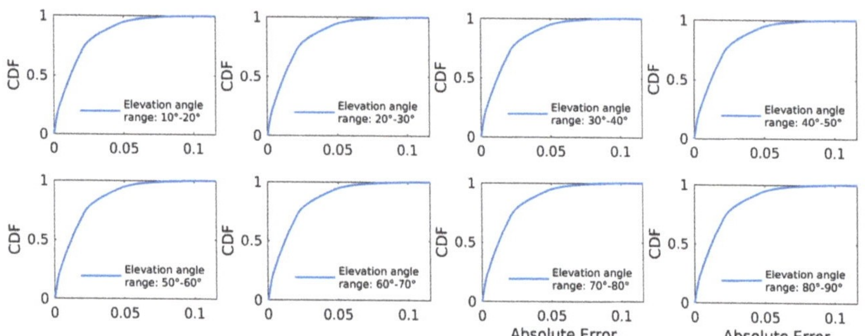

Figure 17. CDF's of the absolute error by elevation angle range for all the simulated orbits and ES's.

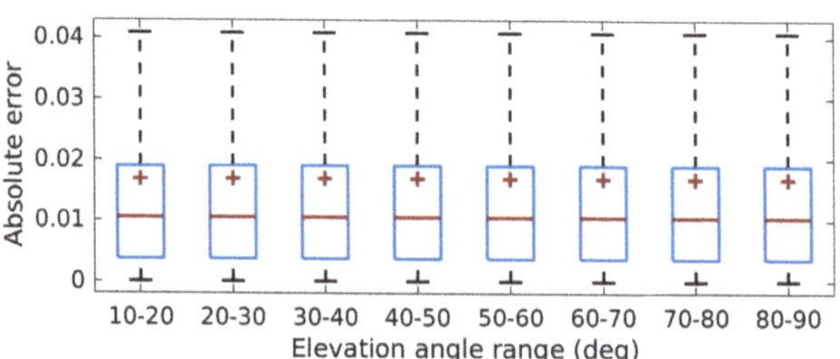

Figure 18. Boxplots of the absolute error by elevation angle range. Simulation results vs proposed methodology.

5. Conclusions

This article demonstrated the feasibility of using a gamma random variable to characterize the elevation angle PDF and CDF. The proposed methodology allowed PDF and CDF calculations of the elevation angle for LEO satellites for altitudes between 1000 km and 2000 km. The characterization of elevation angle through a gamma random variable allowed an easy computation of first and second order statistics, which, to the best of our knowledge, have not been previously addressed in the literature.

The proposed methodology was validated with an error analysis against the empirical CDF's. The results showed that errors are low, with a mean absolute error much below 5% in most of the cases, and with around 95% of the total errors being below 5%.

Furthermore, the proposed methodology allows an easy comparison between multiple orbits in order to determine the most suitable orbital parameters to provide coverage above a minimum elevation angle at a particular latitude.

Author Contributions: Conceptualization, J.M.G.-T. and C.V.-R.; methodology and software, J.M.G.-T.; validation, A.A.-Z. and R.V.-H.; formal analysis, A.A.-Z. and R.V.-H.; investigation, A.A.-Z. and R.V.-H.; writing—original draft preparation, J.M.G.-T.; writing—review and editing, C.V.-R., A.A.-Z. and R.V.-H.; visualization, J.M.G.-T.; supervision, J.M.G.-T., C.V.-R. and A.A.-Z.; project administration, C.V.-R. and R.V.-H. All authors have read and agreed to the published version of the manuscript.

Funding: This research received no external funding.

Institutional Review Board Statement: Not applicable.

Informed Consent Statement: Not applicable.

Data Availability Statement: Not applicable.

Acknowledgments: We would like to acknowledge the support from the Smart Digital Technologies and Infrastructure Research Group, the project "Digital Technologies to Create Adaptive Smart Cities" as part of the Challenge-Based Research Funding Program, and the School of Engineering and Sciences at Tecnologico de Monterrey for providing the means to develop this collaborative work.

Conflicts of Interest: The authors declare no conflict of interest.

Abbreviations

The following abbreviations are used in this manuscript:

CDF	Cumulative distribution function
CCDF	Complementary cumulative distribution function
ES	Earth station
GEO	Geostationary Earth orbit
LEO	Low Earth orbit
LMS	Land mobile satellite
LOS	Line-of-sight
NGEO	Non-Geostationary Earth orbit
NGSO	Non-Geosynchronous orbit
NLOS	Non-line-of-sight
PDF	Probability density function

Appendix A

Figure A1a–f contain the mean error for each elevation angle CDF. Those errors were obtained by comparing the proposed methodology results against the results obtained through simulations. The same format than in Figure 9c,d is applied here; only showing the ES's located below the satellite orbit inclinations at each column.

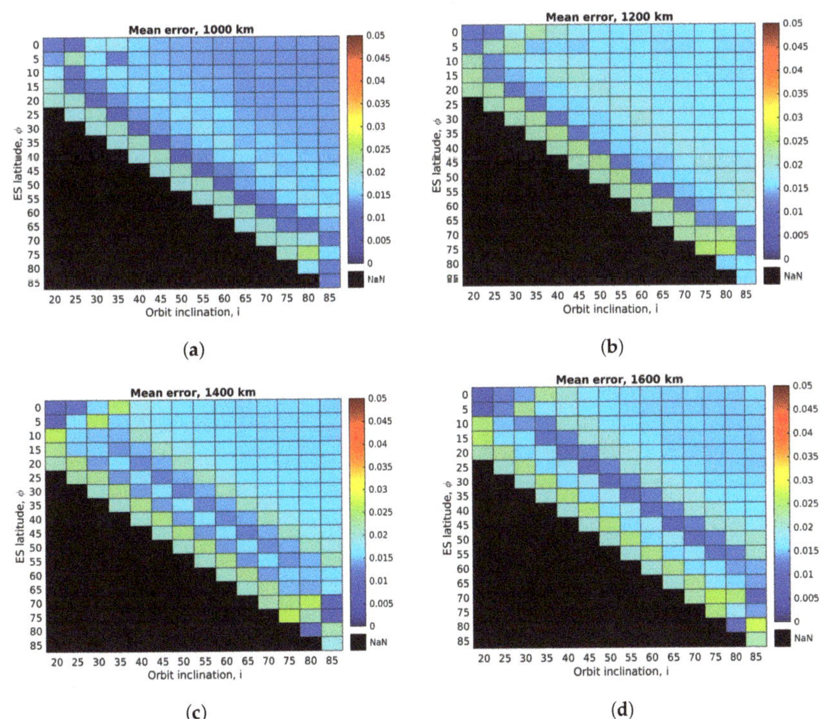

Figure A1. *Cont.*

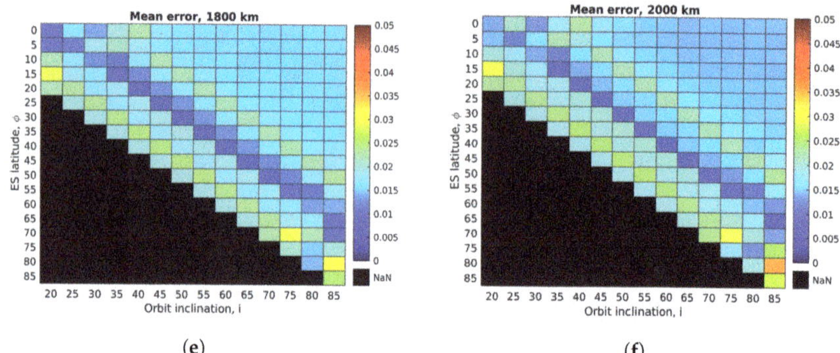

(e) (f)

Figure A1. Mean error obtained with the proposed methodology, as compared with the simulation results for all the simulated orbit inclinations, i_{SAT}, ES's latitudes, ϕ_{ES}, and altitudes, h_k. (**a**) Proposed methodology vs simulations results at 1000 km; (**b**) Proposed methodology vs simulations results at 1200 km; (**c**) Proposed methodology vs simulations results at 1400 km; (**d**) Proposed methodology vs simulations results at 1600 km; (**e**) Proposed methodology vs simulations results at 1800 km; (**f**) Proposed methodology vs simulations results at 2000 km.

Appendix B

Tables A1–A6 contain the shape, a, and rate, b, parameters for $f_\Theta(\theta)$ at the simulated values of i and ϕ. With those results, matrices A_h and B_h are formed as defined in (17) and depicted in Figure 9d. Also, Table A7 contains the numerical values for w_a and w_b.

Table A1. Shape, a, and rate, b parameter for $f_\Theta(\theta)$ and $F_\Theta(\theta)$ for $h_k = 1000$ km.

Shape parameter, a

Orbit inclination, i

ϕ	20°	25°	30°	35°	40°	45°	50°	55°	60°	65°	70°	75°	80°	85°
0°	1.6063	1.5846	1.4659	1.4345	1.4584	1.4742	1.4817	1.4854	1.4864	1.4976	1.4981	1.4990	1.4954	1.5031
5°	1.7662	1.4853	1.4151	1.4548	1.4667	1.4770	1.4859	1.4898	1.4966	1.4968	1.4978	1.5015	1.5017	1.5040
10°	1.5036	1.6270	1.6291	1.4689	1.4303	1.4646	1.4756	1.4785	1.4779	1.4956	1.4867	1.4936	1.4935	1.4983
15°	1.3946	1.5570	1.6595	1.6390	1.4649	1.4340	1.4618	1.4690	1.4724	1.4938	1.4921	1.4914	1.4881	1.4920
20°	1.3749	1.4220	1.5688	1.6689	1.6382	1.4680	1.4307	1.4545	1.4573	1.4891	1.4848	1.4851	1.4828	1.4832
25°	NaN	1.3944	1.4332	1.5776	1.6720	1.6477	1.4676	1.4252	1.4474	1.4854	1.4892	1.4803	1.4776	1.4816
30°	NaN	NaN	1.3925	1.4369	1.5763	1.6771	1.6428	1.4573	1.4084	1.4733	1.4677	1.4715	1.4702	1.4730
35°	NaN	NaN	NaN	1.3925	1.4347	1.5815	1.6756	1.6354	1.4426	1.4446	1.4608	1.4641	1.4622	1.4682
40°	NaN	NaN	NaN	NaN	1.3905	1.4380	1.5786	1.6669	1.6189	1.4831	1.4273	1.4526	1.4554	1.4612
45°	NaN	NaN	NaN	NaN	NaN	1.3940	1.4356	1.5708	1.6451	1.6605	1.4663	1.4221	1.4444	1.4527
50°	NaN	NaN	NaN	NaN	NaN	NaN	1.3910	1.4271	1.5550	1.6894	1.6367	1.4608	1.4121	1.4398
55°	NaN	NaN	NaN	NaN	NaN	NaN	NaN	1.3824	1.4100	1.5894	1.6625	1.6276	1.4507	1.4099
60°	NaN	NaN	NaN	NaN	NaN	NaN	NaN	NaN	1.3660	1.4435	1.5637	1.6466	1.6055	1.4458
65°	NaN	NaN	NaN	NaN	NaN	NaN	NaN	NaN	NaN	1.4001	1.4173	1.5411	1.6072	1.5573
70°	NaN	NaN	NaN	NaN	NaN	NaN	NaN	NaN	NaN	NaN	1.3734	1.3893	1.4656	1.5816
75°	NaN	NaN	NaN	NaN	NaN	NaN	NaN	NaN	NaN	NaN	NaN	1.3180	1.3604	1.6773
80°	NaN	NaN	NaN	NaN	NaN	NaN	NaN	NaN	NaN	NaN	NaN	NaN	1.5359	1.6471
85°	NaN	NaN	NaN	NaN	NaN	NaN	NaN	NaN	NaN	NaN	NaN	NaN	NaN	1.6787

Rate parameter, $b \times 10^{-2}$

Orbit inclination, i

ϕ	20°	25°	30°	35°	40°	45°	50°	55°	60°	65°	70°	75°	80°	85°
0°	9.3564	9.7586	9.5066	8.9569	8.8711	8.8567	8.8380	8.8193	8.8029	8.8260	8.8154	8.8086	8.7865	8.8199
5°	10.5403	10.0061	9.0549	8.9535	8.8763	8.8544	8.8469	8.8317	8.8354	8.8231	8.8158	8.8208	8.8155	8.8212
10°	7.7606	8.9977	9.8007	9.4661	8.8894	8.8741	8.8451	8.8116	8.7805	8.8244	8.7752	8.7949	8.7836	8.8062
15°	6.6219	7.7633	9.0428	9.8124	9.4342	8.8982	8.8559	8.8128	8.7869	8.8387	8.8202	8.8029	8.7816	8.7902
20°	6.6564	6.6032	7.7437	9.0479	9.7937	9.4531	8.8882	8.8237	8.7689	8.8436	8.8078	8.7884	8.7699	8.7666
25°	NaN	6.6326	6.5841	7.7402	9.0362	9.8353	9.4583	8.8648	8.8023	8.8671	8.8434	8.7874	8.7628	8.7717
30°	NaN	NaN	6.5839	6.5680	7.7144	9.0589	9.8236	9.4244	8.8083	8.8931	8.8074	8.7832	8.7553	8.7587
35°	NaN	NaN	NaN	6.5567	6.5410	7.7312	9.0548	9.7985	9.3706	8.9406	8.8534	8.7989	8.7558	8.7643
40°	NaN	NaN	NaN	NaN	6.5304	6.5513	7.7229	9.0267	9.7458	9.5196	8.8811	8.8282	8.7753	8.7717
45°	NaN	NaN	NaN	NaN	NaN	6.5434	6.5459	7.7016	8.9585	9.8790	9.4584	8.8695	8.8088	8.7893
50°	NaN	NaN	NaN	NaN	NaN	NaN	6.5410	6.5295	7.6665	9.0970	9.7983	9.4411	8.8441	8.8240
55°	NaN	NaN	NaN	NaN	NaN	NaN	NaN	6.5306	6.4975	7.7688	9.0234	9.7734	9.4087	8.8794
60°	NaN	NaN	NaN	NaN	NaN	NaN	NaN	NaN	6.5095	6.6064	7.7225	8.9947	9.7121	9.4440
65°	NaN	NaN	NaN	NaN	NaN	NaN	NaN	NaN	NaN	6.6420	6.5845	7.7079	8.9262	9.6598
70°	NaN	NaN	NaN	NaN	NaN	NaN	NaN	NaN	NaN	NaN	6.6549	6.6089	7.6339	9.2649
75°	NaN	NaN	NaN	NaN	NaN	NaN	NaN	NaN	NaN	NaN	NaN	6.7200	6.9875	8.5786
80°	NaN	NaN	NaN	NaN	NaN	NaN	NaN	NaN	NaN	NaN	NaN	NaN	7.7001	7.3090
85°	NaN	NaN	NaN	NaN	NaN	NaN	NaN	NaN	NaN	NaN	NaN	NaN	NaN	6.6598

Table A2. Shape, a, and rate, b parameter for $f_\Theta(\theta)$ and $F_\Theta(\theta)$ for $h_k = 1200$ km.

		Shape parameter, a												
							Orbit inclination, i							
ES latitude, ϕ		20°	25°	30°	35°	40°	45°	50°	55°	60°	65°	70°	75°	80°	85°
	0°	1.9367	1.8476	1.7839	1.6868	1.6955	1.7417	1.7409	1.7492	1.7447	1.7597	1.7591	1.7601	1.7634	1.7602
	5°	2.1177	1.8521	1.6327	1.7080	1.7098	1.7486	1.7488	1.7558	1.7608	1.7694	1.7714	1.7650	1.7716	1.7719
	10°	1.6926	1.8959	1.9305	1.8002	1.6702	1.7141	1.7376	1.7431	1.7430	1.7589	1.7569	1.7578	1.7570	1.7563
	15°	1.6207	1.7863	1.9154	1.9534	1.7777	1.6495	1.7212	1.7317	1.7364	1.7557	1.7562	1.7539	1.7587	1.7589
	20°	1.6095	1.6490	1.8163	1.9346	1.9398	1.7980	1.6558	1.7163	1.7235	1.7514	1.7477	1.7507	1.7465	1.7494
	25°	NaN	1.6232	1.6627	1.7997	1.9360	1.9544	1.7983	1.6558	1.7068	1.7490	1.7432	1.7479	1.7488	1.7452
	30°	NaN	NaN	1.6252	1.6872	1.8215	1.9492	1.9548	1.7931	1.6375	1.7321	1.7337	1.7392	1.7343	1.7354
	35°	NaN	NaN	NaN	1.6447	1.6618	1.8221	1.9537	1.9512	1.7807	1.6748	1.7177	1.7272	1.7325	1.7302
	40°	NaN	NaN	NaN	NaN	1.6225	1.6601	1.8240	1.9460	1.9366	1.8135	1.6560	1.7160	1.7235	1.7234
	45°	NaN	NaN	NaN	NaN	NaN	1.6254	1.6633	1.8164	1.9322	1.9702	1.7978	1.6528	1.7079	1.7157
	50°	NaN	NaN	NaN	NaN	NaN	NaN	1.6238	1.6543	1.8021	1.9626	1.9488	1.7952	1.6478	1.6998
	55°	NaN	NaN	NaN	NaN	NaN	NaN	NaN	1.6163	1.6431	1.8301	1.9366	1.9394	1.7864	1.6401
	60°	NaN	NaN	NaN	NaN	NaN	NaN	NaN	NaN	1.6051	1.6666	1.8028	1.9230	1.9178	1.7669
	65°	NaN	NaN	NaN	NaN	NaN	NaN	NaN	NaN	NaN	1.6296	1.6402	1.7810	1.8767	1.8103
	70°	NaN	NaN	NaN	NaN	NaN	NaN	NaN	NaN	NaN	NaN	1.6007	1.6094	1.6590	1.9132
	75°	NaN	NaN	NaN	NaN	NaN	NaN	NaN	NaN	NaN	NaN	NaN	1.5039	1.6502	1.9769
	80°	NaN	NaN	NaN	NaN	NaN	NaN	NaN	NaN	NaN	NaN	NaN	NaN	1.8388	1.9151
	85°	NaN	NaN	NaN	NaN	NaN	NaN	NaN	NaN	NaN	NaN	NaN	NaN	NaN	1.9243
		Rate parameter, $b \times 10^{-2}$													
							Orbit inclination, i							
ES latitude, ϕ		20°	25°	30°	35°	40°	45°	50°	55°	60°	65°	70°	75°	80°	85°
	0°	9.8678	10.1172	10.1891	9.5261	9.2793	9.3729	9.3072	9.3066	9.2476	9.2919	9.2763	9.2667	9.2673	9.2552
	5°	10.9490	10.7976	9.5396	9.4617	9.4702	9.4077	9.3339	9.3232	9.3145	9.3424	9.3382	9.2878	9.3186	9.3206
	10°	7.9755	9.3073	10.1848	10.2797	9.5051	9.3438	9.3330	9.3051	9.2626	9.2973	9.2773	9.2654	9.2527	9.2481
	15°	7.0189	8.0187	9.2511	10.2487	10.0220	9.3057	9.3599	9.3083	9.2744	9.3104	9.3002	9.2706	9.2890	9.2896
	20°	7.1070	6.9667	8.0353	9.3656	10.2409	10.1571	9.3472	9.3401	9.2772	9.3222	9.2813	9.2749	9.2438	9.2536
	25°	NaN	7.0470	6.9514	7.9165	9.3034	10.2353	10.1621	9.3559	9.2997	9.3585	9.2960	9.2845	9.2656	9.2505
	30°	NaN	NaN	7.0018	6.9659	7.9953	9.3214	10.2443	10.1469	9.2910	9.3912	9.3140	9.2875	9.2399	9.2356
	35°	NaN	NaN	NaN	7.0394	6.8984	8.0022	9.3392	10.2286	10.1082	10.2072	9.3499	9.2977	9.2695	9.2639
	40°	NaN	NaN	NaN	NaN	6.9521	6.8918	7.9991	9.3174	10.1895	10.2072	9.3678	9.3487	9.2934	9.2639
	45°	NaN	NaN	NaN	NaN	NaN	6.9573	6.9004	7.9858	9.2827	10.2757	10.1639	9.3584	9.3328	9.2961
	50°	NaN	NaN	NaN	NaN	NaN	NaN	6.9589	6.8851	7.9537	9.3612	10.2210	10.1526	9.3583	9.3438
	55°	NaN	NaN	NaN	NaN	NaN	NaN	NaN	6.9545	6.8728	8.0376	9.3045	10.1982	10.1379	9.3821
	60°	NaN	NaN	NaN	NaN	NaN	NaN	NaN	NaN	6.9538	6.9553	7.9965	9.2952	10.1634	10.1416
	65°	NaN	NaN	NaN	NaN	NaN	NaN	NaN	NaN	NaN	7.0605	6.9440	8.0014	9.2414	9.9957
	70°	NaN	NaN	NaN	NaN	NaN	NaN	NaN	NaN	NaN	NaN	7.0826	6.9882	7.8742	9.7894
	75°	NaN	NaN	NaN	NaN	NaN	NaN	NaN	NaN	NaN	NaN	NaN	7.1230	7.5669	8.9129
	80°	NaN	NaN	NaN	NaN	NaN	NaN	NaN	NaN	NaN	NaN	NaN	NaN	8.1780	7.6105
	85°	NaN	NaN	NaN	NaN	NaN	NaN	NaN	NaN	NaN	NaN	NaN	NaN	NaN	6.9258

Table A3. Shape, a, and rate, b parameter for $f_\Theta(\theta)$ and $F_\Theta(\theta)$ for $h_k = 1400$ km.

		Shape parameter, a													
								Orbit inclination, i							
ES latitude, ϕ		20°	25°	30°	35°	40°	45°	50°	55°	60°	65°	70°	75°	80°	85°
	0°	2.0032	1.8456	1.8402	1.6716	1.7100	1.7340	1.7449	1.7511	1.7543	1.7637	1.7632	1.7651	1.7702	1.7679
	5°	2.1655	1.9754	1.6522	1.6934	1.7231	1.7409	1.7490	1.7540	1.7591	1.7614	1.7649	1.7663	1.7667	1.7666
	10°	1.6677	1.8992	1.9435	1.8657	1.6721	1.7127	1.7353	1.7436	1.7499	1.7599	1.7588	1.7621	1.7586	1.7636
	15°	1.6088	1.7424	1.9101	1.9636	1.8729	1.6705	1.7134	1.7323	1.7393	1.7562	1.7543	1.7568	1.7675	1.7563
	20°	1.6203	1.6962	1.7952	1.9295	1.9741	1.8764	1.6701	1.7079	1.7269	1.7506	1.7481	1.7527	1.7548	1.7517
	25°	NaN	1.5976	1.6651	1.8075	1.9402	1.9817	1.8801	1.6663	1.7024	1.7427	1.7421	1.7498	1.7327	1.7498
	30°	NaN	NaN	1.6369	1.6635	1.8108	1.9413	1.9814	1.8736	1.6573	1.7178	1.7295	1.7404	1.7402	1.7417
	35°	NaN	NaN	NaN	1.6445	1.6715	1.8127	1.9434	1.9763	1.8665	1.6778	1.7050	1.7302	1.7423	1.7363
	40°	NaN	NaN	NaN	NaN	1.6456	1.6718	1.8117	1.9378	1.9673	1.8869	1.6650	1.7061	1.7206	1.7275
	45°	NaN	NaN	NaN	NaN	NaN	1.6449	1.6710	1.8075	1.9279	1.9885	1.8715	1.6685	1.7048	1.7163
	50°	NaN	NaN	NaN	NaN	NaN	NaN	1.6440	1.6645	1.7968	1.9466	1.9668	1.8700	1.6689	1.6933
	55°	NaN	NaN	NaN	NaN	NaN	NaN	NaN	1.6368	1.6538	1.8118	1.9217	1.9582	1.8582	1.6591
	60°	NaN	NaN	NaN	NaN	NaN	NaN	NaN	NaN	1.6265	1.6684	1.7868	1.9054	1.9326	1.8204
	65°	NaN	NaN	NaN	NaN	NaN	NaN	NaN	NaN	NaN	1.6403	1.6414	1.7611	1.8477	1.8296
	70°	NaN	NaN	NaN	NaN	NaN	NaN	NaN	NaN	NaN	NaN	1.6074	1.5970	1.6574	1.9844
	75°	NaN	NaN	NaN	NaN	NaN	NaN	NaN	NaN	NaN	NaN	NaN	1.5302	1.7400	2.0124
	80°	NaN	NaN	NaN	NaN	NaN	NaN	NaN	NaN	NaN	NaN	NaN	NaN	1.9051	1.9381
	85°	NaN	NaN	NaN	NaN	NaN	NaN	NaN	NaN	NaN	NaN	NaN	NaN	NaN	1.9298

Table A3. *Cont.*

							Rate parameter, $b \times 10^{-2}$							
							Orbit inclination, i							
		20°	25°	30°	35°	40°	45°	50°	55°	60°	65°	70°	75°	80°	85°
ES latitude, ϕ	0°	9.4854	9.6587	9.9384	9.3116	9.0728	9.0317	8.9969	8.9756	8.9583	8.9661	8.9488	8.9440	8.9534	8.9475
	5°	10.3295	10.6508	9.6070	9.1546	9.0625	9.0309	8.9989	8.9776	8.9675	8.9546	8.9515	8.9460	8.9355	8.9351
	10°	7.6674	8.8435	9.6689	9.9236	9.2595	9.0497	9.0065	8.9747	8.9588	8.9634	8.9429	8.9396	8.9208	8.9386
	15°	6.7932	7.5227	8.7334	9.6790	9.9223	9.2504	9.0394	8.9900	8.9562	8.9731	8.9424	8.9348	8.9586	8.9182
	20°	6.9480	6.8614	7.6001	8.7492	9.6886	9.9382	9.2493	9.0196	8.9703	8.9873	8.9450	8.9408	8.9327	8.9139
	25°	NaN	6.7756	6.7266	7.5953	8.7531	9.7130	9.9496	9.2345	9.0001	9.0167	8.9594	8.9542	8.8746	8.9231
	30°	NaN	NaN	6.8307	6.6879	7.5795	8.7557	9.7114	9.9298	9.2092	9.0479	8.9779	8.9616	8.9292	8.9203
	35°	NaN	NaN	NaN	6.8201	6.6873	7.5802	8.7575	9.6934	9.9105	9.2742	9.0104	8.9880	8.9710	8.9328
	40°	NaN	NaN	NaN	NaN	6.8046	6.6837	7.5755	8.7420	9.6708	9.9675	9.2362	9.0237	8.9654	8.9487
	45°	NaN	NaN	NaN	NaN	NaN	6.8013	6.6834	7.5681	8.7196	9.7269	9.9219	9.2539	9.0292	8.9823
	50°	NaN	NaN	NaN	NaN	NaN	NaN	6.8058	6.6760	7.5510	8.7727	9.6706	9.9225	9.2661	9.0351
	55°	NaN	NaN	NaN	NaN	NaN	NaN	NaN	6.8030	6.6671	7.6017	8.7233	9.6600	9.8991	9.2876
	60°	NaN	NaN	NaN	NaN	NaN	NaN	NaN	NaN	6.8055	6.7306	7.5743	8.7167	9.6255	9.8749
	65°	NaN	NaN	NaN	NaN	NaN	NaN	NaN	NaN	NaN	6.8898	6.7259	7.5850	8.6605	9.6210
	70°	NaN	NaN	NaN	NaN	NaN	NaN	NaN	NaN	NaN	NaN	6.9155	6.7657	7.6365	9.4222
	75°	NaN	NaN	NaN	NaN	NaN	NaN	NaN	NaN	NaN	NaN	NaN	7.1459	7.5283	8.4633
	80°	NaN	NaN	NaN	NaN	NaN	NaN	NaN	NaN	NaN	NaN	NaN	NaN	7.9448	7.2608
	85°	NaN	NaN	NaN	NaN	NaN	NaN	NaN	NaN	NaN	NaN	NaN	NaN	NaN	6.6159

Table A4. Shape, a, and rate, b parameter for $f_\Theta(\theta)$ and $F_\Theta(\theta)$ for $h_k = 1600$ km.

							Shape parameter, a							
							Orbit inclination, i							
		20°	25°	30°	35°	40°	45°	50°	55°	60°	65°	70°	75°	80°	85°
ES latitude, ϕ	0°	2.0512	1.9235	1.8537	1.7406	1.6844	1.7333	1.7450	1.7537	1.7570	1.7618	1.7630	1.7656	1.7689	1.7680
	5°	2.1942	2.0607	1.7717	1.6601	1.7164	1.7388	1.7508	1.7558	1.7613	1.7647	1.7677	1.7701	1.7681	1.7687
	10°	1.7309	1.8164	1.9409	1.9105	1.7446	1.6908	1.7336	1.7464	1.7500	1.7590	1.7580	1.7620	1.7657	1.7720
	15°	1.5853	1.7549	1.8968	1.9701	1.9155	1.7528	1.6897	1.7309	1.7402	1.7591	1.7563	1.7607	1.7604	1.7762
	20°	1.6239	1.6532	1.7886	1.9184	1.9776	1.9304	1.7523	1.6878	1.7279	1.7522	1.7538	1.7591	1.7573	1.7647
	25°	NaN	1.6489	1.6724	1.7993	1.9237	1.9910	1.9325	1.7512	1.6846	1.7395	1.7450	1.7540	1.7550	1.7582
	30°	NaN	NaN	1.6547	1.6754	1.7984	1.9330	1.9935	1.9317	1.7473	1.6970	1.7291	1.7440	1.7478	1.7502
	35°	NaN	NaN	NaN	1.6560	1.6715	1.8062	1.9326	1.9893	1.9235	1.7578	1.6838	1.7267	1.7376	1.7407
	40°	NaN	NaN	NaN	NaN	1.6526	1.6789	1.8052	1.9302	1.9815	1.9365	1.7465	1.6836	1.7237	1.7222
	45°	NaN	NaN	NaN	NaN	NaN	1.6606	1.6794	1.8026	1.9224	1.9943	1.9211	1.7483	1.6818	1.7154
	50°	NaN	NaN	NaN	NaN	NaN	NaN	1.6590	1.6753	1.7936	1.9316	1.9736	1.9156	1.7457	1.6728
	55°	NaN	NaN	NaN	NaN	NaN	NaN	NaN	1.6550	1.6660	1.8010	1.9081	1.9610	1.9026	1.7332
	60°	NaN	NaN	NaN	NaN	NaN	NaN	NaN	NaN	1.6450	1.6730	1.7752	1.8866	1.9276	1.8312
	65°	NaN	NaN	NaN	NaN	NaN	NaN	NaN	NaN	NaN	1.6509	1.6425	1.7422	1.7973	1.9083
	70°	NaN	NaN	NaN	NaN	NaN	NaN	NaN	NaN	NaN	NaN	1.6107	1.5687	1.7232	2.0355
	75°	NaN	NaN	NaN	NaN	NaN	NaN	NaN	NaN	NaN	NaN	NaN	1.6165	1.8079	2.0373
	80°	NaN	NaN	NaN	NaN	NaN	NaN	NaN	NaN	NaN	NaN	NaN	NaN	1.9495	1.9564
	85°	NaN	NaN	NaN	NaN	NaN	NaN	NaN	NaN	NaN	NaN	NaN	NaN	NaN	1.9366

							Rate parameter, $b \times 10^{-2}$							
							Orbit inclination, i							
		20°	25°	30°	35°	40°	45°	50°	55°	60°	65°	70°	75°	80°	85°
ES latitude, ϕ	0°	9.1342	9.5174	9.5944	9.3135	8.7818	8.7807	8.7426	8.7223	8.6959	8.6881	8.6741	8.6742	8.6742	8.6766
	5°	9.7966	10.3752	9.7881	8.8842	8.8110	8.7746	8.7474	8.7192	8.7038	8.6957	8.6905	8.6867	8.6685	8.6762
	10°	7.6274	8.2279	9.2022	9.6381	9.2692	8.7674	8.7567	8.7267	8.6895	8.6886	8.6647	8.6670	8.6692	8.6931
	15°	6.6080	7.3062	8.2988	9.2230	9.6287	9.2905	8.7538	8.7376	8.6938	8.7173	8.6821	8.6788	8.6626	8.7096
	20°	6.8179	6.5901	7.3007	8.3061	9.2175	9.6659	9.2847	8.7465	8.7247	8.7338	8.7024	8.6934	8.6663	8.6864
	25°	NaN	6.7566	6.5695	7.2854	8.3000	9.2485	9.6723	9.2833	8.7361	8.7599	8.7128	8.7018	8.6785	8.6748
	30°	NaN	NaN	6.7136	6.5383	7.2623	8.3150	9.2536	9.6715	9.2719	8.7749	8.7319	8.7107	8.6841	8.6771
	35°	NaN	NaN	NaN	6.6823	6.5095	7.2720	8.3100	9.2416	9.6468	9.3038	8.7408	8.7284	8.6969	8.6821
	40°	NaN	NaN	NaN	NaN	6.6578	6.5203	7.2661	8.3021	9.2209	9.6820	9.2726	8.7454	8.7268	8.6719
	45°	NaN	NaN	NaN	NaN	NaN	6.6751	6.5234	7.2610	8.2885	9.2550	9.6393	9.2808	8.7496	8.7354
	50°	NaN	NaN	NaN	NaN	NaN	NaN	6.6781	6.5207	7.2526	8.3177	9.2060	9.6285	9.2832	8.7699
	55°	NaN	NaN	NaN	NaN	NaN	NaN	NaN	6.6827	6.5200	7.2868	8.2773	9.1899	9.6123	9.3061
	60°	NaN	NaN	NaN	NaN	NaN	NaN	NaN	NaN	6.6921	6.5679	7.2625	8.2650	9.1523	9.5221
	65°	NaN	NaN	NaN	NaN	NaN	NaN	NaN	NaN	NaN	6.7600	6.5619	7.2685	8.1612	9.4608
	70°	NaN	NaN	NaN	NaN	NaN	NaN	NaN	NaN	NaN	NaN	6.7813	6.5664	7.5990	9.0780
	75°	NaN	NaN	NaN	NaN	NaN	NaN	NaN	NaN	NaN	NaN	NaN	7.2701	7.4458	8.0980
	80°	NaN	NaN	NaN	NaN	NaN	NaN	NaN	NaN	NaN	NaN	NaN	NaN	7.7161	6.9900
	85°	NaN	NaN	NaN	NaN	NaN	NaN	NaN	NaN	NaN	NaN	NaN	NaN	NaN	6.3842

Table A5. Shape, a, and rate, b parameter for $f_\Theta(\theta)$ and $F_\Theta(\theta)$ for $h_k = 1800$ km.

| | | Shape parameter, a | | | | | | | | | | | | |
| | | Orbit inclination, i | | | | | | | | | | | | |
		20°	25°	30°	35°	40°	45°	50°	55°	60°	65°	70°	75°	80°	85°
ES latitude, ϕ	0°	2.0864	1.9884	1.8169	1.7943	1.6412	1.7248	1.7438	1.7550	1.7609	1.7664	1.7714	1.7707	1.7703	1.7721
	5°	2.2116	2.1213	1.8764	1.5897	1.7065	1.7368	1.7513	1.7599	1.7645	1.7665	1.7747	1.7742	1.7748	1.7745
	10°	1.7923	1.7348	1.9270	1.9337	1.8142	1.6412	1.7269	1.7468	1.7539	1.7655	1.7619	1.7680	1.7676	1.7691
	15°	1.5297	1.7356	1.8821	1.9664	1.9519	1.8180	1.6389	1.7278	1.7445	1.7576	1.7543	1.7642	1.7655	1.7646
	20°	1.6239	1.6565	1.7779	1.9041	1.9797	1.9592	1.8191	1.6388	1.7249	1.7473	1.7510	1.7577	1.7601	1.7588
	25°	NaN	1.6591	1.6767	1.7915	1.9167	1.9911	1.9649	1.8231	1.6361	1.7307	1.7462	1.7516	1.7553	1.7561
	30°	NaN	NaN	1.6676	1.6810	1.7985	1.9204	1.9936	1.9658	1.8162	1.6417	1.7218	1.7392	1.7478	2.1573
	35°	NaN	NaN	NaN	1.6709	1.6870	1.8007	1.9238	1.9959	1.9608	1.8265	1.6329	1.7194	1.7363	2.6343
	40°	NaN	NaN	NaN	NaN	1.6724	1.6879	1.8011	1.9240	1.9903	1.9688	1.8155	1.6332	1.7164	1.7308
	45°	NaN	NaN	NaN	NaN	NaN	1.6746	1.6876	1.8010	1.9156	1.9949	1.9541	1.8130	1.6334	1.7091
	50°	NaN	NaN	NaN	NaN	NaN	NaN	1.6731	1.6855	1.7926	1.9201	1.9743	1.9451	1.8060	1.6291
	55°	NaN	NaN	NaN	NaN	NaN	NaN	NaN	1.6700	1.6776	1.7929	1.8943	1.9580	1.9234	1.7826
	60°	NaN	NaN	NaN	NaN	NaN	NaN	NaN	NaN	1.6603	1.6760	1.7682	1.8697	1.9093	1.7970
	65°	NaN	NaN	NaN	NaN	NaN	NaN	NaN	NaN	NaN	1.6585	1.6442	1.7219	1.7171	1.9728
	70°	NaN	NaN	NaN	NaN	NaN	NaN	NaN	NaN	NaN	NaN	1.6097	1.5173	1.7822	2.0709
	75°	NaN	NaN	NaN	NaN	NaN	NaN	NaN	NaN	NaN	NaN	NaN	1.6951	1.8594	2.0537
	80°	NaN	NaN	NaN	NaN	NaN	NaN	NaN	NaN	NaN	NaN	NaN	NaN	1.9798	1.9692
	85°	NaN	NaN	NaN	NaN	NaN	NaN	NaN	NaN	NaN	NaN	NaN	NaN	NaN	1.9458

| | | Rate parameter, $b \times 10^{-2}$ | | | | | | | | | | | | |
| | | Orbit inclination, i | | | | | | | | | | | | |
		20°	25°	30°	35°	40°	45°	50°	55°	60°	65°	70°	75°	80°	85°
ES latitude, ϕ	0°	8.8196	9.3308	9.1861	9.2502	8.5489	8.5670	8.5293	8.5090	8.4907	8.4811	8.4812	8.4660	8.4518	8.4567
	5°	9.3468	10.0676	9.8322	8.6449	8.6084	8.5685	8.5380	8.5164	8.4966	8.4795	8.4907	8.4766	8.4677	8.4626
	10°	7.5795	7.7597	8.7975	9.3397	9.2387	8.5089	8.5396	8.5159	8.4881	8.4885	8.4597	8.4644	8.4493	8.4529
	15°	6.4075	7.0547	7.9508	8.8251	9.3577	9.2367	8.4880	8.5318	8.5000	8.4924	8.4577	8.4679	8.4561	8.4484
	20°	6.7047	6.4674	7.0536	7.9492	8.8321	9.3703	9.2355	8.4831	8.5173	8.5034	8.4701	8.4654	8.4547	8.4423
	25°	NaN	6.6531	6.4384	7.0389	7.9545	8.8556	9.3818	9.2454	8.4746	8.5307	8.5031	8.4748	8.4606	8.4516
	30°	NaN	NaN	6.6111	6.4100	7.0327	7.9549	8.8569	9.3822	9.2265	8.4932	8.5107	8.4837	8.4704	11.6161
	35°	NaN	NaN	NaN	6.5857	6.4036	7.0290	7.9578	8.8601	9.3696	9.2579	8.4709	8.5067	8.4846	17.2642
	40°	NaN	NaN	NaN	NaN	6.5714	6.4000	7.0262	7.9577	8.8481	9.3920	9.2259	8.4777	8.5106	8.4929
	45°	NaN	NaN	NaN	NaN	NaN	6.5756	6.3990	7.0300	7.9416	8.8626	9.3530	9.2231	8.4927	8.5222
	50°	NaN	NaN	NaN	NaN	NaN	NaN	6.5769	6.4042	7.0208	7.9641	8.8185	9.3373	9.2172	8.5267
	55°	NaN	NaN	NaN	NaN	NaN	NaN	NaN	6.5872	6.4044	7.0426	7.9226	8.7991	9.3054	9.2177
	60°	NaN	NaN	NaN	NaN	NaN	NaN	NaN	NaN	6.5948	6.4369	7.0282	7.9128	8.7370	9.1203
	65°	NaN	NaN	NaN	NaN	NaN	NaN	NaN	NaN	NaN	6.6509	6.4375	7.0166	7.7056	9.2740
	70°	NaN	NaN	NaN	NaN	NaN	NaN	NaN	NaN	NaN	NaN	6.6702	6.3810	7.5445	8.7644
	75°	NaN	NaN	NaN	NaN	NaN	NaN	NaN	NaN	NaN	NaN	NaN	7.3375	7.3430	7.7934
	80°	NaN	NaN	NaN	NaN	NaN	NaN	NaN	NaN	NaN	NaN	NaN	NaN	7.5042	6.7684
	85°	NaN	NaN	NaN	NaN	NaN	NaN	NaN	NaN	NaN	NaN	NaN	NaN	NaN	6.2083

Table A6. Shape, a, and rate, b parameter for $f_\Theta(\theta)$ and $F_\Theta(\theta)$ for $h_k = 2000$ km.

| | | Shape parameter, a | | | | | | | | | | | | |
| | | Orbit inclination, i | | | | | | | | | | | | |
		20°	25°	30°	35°	40°	45°	50°	55°	60°	65°	70°	75°	80°	85°
ES latitude, ϕ	0°	2.1133	2.0407	1.8663	1.8277	1.6851	1.7151	1.7434	1.7543	1.7634	1.7690	1.7730	1.7757	1.7747	1.7801
	5°	2.2236	2.1664	1.9608	1.6623	1.6910	1.7328	1.7511	1.7601	1.7679	1.7742	1.7770	1.7779	1.7768	1.7844
	10°	1.8464	1.7634	1.9054	1.9463	1.8627	1.6881	1.7180	1.7460	1.7568	1.7711	1.7679	1.7719	1.7718	1.7758
	15°	1.5715	1.7152	1.8664	1.9610	1.9725	1.8702	1.6873	1.7189	1.7443	1.7574	1.7629	1.7669	1.7658	1.7698
	20°	1.6176	1.6582	1.7717	1.8939	1.9824	1.9815	1.8733	1.6862	1.7153	1.7485	1.7528	1.7588	1.7612	1.7578
	25°	NaN	1.6675	1.6827	1.7860	1.9111	1.9917	1.9878	1.8782	1.6867	1.7171	1.7421	1.7518	1.7559	1.7612
	30°	NaN	NaN	1.6781	1.6874	1.7972	1.9143	1.9936	1.9895	1.8745	1.6893	1.7104	1.7379	1.7453	1.7527
	35°	NaN	NaN	NaN	1.6799	1.6963	1.7986	1.9170	1.9978	1.9867	1.8777	1.6819	1.7081	1.7317	1.7418
	40°	NaN	NaN	NaN	NaN	1.6883	1.6955	1.7993	1.9211	1.9909	1.9881	1.8674	1.6807	1.7453	1.7237
	45°	NaN	NaN	NaN	NaN	NaN	1.6861	1.6955	1.7996	1.9118	1.9915	1.9746	1.8627	1.6781	1.6952
	50°	NaN	NaN	NaN	NaN	NaN	NaN	1.6849	1.6942	1.7925	1.9092	1.9715	1.9623	1.8499	1.6713
	55°	NaN	NaN	NaN	NaN	NaN	NaN	NaN	1.6842	1.6870	1.7886	1.8857	1.9512	1.9312	1.8113
	60°	NaN	NaN	NaN	NaN	NaN	NaN	NaN	NaN	1.6740	1.6828	1.7609	1.8523	1.8820	1.8489
	65°	NaN	NaN	NaN	NaN	NaN	NaN	NaN	NaN	NaN	1.6644	1.6463	1.6979	1.7474	2.0230
	70°	NaN	NaN	NaN	NaN	NaN	NaN	NaN	NaN	NaN	NaN	1.6026	1.5620	1.8330	2.0953
	75°	NaN	NaN	NaN	NaN	NaN	NaN	NaN	NaN	NaN	NaN	NaN	1.7623	1.8999	2.0647
	80°	NaN	NaN	NaN	NaN	NaN	NaN	NaN	NaN	NaN	NaN	NaN	NaN	2.0023	1.9790
	85°	NaN	NaN	NaN	NaN	NaN	NaN	NaN	NaN	NaN	NaN	NaN	NaN	NaN	1.9584

Table A6. *Cont.*

| | | **Rate parameter,** $b \times 10^{-2}$ | | | | | | | | | | | | |
| | | Orbit inclination, i | | | | | | | | | | | | |
		20°	25°	30°	35°	40°	45°	50°	55°	60°	65°	70°	75°	80°	85°
ES latitude, ϕ	0°	8.5432	9.1399	9.0974	9.1202	8.5964	8.3881	8.3606	8.3318	8.3159	8.3014	8.2988	8.2944	8.2826	8.2926
	5°	8.9698	9.7735	9.7793	8.8345	8.4312	8.3942	8.3640	8.3383	8.3238	8.3219	8.3099	8.3004	8.2846	8.3080
	10°	7.5197	7.6751	8.4453	9.0607	9.1342	8.5590	8.3611	8.3452	8.3173	8.3285	8.2929	8.2901	8.2799	8.2844
	15°	6.4748	6.8508	7.6660	8.4985	9.0872	9.1384	8.5429	8.3518	8.3260	8.3161	8.2966	8.2894	8.2711	8.2777
	20°	6.6023	6.3675	6.8675	7.6707	8.5195	9.1022	9.1422	8.5359	8.3333	8.3363	8.2948	8.2840	8.2721	8.2523
	25°	NaN	6.5665	6.3426	6.8486	7.6805	8.5340	9.1141	9.1546	8.5354	8.3373	8.3136	8.2950	8.2791	8.2811
	30°	NaN	NaN	6.5273	6.3099	6.8477	7.6785	8.5321	9.1163	9.1432	8.5477	8.3180	8.3065	8.2834	8.2823
	35°	NaN	NaN	NaN	6.4935	6.3074	6.8422	7.6793	8.5404	9.1072	9.1546	8.5241	8.3179	8.2998	8.2902
	40°	NaN	NaN	NaN	NaN	6.4955	6.3003	6.8407	7.6895	8.5230	9.1128	9.1241	8.5273	8.2834	8.2992
	45°	NaN	NaN	NaN	NaN	NaN	6.4891	6.3005	6.8438	7.6709	8.5293	9.0800	9.1167	8.5360	8.3301
	50°	NaN	NaN	NaN	NaN	NaN	NaN	6.4918	6.3065	6.8397	7.6762	8.4912	9.0590	9.1003	8.5634
	55°	NaN	NaN	NaN	NaN	NaN	NaN	NaN	6.5068	6.3106	6.8529	7.6475	8.4680	9.0132	9.0711
	60°	NaN	NaN	NaN	NaN	NaN	NaN	NaN	NaN	6.5164	6.3391	6.8386	7.6242	8.3726	9.0403
	65°	NaN	NaN	NaN	NaN	NaN	NaN	NaN	NaN	NaN	6.5571	6.3402	6.8044	7.6273	9.0766
	70°	NaN	NaN	NaN	NaN	NaN	NaN	NaN	NaN	NaN	NaN	6.5688	6.4544	7.4789	8.4834
	75°	NaN	NaN	NaN	NaN	NaN	NaN	NaN	NaN	NaN	NaN	NaN	7.3598	7.2366	7.5378
	80°	NaN	NaN	NaN	NaN	NaN	NaN	NaN	NaN	NaN	NaN	NaN	NaN	7.3162	6.5865
	85°	NaN	NaN	NaN	NaN	NaN	NaN	NaN	NaN	NaN	NaN	NaN	NaN	NaN	6.0765

Table A7. Data for weights matrices, w_a, and w_b.

| | | **Weight matrix,** w_a | | | | | | | | | | | | |
| | | Orbit inclination, i | | | | | | | | | | | | |
		20°	25°	30°	35°	40°	45°	50°	55°	60°	65°	70°	75°	80°	85°
ES latitude, ϕ	0°	1	1	1	1	1	1	1	1	1	1	1	1	1	1
	5°	1	0.9434	1	0.9259	1	1	1	1	1	1	1	1	1	1
	10°	1	1	1	1	1	1	1	1	1	1	1	1	1	1
	15°	1	1	1	1	1	1	1	1	1	1	1	1	1	1
	20°	1	1	1	1	1	1	1	1	1	1	1	1	1	1
	25°	1	1	1	1	1	1	1	1	1	1	1	1	1	1
	30°	1	1	1	1	1	1	1	1	1	1	1	1	1	1
	35°	1	1	1	1	1	1	1	1	1	1	1	1	1	1
	40°	1	1	1	1	1	1	1	1	1	1	1	1	1	1
	45°	1	1	1	1	1	1	1	1	1	1	1	1	1	1
	50°	1	1	1	1	1	1	1	1	1	1	1	1	1	1
	55°	1	1	1	1	1	1	1	1	1	1	1	1	1	1
	60°	1	1	1	1	1	1	1	1	1	1	1	1	1	1
	65°	1	1	1	1	1	1	1	1	1	1	1	1	1	0.9524
	70°	1	1	1	1	1	1	1	1	1	1	1	1	1	1
	75°	1	1	1	1	1	1	1	1	1	1	1	0.9524	1	1
	80°	1	1	1	1	1	1	1	1	1	1	1	1	1	1.1765
	85°	1	1	1	1	1	1	1	1	1	1	1	1	1	1.2500

| | | **Weight matrix,** w_b | | | | | | | | | | | | |
| | | Orbit inclination, i | | | | | | | | | | | | |
		20°	25°	30°	35°	40°	45°	50°	55°	60°	65°	70°	75°	80°	85°
ES latitude, ϕ	0°	1	1	1	1	1	1	1	1	1	1	1	1	1	1
	5°	1	1.0526	1	0.9524	1	1	1	1	1	1	1	1	1	1
	10°	1	1	1	1	1	1	1	1	1	1	1	1	1	1
	15°	1	1	1	1	1	1	1	1	1	1	1	1	1	1
	20°	1	1	1	1	1	1	1	1	1	1	1	1	1	1
	25°	1	1	1	1	1	1	1	1	1	1	1	1	1	1
	30°	1	1	1	1	1	1	1	1	1	1	1	1	1	1
	35°	1	1	1	1	1	1	1	1	1	1	1	1	1	1
	40°	1	1	1	1	1	1	1	1	1	1	1	1	1	1
	45°	1	1	1	1	1	1	1	1	1	1	1	1	1	1
	50°	1	1	1	1	1	1	1	1	1	1	1	1	1	1
	55°	1	1	1	1	1	1	1	1	1	1	1	1	1	1
	60°	1	1	1	1	1	1	1	1	1	1	1	1	1	1
	65°	1	1	1	1	1	1	1	1	1	1	1	1	1	1
	70°	1	1	1	1	1	1	1	1	1	1	1	1	1	1
	75°	1	1	1	1	1	1	1	1	1	1	1	1	1	1
	80°	1	1	1	1	1	1	1	1	1	1	1	1	1	1
	85°	1	1	1	1	1	1	1	1	1	1	1	1	1	1.1111

Appendix C

This appendix contains the well-known four-point-mesh interpolation method, which was applied in this paper to determine the shape, a, and rate, b, parameters from matrices A_h and B_h for a given set of coordinates (i, ϕ). More interpolation methods are widely available in the literature.

Interpolation for a Four-Point Mesh

The bilinear method is proposed as the method to find the query point inside the mesh since it is a well-known method and widely applied for interpolation.

We can obtain the values for a query point as in Figure A2 for a shape parameter value a using bilinear interpolation as follows

$$a(\phi, i) = w_1 a(\phi_2, i_1) + w_2 a(\phi_1, i_1) + w_3 a(\phi_2, i_2) + w_4 a(\phi_1, i_2), \tag{A1}$$

similarly, we can obtain the scale parameter, b, using bilinear interpolation as follows

$$b(\phi, i) = w_1 b(\phi_2, i_1) + w_2 b(\phi_1, i_1) + w_3 b(\phi_2, i_2) + w_4 b(\phi_1, i_2), \tag{A2}$$

where the wights w_1, w_2, w_3 and w_4 are given by

$$w_1 = (i_2 - i)(\phi_1 - \phi)/[(i_2 - i_1)(\phi_1 - \phi_2)] \tag{A3}$$

$$w_2 = (i_2 - i)(\phi - \phi_2)/[(i_2 - i_1)(\phi_1 - \phi_2)] \tag{A4}$$

$$w_3 = (i - i_1)(\phi_1 - \phi)/[(i_2 - i_1)(\phi_1 - \phi_2)] \tag{A5}$$

$$w_4 = (i - i_1)(\phi_1 - \phi_2)/[(i_2 - i_1)(\phi_1 - \phi_2)] \tag{A6}$$

where the known values for $a(i_n, \phi_m), \ldots, a(i_{n+1}, \phi_{m+1})$ and $b(i_n, \phi_m), \ldots, b(i_{n+1}, \phi_{m+1})$, can be obtained from matrices A_h and B_h.

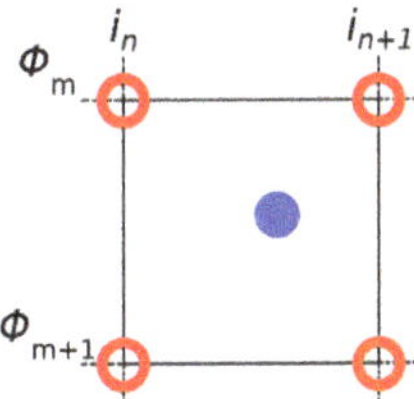

Figure A2. $F_\theta(\Theta)$ for two orbit configurations.

References

1. Crowe, K.; Raines, R. A model to describe the distribution of transmission path elevation angles to the Iridium and Globalstar satellite systems. *IEEE Commun. Lett.* **1999**, *3*, 242–244. [CrossRef]
2. Li, S.Y.; Liu, C. An analytical model to predict the probability density function of elevation angles for LEO satellite systems. *IEEE Commun. Lett.* **2002**, *6*, 138–140. [CrossRef]
3. Corazza, G.; Vatalaro, F. A statistical model for land mobile satellite channels and its application to nongeostationary orbit systems. *IEEE Trans. Veh. Technol.* **1994**, *43*, 738–742. [CrossRef]
4. Davoli, F.; Kouroglorgas, C.; Marchese, M.; Panagopoulos, A.; Patrone, F. Small satellites and CubeSats: Survey of structures, architectures, and protocols. *Int. J. Satell. Commun. Netw.* **2019**, *37*, 343–359. [CrossRef]
5. del Portillo, I.; Cameron, B.G.; Crawley, E.F. A technical comparison of three low earth orbit satellite constellation systems to provide global broadband. *Acta Astronaut.* **2019**, *159*, 123–135. [CrossRef]

6.   Popescu, O. Power Budgets for CubeSat Radios to Support Ground Communications and Inter-Satellite Links. *IEEE Access* **2017**, *5*, 12618–12625. [CrossRef]

7.   Hu, J.; Li, G.; Bian, D.; Gou, L.; Wang, C. Optimal Power Control for Cognitive LEO Constellation With Terrestrial Networks. *IEEE Commun. Lett.* **2020**, *24*, 622–625. [CrossRef]

8.   Chobotov, V.A. *Orbital Mechanics*, 3rd ed.; AIAA: Reston, VA, USA 2002.

9.   Gongora-Torres, J.M.; Vargas-Rosales, C.; Aragón-Zavala, A.; Villalpando-Hernandez, R. Link Budget Analysis for LEO Satellites Based on the Statistics of the Elevation Angle. *IEEE Access* **2022**, *10*, 14518–14528. [CrossRef]

10.   Kolawole, M.O. *Satellite Communication Engineering*; Marcel Dekker, Inc.: New York, NY, USA, 2002; Chapter 2.

11.   Zhengsheng, C.; Qinghua, Z.; Xuerui, L.; Dashuang, S.; Hao, L.; Runtao, Z.; Jinlong, C. A reference satellite selection method based on maximal elevation angle during the observation period. In Proceedings of the 2017 Forum on Cooperative Positioning and Service, CPGPS 2017, 2017, number 2017 Forum on Cooperative Positioning and Service, CPGPS 2017, Harbin, China, 19–21 May 2017; pp. 268–272.

12.   Seyedi, Y. A Trace-Time Framework for Prediction of Elevation Angle Over Land Mobile LEO Satellites Networks. *Wirel. Pers. Commun.* **2012**, *62*, 793–804. [CrossRef]

13.   Su, Y.; Liu, Y.; Zhou, Y.; Yuan, J.; Cao, H.; Shi, J. Broadband LEO Satellite Communications: Architectures and Key Technologies. *IEEE Wirel. Commun.* **2019**, *26*, 55–61. [CrossRef]

14.   Qu, Z.; Zhang, G.; Cao, H.; Xie, J. LEO Satellite Constellation for Internet of Things. *IEEE Access* **2017**, *5*, 18391–18401. [CrossRef]

15.   Lopez-Salamanca, J.J.; Seman, L.O.; Berejuck, M.D.; Bezerra, E.A. Finite-State Markov Chains Channel Model for CubeSats Communication Uplink. *IEEE Trans. Aerosp. Electron. Syst.* **2020**, *56*, 142–154. [CrossRef]

16.   Tropea, M.; De Rango, F. A Comprehensive Review of Channel Modeling for Land Mobile Satellite Communications. *Electronics* **2022**, *11*, 820. [CrossRef]

17.   Saeed, N.; Elzanaty, A.; Almorad, H.; Dahrouj, H.; Al-Naffouri, T.Y.; Alouini, M.S. CubeSat Communications: Recent Advances and Future Challenges. *IEEE Commun. Surv. Tutor.* **2020**, *22*, 1839–1862. [CrossRef]

18.   Kodheli, O.; Lagunas, E.; Maturo, N.; Sharma, S.K.; Shankar, B.; Montoya, J.F.M.; Duncan, J.C.M.; Spano, D.; Chatzinotas, S.; Kisseleff, S.; et al. Satellite Communications in the New Space Era: A Survey and Future Challenges. *IEEE Commun. Surv. Tutor.* **2021**, *23*, 70–109. [CrossRef]

19.   Arapoglou, P.D.; Liolis, K.; Bertinelli, M.; Panagopoulos, A.; Cottis, P.; De Gaudenzi, R. MIMO over Satellite: A Review. *IEEE Commun. Surv. Tutor.* **2011**, *13*, 27–51. [CrossRef]

20.   Baeza, V.M.; Lagunas, E.; Al-Hraishawi, H.; Chatzinotas, S. An Overview of Channel Models for NGSO Satellites. In Proceedings of the 2022 IEEE 96th Vehicular Technology Conference (VTC2022-Fall), London, UK, 26–29 September 2022; pp. 1–6. [CrossRef]

21.   Loo, C. A statistical model for a land mobile satellite link. *IEEE Trans. Veh. Technol.* **1985**, *34*, 122–127. [CrossRef]

22.   Lutz, E.; Cygan, D.; Dippold, M.; Dolainsky, F.; Papke, W. The Land Mobile Satellite Communication Channel-Recording, Statistics, and Channel Model. *IEEE Trans. Veh. Technol.* **1991**, *40*, 375–386 [CrossRef]

23.   Abdi, A.; Lau, W.C.; Alouini, M.S.; Kaveh, M. A new simple model for land mobile satellite channels: First- and second-order statistics. *IEEE Trans. Wirel. Commun.* **2003**, *2*, 519–528. [CrossRef]

24.   Fontan, F.; Vazquez-Castro, M.; Cabado, C.; Garcia, J.; Kubista, E. Statistical modeling of the LMS channel. *IEEE Trans. Veh. Technol.* **2001**, *50*, 1549–1567. [CrossRef]

25.   National Aeronautics and Space Administration. General Mission Analysis Tool (GMAT) Version R2022a. Available online: http://sourceforge.net/projects/gmat/ (accessed on 15 March 2022).

26.   Gongora-Torres, J.M. Time Series for LEO Available online:.. (accessed on 17 March 2022). [CrossRef]

27.   Al-Hraishawi, H.; Chougrani, H.; Kisseleff, S.; Lagunas, E.; Chatzinotas, S. A Survey on Nongeostationary Satellite Systems: The Communication Perspective. *IEEE Commun. Surv. Tutor.* **2022**, *25*, 101–132. [CrossRef]

28.   Papoulis, A. *Probability, Random Variables, and Stochastic Processes*; McGraw-Hill: New York, NY, USA, 1984. [CrossRef]

Article

Closed-Form Method for Atmospheric Correction (CMAC) of Smallsat Data Using Scene Statistics

David P. Groeneveld [1,*], Timothy A. Ruggles [1] and Bo-Cai Gao [2]

[1] Advanced Remote Sensing, Inc., Hartford, SD 57033, USA
[2] Remote Sensing Division, Naval Research Laboratory, Washington, DC 20375, USA
* Correspondence: david@advancedremotesensing.com; Tel.: +1-(505)-690-6864

Featured Application: CMAC software provides reliable and accurate conversion of degraded top-of-atmosphere imagery to surface reflectance. Accomplished in near real-time using only scene statistics, CMAC can reside in-satellite to support low-latency corrected image output to support smallsat's emerging role for intelligence, surveillance, and reconnaissance.

Abstract: High-cadence Earth observation smallsat images offer potential for near real-time global reconnaissance of all sunlit cloud-free locations. However, these data must be corrected to remove light-transmission effects from variable atmospheric aerosol that degrade image interpretability. Although existing methods may work, they require ancillary data that delays image output, impacting their most valuable applications: intelligence, surveillance, and reconnaissance. Closed-form Method for Atmospheric Correction (CMAC) is based on observed atmospheric effects that brighten dark reflectance while darkening bright reflectance. Using only scene statistics in near real-time, CMAC first maps atmospheric effects across each image, then uses the resulting grayscale to reverse the effects to deliver spatially correct surface reflectance for each pixel. CMAC was developed using the European Space Agency's Sentinel-2 imagery. After a rapid calibration that customizes the method for each imaging optical smallsat, CMAC can be applied to atmospherically correct visible through near-infrared bands. To assess CMAC functionality against user-applied state-of-the-art software, Sen2Cor, extensive tests were made of atmospheric correction performance across dark to bright reflectance under a wide range of atmospheric aerosol on multiple images in seven locations. CMAC corrected images faster, with greater accuracy and precision over a range of atmospheric effects more than twice that of Sen2Cor.

Keywords: smallsat; atmospheric correction; atmospheric aerosol; empirical basis; scene statistics; near real-time

Citation: Groeneveld, D.P.; Ruggles, T.A.; Gao, B.-C. Closed-Form Method for Atmospheric Correction (CMAC) of Smallsat Data Using Scene Statistics. *Appl. Sci.* **2023**, *13*, 6352. https://doi.org/10.3390/app13106352

Academic Editors: Simone Battistini, Filippo Graziani and Mauro Pontani

Received: 29 March 2023
Revised: 4 May 2023
Accepted: 10 May 2023
Published: 22 May 2023

1. Introduction

Atmospheric correction is a satellite image processing step that clarifies images for viewing and converts the digital signal to surface reflectance in preparation for artificial intelligence and other machine analysis. Conversion to surface reflectance removes the effect of variable concentrations of atmospheric aerosol particles such as smoke and dust that degrade images by changing the digital signal and obscuring ground features under haze. Atmospheric correction (AC) provides estimates of surface reflectance that support more robust, accurate, and actionable analytics. AC is required for all optical satellites that measure reflected light in spectral wavelengths from visible through near infrared (VNIR).

Prior to the early 1990s, simplified procedures for the retrieval of land surface reflectance factors were developed mainly from Landsat data. A summary of the methods, including radiative transfer modeling [1,2] and image-based dark object subtraction methods [3,4], were given by Moran et al. [5]. More recently, additional algorithms have been developed for multispectral atmospheric correction; Zhang et al. [6] have provided an extensive citation of AC algorithms, some based on radiative transfer modeling [7], others based on dense dark vegetation [8–10]. Other AC methods have used multi-temporal datasets for aerosol and surface reflectance retrievals [11–13]. However, such datasets are inherently unreliable because conditions may change profoundly between dataset acquisitions, as shown in Figure 1. Few of these algorithms can work in near real-time, and many are not readily automatable. Near real-time corrected image output with accuracy and precision is needed to serve the expected future massive data volume generated by smallsats.

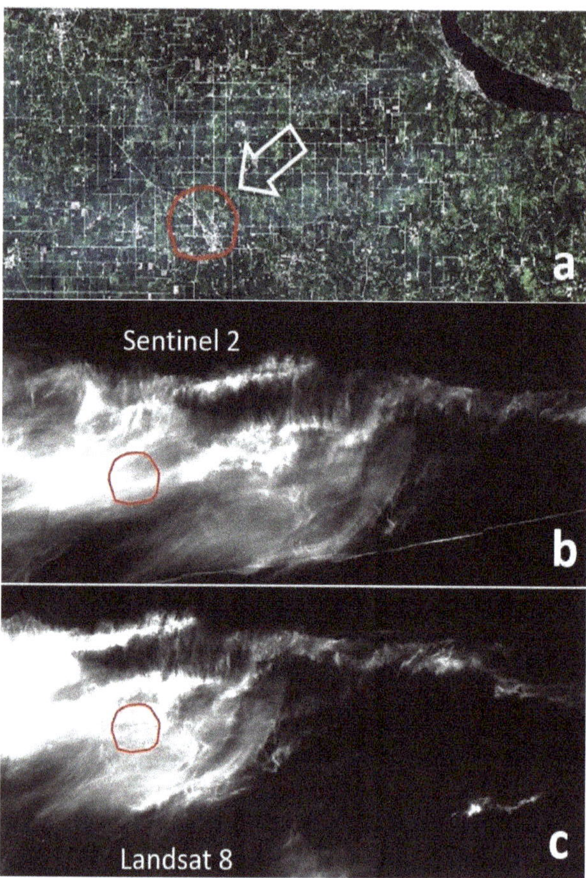

Figure 1. Inexact synchronicity is a source of error for ancillary image application as shown in the following example: (**a**) 14 August 2021 Sentinel 2 TOAR RGB of southern Minnesota, (**b**) the cirrus band (B10) of the same scene, and (**c**) the cirrus band (B09) of Landsat 8 taken about 18 min before. Cirrus affects visible bands as in (**a**).

Closed-form Method for Atmospheric Correction (CMAC) is a software package that was developed using Sentinel-2 (S2) image data specifically for application to smallsats. CMAC works in near real-time to convert top-of-atmosphere reflectance (TOAR) directly to surface reflectance, uniquely using only the statistics from the image. In contrast, automated AC methods in current use are all based upon radiative transfer (RadTran), which requires knowledge of sensor radiometry. This is problematic for smallsats that omit onboard radiometric calibration equipment to conserve weight and size. To maintain AC accuracy for any method, sensor radiometry must be updated periodically to counter changes in sensor response that are well-known to degrade after launch and while in orbit [14]. Hence, for the ideal application of RadTran, sensor radiometry must be tracked for each smallsat. Also required for the RadTran workflow are ancillary data collected by other satellites for the evaluation of atmospheric conditions. Such ancillary data delays AC, thus increasing the latency of the corrected output in direct opposition to the principal value of smallsats: near real-time data for intelligence, surveillance, and reconnaissance afforded by the rapid image cadence from flocks of multiple smallsats.

Although delayed image processing due to ancillary data is problematic, a potentially greater impact may arise from frequently mismatched conditions between datasets caused by atmospheric changes in the elapsed time between image captures (Figure 1). CMAC uses only the statistics from the scene to be corrected, enabling image correction immediately upon download while eliminating correction errors arising through temporal data mismatch.

CMAC development sought a workflow that uses scene statistics for AC rather than applying ancillary data required for RadTran inputs. Examples of ancillary data include aerosol optical thickness (AOT) and water vapor from external satellites such as MODIS [15]. Current RadTran-based methods for smallsats include emulating the radiance of other research-grade satellites such as Landsat 8 and 9 (L8/9) and S2 through the use of datasets that are "harmonized" to support the vicarious calibration of smallsat radiometry [16–18]. Cross-calibration with harmonized datasets provides the basis for estimating smallsat sensor radiance, thereby enabling RadTran application. However, that workflow is inconvenient at least and potentially adds uncertainty.

CMAC operates in near real-time because it uses scene statistics. The accuracy and stability of CMAC version 1.1 output are examined here in relation to that of Sen2Cor, the European Space Agency's (ESA) accepted method for S2 AC that is available as user-applied software [19]. This CMAC version was developed for AC of four-band visual VNIR data represented by S2's blue (B02), green (B03), red (B04) and near NIR (B08a) bands.

Section 2 describes CMAC development to address the AC problem and describes the experimental design to test CMAC. Statistical testing is purposely kept simple to allow expression of the results in clearly understandable terms. Section 3 provides CMAC surface reflectance estimates in comparison to Sen2Cor. This comparison provides the context to judge CMAC accuracy, precision and limitations using both graphical and statistical data representation. The results of both methods are also compared to the uncorrected TOAR data to understand the statistical improvement provided by AC. The data are made available through extensive tables and figures in "live" spreadsheets that are downloadable through Supplementary Materials. Section 4 provides further analysis and context for a better understanding of the AC correction problem and the potential value of CMAC to serve the smallsat industry in its role of providing intelligence, surveillance, and reconnaissance. Section 5 lists five promotional aspects for AC of smallsats resulting from this paper.

2. Materials and Methods

To simplify CMAC development, Level 1C-processed images from Sentinel-2A and Sentinel-2B were considered together, as their responses were not significantly different. The quality and cadence of S2 data, every five days or less, were highly promotional for CMAC empirical testing and development.

CMAC does not differentiate among causal aerosol particles, whether due to smoke, dust, thin clouds, etc., but instead corrects imagery according to the output from an atmospheric index model that is mapped directly from each image's spectral band responses. Aerosol particles are the dominant atmospheric effect resulting in the change from VNIR surface reflectance to the TOAR measured by the satellite [20]. While water vapor is another significant variable that affects atmospheric light transmission, the visible spectral bands are well known to be free of effects from water vapor influences [21], and the S2 NIR B8A in CMAC v1.1 is positioned to avoid water vapor absorption [22]. The B8A has 20 m pixel spatial resolution, while the visible bands have 10 m resolution. This difference has posed no problems for correction or for applications; however, hypothetically, the higher the resolution, the better CMAC can accurately estimate true target surface reflectance because of less reflectance mixing and finer differentiation of atmospheric effects.

2.1. Developing an Atmospheric Index, Atm-I, to Scale the Correction

The VNIR atmospheric effect in imagery is treated in CMAC as a single lump-sum variable. TOAR band statistics are measured across each image within discrete non-overlapping square grid cells. An atmospheric index, Atm-I, is calculated for the S2 blue band (B02) based on VNIR band statistics. This approach resulted from the choice of vegetation as a reference against which atmospheric image effects could be measured. Vegetation is also used as a reference for MODIS, S2 and Landsat as "dense dark vegetation" (DDV) to assess atmospheric aerosol content pursuant to AC application [23]. B02 was selected for Atm-I rather than the shorter-wavelength coastal aerosol band, B01, which saturates at much lower levels of visible haze, thus limiting the potential range for AC. Further reference to the S2 blue band is made specifically to B02.

Development of a statistical model to assess Atm-I as a scalar for CMAC AC began with spectral measurements of continuous DDV plant canopies using an ASD field spectrometer. These data were normalized to S2 responses using the published relative spectral responses for each band [19]. Test applications of this surface reflectance reference measured the lowest non-water blue band values discriminated by NIR threshold in grid cells arrayed across images dominated by DDV from the Amazon Basin and the intensive mechanized farming of the American Midwest. This testing confirmed that a workflow employing vegetation as a surface reflectance reference can produce highly sensitive grayscale maps that emulate visible haze. The remarkable stability of blue reflectance across many plant species is a product of physiologic mechanisms that optimize photosynthesis. Plant carotenoid pigments permit the absorbance of blue light for photosynthesis safely across a range of insolation from shade or cloudy days while avoiding damage during cloudless summer days by dissipating the excess energy as heat [24–26]. Against this known surface reflectance, the atmospheric effect can be readily estimated as the increased reflectance due to backscatter.

Though proving to be an excellent reference for measuring atmospheric effect, a method based solely on vegetation would restrict application only to areas where such plant cover is present [27]. However, DDV can also provide the means to train a predictive statistical model to assess atmospheric effects based solely on scene statistics, regardless of the surface conditions or presence of DDV. The following workflow was used to establish a model to measure the atmospheric effect of images spatially in the form of a single index, Atm-I, from the spectral responses of the four VNIR bands included on virtually all optical smallsats. Translating the vegetation reference would require an enormous amount of ground truth data that did not exist; however, by assuming the surface reflectance of a calibrated vegetation type, Atm-I model development could proceed. Model development was also enabled through the quality, cadence, and resolution of the S2 data. Standardization of the surface reflectance of cultivated crops to serve as DDV provided the means to assume the surface reflectance of target DDV identified on S2 images for model calibration. This workflow proceeded through the following steps:

(1) ASD spectrometer measurements of cultivated crops demonstrated consistent surface reflectance usable as a surrogate blue surface reflectance for the crop when identified in imagery.

(2) A DDV target crop, alfalfa, was adopted as a surrogate estimator of surface reflectance when found in healthy closed canopies. Alfalfa was chosen specifically because it is commonly grown under both irrigated and rainfed conditions. This offered the opportunity to measure spectral responses from DDV index plots and adjacent subplots of variable cover ranging from sparse desert through full DDV expression in humid climates.

(3) Numerous S2 images obtained within two months of summer solstice were assembled that contained the target crop in arid to humid climates and across a range of

atmospheric conditions from clear to hazy. The sampling period was chosen to reduce solar zenith angle effects.

(4)   One or more AOIs were chosen for each image that appeared clearly. For AOIs where haze was present, such haze was examined to assure that it was expressed evenly across the AOI.

(5)   AOIs were chosen to contain an index plot of the most vigorous expression of the alfalfa crop identified through examination of TOAR NDVI. Such plots were chosen as the index samples to serve as the dependent variable that would represent Atm-I.

(6)   Subplots of homogeneous cover were chosen around each index plot to represent various levels of vegetation response ranging from bare ground to DDV across many AOIs set in widely divergent environments.

(7)   The extent of each AOI containing index plots and subplots was restricted to within a diameter of about 5 km to further control variation in Atm-I.

(8)   Values of the four VNIR spectral bands were extracted from the index plots and associated subplots characterized through their maximum, minimum, and median values.

(9)   The lowest consistent TOAR blue band values of the target crop were chosen to represent the surrogate Atm-I of the index plots.

(10)   The extracted VNIR spectral data from the index plots and their associated subplots were pooled and then modeled by multiple linear regression and assuming a negative binomial distribution. Model iterations were tested to determine the best data combinations to predict the surrogate Atm-I blue band TOAR of the index plots from the paired TOAR spectral data.

(11)   The blue and red band reflectance levels were found to be highly significant for the model function, while the green and NIR bands were found to add little. The minimum values for the red and blue bands were chosen as the inputs to the final Atm-I model. Since a binomial distribution was assumed, the estimated value from the Atm-I model was based on a logarithmic scale; the estimated Atm-I level was obtained by applying the exponential function to the model value.

Application of the Atm-I model is made using a non-overlapping grid to select the lowest values of the red and blue bands in each grid cell. The output forms a grayscale, with the brightness constituting the scalar Atm-I levels needed for e AC of the image. Minor smoothing is applied to the Atm-I grayscale image to control pixelation effects due to abrupt changes in estimated Atm-I levels between grid cells. An example Atm-I grayscale output is shown in Figure 2.

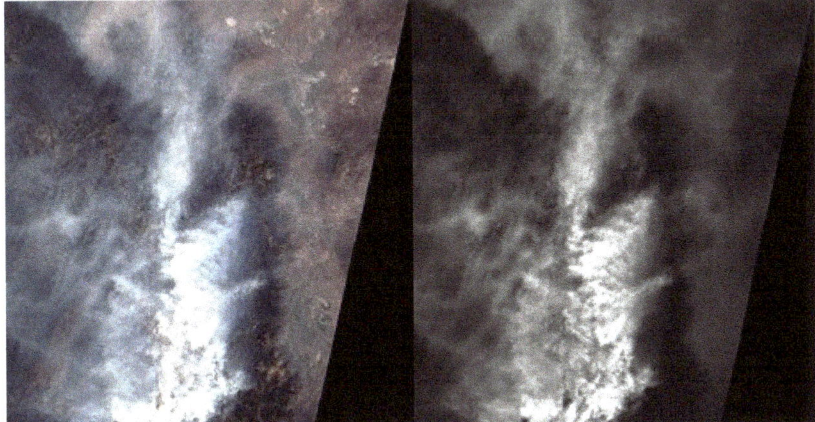

Figure 2. An example 100 m resolution (10 × 10 pixel grid cell) Atm-I grayscale for the 8-22-21 S2 tile over Lake Tahoe, CA, USA. At least some ground signal must remain for correction (exceeded in portions of this image).

The Atm-I model was confirmed by comparison to observable patterns in the grayscale output for images with significant haze mapped using the pattern of blue reflectance over continuous areas of DDV (Figure 2). Because the Atm-I model works directly with the red and blue band responses across the scene, it produces a surrogate spatial estimate for what direct application of dark vegetation would produce were it present in the image. No attempt was made to interrelate Atm-I output to a similar measure used to estimate aerosol particle loading, aerosol optical depth (AOD), which ranges from 0.1 for "clear" atmospheric conditions to 1.0 representing "very hazy" conditions [28] and is derived using sun photometry [29]. In the CMAC application, the Atm-I model output represents the predicted blue band reflectance from DDV and is measured as a digital number—DN (reflectance × 10,000)—that represents a correctable range from the high 700s (extremely clear) to over 1700 (features obscured in a dense haze). The abbreviation DN is used throughout to designate this rescaled reflectance.

2.2. Translating an Observation into the CMAC Equation

The final step in the CMAC workflow reverts TOAR to an estimate of surface reflectance by removing the atmospheric effect measured as Atm-I. This workflow is based on a graphical conceptual model representing the effect from a single Atm-I level as a line in cartesian space defined by surface reflectance and TOAR. Development of this approach began by comparing reflectance between clear and hazy conditions for a given AOI. For images with essentially the same surface reflectance, higher atmospheric aerosol increased dark reflectance from backscatter and decreased bright reflectance through attenuation. This observation was further translated into a graphic conceptual model through inversion of the empirical line method [30], which was adjusted further by placing surface reflectance in both axes (Figure 3). For a single level of Atm-I, this formulation represents the upward reflectance deviation for dark targets and the attenuation of reflectance for bright targets as a single TOAR deviation line (TDL). All VNIR bands were found to conform to this conceptual model.

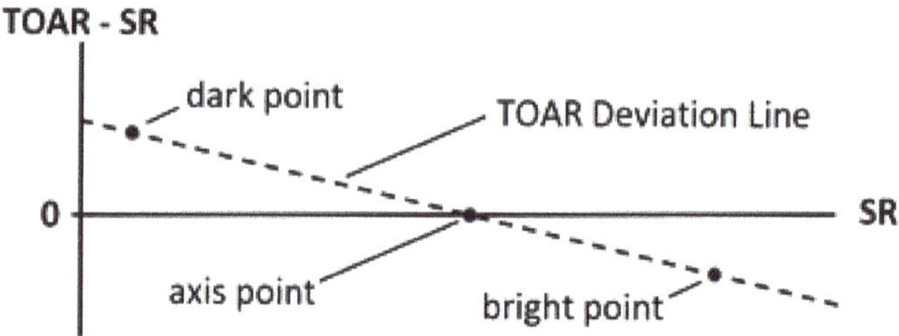

Figure 3. CMAC conceptual model illustrated as a dashed line expressing the effect upon any pixel, dark to bright, from a single level of atmospheric aerosol. SR is surface reflectance. The TDL crosses the x-axis at the axis point.

Precedence for the CMAC conceptual model can be found in Figure 2 of a paper by Fraser and Kaufmann [31] and is reproduced here in Figure 4. The linear relationships are similar to those shown in Figure 3; however, the y-axis in Figure 4 is functionally TOAR. In the context of AC, the parameter L represents radiance that necessarily includes sensor response that is the basis for RadTran applications. Including radiance complicates the calculation of surface reflectance. In the CMAC automated workflow, calibration removes the central importance of true radiance for AC.

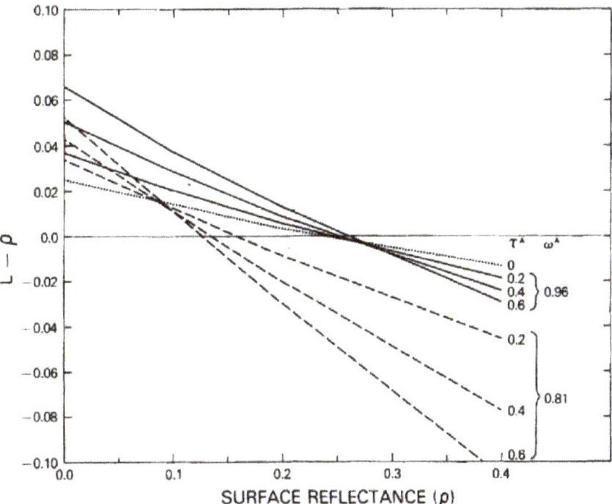

Figure 4. Figure 2 reproduced from Fraser and Kaufmann [31].

The TDL of Figure 3 represents the deviation from surface reflectance caused by the imposed atmospheric effects measured by Atm-I. As Atm-I increases and backscatter and attenuation increase, displacing the intercept upward and steepening the slope. This conceptual model was translated into the CMAC Equation, which converts TOAR to surface reflectance for each pixel differentially across the image according to the slope (m) and the offset (b) for the TDL imposed by the overlying atmosphere. The conversion is rapid due to the efficiency of the closed-form CMAC rather than the less-efficient alternative of iteration and potentially less accurate lookup tables employed in RadTran-based methods [32,33].

$$\text{CMAC Equation: SR} = (\text{TOAR} - b)/(m + 1)$$

The conceptual model and the observations supporting it underscore that the CMAC approach rescales the TOAR to surface reflectance. These changes narrow the reflectance window, a behavior that can be measured for increasing levels of Atm-I (Figure 5). Given the behavior of reflectance to increasing atmospheric aerosol concentration, the often-used dark object subtraction method [3,4] subtracts the lowest reflectance over dark targets (e.g., water bodies) from the entire reflectance distribution. This does little to correct the reflectance distribution while introducing additional error.

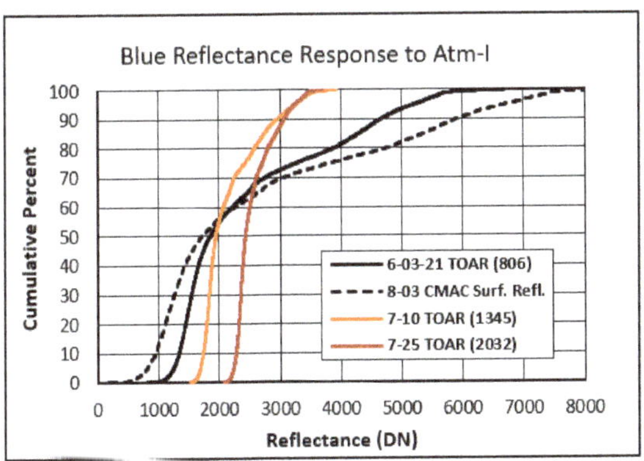

Figure 5. Data extracted from S2 images from 2021 over an area of interest with consistent reflectance in Reno, Nevada that experienced wide swings of aerosol concentration from regional wildfire smoke. The application of such invariant locations is described further in Section 2.4 below. DN refers to reflectance scaled by 10,000.

2.3. Forward Scatter Effects and Calibration

Two parameters per band are needed to apply the CMAC Equation: the TDL slope and offset. Through calibration, these parameters can be predicted from Atm-I for reversing the TOAR snapshot that captured the result from atmospheric degradation of the original surface reflectance. Each smallsat spectral band requires calibration to accommodate differing sensor responses to Atm-I, thereby customizing CMAC for that smallsat's sensor responses. A calibration target is an obvious choice for a precise and convenient establishment of the bright and dark points of the TDL, provided that special concerns are first addressed in the design of such targets.

Through R&D analyses, an additional atmospheric phenomenon became apparent that results from the interactive effects of ground-target reflectance and atmospheric aerosol particles. As Atm-I increases, the higher reflective energy of bright targets illuminates aerosols from below, thus brightening and scattering the Atm-I responses captured by the imager. This phenomenon affects the relationship of slope and offset for the application of the CMAC Equation and may prevent the use of existing targets that combine adjacent dark and bright panels in a checkerboard pattern (Figure 6). Due to forward scatter effects, such checkerboard targets can provide measurable TOAR only when (1) employed under the cleanest of atmospheric conditions, (2) the panels are physically separated, and (3) the panels are large in relation to the imager's spatial resolution. Forward scatter affects calibration curves beyond what is captured by the conceptual model in Figure 3; fortunately, this can be accounted for intrinsically through calibration.

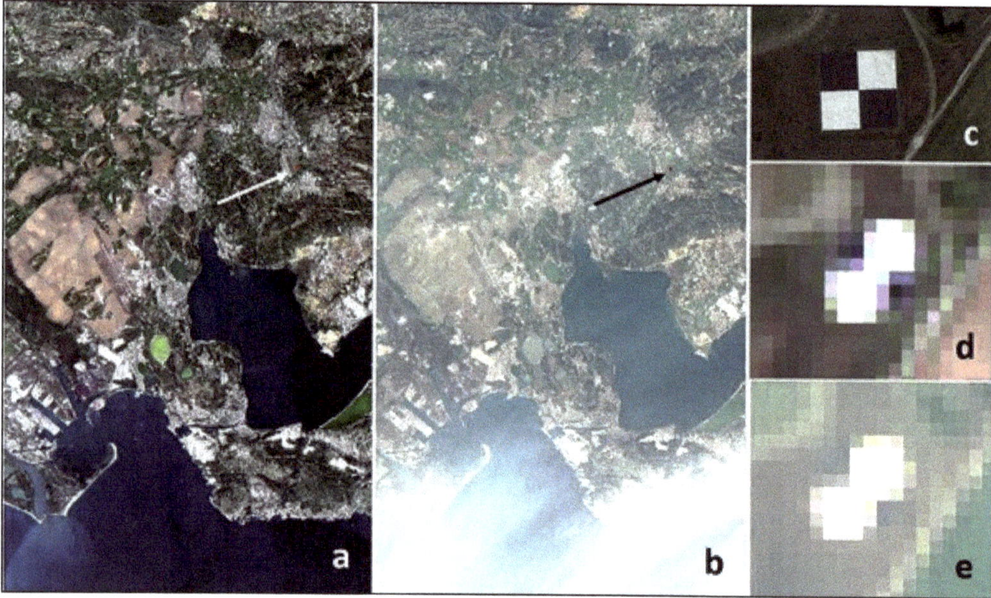

Figure 6. Salon de Provence, France region: a calibration target (arrows) in S2 TOAR regional images 16 June 2021 under light haze (**a,d**) and 8 March 2021 under moderate haze from wildfire smoke (**b,e**). A Google Earth image (**c**) of the target shows the 30 m × 30 m black and white panels.

An appropriate calibration target with accompanying precise and recently measured ground truth was not available; therefore, to provide for CMAC development, slope and offset were calibrated vicariously through image comparisons between clear and hazy conditions, with additional guidance from ground truth data collected by ASD field spectrometer. CMAC results for this calibration compared well with Sen2Cor results developed for the same images. This low Atm-I calibration was then adjusted sequentially for higher Atm-I by matching the surface reflectance estimates of the higher to the lower

Atm-I image through the selection of appropriate slopes and offsets. The slopes and offsets derived in this manner were recorded for each measured Atm-I on multiple clear to hazy images to establish the calibration relationships for surface reflectance estimation across the range of Atm-I that can be encountered operationally.

During previous method proofing, CMAC v1.1 was tested on multiple Sentinel 2 images from diverse environments that included low spectral diversity such as oceans and deserts as well as images degraded from haze due to smoke, fog and, to a more limited extent, dust. CMAC was found to correct all images and remove profound levels of haze, thus demonstrating applicability in all environments and at least to first order, doing so irrespective of aerosol type. This testing also indicated that the corrected VNIR bands are correct relationally and therefore represent close estimates of surface reflectance.

To verify the mathematics of the correction for this investigation, CMAC v1.1 (Advanced Remote Sensing, Inc., Hartford, SD, USA) was tested against Sen2Cor v2.11 (European Space Agency, Paris, France). Performance was judged by estimation of VNIR surface reflectance for seven AOIs exhibiting consistent reflectance throughout the summer. Such AOIs were given the name "quasi invariant areas" (QIAs) in recognition of the special role they can play as targets to assess AC performance under variable atmospheric effects. QIAs provide the opportunity to objectively evaluate accuracy, stability and AC correction limits by the CMAC and Sen2Cor software. Data presented for the comparisons presented here are the output from the processing software, alone, with no other treatment except the statistical analyses documented in downloadable Excel spreadsheets (see Supplementary Materials).

While the investigation of AC here is relative to S2, a research-grade satellite that includes onboard radiometry, CMAC is applicable to smallsats without additional processing or other relationships beyond the Atm-I model and CMAC Equation. Hence, the level of accuracy achievable for AC of S2 data serves as a direct indication of the AC accuracy achievable for smallsat data.

2.4. Experimental Design: Testing CMAC against State-of-the-Art Sen2Cor

This investigation consists of two analyses to evaluate the accuracy, stability and limits for AC applications by CMAC and Sen2Cor. Analysis 1 focused on the performance of these two methods for the correction of 18 images that encompass a range of Atm-I from clear to extremely hazy of a QIA northeast of the Reno, Nevada Airport. Analysis 2 developed AC statistics for 28 visibly clear images of a Southern California (SoCal) metropolitan area within an Atm-I range expected to be more commonly encountered in operations.

Vegetation and surface wetting are two significant influences on the temporal variability of the true surface reflectance. To minimize these effects, warehouse districts dominated by flat roofs and paved interspaces were chosen as targets. Such AOIs have remarkably stable reflectance and hence form desirable QIAs because (1) the AOI boundaries can be mapped to either exclude vegetation or include minor areas of irrigated cover, and (2) the exposed flat roofs and paved surfaces assure rapid drainage and surface drying. All seven QIAs, located in regions with Mediterranean-type climates, were evaluated during June and through early August when vegetation expression was expected to be relatively static. Weather records indicate that both QIAs received no significant rainfall during multiple days before each image acquisition date.

The area of each QIA was variable but sufficiently large such that the influence upon reflectance from variations in numbers of parked cars, semi-trailers, roof repair, etc., constitute a minor stochastic variation. The smallest QIA, Highgrove in southern California with an area of about 1.5 km^2, contains ~15,000 10 m pixels, and the largest, Chino, contains about 6.4 km^2.

Analysis 1: The Reno QIA shown in Figure 7 is located in a sidelap region between adjacent S2 tiles, thus potentially doubling the number of overpasses otherwise available. Eighteen images were selected for analysis from June through to 22 August 2021, a period that initially captured extremely clear conditions followed by periods of intermittent, occasionally severe smoke haze from regional forest fires. The selected images were confirmed to lack cirrus cover through visual inspection of S2 B10, and the Atm-I mapped across the QIA boundaries was confirmed to be homogeneous with only minor variation, generally between two to three percent.

Figure 7. The Reno QIA outlined in red on this S2 image from 6 March 2021 is located just northeast of the Reno, NV airport. The polygon was drawn to exclude vacant lots that might harbor unmanaged vegetation.

Analysis 2: Six warehouse districts were chosen in a metropolitan region of southern California east of Los Angeles covered by a single S2 tile (Figures 8 and 9). Twenty-eight S2 images were selected that met two criteria: (1) collection in June through early August in 2020, 2021 and 2022; and (2) lacking cloud cover, including cirrus determined through visual inspection of S2 band B10. The images exhibited no visibly discernible haze.

Figure 8. Map showing locations of six QIAs located east of Los Angeles, CA. (Source: Maxar, Earthstar Geographics and the GIS User Community). QIA locations are designated as follows: Chino (a); Ontario (b); Highgrove (c); Fontana (d); Redlands (e); and Rochester (f).

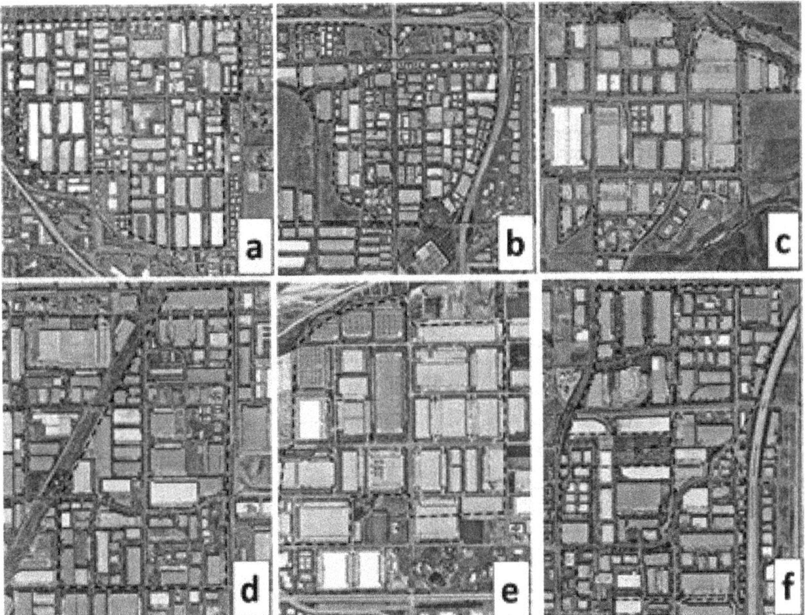

Figure 9. Google Earth closeup images of six QIAs: (**a**) Chino; (**b**) Ontario; (**c**) Highgrove; (**d**) Fontana; (**e**) Redlands; and (**f**) Rochester.

The restriction for the collection period to about three months controlled for reflectance changes of senescing unirrigated weedy plants and restricted the solar zenith angle (SZA) to remain within five degrees of the minimal angle around the summer solstice. These AOIs were accepted as QIAs, and polygons were drawn around each to exclude vacant land and non-irrigated plant cover. Among the 28 images selected, eleven were from 2020, eight were from 2021 and nine were from 2022. This provided 168 separate Atm-I levels for comparison when viewed across the six AOIs (6 QIAs × 28 images). Though the six QIAs are in the same region, measured Atm-I varied among the QIAs even on the same day.

2.5. Data Analysis

CMAC and Sen2Cor corrections were run for the area within each QIA polygon for each image tile. Pixel values were extracted for the four VNIR bands from each QIA for three treatments: TOAR, CMAC-corrected and Sen2Cor-corrected. To provide easy-to-follow documentation, all statistical analyses were performed in Excel spreadsheets. For both analyses, statistical distributions for these treatments were recorded, by band, for each image as cumulative percentiles starting at 0.1, 1, 3 and 5, then continuing in five-percentile increments to the cumulative reflectance at the 95th percentile. The suite of 22 separate statistics for each QIA permitted assessment of AC performance across the range from dark to bright reflectance. Statistical analyses were treated separately for each QIA because the reflectance within each QIA was internally consistent, but variable compared to the other QIAs. Statistical dispersion (therefore, also its antonym, precision) was evaluated for each QIA dataset using the coefficient of variation expressed as a percent (CV%) and calculated from the ratio of standard deviation to the mean for the suite of 28 images.

Estimation error was evaluated for Analysis 2 in an initial step assessed per QIA by averaging the percentile-wise values of the three lowest Atm-I images. The resulting averages per QIA were treated as the standard to estimate error for each percentile of each image within that QIA according to

$$\% \text{ error} = 100 \times (\text{value} - \text{standard})/\text{standard}$$

The resulting values of % error for each band were then pooled from the six QIAs to evaluate error statistics for the operational range of Atm-I and from dark to bright reflectance for both CMAC and Sen2Cor. These error statistics were tracked for each cumulative percentile for each image and were included in the pooled sample that was then ranked from low to high to provide for graphical assessment of CMAC and Sen2Cor error distribution. The pooled sample resulted in suites of $n = 3696$ (6 QIAs × 28 images × 22 percentiles) error estimates for each band of both treatments.

Data extracted from the 46 total S2 images were exported to and analyzed in Excel spreadsheets where figures were generated. A list of these images is presented in Appendix A. All Atm-I data quoted in this paper are median values for the region within each QIA boundary. The data presented in this section are excerpted from over 4000 individual extracted median estimates of surface reflectance from 22 cumulative percentile steps of the reflectance distributions for each of three treatments: TOAR, CMAC and Sen2Cor. These data were examined for the 18 images of the Reno QIA (Analysis 1) and the 28 images of the six Southern California QIAs (Analysis 2).

The workflow for this assessment was comprised of the following steps: (1) clipping the Atm-I, TOAR, Sen2Cor, and CMAC-corrected image bands from the image tiles to each QIA boundary and extracting the median statistics from the clipped areas; (2) aggregating the treatment data for each QIA, sorting it by band and in order of increasing Atm-I; (3) generating Excel spreadsheets containing the aggregated data and (4) analyzing the spreadsheet data within Excel to generate the summary plots and tables. The interested reader is invited to download the Excel data analysis spreadsheets that are available in Supplementary Materials and to test correct 4-band VNIR Sentinel 2 images using the S2 Cloud Pipeline. Instructions for access are given in Supplementary Materials.

3. Results

On a desktop PC (64 MB Ram, Windows 10, C++-based code), CMAC corrected the four VNIR bands of full S2 image tiles (~120 million image pixels) in an average of 91 s. The python-based desktop Sen2Cor processed the same tiles in an average of about 20 min, though for the 13 bands included in S2 ".SAFE" image packages.

Analysis 1: Reno QIA statistics were extracted and analyzed from eighteen images. A summary table presented for the Reno QIA available in the Supplementary Materials provides CV% values that are noticeably elevated above CV% for Analysis 2, also available in the Supplementary Materials. This was due to the inclusion of three high Atm-I images that proved challenging for both methods. This conclusion can also be reached through the examination of Figure 10 in which seven curves are plotted showing the trend in correction as judged by Atm-I from unusually clean atmospheric conditions to the extreme haze from wildfire smoke. Scatter increased with Atm-I. To reduce the number of curves for clarification of trends responding to increasing Atm-I, data of similar Atm-I ranges were averaged ($n = 3$ or $n = 4$) for each percentile of reflectance within each treatment. Sen2Cor can be observed to fail at Atm-I levels of around 1345 and above. Data from the three extreme Atm-I images also show increased CMAC CV% as indicated in Figure 10 by the slight deviation from the statistical weight of the curves generated by results for Atm-I levels of 962 DN and lower. These calculations and the complete dataset are available for "Reno.xlsx" through a link in Supplementary Materials.

As shown in Figure 10, CMAC surface reflectance estimates are roughly commensurate with Sen2Cor for images with lower Atm-I—these curves are displayed in color. The average of the three unusually clear images with the lowest average Atm-I (813) displayed in red was taken to be the best "surrogate" approximation of true surface reflectance. Above Atm-I levels of approximately 2000 DN, the CMAC AC curves significantly diverge above the corresponding surrogate surface reflectance curves. However, this represents a net decrease in estimated surface reflectance.

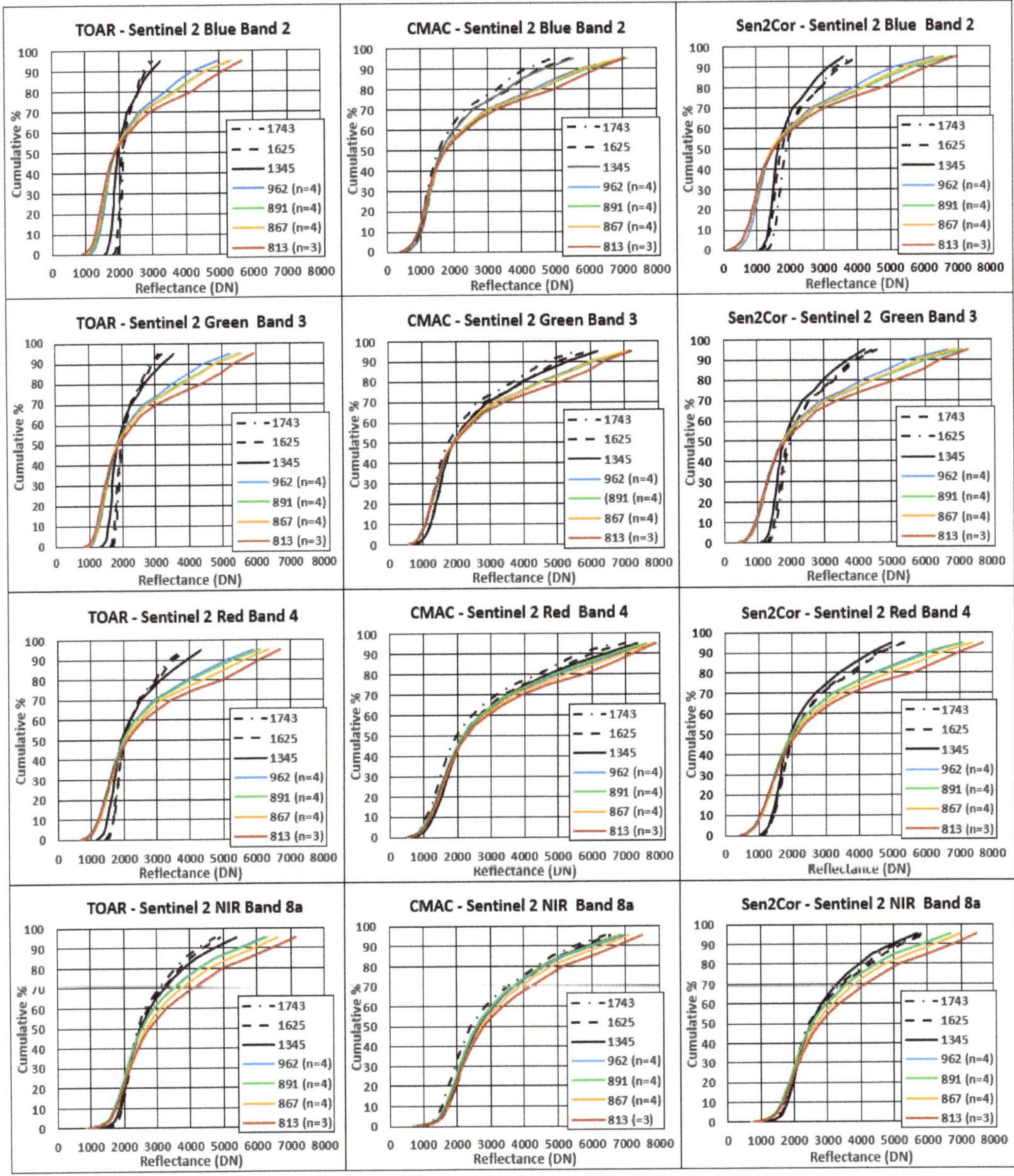

Figure 10. Reno QIA reflectance curves plotted for the four VNIR bands of S2 (rows). Colored curves were derived from $n = 3$ or $n = 4$ percentile averages for each band and treatment for the low-Atm-I images (clear-appearing, lacking haze). Legend values are Atm-I or average Atm-I. Curves in black are for single images that exceed Sen2Cor AC capability. Though not as accurately, CMAC corrected the extremely high Atm-I curves for the visible bands. CMAC curves are tighter (more precise) than Sen2Cor in all bands.

Though CMAC and Sen2Cor AC methods are very different, both captured the trend of decreasing bright reflectance at higher Atm-I levels. Hypothetically, this divergence indicates an atmospheric phenomenon of diffuse shading by aerosol particles that reduces TOAR reflectance. As the CMAC Equation predicts, this results in a lower surface reflectance estimate. This divergence is systematic, so it can potentially be adjusted by empirically fitted relationships for each band if the estimation of the true surface reflectance for brighter targets is desired and the added processing time is not problematic. As discussed later, the true reflectance of bright targets may not be of direct interest for most analyses.

Analysis 2: The six SoCal QIAs were evaluated to test AC performance within an Atm-I range expected to be encountered in typical operation. Figure 11 presents seven curves derived for the Rochester QIA, each being an $n = 4$ average calculated from within similar Atm-I ranges.

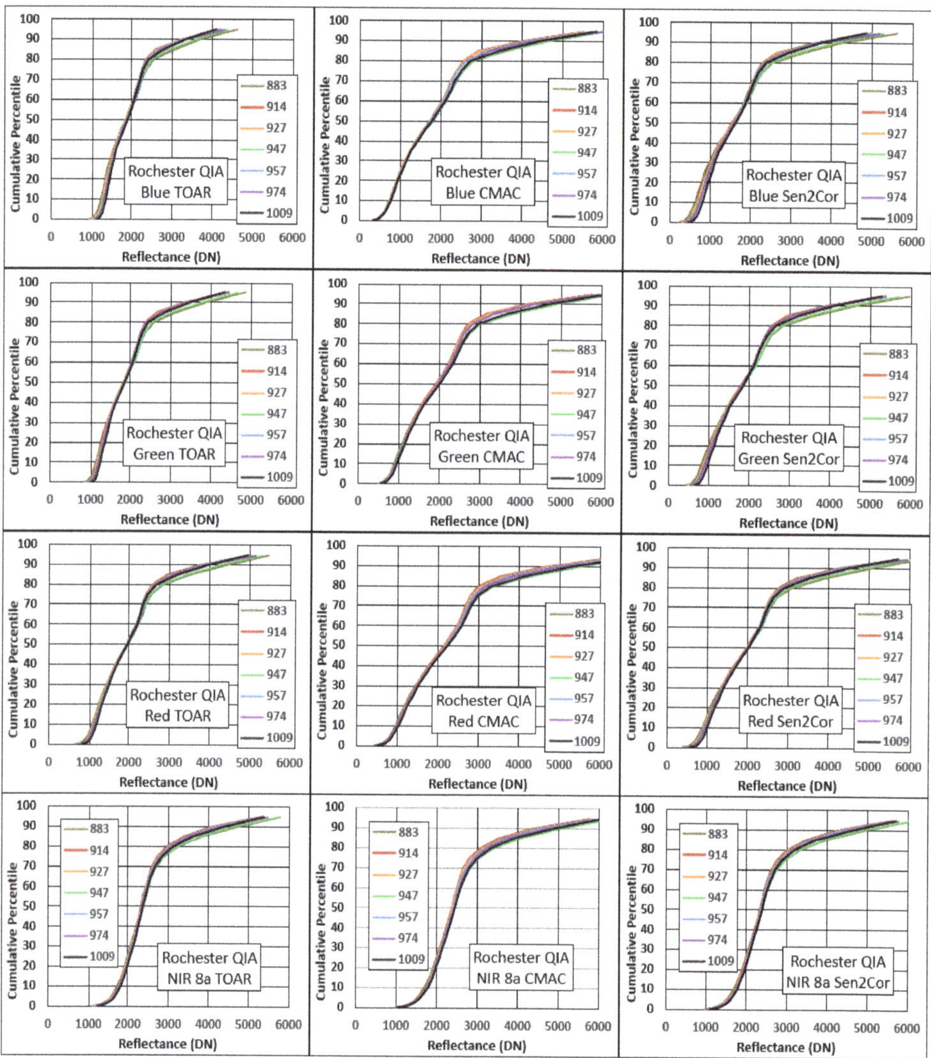

Figure 11. Seven reflectance curves for the Rochester QIA by treatment (columns) and bands (rows). Each curve represents an average of four images with similar average Atm-I. Dispersion is notable for Sen2Cor in the lower limb of visible bands where high precision is needed to support applications such as precision agriculture and AI feature extraction. In CMAC, the lower limb of reflectance is comparatively precise. NIR 8A curves are virtually identical between the two methods. Rochester experienced the highest range of Atm-I levels among the SoCal QIAs.

As previously mentioned, each of the 28 images analyzed for the SoCal QIAs appeared visually clear across the entire range of median Atm-I recorded for the images. Of the 168 unique image-QIA combinations, only four images exceeded an Atm-I of 1050 DN, and those by less than 17 DN. Summary tables for the statistics are presented in for the six QIAs available through a link in Supplementary Materials. The precision of the curves in the Figure 11 example QIA is indicated by how tightly the curves plot and is comparatively better for the suite of SoCal QIAs due to the more expected dynamic range of Atm-I in contrast to the Reno QIA analysis, which included three extremely hazy images. From the analysis of hundreds of images, most atmospheric correction is performed within an approximate Atm-I range of 800 to 1050 DN. We have observed alarming exceptions that occurred during recent summers in the United States from the west coast to the Great Lakes region due to wildfires in Canada and the western United States that drove Atm-I imagery above 1050 DN over millions of hectares of farmed land for extended periods, adversely impacting image uses for precision agriculture and other applications.

CMAC's high accuracy in the lower limb of visible-band surface reflectance, generally less than about 2000 DN, is crucial for precision agriculture and AI feature extraction because these darker features define the analytics of interest in vegetation indices and in prominent observable features [34–36]. The suite of 12 curves as shown in Figure 11 is provided in Excel spreadsheets for each of the six SoCal QIAs are available through the link in Supplementary Materials.

3.1. Precision

Bandwise average CV% for all 28 images and for each of the 22 cumulative percentile steps of the six SoCal QIAs were combined into one dataset to examine how precision is affected by increasing reflectance for both CMAC and Sen2Cor output (in Supplementary Materials as "SoCalCV%.xlsx"). These cumulative percentile-step averages were plotted against TOAR to represent the unbiased trend from low to high reflectance for both methods (Figure 12).

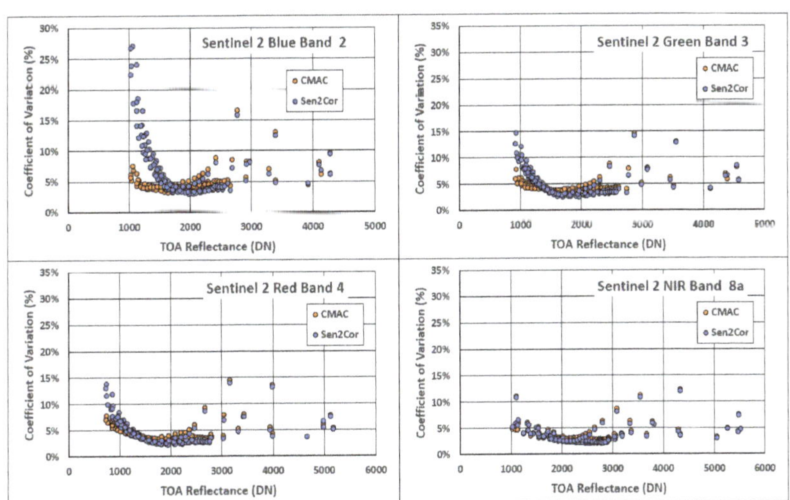

Figure 12. Average CV% distribution for the 22 percentile steps combined for the six QIAs; (*n* = 132) of CMAC and Sen2Cor. Though approached very differently, both methods show similar trends.

The resulting pattern of bandwise CV% is instructive. Moving rightward from the origin toward higher TOAR, CV% is initially high, decreases to a trough and then increases in a quadratic-like manner. This behavior is exhibited in all four bands for both CMAC and Sen2Cor despite very different workflows and data inputs. This result provides support that the CMAC conceptual model, as expressed in Figure 3 with its teeter-totter-like TDL response, correctly portrays the phenomenology of atmospheric aerosol effects. According to the CMAC model, the greatest precision is expected to occur at the axis point where the TDL crosses the x-axis and the affected image's surface and TOAR are equivalent. The statistical distribution in the region around the axis point will, therefore, tend to have minimal scatter. It should be noted, however, that the axis point is a concept and not a fixed value. The axis point migrates rightward as Atm-I increases; hypothetically, this trend is caused by forward scatter.

In Figure 12 the trend in CV% can be seen to increase above a scaled TOAR reflectance level roughly between 2300 to 2800 DN depending upon band and method, but the values become widely spaced and, for each dataset, highly scattered. Uncertainty increases at higher reflectance, as can be seen in the spreadsheets and tables that summarize the QIA CV% data available through the link supplied in the Supplementary Materials.

3.2. Estimation Error

As noted in Figure 10, AC results for the lowest Atm-I (clear) images are expected to be closest to true surface reflectance since the correction requires minimal adjustment for its estimation. An assumption that the surface reflectance estimates of CMAC and Sen2Cor for low Atm-I images are surrogates for true surface reflectance offers a means to examine the propagation of error in each AC method. Following this logic, % error was estimated using the average of the three lowest Atm-I images for the 28 images of each SoCal QIA and used as surrogate true surface reflectance values. These calculations were made per the 22 cumulative percentile steps so that the statistical examination could further define the relationship of the error to reflectance brightness.

Because each of the QIAs was chosen for its quasi-invariant reflectance, AC results for the collection of QIA images are expected to yield reflectance values that remain similarly invariant. Examination of the values for each image indicated that there was no apparent relationship with time for either method. For CMAC, there was also no striking relationship for the magnitude of surface reflectance estimates with Atm-I. For Sen2Cor, however, a trend for surface reflectance error increasing as Atm-I increased was readily discernable.

Error estimates were calculated for CMAC and Sen2Cor for each QIA from their individual surrogate true surface reflectance (CMAC with CMAC, Sen2Cor with Sen2Cor). The values for all six QIAs were combined to robustly represent the error distribution for each band (n = 3696: 6 QIAs × 28 images × 22 percentiles). The error estimates were ranked from lowest (under-estimation error) to highest (over-estimation error) and plotted to compare CMAC to Sen2Cor per band for the entire SoCal dataset (Figure 13). These calculations, as well as the trend in the data output, can be traced in the individual downloadable QIA spreadsheets that fed statistics to the downloadable available in Supplementary Materials as "SoCal%Error.xlsx".

The ranked percent error curves for CMAC and Sen2Cor in Figure 13 agreed well through most of the bandwise % error distributions with the notable exception for the higher ranked values in the visible bands. The under-estimation error was well-constrained and roughly equivalent between the two methods. The upward deviating curves represent an over-estimation error that was much greater for Sen2Cor, especially at shorter band wavelengths—the blue band was more severely affected than the green band. The estimated error was the lowest for NIR 8A.

Over-estimation of visible-band reflectance in Figure 13 was confirmed to be the origin of the greater statistical dispersion of low reflectance observed in the plots of Figure 11 and in the CV% tables and the "raw data" across the six Excel files available through Supplementary Materials. The data in the summary tables show this trend but under-represent the severity of Sen2Cor instability at higher Atm-I because the most widely divergent values were masked by averaging with the majority of comparatively low Atm-I images in the larger image pool. Some Sen2Cor CV% values exceeded 100% for individual images with higher Atm-I, a trend that can be seen in the individual downloadable QIA spreadsheets. The trend for overestimation errors in visible bands is evident in Figure 13 for Sen2Cor. This can be viewed as instability. In contrast to visible bands, NIR 8A error estimates by CMAC and Sen2Cor closely agree.

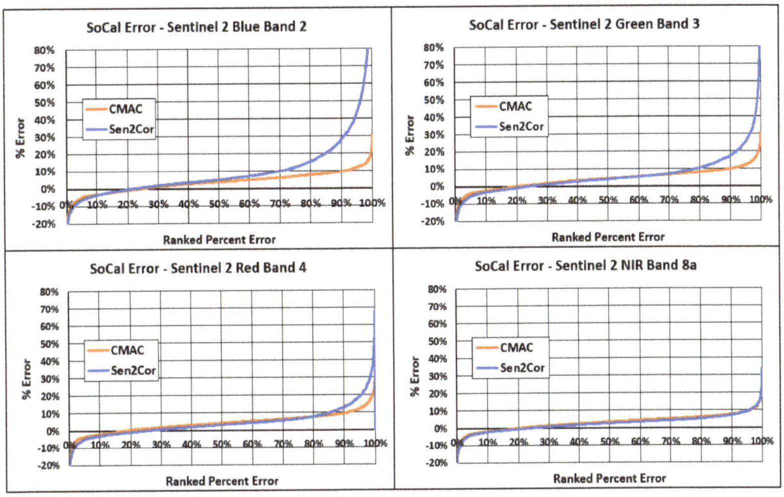

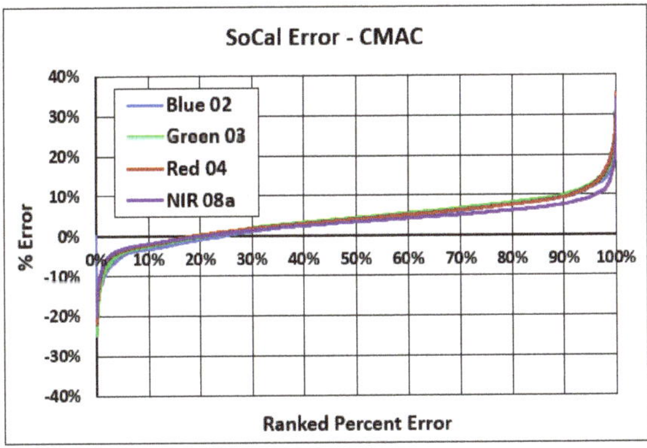

Figure 13. Percent error distribution for CMAC and Sen2Cor plotted according to the rank for the 1st through 3696th estimated error values.

The CMAC % error curves have roughly equivalent shapes and magnitudes. In Figure 14, the estimates for the visible bands plot nearly atop one another. The agreement of CMAC error curves indicates greater stability in VNIR band correction.

Figure 14. CMAC estimated percent error plotted together for all four VNIR bands.

Toward understanding the range of Atm-I that can be corrected accurately by both methods, the low, dark end of reflectance was explored further using averages of the low reflectance % error values as an index to compare to the median QIA Atm-I measured for each image. For this final test, the absolute values of percent error were averaged for the 0.1, 1, 3, 5, 10, 15 and 20th percentiles to form an index, one per image, to compare to the range of the recorded median image Atm-I. The % error values were combined for all six QIAs and plotted (Figure 15). The corresponding reflectance range encompasses up to 2000 DN for NIR 8a and about 1000 to 1200 DN for visible bands. These calculations can be viewed in Supplementary Materials as "SoCalLowRefl%ErrorDistr.xlsx".

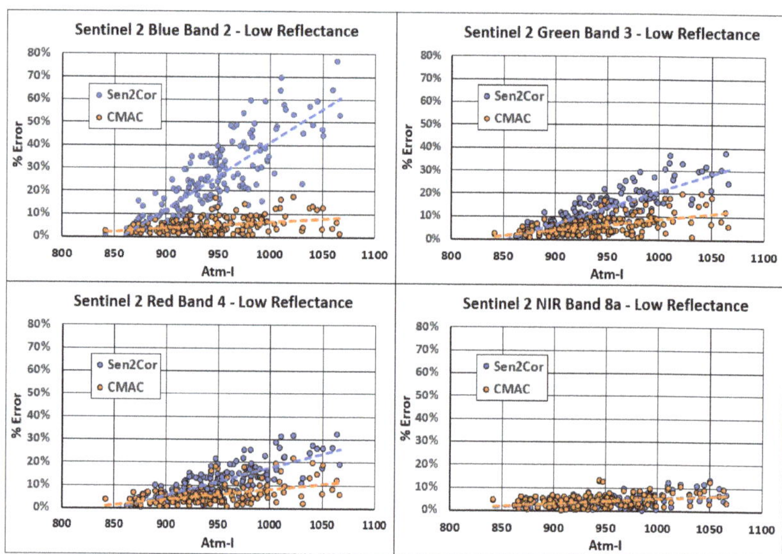

Figure 15. Percent error in Sen2Cor low surface reflectance estimates rapidly increase with Atm-I in all bands except NIR 8A. CMAC error shows a more gradual trend of increased error with increasing Atm-I level.

Because the suite of 28 images for each QIA expressed different Atm-I values, the test in Figure 15 provided 168 unique independent values for comparison to the low reflectance % error index in each band. To place the low reflectance measured over the SoCal QIAs into context, the reflectance level for the visible bands measured at the 0.1 cumulative percentile for five of the six QIAs corresponds to the approximate reflectance typical of the dense dark vegetation used to calibrate the Atm-I model. Hence, a percentile range from 0.1% to 20% is considered to be a valid range for examination of how the low reflectance limb of the surface reflectance distribution is impacted by an error that would affect image utility whether for precision agriculture or AI feature extraction.

Figure 15 illustrates a marked decrease in Sen2Cor's visible band stability as Atm-I increases above about 900 DN. CMAC exhibits a much shallower error trend as Atm-I increases, starting at an Atm-I level of approximately 850 DN. Figure 15 demonstrates that CMAC outperforms Sen2Cor accuracy for AC in the low reflectance portion of visible bands for the most common range of Atm-I that will be encountered in routine AC. As noted in Figure 10, CMAC can still provide useful results even at "extreme" levels of Atm-I exceeding 1700 DN, well beyond the atmospheric effects that can be corrected by Sen2Cor. These graphics form a milestone for judging future R&D efforts to enhance CMAC accuracy.

As mentioned earlier, CMAC v1.1 has been tested on images from diverse environments around the Earth. While these tests demonstrate that CMAC can clear images with profound haze, there has been no direct means to confirm that the cleared images represent true surface reflectance. The data developed for this investigation have provided such direct confirmation. Returning to Figure 10, the most extreme image haze evaluated in this paper, graphed for the Atm-I = 1743 image, also provided results close to surrogate surface reflectance that was confirmed to provide image clearing (Figure 16). Hence, when corrected by CMAC, images that have heavy haze cleared from the image can provide results that are close to the true surface reflectance. This is a valuable result that will help push the upper limit of correction to even higher Atm-I levels.

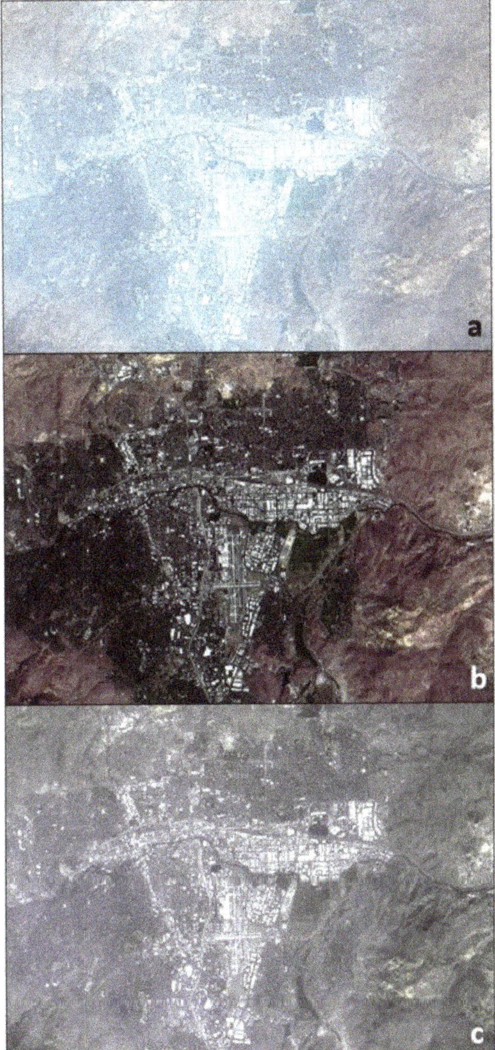

Figure 16. A clip from the S2 Reno image whose data are shown in Figure 10 statistics for Atm-I = 1743 and as the 22-8-2021 statistics in Supplementary Materials as "Reno QIA Curves.xlsx". As in all other image displays in this paper, these examples are screenshots from QGIS display. (**a**) TOAR representation made from the full tile stretch. A full tile image stretch can be taken to visually represent the degraded TOAR mathematics of the image. The alternative, a clipped image stretch, does not appropriately represent the unbiased mathematics of the image that confronts AI and other machine analysis; such clip stretches may visibly clear some haze (but are typically accompanied by color balance problems). Color balance is a valuable indicator of potential problems that could occur through use of machine analyses. (**b**) CMAC clearing of the image provides the color balance conferred by the TOAR image. The features within both are darkened through hypothesized diffuse shading from aerosol particles. (**c**) Sen2Cor correction of the same clip displaying color balance problems and residual haze.

4. Discussion

Smallsats are unequivocally the future for remote sensing intelligence, surveillance and reconnaissance (ISR) applications. CMAC was developed to support this mission by enabling smallsat AC to be independent of ancillary data and operate in near real-time. Because CMAC correction is generated exclusively from scene statistics, TOAR can be converted to surface reflectance directly upon image download without waiting for ancillary data. Application of scene statistics, therefore, enables AC of smallsat data to support the most time-critical applications for ISR.

Statistical analyses show that CMAC suffers no performance loss through the use of scene statistics and instead provided much greater precision and accuracy over a much wider range of atmospheric effects in comparison to Sen2Cor. This can be attributed to the CMAC conceptual model that captures the effect upon light from atmospheric transmittance. In comparison, Sen2Cor, like all competing automated methods, is based upon RadTran calculations. For existing smallsat AC applications, similar RadTran workflows must apply radiance calibration from a harmonized set of S2 and L8/9 data. Therefore, such secondary application to smallsats can perform no better than that recorded here for Sentinel 2.

CMAC performs within the dark-to-bright reflectance distribution of most concern: lower reflectance in the visible spectrum of less than 2000 DN. This low reflectance carries the signals of greatest interest for remotely sensed ISR, whether for national security, precision agriculture or any other machine analysis. In these tests, Sen2Cor performed relatively poorly at the low end of visible reflectance and did not reliably estimate surface reflectance unless the image was already virtually clear of atmospheric effects (Figure 15). Smallsats employing RadTran-based correction calibrated with harmonized data can be expected to face these same issues at least.

While CMAC can provide enhanced precision and accuracy, its application to smallsats also offers significant convenience through the adoption of a simple two-step procedure for calibration. In the first step, by assuming pre-launch sensor radiance, the digital sensor output can be converted to an assumed TOAR following well-established procedures. Any change of sensor response from benchtop calibration after orbital launch can then be accommodated in a second step through a rapid calibration procedure that customizes the conversion of the assumed TOAR to surface reflectance for the smallsat. In this way, the uncertainty induced by assuming pre-launch calibration is removed through a post-launch calibration step. Once calibrated, CMAC software produces estimates of surface reflectance from that smallsat's assumed TOAR from then on but is subject to periodic recalibration to compensate for sensor response drift. Recalibration can be performed two to three times per year to maintain the highest quality output. CMAC calibration methods and the associated science for their application are undergoing further development and promise to be automatable, rapid, highly precise and applicable to any VNIR spectral band.

While CMAC proved accurate and precise for the lowermost limb of reflectance in all four VNIR bands below a surface reflectance of about 2000 DN. CMAC underpredicts surface reflectance values above 2000 DN, which is a behavior also exhibited by Sen2Cor. This intrinsic divergence is hypothesized to result from the alteration of TOAR reflectance through diffuse shading by aerosol particles. As seen in the CMAC data from all seven QIAs, even for extreme levels of Atm-I, CMAC estimates of true surface reflectance remain relatively stable in the critical reflectance region below 2000 DN. With the exception of beach sand, dry lake playas, ice and snow, most targets exceeding a surface reflectance of 2000 DN are manmade. Depending upon the intent of analysis, highly reflective manmade surfaces subject to such divergence from true surface reflectance may not be of interest, though their brightness certainly could be. Perhaps a better representation for such highly reflective targets is simply to map the targets that exceed some set threshold. This solution could be especially robust for machine analyses and AI while alleviating the need to adjust bright target reflectance to compensate for this systematic divergence. Such compensation can be made, but only at the cost of additional processing time.

Future technical upgrades are planned for CMAC. A version that includes additional bands of interest for agriculture (such as S2's red edge bands B05-B07 and 10m NIR B08 is under development. As previously mentioned, this paper examined image data through periods selected to constrain solar zenith angle (SZA) to yield the most robust comparison possible. Otherwise, significant errors in surface reflectance estimation can occur as SZA increases in the presence of high levels of Atm-I. Such conditions force incident sunlight to travel a greater path length through layers of atmospheric aerosol, thus causing more light attenuation (i.e., reduction of signal) than from Atm-I alone. Current efforts for CMAC R&D include measuring the interactive effect of SZA and Atm-I and then formulating a bivariate model accounting for both; this is a required step to prepare AC of smallsat imagery to address all conditions that can be encountered. This advancement will be made applicable to all smallsats as a last step in CMAC processing.

Pointable smallsats will open a new world of ISR application for areas of conflict to serially gather data nearly continuously during daylight. An additional application of interest involves the development of systematic adjustment for AC of data gathered from far off-nadir viewing by pointable smallsats. Through edge applications with CMAC software operating on board, pointable smallsats may effectively achieve real-time AC by correcting lines of push broom-scanned image data after a few-second lag. Such AC-corrected data could then be forwarded to AI feature extraction and on to first responders or warfighters within minutes of collection. CMAC confers promotional benefits for solving all of these problems through its simplicity, i.e., nothing is black box, and the correction is made only with scene statistics. Simply stated, this will be a rules-based "see it, correct it" approach for smallsat ISR.

CMAC is currently applicable for nadir-look four-band smallsats and is ready for test application to other bands in the VNIR spectrum. In principle, CMAC AC should be applicable to hyperspectral image datasets. Perhaps of greater interest and utility are completed algorithms for AC application over oceans, where CMAC can produce excellent results with respect to appearance (Figure 17). AC over large water bodies is inherently more difficult than over land because the air-water interface is highly reflective and can affect AC output due to wave effects and the geometry of image capture relative to the solar angle. AC over water deserves focused R&D effort.

Figure 17. S2 image, 5 March 2021 closeup of the Mexican Gulf Coast north of Veracruz: (**a**) TOAR with smoke effects from fields burned before planting; (**b**) CMAC v1.1 corrected.

While CMAC was developed specifically for smallsats it was enabled by the consistent quality and cadence of free data from the Copernicus S2 program that flies two research-grade satellites. For convenience, this paper applies a statistical comparison of CMAC output to the existing method for the AC of S2 data, Sen2Cor. CMAC calibration is readily translatable to other satellites. Figure 18 displays Atm-I and CMAC for Landsat 8/9 data for comparison to TOAR and the RadTRan-based method, LaSRC, that was developed for the correction of L8/9. An example of L8 correction is presented in Figure 18, where problems are visible for the LaSRC correction including significant remaining haze, a bluish shift for green vegetation (southwest corner) and dark artifacts in the ocean. The CMAC correction is clear, and the ocean is portrayed with the expected deep blue color balance. Recapping the results presented earlier in this paper and with reference to Figures 10 and 16, the restored color is likely close to the actual surface reflectance.

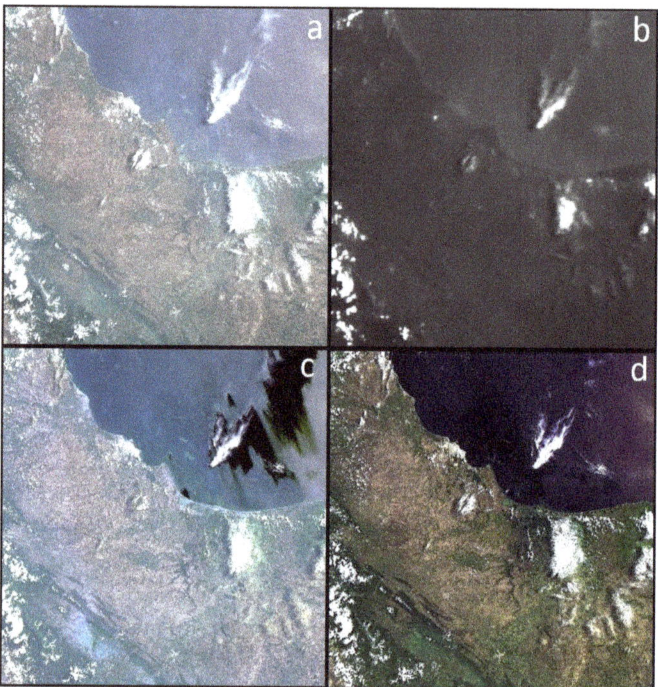

Figure 18. L8 full tile of the Mexican Gulf Coast from 5 December 2021: (**a**) TOAR, (**b**) Atm-I, (**c**) LaSRC correction and (**d**) CMAC correction. The images were rotated from their collection angle to fit squarely.

Harmonized radiance products from L8/9 and S2 are meant to prepare smallsats for the application of RadTran for AC. The results from this paper call into question the application of RadTRan methods to smallsats since the software performance is limited even when applied, using the software designed specifically for S2 and L8/9.

As mentioned above, CMAC was developed to work with the four-band VNIR data tested in this paper for application to S2 data. Hence, as shown by the application to L8 in Figure 18, CMAC should also be applicable to smallsats. This was tested with a limited dataset using experimental procedures. As with the L8/9 application, conclusive testing has not yet been completed. However, using the qualitative criteria of correct color balance and image clearing as an indication that the output is close to true surface reflectance, application for smallsat flocks is expected to work as planned. This initial testing was conducted with four-band VNIR Planet Dove data that yielded clear images with correct color balance, confirming that CMAC will work as designed for smallsat application (Figure 19). The major challenge for this application is that smallsat systems will likely contain members launched at different times with different sensors, optics, and designs, thus raising the question of whether each individual smallsat will require its own exclusive calibration. A partial answer was confirmed through analysis of the data from two cohorts bearing file names PS2.SD and PSB.SD, which required separate calibration, but further testing was not possible with the small data set available. Anticipating that each smallsat will need individual calibration is likely, so a CMAC calibration workflow has been designed that promises to be both automated and highly precise.

Figure 19. TOAR (**a**) and CMAC (**b**) views of 8-18-2021 Planet Labs Dove satellite image over Fargo, North Dakota (20210818_175123_23_105a_3B_AnalyticMS_clip.tif cohort PS2.SD).

5. Conclusions

CMAC is a new AC method developed to serve the smallsat remote sensing industry. From this investigation using S2 data as the testbed, it is apparent that CMAC has an accuracy that exceeds the existing state-of-the-art software, while requiring a fraction of the processing time. In other tests, CMAC utility has proven readily transferable to smallsat data through automated calibration.

CMAC is new to remote sensing's AC science. The following points are contributions by CMAC and this investigation:

1. Application of quasi-invariant areas (QIAs) for proofing atmospheric correction: QIAs are large-scale spectrally diverse targets whose surface reflectance remains sufficiently stable for testing methods across variable levels of atmospheric effects from clean to extremely hazy conditions through time.
2. The CMAC method in its entirety: Based on the analysis results presented here, CMAC correction agrees well with the accepted method for S2 AC and showed greater statistical stability, particularly for visible band reflectance below 2000 DN, the region of reflectance distribution that defines image quality and utility.
3. The CMAC conceptual model: The atmospheric correction algorithm is based upon empirical observations represented by a linear conceptual model. Statistical dispersion measured for 28 images across six QIAs provided evidence supporting that the simple CMAC conceptual model mathematically captures the atmospheric effects upon transmitted light.
4. Treatment of atmospheric effect as a single lump-sum variable: RadTRan methods attempt to correct for multiple effects from the atmosphere; however, their interactive effects are largely unknown. CMAC is a simplification that performs AC based upon a single estimated variable: scatter in the S2 blue band calibrated to vegetation responses. This simplification resulted in comparatively higher accuracy and stability than the state-of-the-art Sen2Cor.
5. The simplicity of CMAC is promotional—nothing is black box: The simplification for estimation of surface reflectance through the pathway of reflectance, not radiance, enables rapid calibration of any smallsat and accurate results. The resulting CMAC workflow led to the discovery of the important role that forward scatter plays in image effects from the atmospheric transmission and promises to solve reliable AC over the ocean for SZA-Atm-I interactive effects as well as for other issues.

6. Patents

Currently, one CMAC patent is granted but is not yet in print. Two additional patent applications are pending before the US Patent Trade Office with one of these filed internationally through the Patent Cooperation Treaty (language approved as filed per International Search Report).

Supplementary Materials: The following supporting information can be downloaded from https://www.mdpi.com/article/10.3390/app13106352/s1: "live" Excel workbooks referenced in Section 3 and the Python computer code for the S2 4-band version 1.1. An online application for free test-correction of 30 Sentinel 2 images that can be browsed and selected by the user is available on https://strato.advancedremotesensing.com/app (accessed on 9 May 2023). If more images are needed for your research or work, please make contact through https://advancedremotesensing.com (accessed on 9 May 2023).

Author Contributions: Conceptualization, D.P.G., B.-C.G. and T.A.R.; methodology, D.P.G. and T.A.R.; software, T.A.R. and D.P.G.; Formal analysis, D.P.G.; data curation, T.A.R. and D.P.G.; visualization, D.P.G. and B.-C.G.; original draft, D.P.G.; review and editing, D.P.G., T.A.R. and B.-C.G.; resources, D.P.G. and B.-C.G.; project administration, D.P.G. All authors have read and agreed to the published version of the manuscript.

Funding: This research was funded by the U.S. National Science Foundation Small Business Innovation Research Program through grants #1840196 and #1950746.

Institutional Review Board Statement: Not applicable.

Informed Consent Statement: Not applicable.

Data Availability Statement: Please see Supplementary Materials for accessing the analyzed data and Appendix A for the list of images that were analyzed.

Acknowledgments: We thank the European Space Agency's Sentinel 2 program for the steady stream of free, high-quality imagery and its excellent documentation and celebrate Sentinel 2 as a shining example of a major societal benefit with a global scale impact. Our heartfelt thanks in memory of Thomas Loveland (dec. 13 May 2022), a central figure in Landsat applications, for his friendship, encouragement and insights early in our R&D process.

Conflicts of Interest: The authors declare no conflict of interest.

Appendix A. Sentinel 2 Images Used for Analyses

Table A1. 18 Reno QIA images used for Analysis 1—all 2021.

T10SGJ_20210603T184919	T11SKD_20210608T184921
T10SGJ_20210613T184919	T11SKD_20210615T183921
T11SKD_20210620T183919	T10SGJ_20210625T183921
T10SGJ_20210628T184921	T10SGJ_20210630T183919
T11SKD_20210705T183921	T10SGJ_20210710T183919
T11SKD_20210715T183921	T11SKD_20210720T183919
T10SGJ_20210723T184919	T10SGJ_20210728T184921
T10SGJ_20210804T183921	T10SGJ_20210807T184921
T11SKD_20210809T183919	T10SGJ_20210822T184919

Table A2. Sentinel 2 images used for Analysis 2—southern California QIAs.

2020	T11SMT_20200612T182919	T11SMT_20200622T182919
	T11SMT_20200627T182921	T11SMT_20200702T182919
	T11SMT_20200707T182921	T11SMT_20200712T182919
	T11SMT_20200717T182921	T11SMT_20200722T182919
	T11SMT_20200727T182921	T11SMT_20200801T182919
	T11SMT_20200806T182921	
2021	T11SMT_20210612T182921	T11SMT_20210627T182919
	T11SMT_20210702T182921	T11SMT_20210707T182919
	T11SMT_20210717T182919	T11SMT_20210727T182919
	T11SMT_20210801T182921	T11SMT_20210806T182919
2022	T11SMT_20220612T182919	T11SMT_20220617T182931
	T11SMT_20220627T182931	T11SMT_20220702T182919
	T11SMT_20220707T182931	T11SMT_20220712T182919
	T11SMT_20220717T182931	T11SMT_20220727T182931
	T11SMT_20220801T182919	T11SMT_20220806T182931

References

1. Richter, R. A fast atmospheric correction algorithm applied to Landsat TM images. *Int. J. Remote Sens.* **1990**, *11*, 159–166. [CrossRef]
2. Dozier, J.; Frew, J. Atmospheric corrections to satellite radiometric data over rugged terrain. *Remote Sens. Environ.* **1981**, *11*, 191–205. [CrossRef]
3. Chavez, P.Z. An improved dark-object subtraction technique for atmospheric scattering correction of multispectral data. *Remote Sens. Environ.* **1988**, *24*, 459–479. [CrossRef]
4. Ahern, F.J.; Goodenough, D.G.; Rao, V.R.; Rochon, G. Use of clear lakes as standard reflectors for atmospheric measurements. In Proceedings of the Eleventh International Symposium on Remote Sensing of Environment, Ann Arbor, MI, USA, 13–16 October 1977; pp. 731–755.
5. Moran, M.S.; Jackson, R.D.; Slater, P.N.; Teillet, P.M. Evaluation of simplified procedures for retrieval of land surface reflectance factors from satellite sensor output. *Remote Sens. Environ.* **1992**, *41*, 169–184. [CrossRef]
6. Zhang, H.; Yan, D.; Zhang, B.; Fu, Z.; Li, B.; Zhang, S. An Operational Atmospheric Correction Framework for Multi-Source Medium-High-Resolution Remote Sensing Data of China. *Remote Sens.* **2022**, *14*, 5590. [CrossRef]
7. Liang, S.; Fang, H.; Chen, M. Atmospheric correction of Landsat ETM+ land surface imagery—Part I: Methods. *IEEE Trans. Geosc. Remote Sens.* **2001**, *39*, 2409–2498.
8. Kaufman, Y.J.; Wald, A.; Remer, L.A.; Gao, B.-C.; Li, R.R.; Flynn, L. The MODIS 2.1-µm channel-Correlation with visible reflectance for use in remote sensing of aerosol. *IEEE Trans. Geos. Remote Sens.* **1997**, *35*, 1286–1298.
9. Vermote, E.F.; Kotchenova, S. Atmospheric correction for the monitoring of land surfaces. *J. Geophys. Res.* **2008**, *113*, D23S90. [CrossRef]
10. Remer, L.A.; Kaufman, Y.J.; Tanre, D.; Mattoo, S.; Chu, D.A.; Martins, J.V.; Li, R.-R.; Ichoku, C.; Levy, R.C.; Kleidman, R.G.; et al. The MODIS Aerosol Algorithm, Products, and Validation. *J. Atmos. Sci.* **2005**, *62*, 947–973. [CrossRef]
11. Lyapustin, A.Y.; Wang, I.; Laszlo, R.; Kahn, R.; Korkin, S.; Remer, L.; Levy, J.; Reid, S. Multiangle implementation of atmospheric correction (MAIAC): 2. Aerosol algorithm. *J. Geophys. Res.* **2011**, *116*, D03211. [CrossRef]
12. Hsu, N.C.; Tsay, S.C.; King, M.D.; Herman, J.R. Deep blue retrievals of Asian aerosol properties during ACE-Asia. *IEEE Trans. Geosci. Remote Sens.* **2006**, *44*, 3180–3195.
13. Hagolle, O.; Dedieu, G.; Mougenot, B.; Debaecker, V.; Duchemin, B.; Meygret, A. Correction of aerosol effects on multi-temporal images acquired with constant viewing angles: Application to Formosat-2 images. *Remote Sens. Environ.* **2007**, *112*, 1689–1701. [CrossRef]
14. Kabir, S.; Leigh, L.; Helder, D. Vicarious Methodologies to Assess and Improve the Quality of the Optical Remote Sensing Images: A Critical Review. *Remote Sens.* **2020**, *12*, 4029. [CrossRef]
15. NASA. MODIS, Moderate Resolution Imaging Spectroradiometer. Available online: https://modis.gsfc.nasa.gov/ (accessed on 13 March 2023).
16. Planet Labs. Planet Fusion Monitoring Technical Specification Version 1.0.0. 2022. Available online: https://assets.planet.com/docs/Fusion-Tech-Spec_v1.0.0.pdf (accessed on 7 January 2023).

17. Claverie, M.; Ju, J.; Masek, J.G.; Dungan, J.L.; Vermote, E.F.; Roger, J.-C.; Skakun, S.V.; Justice, C. The Harmonized Landsat and Sentinel-2 surface reflectance data set. *Remote Sens. Environ.* **2018**, *219*, 145–161. [CrossRef]
18. Zhang, M.; Zhu, D.; Su, W.; Huang, J.; Zhang, X.; Liu, Z. Harmonizing Multi-Source Remote Sensing Images for Summer Corn Growth Monitoring. *Remote Sens.* **2019**, *11*, 1266. [CrossRef]
19. ESA. Level-2A Algorithm Overview. Available online: https://sentinels.copernicus.eu/web/sentinel/technical-guides/sentinel-2-msi/level-2a-algorithms-products (accessed on 7 January 2023).
20. Lee, K.H.; Won, M.S.; Kim, K.; Park, S.S. Analytical approach to estimating aerosol extinction and visibility from satellite observations. *Atmos. Environ.* **2014**, *91*, 127–136. [CrossRef]
21. Pope, R.M.; Fry, E.S. Absorption spectrum (380–700 nm) of pure water II Integrating cavity measurements. *Appl. Opt.* **1997**, *36*, 8710. [CrossRef]
22. ESA. Sentinel 2 Document Library Sentinel-2 Spectral Response Functions (S2-SRF). Available online: https://sentinels.copernicus.eu/web/sentinel/user-guides/sentinel-2-msi/document-library/-/asset_publisher/Wk0TKajiISaR/content/sentinel-2a-spectral-responses (accessed on 25 January 2023).
23. Vermote, E.; Roger, J.C.; Franch, B.; Skakun, S. LaSRC (Land Surface Reflectance Code): Overview, application and validation using MODIS, VIIRS, LANDSAT and Sentinel 2 data's. In Proceedings of the IGARSS 2018—2018 IEEE International Geoscience and Remote Sensing Symposium, Valencia, Spain, 22–27 July 2018; pp. 8173–8176. [CrossRef]
24. Guidi, L.; Tattini, M.; Landi, M. How Does Chloroplast Protect Chlorophyll Against Excessive Light? In *Chlorophyll*; Jacob-Lopes, E., Zepka, L.Q., Queiroz, M.I., Eds.; IntechOpen: London, UK, 2017. Available online: https://www.intechopen.com/chapters/54493 (accessed on 7 January 2023).
25. Kume, A. Importance of the green color, absorption gradient, and spectral absorption of chloroplasts for the radiative energy balance of leaves. *J. Plant Res.* **2017**, *130*, 501–514. [CrossRef]
26. Son, M.; Pinnola, A.; Gordon, S.C.; Bassi, R.; Schlau-Cohen, G.S. Observation of dissipative chlorophyll-to-carotenoid energy transfer in light-harvesting complex II in membrane nanodiscs. *Nat. Commun.* **2020**, *11*, 1295. [CrossRef]
27. Gillingham, S.S.; Flood, N.; Gill, T.K.; Mitchell, R.M. Limitations of the dense dark vegetation method for aerosol retrieval under Australian conditions. *Remote Sens. Lett.* **2012**, *3*, 67–76. [CrossRef]
28. NASA. Earth Observatory: Aerosol Optical Depth. Available online: https://earthobservatory.nasa.gov/global-maps/MODAL2_M_AER_OD (accessed on 7 January 2023).
29. NASA. How Aerosols Are Measured: The Science of Deep Blue. Available online: https://earth.gsfc.nasa.gov/climate/data/deep-blue/science (accessed on 6 February 2023).
30. Karpouzli, E.; Malthus, T. The empirical line method for the atmospheric correction of IKONOS imagery. *Int. J. Remote Sens.* **2003**, *24*, 1143–1150. [CrossRef]
31. Fraser, R.S.; Kaufman, Y.J. The relative importance of aerosol scattering and absorption in remote sensing. *IEEE Trans. Geosci. Remote Sens.* **1985**, *GE-23*, 625–633.
32. Richter, R.; Louis, J.; Muller-Wilm, U. *Sentinel-2 MSI-Level 2A Products Algorithm Theoretical Basis Document*; Telespazio VEGA Deutschland GmbH: Darmstadt, Germany, 2012.
33. Son, M.; Pinnola, A.; Gordon, S.C.; Bassi, R.; Schlau-Cohen, G.S. Validation of a vector version of the 6S radiative transfer code for atmospheric correction of satellite data Part I: Path radiance. *Appl. Opt.* **2006**, *45*, 6762–6774. [CrossRef]
34. Zheng, Q.; Ye, H.; Huang, W.; Dong, Y.; Jiang, H.; Wang, C.; Li, D.; Wang, L.; Chen, S. Integrating Spectral Information and Meteorological Data to Monitor Wheat Yellow Rust at a Regional Scale: A Case Study. *Remote Sens.* **2021**, *13*, 278. [CrossRef]
35. Luo, L.; Chang, Q.; Wang, Q.; Huang, Y. Identification and Severity Monitoring of Maize Dwarf Mosaic Virus Infection Based on Hyperspectral Measurements. *Remote Sens.* **2021**, *13*, 4560. [CrossRef]
36. Romero, A.; Gatta, C.; Camps-Valls, G. Unsupervised Deep Feature Extraction for Remote Sensing Image Classification. *IEEE Trans. Geosci. Remote Sens.* **2016**, *54*, 1349–1362. [CrossRef]

applied
sciences

Article

On-Orbit Magnetometer Data Calibration Using Genetic Algorithm and Interchangeability of the Calibration Parameters

Dulani Chamika Withanage *, Mariko Teramoto and Mengu Cho

Laboratory of Lean Satellite Enterprises and In-Orbit Experiments (LaSEINE), Kyushu Institute of Technology, Kitakyushu 804-8550, Japan; teramoto.mariko418@mail.kyutech.jp (M.T.); cho.mengu801@mail.kyutech.jp (M.C.)
* Correspondence: withanage.dulani-chamika622@mail.kyutech.jp

Abstract: Magnetometers are important sensors with applications in the attitude determination and control systems of satellites. CubeSats have certain limitations related to power, mass, and volume. Due to this, CubeSat magnetometers are not separated from other electrical circuits inside the satellite. Thus, it is important to calibrate the magnetometer, simulating operating conditions while the satellite is running before the launch. However, due to the limited facilities, not every CubeSat is able to calibrate its magnetometers properly on the ground. This study focuses on the calibration of on-orbit magnetometer data observed by BIRDS-3 CubeSats with a genetic algorithm. High oscillations in the total magnetic field were found in the on-orbit magnetic field data measured by magnetometers inside BIRDS-3 CubeSats. Nine unknowns, scaling factors, non-orthogonal angles, and offsets are identified with the genetic algorithm. This paper discusses the factors that affect the high oscillations in the measured total magnetic field data. For the calibration, we used magnetic field data similar to those of a model magnetic field, as the deviation is smaller. This paper presents the accuracy of determining unknowns using the genetic algorithm, as well as the interchangeability of the answers with additional orbit data from the same satellite. This method can be used in the future to calibrate magnetometers inside CubeSats before or after launch.

Keywords: CubeSats; magnetometer; on-orbit data; calibration; genetic algorithm

Citation: Withanage, D.C.; Teramoto, M.; Cho, M. On-Orbit Magnetometer Data Calibration Using Genetic Algorithm and Interchangeability of the Calibration Parameters. *Appl. Sci.* 2023, 13, 6742. https://doi.org/10.3390/app13116742

Academic Editor: Jérôme Morio

Received: 28 April 2023
Revised: 25 May 2023
Accepted: 29 May 2023
Published: 1 June 2023

1. Introduction

The attitude determination and control system (ADCS) is a subsystem in satellites. Magnetometers are sensors commonly used in the ADCS. If the magnetometer data are not calibrated properly, using simulations of on-orbit operating conditions with the running satellite, these data can be affected by residual magnetic fields. CubeSats have limited volume, space, and power. Due to the limitations of CubeSats, their magnetometers are not separated from other electrical circuits installed inside the satellite. Incorrect magnetometer data can lead to malfunction of the attitude determination and control system. However, not every CubeSat developer has the facilities to calibrate the magnetometers precisely. Therefore, not every CubeSat is able to calibrate its magnetometers correctly before flight.

Magnetometers can provide unexpected results due to scaling factors, offsets, and non-orthogonality errors. Due to the residual magnetic field inside the satellite, the magnetometers require an additional calibration. That is, despite being calibrated by the ground test, magnetometers need to be calibrated again with on-orbit data.

Previous studies have demonstrated calibration methods to calibrate on-orbit magnetometer data. Bangert et al. [1] described the performance characteristics of the UWE-3 pico-satellite's attitude determination and control system. According to the on-orbit results, attitude estimation was not accurate. It was found that the magnetometer data were not calibrated properly. Oscillations in total magnetic field, $|B|$, were identified. Bangert's study attempted to calibrate the on-orbit magnetometer data and set the correct parameters using standard minimization algorithms. Gain, cross-axis effects, and offsets were considered in the calibration method. A study presented by Foster [2] shows an extension of the two-step estimation algorithm to calibrate solid-state strap-down magnetometers. Additionally, the authors stated that this algorithm can be applied to any two- or three-axis sensor. This method was not tested for magnetometers installed in a whole vehicle or spacecraft. Pourtakdoust et al. [3] presented the calibration of magnetometers using the hyper least square (HyperLS) method. This method was tested with a Honeywell HMC5883L magnetometer. The authors mentioned that magnetometer outputs were affected by various disturbances due to manufacturing faults and limits as well as environmental impacts, requiring calibration for meaningful application. The authors defined the errors as hard iron errors, soft iron errors, bias, scale factors, and misalignment in their study. Wang et al. presented a calibration method utilizing a back propagation (BP) neural network [4]. The trained BP neural network is able to predict the true magnetic field, referring to the observational data. This method provided an error of less than 10 nT. The authors trained the network with the Levenberg–Marquardt (LMBP) algorithm, considering the non-orthogonal error, measurement noise, and constant drift. This calibration method is verified by a simulation. Previous studies used the genetic algorithm to calibrate magnetometers. Xueliang et al. [5] presented a calibration method for a three-axis magnetometer using the genetic algorithm. In this study, three-axis non-orthogonal error, sensitivity error, and the residual magnetic effect are considered. This method was tested by numerical simulation and experimentation using a fluxgate magnetometer. The author mentioned that it is difficult to achieve high accuracy with methods such as the least square method for model parameter estimation. In our study, we used on-orbit data collected by three satellites, with the evaluation function as the error (defined as in Equation (5) shown in the methodology). Moreover, we focused on low-latitude regions where the deviation of the true magnetic field from a model magnetic field is expected to be small and applied the genetic algorithm.

Chekhov et al. [6] also presented magnetometer calibration with the genetic algorithm for industrial micro-controllers. Remotely operated underwater vehicles use attitude and heading reference systems (AHRSs). AHRSs use magnetometer and accelerometer data to determine the heading of a vehicle. For genetic algorithm calibration, bias errors, gain errors, and misalignment errors were considered. This method was verified with simulated and collected magnetometer data. It is shown that the genetic algorithm shows an improvement over the recursive least square method after testing on a mission. Most research has attempted to use two-step estimation [2], standard minimization [1], the hyper least square method, strict ellipsoid, and the three-step method [3]. However, in every case, the magnetometer calibration conditions are different. The majority of previous authors have used error models. The accuracy of calibration depends on the true magnetic field or the model used. Xuelinag and Chekhov stated that the genetic algorithm shows an improvement in magnetometer calibration.

Genetic algorithms have been applied to many areas successfully [7] (p. 3). Genetic algorithms are suitable for solving problems unable to be solved easily using classical methods such as least square methods [6–8]. Genetic algorithms are known as effective candidates for parameter selection for non-linear, non-continuous problems wherein a large number of unknown parameters must be found while local maxima can exist in the search domain. In order to observe the performance, we used the genetic algorithm for magnetometer calibration.

The purpose of the paper is to use a genetic algorithm to calibrate the on-orbit magnetometer data collected by CubeSats. Our study focuses on calibrating on-orbit magnetometer data using the genetic algorithm. This method is tested with magnetometer data observed by BIRDS-3 CubeSats. This study discusses the accuracy and interchangeability of the answers found using the genetic algorithm for selected regions. Even though there remain limitations, this study demonstrates the level of calibration accuracy that can be achieved using the genetic algorithm for magnetometers installed in CubeSats.

The novelty of the present work concerns applying the genetic algorithm to low-latitude regions (where the true magnetic field is expected to be similar to a model magnetic field) of on-orbit magnetometer data and checking the interchangeability of the parameters with other on-orbit data measured from the same satellite. Moreover, this work considers the reasons for the high oscillations seen in the total magnetic field measured on orbit.

The present work can contribute to calibrating the on-orbit magnetometer data collected by CubeSats, aiding CubeSat developers in overcoming limitations when calibrating magnetometers.

This paper consists of six sections. Section 1 is the introduction; Section 2 describes the materials; Section 3 describes the methodology used in this study; Section 4 describes the results obtained using the aforementioned methodology; Section 5 is the discussion; and Section 6 is the conclusion.

2. Materials

BIRDS-3 is the third project of the BIRDS program [9]. BIRDS-3 is a constellation of three CubeSats deployed to orbit on 17 June 2019 [10]. Figure 1 shows the BIRDS-3 flight models. The external dimensions and weight of each BIRDS-3 CubeSat are 113.5 mm × 100 mm × 100 mm and 1.05 kg, respectively. All three satellites re-entered the atmosphere on October 2021.

Figure 1. BIRDS-3 Flight models.

The BIRDS-3 CubeSats had four missions, namely the Imaging Mission (CAM), the LoRa Demonstration Mission (LDM), the Attitude Determination and Control System (ADCS), and the Software Configurable Backplane Mission (BPB):BIRDS-3 FM CPLD. The BIRDS-3 CubeSats had two mission boards named Mission Board 1 and Mission Board 2. ADCS and CAM were installed on Mission Board 2. BIRDS-3 satellites have three-axis magnetic torquers for ADCS and were implemented in printed circuit boards in order to save space. The magnetometer was installed in Mission Board 2 in the ADCS section. Figure 2 shows the location of Mission Board 2 (ADCS mission board), and Figure 3 shows the location of the magnetometer. The MMC5883MA magnetometer was used in BIRDS-3 CubeSats.

On-orbit magnetometer data were used as part of the BIRDS-3 attitude stabilization system. B-dot control was used to stabilize the CubeSats. The B-dot algorithm [11] is a simple algorithm used to reduce the angular velocity of satellites. It requires the rate of changes in magnetic field flux density, measured by magnetometers. Magnetometer data were the key input of the B-dot algorithm. The attitude stabilization system of BIRDS-3 failed in orbit. It was found that the on-orbit magnetic field data observed by all three magnetometers were different from the expected results when compared to magnetic field models such as the World Magnetic Model (WMM) [12]. Before the flight, a magnetometer calibration test using a Helmholtz coil was conducted on the ground utilizing an engineering model of the magnetometer. This calibration test was performed only for the magnetometer. The whole satellite was not used to calibrate the magnetometer. Moreover, the gain and non-orthogonality angles were not estimated before the flight. Figure 4 shows the on-orbit magnetometer data observed by BIRDS-3 CubeSats after their deployment from the International Space Station. The BIRDS-3 CubeSats were deployed into orbit at 7:15 p.m. (JST) on 17 June 2019. The data were collected after 16 s of deployment. The zero in the time axis corresponds to the 16 s after the deployment. Latitude at deployment was approximately 1.433 degrees north, and longitude was −57.903 degrees.

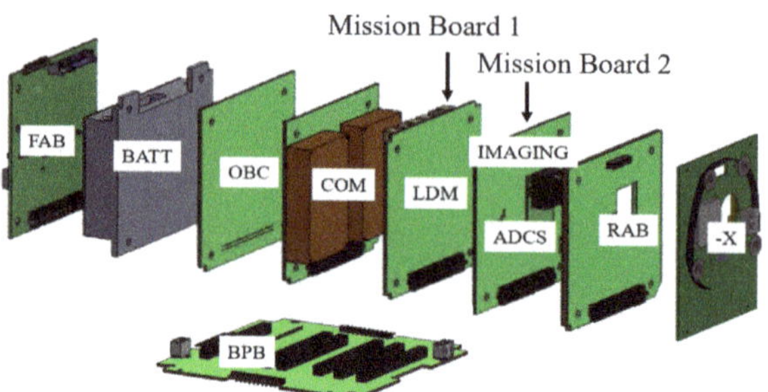

Figure 2. Location of Mission Board 2 in the BIRDS-3 CubeSats.

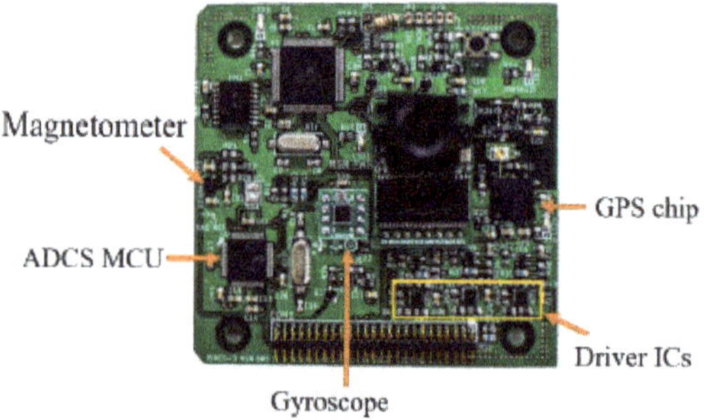

Figure 3. Mission Board 2 and the location of the magnetometer.

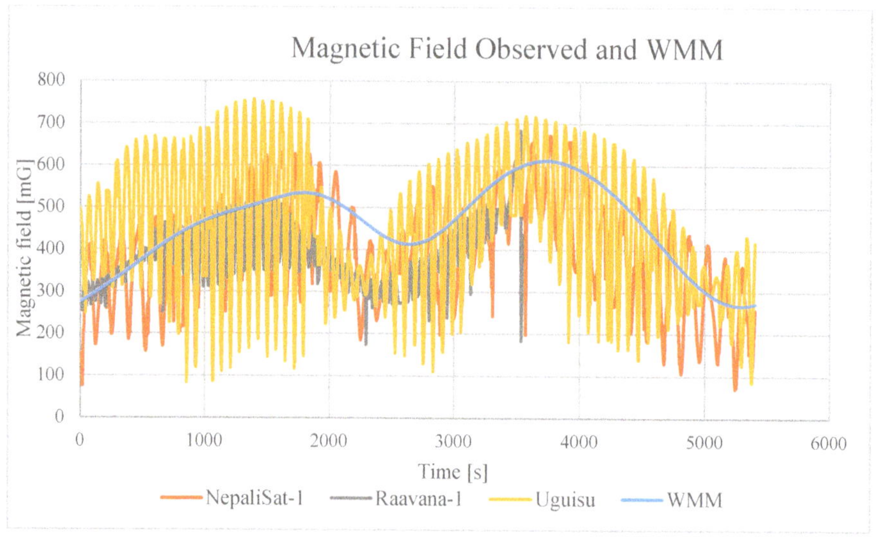

Figure 4. The total magnetic field observed by BIRDS-3 (NepaliSat-1, Raavana-1, and Uguisu) and the model magnetic field (WMM: World Magnetic Model).

In Figure 4, the magnetometer data are compared with the WMM. The orange line indicates the magnetometer data of NepaliSat-1, the gray line indicates the magnetometer data of Raavana-1, the yellow line indicates the magnetometer data of Uguisu, and the blue line indicates the WMM. Figure 4 shows the difference between observed and WMM magnetic fields. There are heavy variations in the observed total magnetic field measured in all three satellites due to the offsets, gains, and non-orthogonal angles in the magnetometers. These errors in the magnetometer data must be corrected.

3. Methodology

3.1. Ground Testing

Before launch, the residual magnetic field of the satellite was not tested. As the magnetic field measured by the magnetometers had high oscillations during orbit, we decided to measure the residual magnetic field of the satellite. This test was conducted within the magnetic shield chamber belonging to ISAS/JAXA (Institute of Space and Astronautical Science/Japan Aerospace Exploration Agency) located in Sagamihara, Japan. It is challenging to perform this experiment in a normal environment, due to the Earth's magnetic field as well as disturbance magnetic fields. Therefore, the experiment was conducted in the magnetic shield room. Figure 5 shows the test setup.

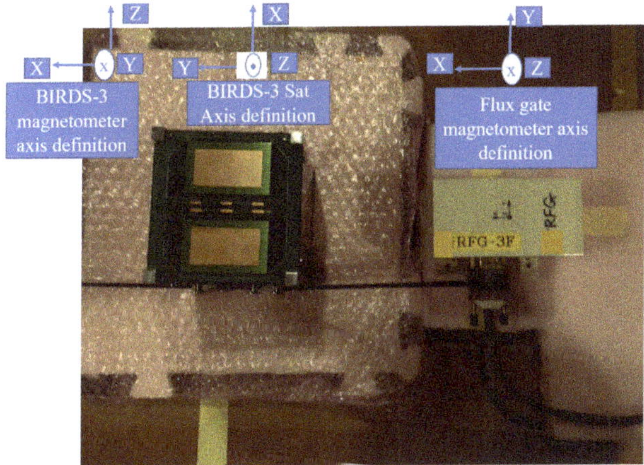

Figure 5. Residual magnetic field test setup.

Different operation modes were provided as input to the Birds-3 FM (Flight Model) backup, and magnetometer readings inside the CubeSats were checked, the difference between the FM and the FM backup being that no solar cells were attached to the FM backup. Normal sampling (NS), high sampling (HS), and ADCS modes were executed. The normal-sampling mode was the nominal mode of the CubeSats. In high-sampling mode, the sensor data were collected every 5 s. The residual magnetic field after deducting the ambient magnetic field in the shield room was as follows: X: 16,570.8 nT (165.70 mG), Y: 1070.8 nT (10.70 mG), and Z: 2283.3 nT (22.83 mG).

We have mentioned three calibration parameters, namely offsets, non-orthogonality errors, and gains. Offsets are affected by residual magnetic fields. Non-orthogonal errors are due to manufacturing errors. Gains are treated as the sensitivity of the magnetometer. Among the parameters, only the offset values are affected due to the residual magnetic field inside the satellite.

Figure 6 shows the experimental results. Blue indicates the X-axis of the magnetometer; orange indicates the Y-axis of the magnetometer; and gray indicates the Z-axis of the magnetometer.

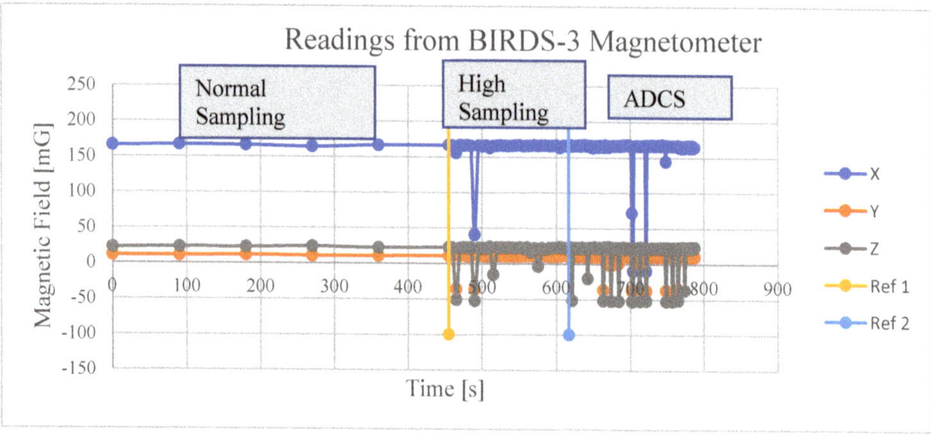

Figure 6. Results of residual magnetic field test at ISAS/JAXA.

3.2. Calibration Method

The genetic algorithm (GA) has been proven to be better in complex finding problems [7]. The genetic algorithm is an algorithm based on natural selection and reproduction [13]. Genetic algorithms can be used to find the best solution to a given problem. The main processes of the genetic algorithm are selection, crossover, and mutation. An initial population was defined in order to find solutions from the genetic algorithm.

The nine unknowns to be found in this study were the three offsets of x, y, and z, gains of a, b, and c, and non-orthogonality angles of λ, ρ, and ϕ. The unknowns are shown in Figure 7. Definitions of the angles are shown in Figure 8 [14]. The unknowns were defined as chromosomes or genes, as shown in Figure 7. One gene was represented by 12 bits.

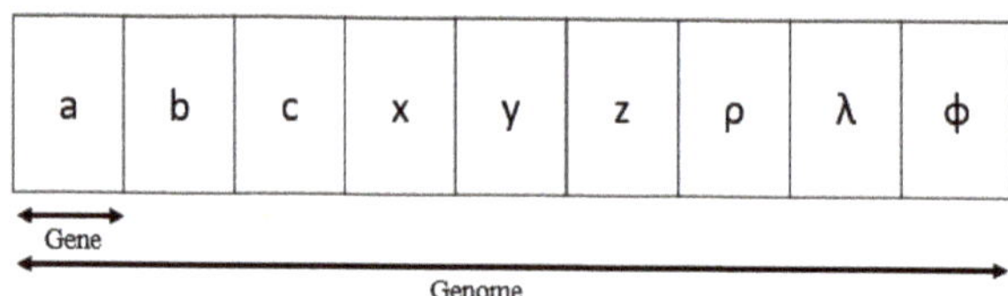

Figure 7. Nine unknowns defined in this study.

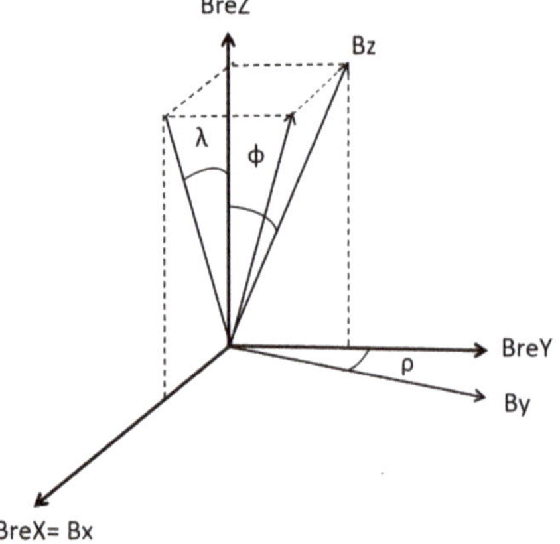

Figure 8. Axis definition.

The genetic algorithm requires a fitness function to find the best match for unknowns. The difference between the observed magnetic field and the model magnetic field was defined as the fitness function. The non-orthogonality angles of ρ, λ, and ϕ are shown in Figure 8. BreX, BreY, and BreZ axes are considered perfect orthogonal axes taken as calibrated values in Equations (1)–(3). Bx, By, and Bz are the non-orthogonal axes of the magnetometer sensors considered as measured values in Equations (1)–(3). ρ is the angle from the nominal Y axis to the X–Y plane. λ and ϕ are the angles from the nominal Z-axis to the X–Z and Z–Y planes, respectively. Equations (1)–(3) show the relationship between the calibrated magnetic field in each axis, the measured values, gain, and offsets, and the non-orthogonal angles.

$$BreX = Bx - x/a \tag{1}$$

$$BreY = \frac{(By - y) - BreX \times \sin(\rho) \times b}{\cos(\rho) \times b} \tag{2}$$

$$BreZ = \frac{(Bz - z) - BreX \times \sin(\lambda) \times c - BreY \times \sin(\phi) \times \cos(\lambda) \times c}{c \times \cos(\phi) \times \cos(\lambda)} \tag{3}$$

$$Bt^2 = BreX^2 + BreY^2 + BreZ^2 \tag{4}$$

The genetic algorithm attempts to minimize the error between the total calibrated magnetic field and the WMM, which is considered to be the true solution. The RMS (root mean square) difference between track line data and the WMM was less than 200 nT [15] (p. 60). Thus, we believed this accuracy to be sufficient for us to use the WMM as the true solution. The calibrated total magnetic field Bt is shown in Equation (4). The error "e" is defined as Equation (5). Bwmm stands for the values taken from the WMM. In this equation, N denotes the number of data available.

$$e = \frac{\Sigma\left(\frac{\sqrt{(Bwmm - Bt)^2}}{Bwmm}\right)}{N} \tag{5}$$

The values used for the parameters in the genetic algorithm are shown in Table 1. The initial population was a set of randomly chosen solutions. In this study, we selected an initial population of 1000. The initial population can be thought of as a set of genes. The number of generations denotes how many generations are iterated to find the best solution. The software stops when the given number of generations is processed. We selected the number of generations as 100 in order to find the unknowns for this study. Next, we processed the initial population to find the most suitable answers for the unknowns. In the selection operation, we selected individuals from the current population. The selection operation allows the genes to pass to the next generation. Best genes are determined using the fitness function. A selection ratio of 0.3 was used in this study. Mutation is another genetic operator. Mutation can change a gene and create a new individual. The mutation ratio is the probability of altering the bits in genes. Successful mutations survive during this process. Precision of the gene is represented by the number of bits. In the genetic algorithm, the genes are represented in binary. In this study, one gene was represented by 12 bits, as mentioned in the previous section.

Table 1. Parameters used in the genetic algorithm.

Parameter	Selected Value
Initial population	1000
Number of generations	100
Selection ratio	0.3
Mutation ratio	0.001
Number of bits	12

We focused on the data collected from low-latitude regions in which the true magnetic field is similar to the model magnetic field. Next, we applied the genetic algorithm to the selected region. After that, we checked the interchangeability of answers between other orbits. For the BIRDS-3 data, the World Magnetic Model was used as the true data. This method was applied to the on-orbit magnetometer data observed by BIRDS-3 CubeSats, and the results are presented in the next section. Please refer to the Supplementary Materials section for the magnetometer data observed by BIRDS-3.

4. Results

Accuracy of Interchangeability of Answers

We selected low-latitude regions to apply the genetic algorithm. This is because the deviation of the true magnetic field from the WMM is expected to be small. Next, the answers given by the genetic algorithm were applied to other available orbit data of the same satellite. This method was applied to the on-orbit magnetic field data from NepaliSat-1, Uguisu, and Raavana-1.

Figure 9 shows the data set used for calibration. The region between Ref 1 and Ref 2 was selected for GA application. The geographic latitude and magnetic latitude are shown in Figure 10. Table 2 shows the range of unknowns used for the GA. The nine parameters were searched to provide the best match with reference to the WMM values, and the calibrated values were calculated using the nine parameters.

Table 3 shows the parameters found by applying the GA to the selected region of Data Set 2 of Raavana-1, shown in Figure 9.

$$e_m = \frac{\Sigma \left(\frac{\sqrt{(Bwmm - Bm)^2}}{Bwmm} \right)}{N} \tag{6}$$

$$e = \frac{\Sigma \left(\frac{\sqrt{(Bwmm - Bt)^2}}{Bwmm} \right)}{N} \tag{7}$$

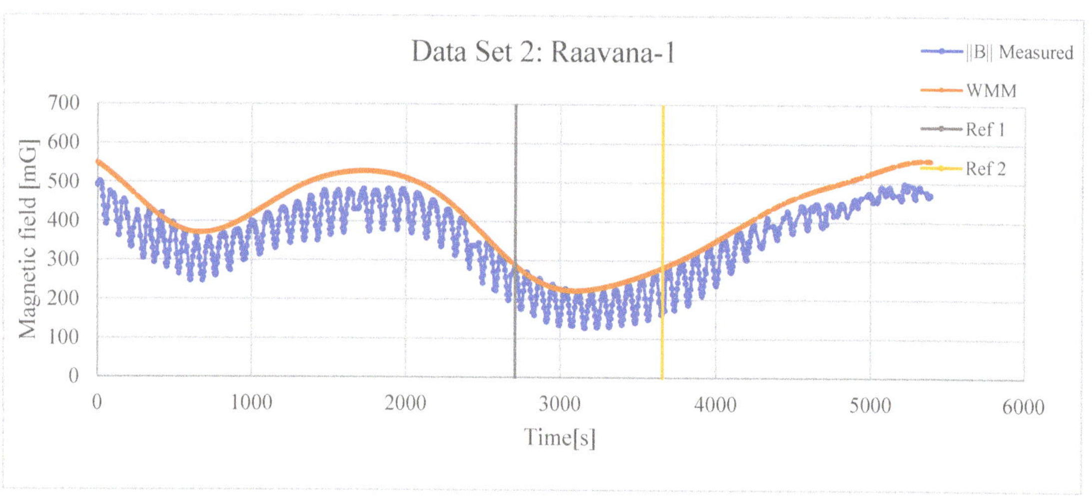

Figure 9. Data Set 2, Raavana-1, measured on 14 February 2021 at 5:23 a.m. UTC and the selected region for the calibration.

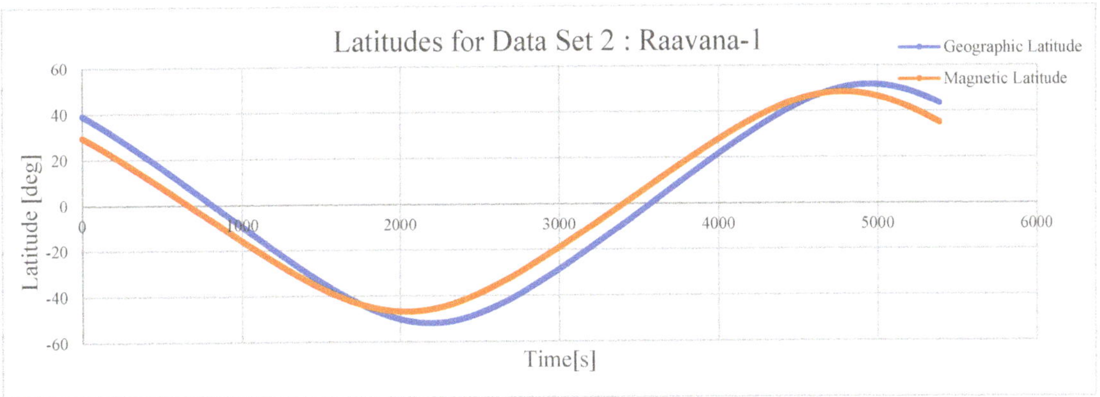

Figure 10. Magnetic Latitude and Geographic Latitude for Data Set 2 of Raavana-1.

Table 2. Range of the parameters used in the genetic algorithm.

Offsets [mG]	Gains	Non-Orthogonal Angles [deg]
X/Y/Z	a/b/c	$\lambda/\rho/\phi$
250, −180	1.5, 0.6	20, −20

Table 3. Unknowns found by the GA for the selected region of Data Set 2, Raavana-1.

Offsets [mG]			Gains			Non-Orthogonal Angles [deg]		
X	Y	Z	a	b	c	λ	ρ	ϕ
16.21	63.96	7.94	0.87	0.84	0.87	−1.25	1.88	−0.47

Figure 11 shows the results of Data Set 2 collected by Raavana-1. First, the genetic algorithm was applied to the selected region, as shown in Figure 9. Next, the answers given by the genetic algorithm (Table 3) were applied to the whole orbit and to other available on-orbit data collection by the same satellite. Table 4 provides a summary of the error values before calibration, calculated according to Equation (6), and the error values after calibration according to Equation (7). In Equation (6), Bm denotes the measured total magnetic field; e_m denotes the error value before calibration; and "e" denotes the error value after calibration in Equation (7). The third column of Table 4 shows the error value for the selected region between Ref 1 and Ref 2 in Figure 9. The fourth column shows the error value for all the data used in the data set.

Using the parameters shown in Table 3, i.e., the parameters found from Data Set 2, we calibrated the magnetometer data of a whole orbit, named Data Set 3. Data Set 2 comprises data collected on 14 February 2021 by Raavana-1. Data Set 3 comprises data collected again on 11 April 2021 by Raavana-1. Figure 12 shows the calibration results obtained in this way. The calibrated magnetic field matches well with the WMM model. The overall error is 0.03, as shown in the fifth column of Table 4, which is as good as 0.02 (the fourth column) of the error for the original data of Data Set 2. We repeated similar steps using Data Set 3. The result is shown in Table 5. The calibration parameters were derived by applying the genetic algorithm to Data Set 3. Then, the parameters were used to calibrate the measurement data of Data Set 2. The overall errors (0.03 and 0.07 in the fourth and fifth columns of Table 5) are not as good as those in Table 4. Therefore, we use the parameters in Table 3 as the calibration parameters of magnetometers onboard Raavana-1.

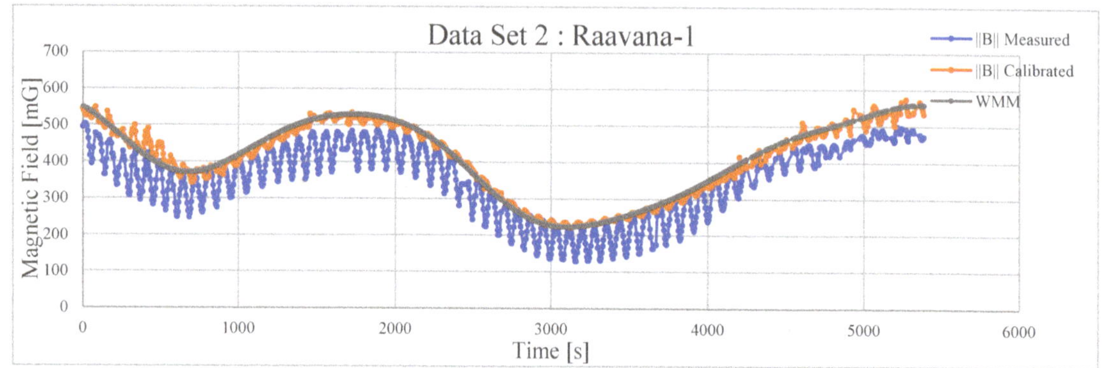

Figure 11. Calibration of Data Set 2, Raavana-1, measured on 14 February 2021 at 5:23 a.m. UTC.

Table 4. Errors before and after calibration by using the parameters derived from Data Set 2 of Ravaana-1.

Data Set Number	Equation Used to Calculate the Error	Error of Selected Region	Error of Whole Orbit	Error of Data Set 3
2	e_m	0.19	0.16	
2	e	0.01	0.02	0.03

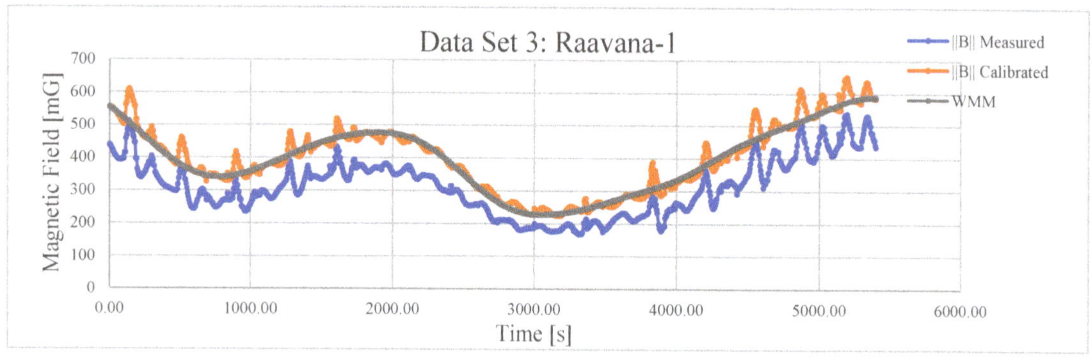

Figure 12. Calibration of Data Set 3, Raavana-1, measured on 11 April 2021 at 5:09 a.m. UTC using the values derived from Data Set 2 measured on 14 February 2021.

Table 5. Errors before and after calibration by using the parameters derived from Data Set 3 of Raavana-1.

Data Set Number	Equation Used to Calculate the Error	Error of Selected Region	Error of Whole Orbit	Error of Data Set 2
3	e_m	0.22	0.21	
3	e	0.04	0.03	0.07

The results of the other two satellites' (Uguisu and NepaliSat-1) magnetometer calibrations are shown below. First, the genetic algorithm was applied to the section of Data Set 2 shown between Ref 1 and Ref 2 in Figure 13, where the magnetic latitude is between −30.61 and 10.28 degrees for Uguisu. Next, the answers were applied to the whole of Data Set 3. The genetic algorithm was applied to the section shown in Figure 14, where the magnetic latitude is between 12.45 and −21.97 degrees for Nepalisat-1. Next, the answers were applied to Data Set 2.

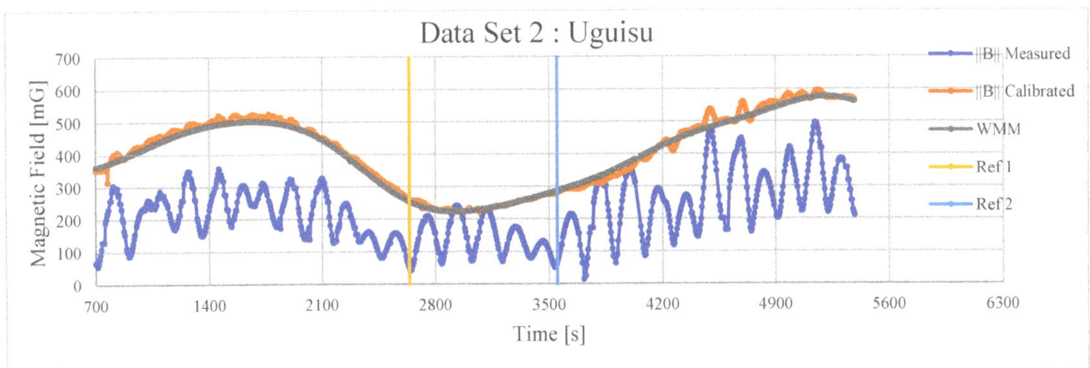

Figure 13. Calibration of Data Set 2, Uguisu, measured on 14 February 2021 at 4:40 a.m. UTC.

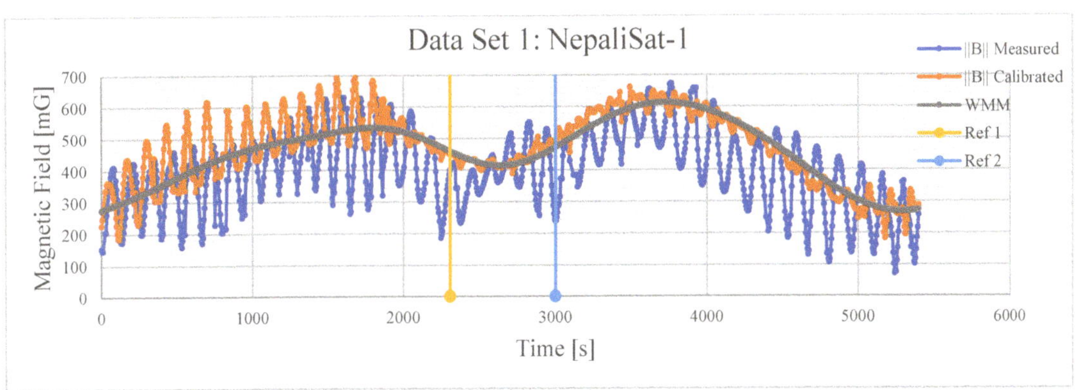

Figure 14. Calibration of Data Set 1, NepaliSat-1, measured on 17 June 2019 at 10:15 a.m. UTC.

Table 6 shows the parameters found by applying the GA to the data shown in Figures 13 and 14 using the selected region shown in the figures. Using those parameters, we evaluated the error of other data sets obtained by the same satellites. The results are shown in Tables 7 and 8.

According to the results shown above using the on-orbit magnetometer data of Raavana-1, Uguisu, and NepaliSat-1, it is proven that the genetic algorithm can be used to calibrate on-orbit magnetometer data, reducing discrepancies from the WMM.

Table 6. Unknowns found by GA for Uguisu and NepaliSat-1.

Satellite	Offsets [mG]			Gains			Non-Orthogonal Angles [deg]		
	X	Y	Z	a	b	c	λ	ρ	ϕ
Uguisu	−145.63	179.57	−40.53	0.98	0.79	0.97	−0.94	9.04	1.11
NepaliSat-1	167.48	51.58	−50.76	0.82	0.79	0.84	−5.33	5.05	−0.23

Table 7. Errors before and after calibration by using the parameters derived from Data Set 2 of Uguisu.

Data Set Number	Equation Used to Calculate the Error	Error of Selected Region	Error of Whole Orbit	Error of Data Set 3
2	e_m	0.42	0.45	0.50
2	e	0.01	0.03	0.04

Table 8. Errors before and after calibration by using the parameters derived from Data Set 1 of NepaliSat-1.

Data Set Number	Equation Used to Calculate the Error	Error of Selected Region	Error of Whole Orbit	Error of Data Set 2
1	e_m	0.17	0.21	0.09
1	e	0.03	0.19	0.02

5. Discussion

It is shown that our results for a selected region can be applied to other orbit data obtained by the same satellite. If the obtained nine parameters are correct, the calibrated result should match with the WMM regardless of when the measurement is performed, as long as the WMM is steady. WMM coefficients are estimated using data measured during magnetically quiet periods. The combined error of the WMM 2020 is 129 nT in the total magnetic field [15] (p. 67). In Table 9 below, we list the Kp-index of when the measurement was performed. From this table, we can say that the magnetic field was quiet, and the WMM is an adequate "solution" for the genetic algorithm.

Table 9. Kp-index when the magnetometer data were obtained [16–18].

Satellite	Data Set	Date	Measurement Time (UTC) Start	End	Kp-Index
Raavana-1	1	17 June 2019	10:15	11:12	0.667
	2	14 February 2021	05:23	06:52	1.667/2.000
	3	11 April 2021	05:09	06:39	2.333/1.667
Uguisu	1	17 June 2019	10:15	11:44	0.667
	2	14 February 2021	04:40	06:09	1.333/1.667
	3	11 April 2021	04:41	06:13	2.333/1.667
NepaliSat-1	1	17 June 2019	10:15	11:44	0.667
	2	14 February 2021	05:13	06:41	1.667/2.000

Although the calibrated results obtained by parameters derived from an additional data set provided reduced error values, as shown in Tables 4, 7 and 8, we still could not reduce the oscillation to the degree we desired. To determine the measured source of oscillation, we ran the GA simulation by dropping some of the parameters from the subject of the search. We applied the GA for specific low-latitude regions of each data set. In each simulation, only six parameters, instead of nine, were searched. Table 10 shows the results. In the table, without gain means that we searched only the offset (X, Y, and Z) and non-orthogonal angles (λ, ρ, ϕ) while fixing the gain values (a, b, c) to 1. The errors when gain and offsets were not considered were larger than the error when non-orthogonal angles were not considered. According to the results, the main reasons for the high oscillations are offsets and gain.

Table 10. Error obtained by dropping 3 of 9 parameters from the search in the GA.

Satellite	Data Set Number	Without Gain	Without Offset	Without Non-Orthogonal Angles
Ravaana-1	1	0.08	0.09	0.06
Ravaana-1	2	0.08	0.08	0.02
Ravaana-1	3	0.04	0.04	0.04
Uguisu	1	0.15	0.21	0.13
Uguisu	2	0.03	0.25	0.02
Uguisu	3	0.03	0.19	0.02
NepaliSat-1	1	0.13	0.15	0.07
NepaliSat-1	2	0.03	0.04	0.02

Next, the solar cell output pattern and the magnetometer output pattern for each axis were checked to confirm the reason for the high oscillations observed. The solar cell graph pattern and the magnetometer graph pattern have similar behavior, proving that each axis' oscillation of magnetometers stems from the satellites' rotation. One example is shown in Figure 15; the solar cell current is shown in mA, and the magnetic field is shown in mG. Each axis of the magnetometer data has an oscillation due to satellite rotation. However, we should not observe an oscillation in the total magnetic field. Therefore, the oscillation remaining after calibration is because the parameters obtained by GA are not yet perfectly matched.

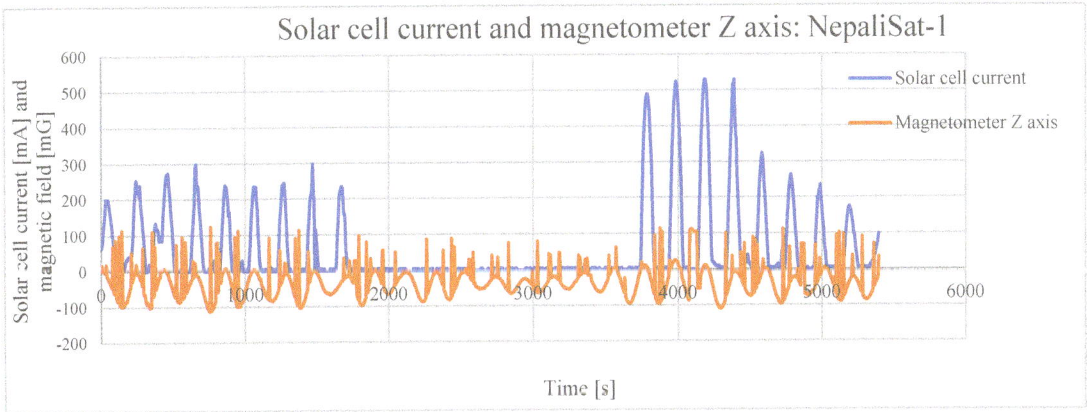

Figure 15. Solar cell current and magnetometer measurements.

As we still observed high oscillations in the calibrated result, we attempted to use the frequency amplitude contained in the measured data to reduce the high oscillations. From Figure 15, we can see that the high-frequency oscillation is due to a combination of the satellite's rotation and the mismatch of calibration parameters. From Figure 15, we determine the satellite rotates at about 0.005 Hz (1.8 deg/s). In Figure 16, we show the frequency spectrum of Data Set 1 of NepaliSat-1. It shows a peak at around 0.007 Hz. Instead of using the error defined in Equation (5) as the evaluation function of the GA, we used the sum of the magnitude of frequency amplitude and WMM data. The smaller the amplitude, the better the ranking is. In this search, we used the entire data sets of whole orbits rather than a selected region within low-latitude regions.

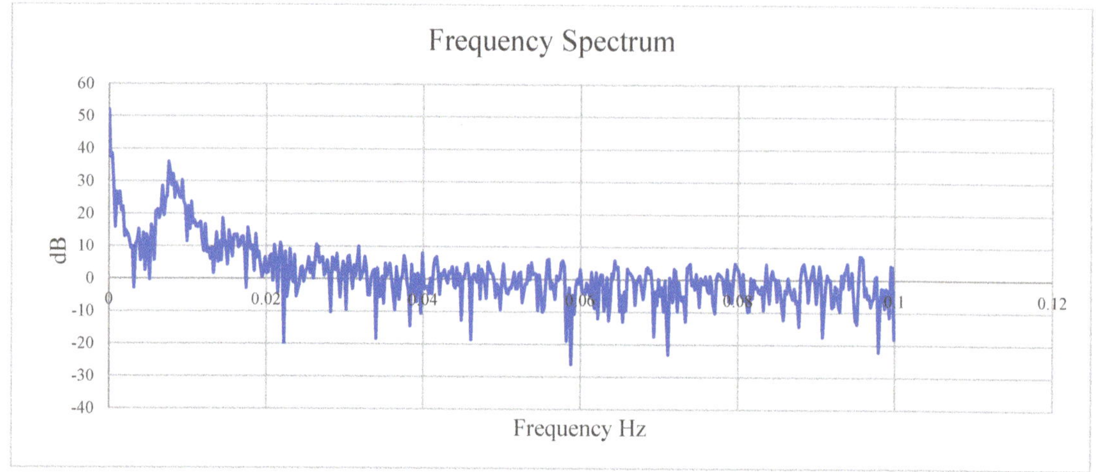

Figure 16. Frequency spectrum of measured magnetic field by NepaliSat-1 (Data Set 1).

The following Equations (8)–(10) were used to apply genetic algorithms in this case. Equation (10) was used as the evaluation function. E1 and E2 were chosen to make e/E1 and e2/E2 both approximately one at the first generation.

$$e = \frac{\Sigma\left(\frac{\sqrt{(Bwmm - Bt)^2}}{Bwmm}\right)}{N} \tag{8}$$

$$e2 = \Sigma(\text{amplitude of frequency spectrum}) \tag{9}$$

$$e = \left(\frac{e}{E1} + \frac{e2}{E2}\right)/2 \tag{10}$$

Table 11 shows the parameters obtained from the GA. By comparing with Table 6, we can observe that the offset of the X-axis has been reduced. Figure 17 shows the calibrated result. Compared to Figure 14, we can see that the oscillations in the observation were reduced significantly in the first part of the graph, where we noticed high oscillations even after calibration with Equation (5).

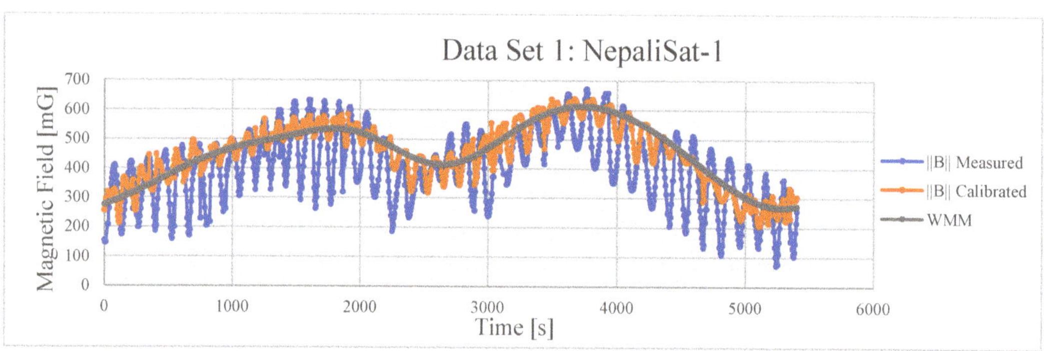

Figure 17. Calibration of Data Set 1, NepaliSat-1, using the oscillation amplitude and WMM as the evaluation function of GA.

Table 11. Calibration parameters obtained by using the frequency amplitude due to satellite rotation as the evaluation function of the GA.

Satellite	Offsets [mG]			Gains			Non-Orthogonal Angles [deg]		
	X	Y	Z	a	b	c	λ	ρ	ϕ
NepaliSat-1	127.91	62.61	-77.12	0.87	0.80	0.88	4.54	-1.80	0.38

6. Conclusions

CubeSats have limitations, as mentioned in the introduction section. Thus, magnetometers are not separated from electrical circuits inside the satellites. Therefore, magnetometers should be calibrated properly before flight. However, most CubeSat developers do not have the facilities to calibrate magnetometers with in-orbit conditions. Magnetometers are important sensors used in attitude determination and control systems. In this study, we investigated the magnetometer data collected by the BIRDS-3 CubeSats Raavana-1, NepaliSat-1, and Uguisu that flew with an orbit of 51.6-degree inclination and approximately 400 km altitude. There was a heavy variation in the magnetometer data collected when compared to a magnetic model.

The genetic algorithm was used in this study to calibrate the magnetometer data. We focused on non-orthogonality angles, offsets, and gains to calibrate the on-orbit magnetometer data. According to the results, applying the genetic algorithm to low-latitude regions and interchanging the answers with other available on-orbit data collected by the same satellite is possible. We defined the error as the difference from the WMM value, regarding the WMM value as a "true" solution. We use the error as the evaluation function of the genetic algorithm; the smaller, the better. According to the results, we could reduce the error with the genetic algorithm compared to the error in the measured magnetometer data. We were able to show that the high oscillations in the total magnetic field are mostly due to the gain and the offset. Next, we discussed using frequency amplitude in the genetic algorithm to reduce the oscillations. In this case, we defined two errors. One was the difference from the WMM value. The other was the oscillation amplitude of frequency, which was higher than expected due to satellite rotation. In this case, we used entire data sets of whole orbits. According to the results, we show that only using the difference from the WMM value for low-latitude data as the evaluation function gives better calibration results than using both the high-frequency oscillation and the difference from the WMM value as the evaluation function.

We tested the genetic algorithm for flight magnetometer data collected by three satellites and confirmed that the method is sufficient in calibrating on-orbit magnetometer data.

The calibration method outlined in this paper still has room for improvement. We could not remove the oscillations in the total magnetic field completely. Finding a better way to reduce oscillations remains a goal of future work.

Supplementary Materials: The following supporting information can be downloaded at Withanage, Dulani (2023), "On-orbit magnetometer data observed by BIRDS-3 CubeSats", Mendeley Data, V1, doi: 10.17632/2b2rvxccyx.1, available online: https://data.mendeley.com/datasets/2b2rvxccyx/1, accessed on 27 April 2023.

Author Contributions: Conceptualization, D.C.W., M.T. and M.C.; methodology, D.C.W., M.T. and M.C.; software D.C.W., M.T. and M.C.; validation D.C.W., M.T. and M.C.; formal analysis, D.C.W.; investigation, D.C.W.; resources, M.T. and M.C.; data curation, D.C.W.; writing—original draft preparation, D.C.W.; writing—review and editing, D.C.W., M.T. and M.C.; visualization, D.C.W.; supervision, M.T. and M.C.; project administration, M.C.; funding acquisition, M.C. All authors have read and agreed to the published version of the manuscript.

Funding: This research was partially funded by the JSPS Core-to-Core Program B: Asia–Africa Science Platforms.

Institutional Review Board Statement: Not applicable.

Informed Consent Statement: Not applicable.

Data Availability Statement: Not applicable.

Acknowledgments: The authors would like to express gratitude to BIRDS-3 team members. Moreover, the authors would like to acknowledge Sangkyun Kim, George Maeda, Hirokazu Masui, and Takashi Yamauchi.

Conflicts of Interest: The authors declare no conflict of interest.

References

1. Bangert, P.; Busch, S.; Schilling, K. Performance characteristics of the uwe-3 miniature attitude determination and control system. In Proceedings of the 2nd IAA Conference on Dynamics and Control of Space Systems (DYCOSS), Rome, Italy, 24–26 March 2014.
2. Foster, C.C.; Elkaim, G.H. Extension of a Two-Step Calibration Methodology to Include Nonorthogonal Sensor Axes. *IEEE Trans. Aerosp. Electron. Syst.* **2008**, *44*, 1070–1078. [CrossRef]
3. Pourtakdoust, S.H.; Kiani, M.; Sheikhy, A.K. Consistent calibration of magnetometers for nonlinear attitude determination. *Measurement* **2015**, *73*, 180–190. [CrossRef]
4. Wang, Y.; Sheng, T.; He, L.; Cheng, Z. Calibration Method of Magnetometer Based on BP Neural Net-work. *J. Comput. Commun.* **2020**, *8*, 31–41. [CrossRef]
5. Xueliang, P.; Pei, J.; Chunsheng, L. The Calibration Method of Three-axis Magnetometer Based on Genetic Algorithm. *Appl. Mech. Mater.* **2015**, *722*, 373–378. [CrossRef]
6. Chekhov, E.L.; Antonov, D.A.; Kolganov, L.A.; Savkin, A.V. Magnetometer Calibration Using Genetic Algorithms. *TEM J.* **2020**, *9*, 907–914. [CrossRef]
7. Coley, D.A. *An Introduction to Genetic Algorithms for Scientists and Engineers*; World Scientific Publishing Co. Pte. Ltd.: Singapore, 1999.
8. Massart, D.L.; Vandeginste, B.G.M.; Buydens, L.M.C.; De Jon, S.; Lewi, P.J.; Smeyers-Verbeke, J. Data Handling in Science and Technology. In *Chapter 27 Genetic Algorithms and Other Global Search Strategies*; Elsevier: Amsterdam, The Netherlands, 1998; Volume 20, Part A; pp. 805–848.
9. Kim, S.; Maeda, G.; Cho, M. BIRDS Program Digital Textbook, CubeSTD-2019-001G. Available online: https://www.birds-project.com (accessed on 2 May 2022).
10. Chamika, W.D.; Cho, M.; Maeda, G.; Kim, S.; Masui, H.; Yamauchi, T.; Panawennage, S.; Shrestha, S.B. BIRDS-3 Satellite Project Including the First Satellites of Sri Lanka and Nepal. In Proceedings of the 70th International Astronautical Congress, Washington, DC, USA, 21–25 October 2019.
11. Böttcher, M.; Eshghi, S.; Varatharajoo, R. Testing and validation of $\dot{B}$ algorithm for cubesat satellites. In *IOP Conference Series: Materials Science and Engineering*; IOP Publishing: Bristol, UK, 2016; Volume 152, p. 012029.
12. The World Magnetic Model. Available online: https://www.ngdc.noaa.gov/geomag/WMM/ (accessed on 3 December 2022).
13. Hassanat, A.; Almohammadi, K.; Alkafaween, E.; Abunawas, E.; Hammouri, A.; Prasath, V.B.S. Choosing Mutation and Crossover Ratios for Genetic Algorithms—A Review with a New Dynamic Approach. *Information* **2019**, *10*, 390. [CrossRef]
14. Springmann, J.C.; Cutler, W.J. Attitude-Independent Magnetometer Calibration with Time-Varying Bias. *J. Guid. Control. Dyn.* **2012**, *35*, 1080–1088. [CrossRef]
15. The US/UK World Magnetic Model for 2020–2025. Available online: https://www.ngdc.noaa.gov/geomag/WMM/data/WMM2020/WMM2020_Report.pdf (accessed on 16 January 2022).
16. GFZ German Research Centre for Geosciences. Available online: https://kp.gfz-potsdam.de/en/ (accessed on 10 January 2022).
17. Matzka, J.; Bronkalla, O.; Tornow, K.; Elger, K.; Stolle, C. Geomagnetic Kp index. V. 1.0. GFZ Data Services. *Space Weather* **2021**. [CrossRef]
18. Matzka, J.; Stolle, C.; Yamazaki, Y.; Bronkalla, O.; Morschhauser, A. The Geomagnetic Kp Index and Derived Indices of Geomagnetic Activity. *Space Weather* **2021**, *19*, e2020SW002641. [CrossRef]

MDPI
St. Alban-Anlage 66
4052 Basel
Switzerland
www.mdpi.com

Applied Sciences Editorial Office
E-mail: applsci@mdpi.com
www.mdpi.com/journal/applsci